W9-APO-967

The Cactus Primer

The Cactus Primer

Arthur C. Gibson
Park S. Nobel

Harvard University Press
Cambridge, Massachusetts
London, England 1986

Printed in the United States of America
10 9 8 7 6 5 4 3 2 1

This book is printed on acid-free paper, and its binding materials have been chosen for strength and durability.

Library of Congress Cataloging-in-Publication Data

Gibson, Arthur C.
The cactus primer.
Bibliography: p.
Includes index.
1. Cactus. I. Nobel, Park S. II. Title.
QK495.C11G5 1986 583'.47 85-24874
ISBN 0-674-08990-1 (alk. paper)

Preface

People around the world and from all walks of life are hopelessly susceptible to a condition called cactophily, the love of cacti. Victims often contract this condition when their first cactus plant, residing on a sunny windowsill, produces its showy flowers. As personal interest in cacti increases, this single plant may become part of a larger collection that can threaten to take over every habitable space. For others, cactophily may begin with an innocent visit to the desert to view these marvelous plants—plants that, unlike animals, do not run and hide from the parching sun.

As familiarity with cacti grows, the owner's curiosity about cacti also grows. We have found that the typical cactophile is interested not only in ways to propagate, grow, and care for cacti but also in technical issues. What is the structure of a cactus, and how does a cactus survive in nature? Where did the cactus family originate, and when and how have the many species come into existence? What are the proper names for the species, and how are they related to one another? How much water does a cactus lose on a hot day? These and many other questions are addressed in scientific journals not generally read by most cactus collectors—or for that matter even by scientists preoccupied with other scientific topics. This book is intended to bridge the gap between the professional cactus specialist, who obtains the basic information through research, and others who wish to have this information in accessible form. A cactus enthusiast with basic training in science should be able to use this book to gain an understanding of the biology of cacti. Likewise, this book will aid teachers in colleges and universities who use cacti in lectures to illustrate desert adaptations, Crassulacean acid metabolism, or convergent evolution. Often such a teacher presents a very brief account, and neither the instructor nor the student sees the whole picture or has a convenient reference in which the biology of cacti is comprehensively discussed. This book has been designed to provide such a complete account.

Typically, books on a special group of plants or animals present separate chapters on morphology, anatomy, development, physiology, ecology, evolution, and systematics. Here we have integrated the materials to show how structure, function, and distribution are interrelated and how and why the many interesting structural features in the cactus family have evolved. The general features of cacti are described in Chapter 1, and the primitive features of certain cacti and early evolutionary trends are discussed in Chapter 2. Following that, a series of chapters is devoted to the special features of a cactus plant: succulence, Crassulacean acid metabolism, areoles and spines, ribs and tubercles, ecological distribution, and growth form and habit. Chapter 9 is a summary of the secondary compounds and other substances found in cacti, and Chapter 10 discusses how information on structure and plant chemistry can be used to reconstruct the phylogeny of cactus lineages. In the last chapter we present the available evidence on evolutionary relationships of the cactus family. A glossary is included to provide definitions and to facilitate independent usage of different sections.

We have cited very little literature in the text itself; instead we have included a list of selected references at the end of each chapter. Naturally we have drawn heavily on our own work on cactus biology and included original data that have not previously been published. Our coverage of cactus biology certainly is not complete or exhaustive, but we do review most of the subjects that have been well studied. Indeed, writing this book together has been a meaningful educational experience for both of us. We hope that our readers find these plants as interesting as we do and that our interest is contagious. We also hope that this book will lead to a deeper appreciation of cacti and perhaps even serve as a springboard for new studies.

This book has benefited from the comments and contributions of many individuals. First, we must thank the National Science Foundation and the Department of Energy for supporting many of our cactus research projects. Permission to reproduce certain published material or to use original photographs was generously granted, as acknowledged in the appropriate figure captions. A draft of this book formed the basis of a lively graduate seminar at UCLA and we thank the participants: William Bond, Augusto Franco, Maureen Gardner, Gary Geller, Brian Henen, Loraine Kohorn, Lynn McMurtry, Gary Peter, Barry Prigge, Debbie Raphael, Grant Steen, and Judy Verbeke. We are also indebted to Ed and Betty Gay and Myron Kimnach for discussions about the manuscript. Line drawings were skillfully done by Hildy Heinkel, Margaret Kowalczyk, and Gretchen North. Marjorie Macdonald did an outstanding job on the typing. Finally, we thank our wives and daughters for their understanding during our period of preoccupation with spines and stems and species.

A.C.G.
P.S.N.
July 15, 1985

Contents

Figure 1.1. Small cactus garden in the Mildred E. Mathias Botanical Garden, University of California at Los Angeles, Westwood, California. Individual cacti are identified on the schematic diagram: 1, *Mammillaria longimamma* var. *uberiformis;* 2, *Obregonia denegrii;* 3, *Ariocarpus trigonus;* 4, *Ferocactus viridescens;* 5, *Leuchtenbergia principis;* 6, *Opuntia invicta;* 7, *Ferocactus ("Echinofossulocactus")* sp.; 8, *Stenocereus eruca;* 9, *Opuntia microdasys;* 10, *Echinopsis* spp.; 11, *Parodia* sp.; 12, *Mammillaria elongata;* 13, *Mammillaria compressa;* 14, *Echinocereus pentalophus;* 15, *Trichocereus* sp.; 16, *Notocactus magnificus;* 17, *Eulychnia iquiquensis;* 18, *Opuntia* sp. (platyopuntia); 19, *Opuntia echios* var. *gigantea;* 20, *Borzicactus* sp.; 21, *Mammillaria zuccariana;* 22, *Echinocactus grusonii;* 23, *Cleistocactus morawitzianus;* 24, *Opuntia* sp. (cholla); 25, Unknown; 26, *Opuntia quimilo;* 27, *Stenocereus thurberi;* 28, *Mammillaria* sp.; 29, *Stenocereus gummosus;* 30, *Trichocereus werdermannianus;* 31, *Samaipaticereus corroanus;* 32, *Opuntia pittieri;* 33, *Pachycereus marginatus* var. *gemmatus;* 34, *Opuntia echinocarpa;* 35, *Espostoa lanata;* 36, *Myrtillocactus geometrizans;* 37, *Cereus peruvianus* 'Monstruosus'; 38, *Pereskia aculeata;* 39, *Cereus* sp.; 40, *Cereus peruvianus;* 41, *Pereskia grandifolia;* 42, *Rhipsalis* aff. *linearis;* 43, *Cryptocereus anthonyanus;* 44, *Rhipsalis paradoxa;* 45, *Austrocylindropuntia subulata;* 46, *Pereskiopsis porteri.*

44
42
37
43
39
37
40
45
29
41
39
36
36
35
47
38
33
34
35
46
46
24
18
35
30
31
20
19
23
24
6
28
10
10
32
9
22
10
21
10
26
22
25
30
17
25
22
27
12
10
10
13
15
16
9
10
10
4
11
4
4
14
8
4
5
4
10
6
7
1
3
2

1

The Cactus Plant

Just outside the building in which we teach botany at UCLA is a small cactus garden (Fig. 1.1). In this 5 m (meter) × 6 m plot we currently cultivate 55 species of the cactus family (Cactaceae). Compared with the size of most cactus collections, this is not very remarkable, because the cactus family consists of about 1600 species and thousands of additional varieties and forms. Neither is this a collection of particularly valuable species, because whenever we add a species that is especially interesting, someone surreptitiously transplants our specimen to a private collection — the great popularity of the group does carry hidden costs.

Although this cactus patch is not large, visitors to campus and students on their way to class stop to examine the plants. Rarely are these specimens in flower, so the attraction of the cacti lies mainly in their bizarre vegetative features.

Diversity of Form

Size and Structure

One of the fascinating aspects of the cactus family is the array of fleshy (succulent) growth forms and habits that are represented. Only some of the many forms are on display in the garden (Fig. 1.1). The largest specimen, *Cereus peruvianus,* is 4 m tall; but in South America the treelike (arborescent) species of *Cereus* may achieve a height of 15 m or more. The 1985 *Guinness Book of World Records* lists saguaro, *Carnegiea gigantea,* a native of the Sonoran Desert of North America, as the "largest" cactus; one plant was measured at 23 m (70 feet). But this accolade is probably not justified. Other large forms, such as *Pachycereus weberi* (Fig. 1.2) of southern Mexico and *Cereus jamacaru* of Brazil, are generally as tall as or taller than most saguaros and are many times more massive, with trunks over a meter in diameter. Consider the weight of one of these trees, which is 90% water. A piece of stem 1 m long and 30 cm (centimeters) in diameter would weigh approximately 70 kg (kilograms) (150 pounds); a plant of *P. weberi,* which is 15 m tall and has over 70 ascending branches plus a thick trunk, would weigh in excess of 25,000 kg, or about 25 tons. This is more than the weight of a large humpback whale.

The arborescent species (Figs. 1.2 – 1.4) are called columnar cacti because they have long, often unbranched, columnlike stems. Following convention, we include in this category only species that are 0.5 m or more in height with fluted (ribbed) stems, which are 10 to 50 times longer than they are wide. Large treelike forms with long, unbranched stems grade into a wide variety of smaller trees and shrubs with erect, ascending, arching, or pendent branches. For example, our garden contains specimens of *Myrtillocactus geometrizans* with highly branched stems having 6 shallow ribs per stem, of *Pachycereus marginatus* with erect unbranched stems having 5 shallow ribs, and of *Espostoa lanata* with erect stems having about 20 moderately deep ribs.

Barrel cacti are ribbed, columnar species that attain a maximum height of 0.5 to 2 m and normally have only one very thick, erect stem. Barrel cacti, such as *Ferocactus* (Figs. 1.1 and 1.5) and *Astrophytum,* are typically cylindrical; but those in the genus *Echinocactus* (Fig. 1.6), such as the golden barrel cactus, *E. grusonii,* are more nearly spherical (globose, or globular) and may occur in clusters. Various historical accounts tell how drinking water can be obtained from a barrel cactus by decapitating a plant and pounding the juicy tissues with a club. But this is a beverage of last resort because many cacti contain distasteful substances, such as mucilage and alkaloids.

Cacti with the smallest growth habits, which constitute over half the species in the family, are naturally the ones favored by collectors. Large

Figures 1.2–1.9. Growth habits of cacti.
1.2 *Pachycereus weberi,* a massive, candelabriform, arborescent cactus in Puebla, Mexico.
1.3. *Cereus* sp., a small, arborescent specimen in Bahia, Brazil.
1.4. *Neobuxbaumia mezcaelensis,* a tall, solitary columnar, arborescent cactus, growing with a species of *Cephalocereus* that is a shorter solitary columnar in Puebla, Mexico.
1.5. *Ferocactus recurvatus,* a cylindrical barrel cactus in the Valley of Tehuacán, Puebla, Mexico.
1.6. *Echinocactus platyacanthus (= E. grandis),* an extremely wide barrel cactus in the Valley of Tehuacán, Puebla, Mexico.
1.7. *Mammillaria (Solisia) pectinata,* a small, globular cactus about 6 cm in diameter and a native of Tehuacán, Puebla, Mexico.
1.8. *Mammillaria compressa,* a tightly packed caespitose cactus in Hidalgo, Mexico.
1.9. *Echinocereus cinerascens,* an open, caespitose cactus in Hidalgo, Mexico.

Figures 1.10–1.15. Ribs and tubercles of small cacti.
1.10. *Echinopsis* sp., a small, cylindrical plant with well-developed ribs.
1.11. *Gymnocalycium mihanovichii* var. *stenogonum,* a globular plant with well-developed ribs. Each rib has a bump or a "chin" between successive areoles, and the position of the fruit *(upper right)* shows where the flower is formed on the upper side of the areole just below the chin.
1.12. *Mammillaria zuccariniana,* a small, cylindrical plant with helically arranged tubercles.
1.13. *Mammillaria longimamma* var. *uberiformis,* a small, globular plant with relatively long tubercles.
1.14. *Lophophora williamsii,* or peyote, a cluster of globular plants with rounded ribs having relatively poorly developed tubercles and no spines on the areoles.
1.15. *Astrophytum asterias,* a small, globular plant with rounded ribs and no visible sign of tubercle development; areoles have many trichomes but no spines, and the small indentations in the skin are deeply sunken stomates.

globular or cylindrical barrel forms grade into smaller plants with solitary stems (Fig. 1.7) or plants with a hemispherical mound of many small stems (caespitose) arising from a common stem base and root system (Figs. 1.8 and 1.9). Even a single genus may include several different diminutive growth forms; *Mammillaria,* for example, contains about 180 small species. Honors for the tiniest cactus belong to the aptly named *Blossfeldia liliputana* of Argentina, a tightly caespitose plant the individual stems of which are often less than 1 cm in diameter or height at maturity.

General Features of the Shoot

What fascinates many collectors is the great variation in features of the shoot (stem and leaves). Like their larger cousins and the ancestors from which they evolved, many small cacti have ribs. Ribs are well developed and universally present in genera such as *Echinocereus* of North America and *Echinopsis* (Fig. 1.10), *Gymnocalycium* (Fig. 1.11), and *Notocactus* of South America. In other genera, for example, *Mammillaria* and *Coryphantha* of North America and *Rebutia* of South America, ribs never occur; instead, the stem is covered by succulent projections (called tubercles) that are arranged in intricate helical patterns (Fig. 1.12). Some very prized species, such as *Ariocarpus trigonus, Mammillaria longimamma* (Fig. 1.13), and *Leuchtenbergia principis,* have exceedingly long tubercles, whereas the hallucinogenic peyote, *Lophophora williamsii* (Fig. 1.14), and *Astrophytum asterias* (Fig. 1.15) have stems that essentially lack ribs and tubercles. Many genera of small cacti have some ribbed species, some tuberculate species, and some transitional species with shallow ribs and strong tubercles. A genus with ribs is often closely related to a genus with tubercles only.

The most distinguishing nonsexual (vegetative) feature of a cactus is the areole (Fig. 1.16) with its spines and hairs (trichomes). Borne atop a tubercle or on the edge of a rib, an areole is actually a lateral bud (axillary bud) that directly produces a cluster of spines. In seed plants this bud often forms a lateral branch, whereas in cacti the number of branches is greatly reduced. But an areole can produce a branch when a stimulus is provided, such as injury to the plant, for example, new plantlets or branches arise from the base of a plant when the top is removed, or an areole can be used in the production of a flower.

The variation in cactus spines is truly amazing: straight to hooked, curved, and twisted; broad to narrow; thick to thin; more than 15 cm long to minute (or absent); steellike to soft or brittle; pale to brightly colored; smooth or variously sculptured; and persistent to deciduous. Moreover, spine structure varies greatly on a single plant, from seedling to adult, and even on a single areole. To make matters more interesting, the spines frequently emerge through a carpet of soft hairs that cover the areole and the shoot tip. The hairs may be long or short, dense or sparse, persistent or deciduous, and variously colored. On our specimen of organ pipe cactus, *Stenocereus thurberi,* from the Sonoran Desert, the areolar hairs are first red and eventually turn black.

An individual cactus plant may pass through several different growth habits and stem features en route to the mature structure. The saguaro, *Carnegiea gigantea,* illustrates this point. The young plant is globose, and its tubercles are not organized into ribs. As the plant grows, it increases in both length and diameter, first becoming a large globular form with many well-defined ribs and then becoming cylindrical. At a height of 1 m, the plant, which is now 20 to 25 years old, is remarkably similar to a mature specimen of a barrel cactus, for example, the compass plant or biznaga, *Ferocactus wislizenii.* Each year the shoot grows 6 to 10 cm, and the plant becomes a solitary column (Fig. 1.17*A*). The first flowers may be produced when *C. gigantea* is 2 m high, or about 30 years old. When the plant is 40 to 50 years old, lateral branches may develop in the middle region of the column, giving rise to the "arms" of the mature, branched, arborescent cactus (Fig. 1.17*B*). When the plant attains a height of 10 m, it is about 150 years old.

To the casual observer, cacti appear to lack leaves, but this is not strictly the case. In the first place, many cacti produce minute leaves a fraction of a millimeter in length at the growing tip of the shoot (Fig. 1.18). Each leaf is formed on the outer side of the areole, and in most species the leaves are ephemeral and quickly shed; thus, they are rarely observed.

Figure 1.16. *Copiapoa cinerea,* a barrel cactus from northern Chile, which has areoles with sharp, dark spines and much soft, light-colored pubescence. This is the top of a young plant.

Figure 1.17. Growth habits of saguaro, *Carnegiea gigantea,* growing in Organ Pipe Cactus National Monument near Lukeville, Arizona. ***(A)*** Solitary specimen about 2 m tall, when flowering begins. ***(B)*** Branching features of mature specimens that are 4 to 6 m tall.

Other species have short, narrow leaves a centimeter or more in length that persist for several weeks until the stem is fully expanded (Figs. 1.19–1.22). But most amazing are the leaf-bearing cacti, which have conspicuous and even fairly large leaves that persist throughout a full growing season. Located at the rear of our cactus patch are two such plants. In the back of the garden are specimens of *Pereskia grandifolia* from South America, a cactus that has thin leaves 7 to 10 cm long and 3 to 4 cm wide (Fig. 1.23), and *P. aculeata,* a small, scrambling (scandent) shrub from the West Indies with somewhat shorter leaves. Next to these lives *Austrocylindropuntia subulata* from South America, which has succulent, narrowly cylindrical leaves 7 to 10 cm in length (Fig. 1.20). People just learning about cacti are initially surprised when they see these forms, although they quickly learn that large, wide leaves in cacti are an evolutionarily primitive feature, a holdover from early cacti that evolved from some other leaf-bearing, flowering plants.

In our garden, we find five cactus plants growing on the fence behind the plot. These are epiphytes, species that do not have to have their roots planted in the ground. A Central American species of *Hylocereus* shows the transition from terrestrial life to true epiphytism; this plant was initially planted in the ground, but it can be grown on a tree branch without having its roots in soil. The other plants, species of *Rhipsalis,* are hanging tropical epiphytes that can live in soil pockets among boulders; but more frequently their stems hang from the crotches of a tropical tree (Fig. 1.24). In *R. baccifera,* the young stems have shallow ribs and soft spines (Fig. 1.24*A*); the mature stems are cylindrical and essentially spineless (Fig. 1.24*B*). There are also several cactus species from tropical North America (for example, *Selenicereus testudo* and *Strophocactus wittii*) that grow tightly appressed to tree trunks, clinging to the bark with numerous roots. Cactus epiphytes either have very few, short spines or lack spines entirely.

Cacti are famous for their ability to root from vegetative cuttings. If a portion of a plant is severed and falls to the ground, the fallen piece can form a callus on the broken surface. In time, roots will emerge from the callused surface or, in certain forms, from areoles on the lower side. These roots are called adventitious roots, a term meaning that they actually arise from stem or leaf tissue. Adventi-

Figures 1.18–1.22. Leaves of succulent cacti.
1.18. *Corryocactus melanotrichus,* with tiny, vestigial leaves *(arrows)* that are characteristic of cereoid cacti; the areole is situated in the axil of the leaf.
1.19. *Opuntia rufida,* a platyopuntia; the upper half of a cladode is shown. It has numerous areoles and a small, cylindrical leaf associated with each glochid-bearing areole.
1.20. *Austrocylindropuntia subulata,* a South American opuntioid with cylindrical stems and long, cylindrical leaves.
1.21. *Tephrocactus zehnderi,* showing small but prominent cylindrical leaves on a young stem.
1.22. *Opuntia fulgida,* jumping cholla, a common cylindropuntia of North America with small, ephemeral leaves.

tious roots can also form on the lower side of a stem that is still attached to the mother plant; or, as we have just mentioned, creeping epiphytes may root against nearly any surface.

Perhaps the most remarkable case of vegetative reproduction in the cactus family is the creeping devil, *Stenocereus eruca* (Fig. 1.25), which is endemic (restricted) to barren sandy flats in the Magdalena region of Baja California. This is truly one of the world's most curious plants. Possessing horizontal stems that lie on the ground (procumbent stems), it does indeed creep, rooting as it goes and dying away at the base.

A plant of great size, such as a large treelike cactus, must be supported by an internal framework of wood. It is not obvious that nearly all cacti, even small or procumbent ones, have wood in their perennial stems and roots. In fact, almost all cacti are long-lived, aboveground perennials and, therefore, are woody plants, not herbs. In the past, researchers

Figure 1.23. *Pereskia grandifolia,* a commonly cultivated species of leaf-bearing cactus that has large, relatively thin leaves and fruits that are borne on short but conspicuous pedicels.

have been misled because the wood in the stem, which is often soft, is obscured by the presence of succulent tissues.

There are, however, some herbaceous cacti, especially the smallest epiphytes. Probably the most familiar herbaceous perennials of Cactaceae are the wilcoxias *(Wilcoxia).* One to several slender stems arise from a set of tuberlike roots. The shoot grows to less than a meter in length and produces a showy flower and spiny fruit before dying back to ground level. This type of life form is called a geophyte (Greek, earth plant). A handful of highly specialized, dwarf species belonging to unrelated lineages of cacti have independently evolved as geophytes, generally in sandy habitats.

Joints

The diversity of form within the cactus family is nowhere better displayed than in the large genus *Opuntia,* which encompasses about 160 species and occurs from southern Canada in the north to Tierra del Fuego in the south and on nearly all islands where cacti are found. Most species of the genus are jointed.

A joint is a shoot segment that arises abruptly from an old stem areole and is clearly demarcated by a relatively narrow base (Figs. 1.20–1.22, 1.26). Hence, a long branch often consists of a series of discrete segments organized in a straight or zigzag manner. These individual units can often detach at their narrow base; they can fall and root where they land or they can be carried involuntarily by an animal to another location where they can produce adventitious roots and begin a new plant. Although joints are best known from opuntias, they also occur in other groups of cacti, for example, the popular Christmas cactus, *Schlumbergera truncata* (Fig. 1.27).

Joints of opuntias may be cylindrical, club-shaped,

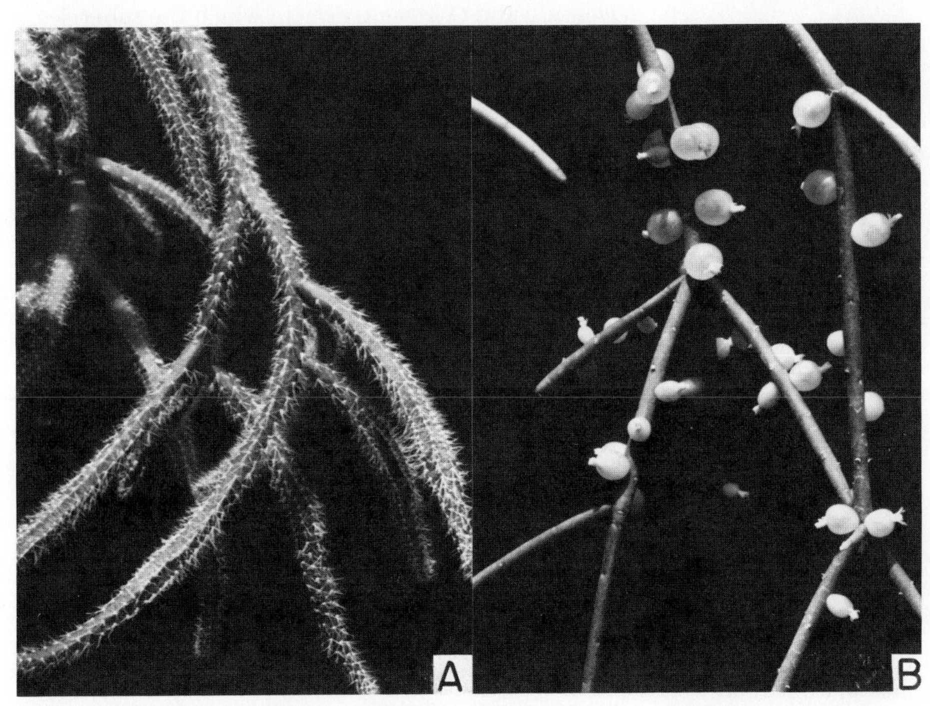

Figure 1.24. External features of *Rhipsalis baccifera.* *(A)* Juvenile stems of the plant, which have low ribs covered with many thin, relatively soft spines. *(B)* Adult stems of the same plant; they are cylindrical and lack conspicuous spines. Spherical structures on the stems are whitish fruits.

Figure 1.25. *Stenocereus eruca,* formerly called *Machaerocereus eruca,* the creeping devil of western Baja California. The stems of this cactus lie on the ground.

Figures 1.26 and 1.27. Joints of cacti.
1.26. *Opuntia prolifera,* a cylindropuntia, which produces a set of joints at the beginning of each growing season.
1.27. *Schlumbergera truncata,* Christmas cactus, which grows by the addition of flattened, two-ribbed joints.

globose, or laterally flattened. Shrubby species of North America with cylindrical stems are called chollas (or cylindropuntias). Cholla joints are not really cylindrical because they have prominent tubercles covering their surface (Fig. 1.22); thus the stem rarely appears circular in cross section. Only in some very thin-stemmed pencil chollas, such as *O. leptocaulis* and *O. ramosissima,* in which the tubercles are very low are the stems actually cylindrical. Club-shaped and even globose joints characterize the South American genus *Tephrocactus* (Fig. 1.21). Finally, many species occurring throughout the New World have flattened stems, called cladodes. In Figure 1.19, which shows the upper portion of the cladode, you can observe the areoles on one broad surface and along the edge. Species with cladodes are called platyopuntias (*platy* means flat or platelike) or prickly pears. These cladodes, which are often called pads, can fool amateurs because they mimic leaves; they have many shapes, ranging from circular to elongate with the narrow end either at the base or at the tip. Cladodes generally lack prominent tubercles.

The tallest species of *Opuntia* are platyopuntias, and they can be as tall as large arborescent, columnar cacti. Many platyopuntia species occur as trees and shrubs in both North and South America. Probably the most famous trees are the photogenic prickly pears of the Galápagos Archipelago, *O. echios* and *O. megasperma,* which have massive trunks and can be over 10 m in height. These cacti were closely observed by Charles Darwin while he was developing his famous evolutionary theory.

The largest chollas have a central trunk; but, unlike the arborescent platyopuntias, these species rarely achieve a height of 4 m. In addition, there are many small caespitose and low mat-forming species of platyopuntias and cylindropuntias; and several species, most notably *Opuntia chaffeyi*, are true herbaceous perennials with a large storage root.

In *Opuntia* there are two types of spines: relatively thin, sharp, persistent spines and short, deciduous, barbed glochids. Glochids are those nasty, irritating hairlike structures one invariably gets in the skin by touching a plant or even the soil beneath it. Glochids are a practical reason why many serious cactus hobbyists do not cultivate *Opuntia*. In our garden, we have positioned our six species in places we do not want people to enter.

People who have looked at a young opuntia pad or joint have seen cactus leaves, although they may not be aware that they have. The leaves are narrowly cylindrical (terete) structures more than a centimeter long. They persist for several weeks before they dry and fall off, usually when the largest spines on the accompanying areole are becoming conspicuous. Thus, for a short while these minute leaves may take up carbon dioxide and make sugars; but in opuntias as in most cacti, the chief photosynthetic organ is the stem.

General Physiological Characteristics

Most cacti live in deserts and semiarid habitats in which rainfall is low and the amount of water present in the soil is very limited for much of the year. Under these dry conditions the key to survival is water management. How does a desert plant use the water available in the soil to grow, flower, and set seed before the soil dries out and growth stops or parts are killed by lack of water? Interestingly, there is no single strategy for all desert plants. For example, some plants are spendthrifts and use water freely, only to die or lose shoots when the supply is exhausted or cannot keep up with the demand. Some plants pace themselves and are more economical in water usage until the soil dries. Others are able to grow slowly or retain their leaves even when the soil is very dry. Still other plants rely upon their own large water reserves, and this strategy, of course, is what has made cacti so famous.

Any cactus enthusiast would be happy to tell you how long a certain cactus plant in the collection survived without any water. To propagate a cactus stem the common practice is to lay the part on a dry bench until the cut end calluses over and adventitious roots begin to form. This process often takes months, but occasionally a specimen is forgotten and remains unplanted for one to several years without lethal consequences. Cacti typically first shrivel and die from the base upward, so that even after several years the shoot apex may be perfectly capable of growing, although all the lower portions of the stem and any lateral branches are dead. Some cacti with a water content as low as 20% can resume growth, whereas a nonsucculent often dies when water content is below 50%.

A remarkable feature of cacti is that, when they are fully hydrated, water constitutes 90 to 94% of the stem. At maximal water-holding capacity the ribs and tubercles are swollen until they are turgid and bowed outward. Over the course of weeks and months, as water is lost through the surface (transpiration), the ribs and tubercles flatten and collapse on themselves, thus forming more narrow valleys and a crinkled surface appearance. Hence, in a ribbed species, the stem can expand and contract seasonally like an accordion.

Figure 1.28 shows the very gradual decrease of water content that occurred in the stem of a large barrel cactus, *Ferocactus acanthodes*,* during an extended drought period in the northwestern section of the Sonoran Desert. Even after transpiring for 2 months following a major rainy period, the stem still had a higher water content than that normally found in a nonsucculent crop plant like peas or tomatoes after 1 day of active transpiration.

Another remarkable feature of cacti is the rapidity with which the stem can recover water. For example, Figure 1.28 shows the water recovery of *F. acanthodes* following drought. Water uptake into the stem is noticeable in 12 hours, and the stem can become fully hydrated in a few days. Water uptake can even occur after a small rainfall of only 7 mm (about $\frac{1}{4}$ inch). In the case of an adult barrel cactus, the amount of water absorbed in a few days from a recently moistened soil can be about 20 liters (just over 5 gallons). In contrast, it would take about 40 days to lose this same amount of water through transpiration.

The rapid uptake of water is even more interesting if one studies the root system. During periods of drought the shallow roots typically have relatively few new lateral branches or other whitish roots that are characteristically involved in most of the water absorption—indeed, nearly the entire root system is covered by a flaky orange-brown bark (periderm). When water is applied to old roots, root buds hidden

* According to Nigel Taylor of the Kew Gardens in England, this well-known barrel cactus from the southwestern United States should be called *Ferocactus cylindraceus* (Engelm.) Orcutt. N. P. Taylor, "Notes on *Ferocactus* B. & R.," *Cactus and Succulent Journal* (London) 41(1979):88–94.

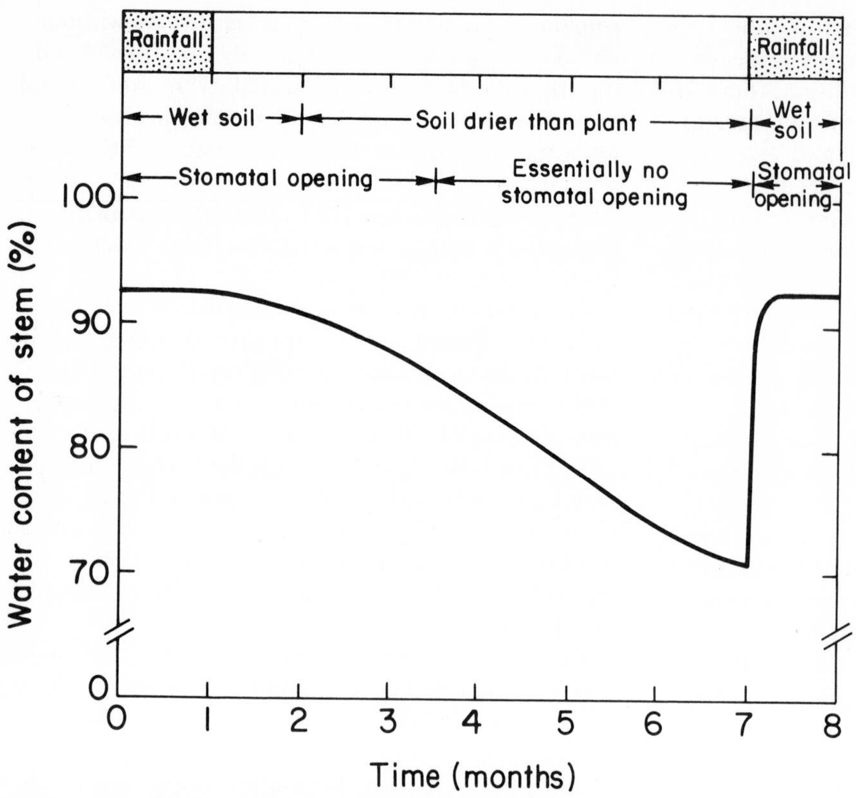

Figure 1.28. Relationships between rainfall, soil water, nocturnal stomatal opening, and stem water content of a barrel cactus, *Ferocactus acanthodes,* in the western Sonoran Desert. The plant was 34 cm tall. (Data are adapted from Nobel, 1977, and unpublished observations of P. S. Nobel.)

inside the root (primordia) elongate rapidly, often without much cell division, to form rain roots—a change that increases the capability for water uptake. Moistening of the soil also leads to rehydration of old roots, which greatly increases their conductance in a matter of hours—these rehydrated roots take up water very readily from the soil. When the soil water content plunges, rain roots shrivel and are shed, the conductance of the old roots decreases, and once again the plant returns to the low-conductance root system appropriate for drought conditions. These changes mean that the precious water in the stem is not readily lost back to the soil.

No detailed study has yet been made on the pathway of water movement in cacti from the roots to the stem and eventually to the areoles at the top. In typical nonsucculent dicotyledons, bulk movement of water occurs via the xylem (secondary xylem is wood), in which there are long, hollow conduits called vessels; this water pathway is fairly simple and direct. In cacti, the vascular network that includes the xylem is quite complex, consisting of the major ring of bundles leading to the shoot tip and to each areole plus, frequently, a separate system in the fleshy cortex (outer tissue including ribs and tubercles) and sometimes also another system in the pith (central, spongy tissue). The cells of the stem cortex are loosely packed, and water can diffuse as a vapor through the intercellular air spaces. The stem tissues of cacti often have large quantities of mucilage, which is hydrophilic (water-loving), and this may affect hydration. Dye experiments and other measurements are needed to demonstrate the actual water course during rehydration of desiccated cacti.

Water management by succulent plants such as cacti only makes sense when it is studied in the context of a special physiological process called Crassulacean acid metabolism (CAM). Plants using the CAM pathway open their stomates (numerous small pores that occur over essentially the entire stem surface; Figs. 1.29–1.31) and take up carbon dioxide mainly at night, when temperatures are low. Stomates are closed during most of the day, when temperatures are high. Yet any opening of stomates inevitably causes loss of water from the plant, although opening stomates primarily at night has important water-conserving consequences. Such timing is in marked contrast to that of most leafy plants (such as peas, tomatoes, corn, and maple trees), which open their stomates during the daytime.

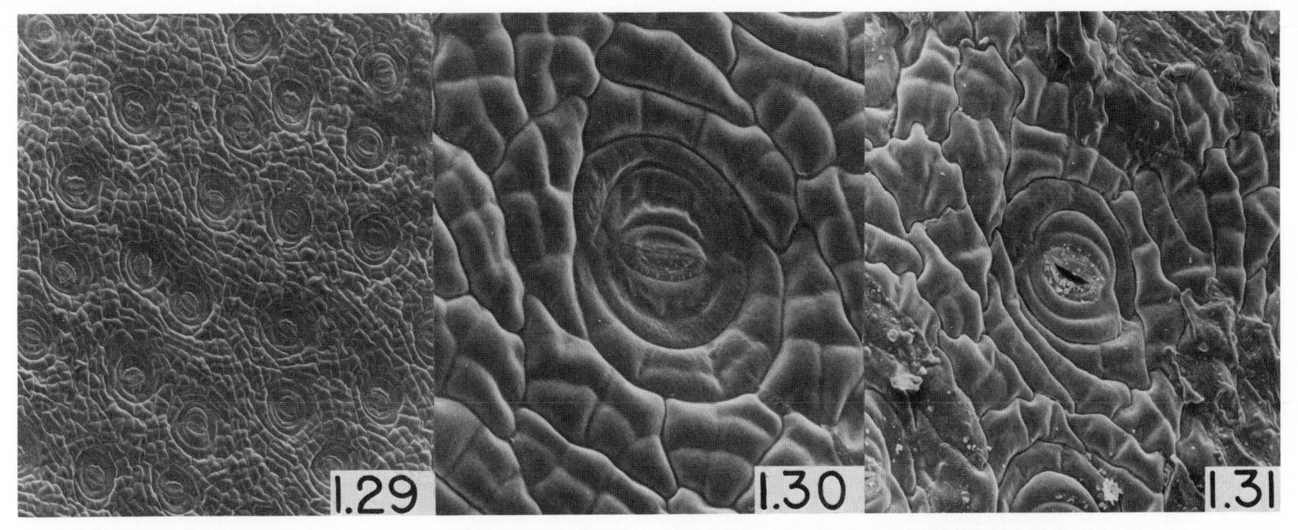

Figures 1.29–1.31. Scanning electron photomicrographs of cactus stems.
1.29. *Stenocereus gummosus,* or agria, from Baja California, Mexico. The rib surface (area of photograph, 0.75 mm^2), which has had the wax removed by chemicals, has slightly raised epidermal cells (called the groundmass) and irregularly spaced stomates with their associated cells.
1.30. A close-up of the agria in Figure 1.29, enlarged five times, to show a single pair of closed guard cells (no stomatal pore) surrounded by crescent-shaped subsidiary cells that are situated parallel to the long axis of the guard cells—a pattern called parallelocytic. The individual epidermal cells, which are clearly discernible, occur in discrete clusters that look like shapes from a butcher's shop. Each cluster of groundmass cells is the product of one initial cell that divides repeatedly; in this way the epidermis of this species expands greatly in surface area.
1.31. *Stenocereus alamosensis,* or cina, from Sonora, Mexico; a close-up of the rib surface at the same magnification as Figure 1.30, showing a partially open stomate, which is about 0.02 mm in length. Once again, the wax was chemically removed to reveal the nature of the epidermal cells, but the wax remains on the corners of the photomicrograph. In this species, the clusters of epidermal cells are less well developed.

The rate of transpiration—the movement of water vapor out of a plant, in this case from the stem—is determined by the amount of opening by the stomatal pores and the difference in water vapor concentration or pressure between the inside of the stem and the air next to the stem. The intercellular air spaces inside the stem are nearly saturated with water vapor, and this saturation vapor concentration is very dependent on temperature. Indeed, the ability of air to hold water increases approximately exponentially with temperature. Thus, if a plant can open its stomates when the temperature is low, as at night, then water savings can be tremendous. We can relate transpiration, which technically is the amount of water leaving a unit area of the stem surface in unit time (symbolized by J_{wv}, where J is the conventional symbol for a flux density and the subscript wv indicates water vapor), to the stomatal conductance for water vapor (g_{wv}), which describes the degree of opening of the stomtal pores, times the drop in water vapor concentration from the stem to the air (Δc_{wv}):

$$
\begin{aligned}
J_{wv} &= g_{wv} \Delta c_{wv} \\
&= g_{wv} (c_{wv}^{ias} - c_{wv}^{air})
\end{aligned}
\qquad (1.1)
$$

where c_{wv}^{ias} is the water vapor concentration in the intercellular air spaces adjacent to the chlorophyll-containing cells of the stem (essentially a water-saturated environment) and c_{wv}^{air} is the water vapor concentration of the ambient air.

Let us consider the case where transpiration occurs into a representative desert atmosphere with a water vapor concentration in the air of 8.1 g m^{-3}, as would occur for 35% relative humidity at 25°C. (Unless there is a change in the weather, water vapor concentration tends to be relatively constant throughout a day, whereas relative humidity can change markedly as the air temperature changes.) We will use three stem temperatures: 40°C, 25°C, and 10°C. For the same degree of stomatal opening (the same g_{wv}, where g_{wv} increases as the stomatal pore becomes larger), transpiration is decreased 65% at 25°C (relative to that at 40°C) and is decreased fully 97% at 10°C (again, relative to that at 40°C). Thus, stomatal opening at night, when stem temperatures tend to be lower, results in a considerable savings of water and allows CAM plants like cacti to exist in habitats where little water is available. Later we shall use Equation 1.1 to calculate actual transpiration rates.

Let us next consider gas exchange of a typical CAM plant over an entire 24-hour period. Beginning at sunset, the following events occur. When the sun goes down, photosynthesis ceases because it requires light. At this time the stomates open, thereby allowing carbon dioxide to enter the chlorophyll-containing cells (chlorenchyma) and be incorporated into simple organic acids, which are stored in the cell vacuole (sac containing mostly water). The vacuole fills with these organic acids, and so its pH becomes very low (for example, pH 3). When the sun rises the next morning, the stomates close, light is trapped by the chlorophyll-containing structures in the cell (chloroplasts), and carbon dioxide is split off the organic acids and combined with other simple organic compounds to make sugars. Hence, unlike the uptake of carbon dioxide by typical green plants with stomatal opening during the daytime, CO_2 uptake by cacti is temporally separated from the manufacture of sugars. Moreover, there is such a large pool of organic acids in the chlorenchyma that sugar production can continue for the entire lighted interval without having the stomates open.

Researchers use water-use efficiency (grams of carbon dioxide fixed in photosynthesis per kilogram water vapor lost by transpiration) as a measure of water management. Indeed, the nocturnal opening of stomates greatly enhances the water-use efficiency of cacti compared to that of non-CAM plants. For instance, the water-use efficiency of *Ferocactus acanthodes* in California can be 14 g of carbon dioxide fixed per kg of water vapor lost over an entire year, whereas values for crop plants are generally 1 to 3 g CO_2/kg water. The high water-use efficiencies for cacti also reflect the fact that the stems can essentially seal themselves off from the surrounding air. Thus, during periods of drought cactus stomates stay closed and very little water is lost by the stem. At such times the small amount of carbon dioxide released by respiration (the production of energy when oxygen is combined with sugars) remains locked within the plant and is refixed by photosynthesis during the daytime. This strategy also enables severed stems to remain alive for many years—until they can place roots into the soil.

There are trade-offs in any design, and a cactus is no exception. First, a cactus stem generally has very little transpiration during light periods, so there can be almost no heat dissipation through evaporative cooling. Hence, the temperature of the sunlit part of a cactus stem can rise to extremely high temperatures, 55°C or more. These temperatures would "cook" most plants by destroying their proteins; but, as we shall see, cacti can tolerate these and higher temperatures without death to the cell contents (protoplast). Second, CAM is a somewhat cumbersome process; therefore, the overall productivity and growth rates of cacti tend to be low, especially in low-light environments. Indeed, maximal carbon dioxide uptake by cacti requires the stems to be in full sunlight without any shading. Certain morphological features of stems, such as the orientation of terminal cladodes on platyopuntias, may be interpreted as attempts to maximize the interception of sunlight (see Chapter 8). In summary, cacti have a number of interesting physiological properties that permit these plants to survive—even thrive—in some very inhospitable, dry locations.

Reproductive Structures

With practice anyone can identify a cactus and not confuse it with species of other plant families that have succulent stems. Generally a cactus can easily be recognized in the vegetative condition by judging the features described earlier in this chapter: growth habit; presence of ribs or tubercles or both; presence of areoles with spines; and absence of conspicuous foliage leaves. Nonetheless, there are many species of stem succulents, especially in Africa, that possess many of these same features, particularly those in the genus *Euphorbia* (family Euphorbiaceae). In fact, throughout this century evolutionists have commonly cited the similarity of the stems of succulent cacti of the Western Hemisphere and the succulent, cactus-like euphorbs of Africa as an example of convergent evolution. One way to distinguish succulent euphorbs from cacti is by the presence of copious white latex. There are, however, some latex-bearing species of cacti, as in certain species of *Mammillaria.* Therefore, the only reliable features for distinguishing cacti from other plants are the reproductive features, namely, the flowers and fruits, the designs of which are unique to each plant family.

In a family as large and as diversified as Cactaceae, it is not easy to find features that are uniformly present or to make generalizations that do not have important exceptions. Therefore, to set the stage for later discussions, we first shall describe the reproductive features of a single species that is "typical" of the family and subsequently shall enumerate the variations.

Our example is organ pipe cactus, *Stenocereus thurberi* (Fig. 1.32), which is the namesake of a national monument located near Lukeville, Arizona. This cactus has showy, radially symmetrical flowers that are 6 to 8 cm in length and are borne near the tip and along the sides of erect branches (Fig. 1.32*A–C*). Flowers are sessile (lacking a basal stalk, or peduncle) and solitary (only one flower produced per areole). After the flower drops from the plant or

Figure 1.32. Reproductive structures of organ pipe cactus, *Stenocereus thurberi.*
(A) Unopened flower bud that is situated on the upper side of an areole; there are a few inconspicuous spines present on the pericarpel.
(B) A flower open at night, showing its whitish perianth parts. Just above the flower is an immature fruit with well-developed spines, which grow after fertilization. The dry remnant of the floral tube persists above the developing fruit. The dark spots above certain stem areoles indicate where flowers have been produced in previous years.
(C) A flower *(center)* closing after dawn and a young bud that will grow much larger before opening in several days. The bracts of the floral tube show the gradual transition to the perianth parts.
(D) Two mature, spiny fruits; on the lower one, the dry floral tube is still present.
(E) A mature fruit *(center)* that has been pecked open by a bird, thus exposing the juicy, red pulp and the small, dark seeds. Above the mature fruit is an immature fruit with a persistent floral tube.
(F) Scanning electron photomicrograph of a seed (1.5 mm in length) that has been magnified many times. The scar at the top of the seed is the hilum, the point of attachment of the seed to the funiculus. On the seed coat (testa) can be observed the individual cells, which are fairly flat in this species.

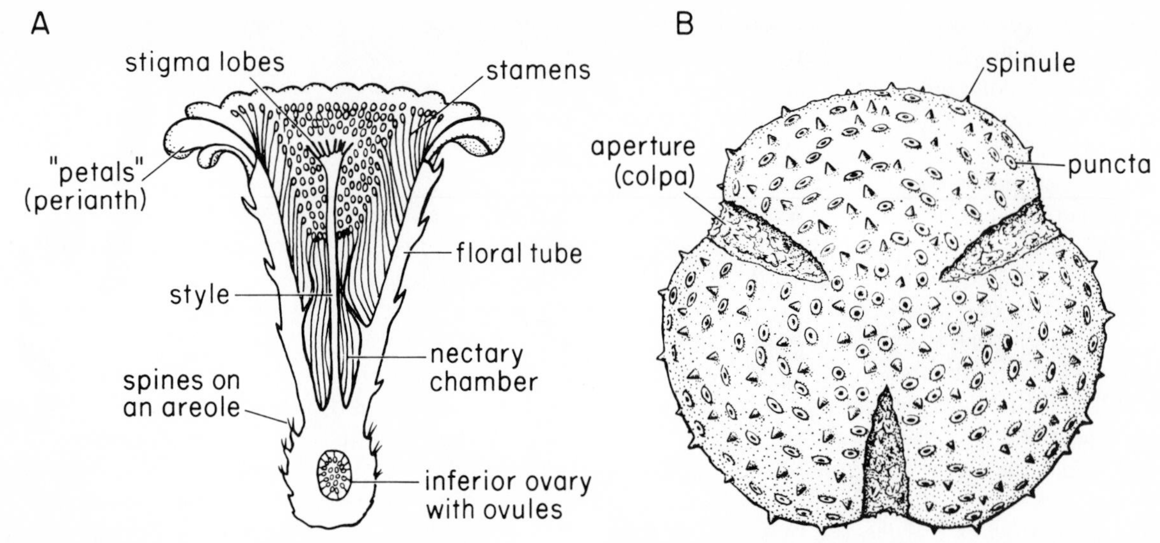

Figure 1.33. Organ pipe cactus, *Stenocereus thurberi.* *(A)* Longitudinal section through a flower, showing the major structures and the inferior position of the ovary. *(B)* A pollen grain, which is about 0.1 mm in diameter, for which the prominent features of the surface (exine) are labeled. (Redrawn from Leuenberger, 1976.)

successfully develops into a fruit, growth of the areole is arrested because the growing tip of the areole (areolar meristem) is used up in the production of the flower.

The most fundamental characteristic of a cactus flower is the inferior ovary (Fig. 1.33*A*)—the ovary (female structure) is positioned below (inferior to) the rest of the flower parts, or, when you reverse your perspective, the stamens (male structures) and showy, leaflike parts (perianth parts) arise from the top of the ovary (epigyny: above the female). An inferior ovary is always the hallmark of a highly specialized group of plants. Other families with inferior ovaries are the sunflowers, orchids, irises, and carrots. Much scientific research has shown that the primitive condition for flowering plants was the superior ovary (hypogyny: below the female); in this condition the stamens and the perianth parts (sepals and petals) arise from the base of the ovary. In most plant groups, the inferior ovary evolved when the base of the perianth formed into a tube or cup (hypanthium) and this became fused with the ovary, either to protect the ovary from mechanical injury or to form an extra coating around the ovary and thereby promote the development of fleshy fruits. In cacti, however, there is substantial evidence to show that this was not the origin of the inferior ovary; rather, the ovary sank into the stem tissues (see Chapter 2). The consequence of this sinking is that areoles of the stem cover the outside of the cactus ovary. Because this covering is a novel structure, the outer wall of the flower is called the pericarpel. Areoles on the pericarpel of an organ pipe cactus flower are located above vestigial leaves and have abundant hairs; and the upper areoles of its flowers may have thin, sharp spines (Fig. 1.33*A*).

Above the ovary–pericarpel complex is a floral tube (Fig. 1.33*A*), which arises as a solid, narrow funnel 2 to 3 cm long. The outside of the floral tube is covered by leaflike structures (Fig. 1.32*C*). At its base the floral tube has areoles, each with a scalelike leaf blade (lamina) and hairs (trichomes); but, progressing upward, one sees that the areoles do not develop and the lamina are longer and more like petals. Forming the entrance to the flower are 25 or more separate (free), pinkish-white "petals." In typical dicotyledons the outer floral parts are called sepals (collectively, the calyx) and the inner ones are called petals (collectively, the corolla); but in cacti, where they form as a graded series, workers merely call both sets perianth parts.

On the inside of the flower, the floral tube is covered by hundreds of white, pollen-bearing stamens, and the bases of the lowest stamens line the base of the floral tube to form a nectary chamber, which is hidden by the stamens. All stamens are approximately the same length (about 2 cm); thus, the uppermost stamens are exserted evenly beyond the perianth (Fig. 1.33*A*), as in an old-fashioned shaving brush. From the top of the ovary and arising through an opening in the staminal brush is the white style, which has 11 or 12 distinct stigma lobes (the areas where the pollen is received).

Flower opening (anthesis) is nocturnal; and the strongly scented, funnel-shaped flowers attract bats. Pollen (Fig. 1.33*B*) from one flower is transferred to the slightly exserted stigma lobes of another flower when a bat buries its head in the flower to extract the nectar from the hidden nectary chamber. The flowers remain open the next morning and may be pollinated then by diurnal animals, such as bees. Over the following days the floral tube and perianth shrivel but remain attached to the developing fruit.

At anthesis, the large central cavity of the ovary (locule) is filled with hundreds of small, white, unfertilized seeds (ovules) that are borne on branched stalks (called funiculi) and thereby attached to the inner ovary wall (parietal placentation; Fig. 1.33*A*). The single locule is in reality a multiple structure (called a multicarpellate, or compound, ovary) in which the walls separating the individual chambers have been evolutionarily lost. When fertilized, the ovules grow to about 1.5 mm in length and form a dark seed coat (testa). These mature seeds are embedded in a sweet, red pulp. Meanwhile, formidable clusters of spines develop on the pericarpel areoles, and at maturity the pericarpel surface also turns red and is fleshy (Fig. 1.32*E*). Hence, this fruit is a many-seeded berry. Red coloration and the rewards of sweet pulp and seeds attract birds, which feed upon the pulp and inadvertently distribute the seeds via their excrement.

A mature seed (Fig. 1.32*F*) contains a markedly curved embryo that encircles the central nutritive tissue. In typical angiosperms the nutritive tissue in the seed is endosperm, composed of triploid (three copies of each type of chromosome, designated $3n$) cells and that is formed by the fusion of a sperm nucleus derived from the male pollen grain with two haploid (one copy of each type of chromosome, designated n) nuclei found in the embryo sac of the female. But in cacti the central food tissue in the seed is called perisperm because it is diploid (two copies of each type of chromosome, designated $2n$) and derived from special cells of the mother plant only.

The number of seeds produced by a single plant of *Stenocereus thurberi* may be enormous, but very few seeds ever chance upon the proper circumstances for germination. Germination occurs when seeds are exposed simultaneously to adequate moisture, sunlight (or its physiological equivalent), and a warm temperature. Within a week under optimal germination conditions, a seed will produce a small seedling with a succulent hypocotyl–root axis and two short, succulent, triangular seed leaves (cotyledons). Many months are required before this seedling develops a cylindrical body with ribs.

Table 1.1 lists some of the noteworthy ways that the reproductive features of cacti depart from the fairly general design we have just described. For example, the flowers of cacti come in many different lengths, shapes, and colors; and each construction attracts specific types of animal pollinators, such as bees, hummingbirds, bats, and nocturnal moths. Consider the flowers of species that are pollinated by hummingbirds (Figs. 1.34 and 1.35), such as cina or rathbunia *(Stenocereus alamosensis)* from Sonora, Mexico, and any of numerous species of *Borzicactus* from Peru, which have long, narrow, tubular flowers that open during the morning and are red. These flowers tend to be produced along the length of the vertical stems. The hummingbird is attracted to red flowers, and it hovers at the flower, inserts its long bill, and extracts the copious nectar at the base of the floral tube. The form of these flowers has evolved specifically to fit the bird's feeding strategy and is specialized enough to exclude organisms that cannot fit in the narrow tube or are not attracted to red or are active only at night. In fact, mechanisms of pollinator attraction have affected flower evolution in cacti (see Chapter 10). Later (Chapter 5) we also shall describe some of the unusual ways that flowers are borne in special cacti, as in a dense terminal head (a cephalium; in *Melocactus* and *Backebergia*), in rings or vertical strips, and even on separate growing points (flower meristems) separated from the spine-producing areole.

Even though the succulent euphorbs mentioned

Table 1.1. Reproductive features in Cactaceae.

Feature	Alternatives or range[a]	Examples
Length of flower	0.5 cm to	*Disocactus rhamulosus* and *Rhipsalis* spp.
	37 cm	*Epiphyllum chrysocardium, E. grandilobum,* and *Mediocactus megelanthus*
Symmetry of flower	**Radial**	—
	Zygomorphic	*Borzicactus, Cleistocactus,* and *Schlumbergera*
Number of flowers per areole	**1** to	—
	Numerous	*Disocactus eichlamii, Lophocereus,* and *Myrtillocactus*
	As chains	*Opuntia fulgida*
	As inflorescences	*Pereskia grandifolia*
Position of flower relative to tubercle	**Apical** to	—
	Basal	*Mammillaria*

Table 1.1 ***(continued)***

Feature	Alternatives or range[a]	Examples
Longevity of flowering areoles	**Single season**	—
	Perennial	*Myrtillocactus* and *Neoraimondia*
Attachment of flower	**Sessile** to	—
	Stalked	*Pereskia*
Position of ovary	**Inferior** to	—
	Almost superior	*Pereskia*
Shape of floral tube	Broadly or narrowly funnel-shaped to	—
	Tubular	*Borzicactus* and *Cleistocactus*
	Bell-shaped	*Mitrocereus*
	Disk-shaped	*Opuntia spinosior*
	Almost absent	*Rhipsalis*
Color of perianth parts	White to	—
	Pink, rose, red, or purple	—
	Yellow or orange	—
	Green	—
	Blue	*Disocactus amazonicus*
Structures on the pericarpel areoles	**Spines**	—
	Hairs	—
	Bristles	*Pachycereus*
Size of the bracts on the pericarpel	Large to	—
	Small or absent	—
Nature of the bracts on the pericarpel	**Flexible** to	—
	Stiff	*Escontria chiotilla*
Number of stamens	Relatively few to	—
	Over 1500	*Carnegiea* and *Selenicereus*
Nature of the nectary chamber	**Open** to hidden	—
Time of flower opening (anthesis)	**Day** to	—
	Night	Many
	Never open	*Frailea* and *Melocactus*
Number of stigma lobes	1 to over 20	—
Number of ovules (and seeds)	Fewer than 20 to	*Pereskia grandifolia*
	Several hundred	—
Occurrence of sexes in flowers	**Hermaphroditic**	—
	Functionally dioecious	*Mammillaria dioica*
	Dioecious	*Opuntia stenopetala*
	Monoecious or gynodioecious	*Selenicereus innesii*
Nature of fruit	**Juicy berry** to	—
	Berry that splits open	*Carnegiea*
	Dry fruit that does not split open	*Opuntia basilaris*
Color of fruit pulp	**Red to purple**	—
	White	*Mitrocereus*
	Clear	*Stenocereus dumortieri*
Size of seeds	Less than 0.5 mm to	*Blossfeldia* and *Rhipsalis*
	Over 6.0 mm	*Backebergia, Opuntia,* and *Pereskia*
Color of seeds	**Black** to	—
	Brown	Many
	Reddish black	*Carnegiea*
	White (with a bony covering, or aril)	*Opuntia*
	Tan	*Pterocactus*

a. Conditions set in boldface occur in at least two-thirds of all species in the family.

Figures 1.34 and 1.35. Hummingbird-pollinated flowers of cacti.
1.34. *Stenocereus alamosensis,* formerly *Rathbunia alamosensis,* a common shrubby species of the Sonoran Desert that is classified in the tribe Pachycereeae and has narrowly tubular, terracotta red flowers, which are used by hovering hummingbirds.

1.35. *Borzicactus fieldianus,* a small, cylindrical species from western South America, that is classified in the tribe Trichocereeae and has narrowly tubular, hummingbird-pollinated flowers that are yellowish orange and bear long, silky hairs on the floral tube.

earlier resemble cacti in vegetative form, the two groups differ in their reproductive structures. Close inspection of fertile organs on a cactuslike species of *Euphorbia* reveals a very unusual structure (Figs. 1.36 and 1.37). What might appear to be a stalked flower is not a single flower but rather an aggregate of numerous flowers, a very highly reduced inflorescence that is given the special name cyathium. A cyathium, which is unique to the Euphorbiaceae, consists of one female (pistillate) flower, often borne on a pedicel, and a number of male (staminate) flowers that frequently have only one stamen and are somewhat hidden within a cuplike structure formed by the fusion of several small, leaflike structures (called bracts). In many species petallike structures around the top of the cup may be brightly colored

Figures 1.36 and 1.37. Reproductive structures of cactuslike species of *Euphorbia.*
1.36. The cyathium (inflorescence) consists of a central female flower surrounded by several stamens, each of which is actually a highly reduced male flower, that are enclosed by five petallike bracts.

1.37. A cluster of the strongly three-lobed, capsular fruits. The persistent "beak" on each fruit is the stigma and style. When mature, the fruit is dry and splits open to release the seeds.

(for example, the red lobes in the crown of thorns, *Euphorbia milii* var. *splendens*). The marked reproductive differences between cacti and cactuslike euphorbs do not stop here but extend into other features. For example, *Euphorbia* has a dry fruit with three chambers (Fig. 1.37), each of which contains only one seed that is attached to the central axis of the fruit (axile placentation). Hence, if one has any fertile material of a specimen, members of these two families can be readily distinguished.

If cacti are not closely related to Euphorbiaceae, then which families are their closest relatives? Many botanists have struggled with this question. Four lines of evidence all indicate that the Cactaceae belong to a group of families called the centrosperms, or more precisely the order Chenopodiales. Cacti share the following characters with about 10 other families: (1) a strongly curved embryo surrounding perisperm in the seed; (2) similar types of pollen grains; (3) unique red and yellow pigments, the betalains, known only from this order; and (4) a special type of protein inclusion in the sugar-translocating tissue (phloem). Families that appear to share many specific characters with the cacti are the portulacas (Portulacaceae), the didiereas (Didiereaceae), and the pokeweeds (Phytolaccaceae and segregates). (See Chapter 11 for a discussion of the origin of the cactus family.)

Classification

For most plant families, the task of naming and classifying species (taxonomy) has had a long and complicated history, and this is especially true for the Cactaceae. In his seminal work *Species Plantarum* (1753), where taxonomy of plants officially began, Carolus Linnaeus named only 22 species of cacti, which he classified in a single genus *Cactus*. In this one group (taxon) he classified arborescent columnars, epiphytes, barrel cacti, dwarf growth forms, platyopuntias, and even leaf-bearing cacti. The very next year Philip Miller, a famous British horticulturist, recognized three more genera, *Pereskia, Cereus,* and *Opuntia*—all pre-Linnaean names—in which he included the species described by Linnaeus. From those humble beginnings we now have over 11,000 published Latin binomials (genus–species names) for cacti, most of which are illegitimate, invalid, or incorrect. For example, the mat-forming platyopuntia of the temperate regions of the United States, *Opuntia compressa,* which was described by Linnaeus as *Cactus opuntia,* has been described under at least 35 different binomials. Few families have had so many duplicate names as the cacti.

Normally a family is studied one genus at a time—an ambitious taxonomist takes an interest in a small taxon and determines the correct names and probable evolutionary relationships of each species in that genus. This research is frequently published as a detailed generic monograph or perhaps as part of a regional flora. The taxonomic history of cacti, however, has comparatively few monographs describing individual genera. Instead, the taxonomy of Cactaceae has been directed primarily by large descriptive catalogs of the entire family. The most famous of these was *The Cactaceae* by the renowned American botanists Nathaniel Lord Britton and Joseph Nelson Rose; it was published in four volumes from 1919 to 1923. Here Britton and Rose made the first major overhaul. They recognized 123 genera and subdivided the large columnar genus *Cereus* of Miller into many smaller and more homogeneous units. Since then, cactus researchers have gradually redefined these genera of Britton and Rose, sometimes lumping them together, sometimes splitting them apart, and sometimes completely reshuffling the species to be included in each genus. The number of genera has varied from a high of 220 to a low around 30, with many intermediate values.

No one has ever shown total mastery of the entire family, and there are good reasons for this. First and foremost, in terms of species Cactaceae is a very large family—one of the largest in the New World and the second largest family that is essentially restricted to the New World (second to the pineapple family, Bromeliaceae). Only the epiphytic cactus *Rhipsalis* occurs naturally in the Old World tropics. In the New World, cacti are found in all countries and on most islands that have the appropriate climate. Given that no one has the time or funds to visit, study, and collect in all these areas, an author will consequently understand the classification of some groups better than that of others. Second, scientific specimens of cacti, unlike specimens of most plant families, are rarely made because they are nasty things to cut up and dry, so that they are poorly represented in herbaria, which typically provide much of the data for generic monographs. Third, herbarium specimens of cacti are often inadequate because they are fragmentary and because dried specimens do not show many of the important taxonomic features. For succulents like cacti, there is a special need to study both fresh and liquid-preserved materials. Fourth, the family has many polymorphic (morphologically variable) species, so that one cannot be certain about the morphological limits of each species without detailed study of natural populations. Finally, in the evolution of this family an incredible amount of morphological convergence has occurred, so that the same growth form, floral designs, and fruit and seed features have

Table 1.2. Genera of family Cactaceae.

Taxonomic grouping and authority[a]	Estimated number of species	Distribution
SUBFAMILY PERESKIOIDEAE		
Maihuenia Phil.	2–3	Andean Chile and Argentina
Pereskia Miller	15–18	Southern Mexico to Argentina and the West Indies
SUBFAMILY OPUNTIOIDEAE		
Austrocylindropuntia Backeb.	ca. 15	Western South America
Opuntia Miller	Over 160	Southern Canada to southern South America, the Galápagos Archipelago, the West Indies, and many other American islands (includes *Brasiliopuntia, Consolea, Corynopuntia, Cylindropuntia, Grusonia, Marenopuntia, Micropuntia,* and probably *Nopalea*)
Pereskiopsis Britt. & Rose	9–11	Mexico and Guatemala
Pterocactus K. Schum.	9	Southern and western Argentina
Quiabentia Britt. & Rose	3–4	Brazil, Bolivia, and Argentina
Tacinga Britt. & Rose	1–2	Northeastern Brazil
Tephrocactus Lem.	ca. 50	Southern and western South America
SUBFAMILY CACTOIDEAE		
TRIBE BROWNINGEAE		
Browningia Britt. & Rose	5–6	Northern Peru to northern Chile (includes *Azureocereus* and *Gymnocereus*)
Castellanosia Card.	1	*C. caineana.* Eastern Bolivia
TRIBE CACTEAE		
Ancistrocactus Britt. & Rose	4	Southern Texas and northern Mexico
Ariocarpus Scheidw.	6	Texas to central Mexico
Astrophytum Lem.	4	Texas and northern and central Mexico
Aztekium Boedek.	1	*A. ritteri.* Nuevo León, Mexico
Coryphantha (Engelm.) Lem.	70	Western Canada, western United States, and Mexico (includes *Cochiseia* and *Escobaria*)
Cumarinia Buxb.	1	*C. odorata.* Tamaulipas and San Luis Potosí, Mexico
Echinocactus Link & Otto	5–6	Southern California, Nevada, and Texas to southern Mexico (includes *Homalocephala?*)
Echinomastus Britt. & Rose	6	Southern United States to central Mexico
Epithelantha Weber ex Britt. & Rose	2	Southern Texas to Coahuila and Nuevo León, Mexico
Ferocactus Britt. & Rose	ca. 25	Southwestern United States to southern Mexico
Gymnocactus Backeb.	9	Northern and central Mexico, especially the Chihuahuan Desert
Lophophora Coult.	2	Southern Texas to San Luis Potosí and Querétaro, Mexico
Leuchtenbergia Hooker	1	*L. principis.* Chihuahuan Desert, Mexico
Mammillaria Haw.	ca. 200	Southwestern United States, Mexico southward to Venezuela and Colombia, and the West Indies (includes *Bartschella, Chilita, Cochemiea, Dolichothele, Mammillopsis, Porfiria,* and *Solisia*)
Neolloydia Britt. & Rose	2	Texas to central Mexico
Obregonia Frič	1	*O. denegrii.* Tamaulipas, Mexico
Ortegocactus Alexand.	1	*O. macdougallii.* Oaxaca, Mexico
Pediocactus Britt. & Rose	8	Western United States (includes *Navajoa, Pilocanthus,* and *Utahia*)
Pelecyphora Ehrenb.	2	Chihuahuan Desert, Mexico
Sclerocactus Britt. & Rose	ca. 10	Western United States (includes *Hamatocactus?*)
Stenocactus (Schum.) Backeb.	5 or more	Mexico
Strombocactus Britt. & Rose	1	*S. disciformis.* Hidalgo and Querétaro, Mexico
Thelocactus (Schum.) Britt. & Rose	ca. 10	Texas to Hidalgo and Querétaro, Mexico

Table 1.2 *(continued)*

Taxonomic grouping and authority[a]	Estimated number of species	Distribution
Toumeya Britt. & Rose	1	*T. papyracantha.* New Mexico
Turbinicarpus (Backeb.) Buxb. & Backeb.	6	Northern and central Mexico
TRIBE CEREEAE		
Arrojadoa Britt. & Rose	5 or more	Eastern Brazil
Austrocephalocereus Backeb.	5 or more	Eastern Brazil
Cereus Miller	ca. 30	Eastern West Indies and South America, east of the Andes (includes *Piptanthocereus;* includes *Subpilocereus?*)
Coleocephalocereus Backeb.	10	Eastern Brazil (includes *Buiningia*)
Melocactus Link & Otto	ca. 35	Mexico and the West Indies to eastern Brazil and Peru
Micranthocereus Backeb.	ca. 5	Eastern Brazil
Monvillea Britt. & Rose	20	Northern and eastern South America, east of the Andes (includes *Brasilicereus*)
Stephanocereus Berger	1	*S. leucostele.* Eastern Brazil
Stetsonia Britt. & Rose	1	*S. coryne.* Central Argentina to northwestern Paraguay and southeastern Bolivia
TRIBE ECHINOCEREEAE		
Echinocereus Engelm.	ca. 50	Southwestern United States and Mexico (includes *Wilcoxia striata?*)
TRIBE HYLOCEREEAE		
Acanthocereus (Berger) Britt. & Rose	9	Mexico, Central America, Florida and the entire Caribbean region, northern and eastern South America (probably includes *Dendrocereus*)
Anisocereus Backeb.	1	*A. lepidanthus.* Southern Mexico and Guatemala (inclusion here unconfirmed, but certainly not a species of *Escontria*)
Aporocactus Lem.	4	Mexico
Bergerocactus Britt. & Rose	1	*B. emoryi.* Southern California and western Baja California, Mexico (inclusion here provisional)
Brachycereus Britt. & Rose	1	*B. nesioticus.* Galápagos Islands (inclusion here provisional)
Cryptocereus Alexand.	1	*C. anthonyanus.* Southern Mexico
Disocactus Lindl.	10–11	Southern Mexico to Costa Rica; one species presently assigned to this genus in Brazil (includes *Bonifazia, Chiapasia, Pseudorhipsalis,* and *Wittia*)
Eccremocactus Britt. & Rose	3	Costa Rica and Ecuador (possibly not a natural genus)
Epiphyllum Haw.	11–12	Mexico to Paraguay, West Indies
Harrisia Britt.	ca. 20	Florida, West Indies, and South America from Brazil to Argentina east of the Andes (includes *Eriocereus* and *Roseocereus*)
Heliocereus (Berger) Britt. & Rose	4	Sinaloa, Mexico to Nicaragua
Hylocereus Britt. & Rose	19–20	Mexico northern South America, West Indies, and Florida
Mediocactus Britt. & Rose	3–4	Peru, Brazil to Argentina
Morangaya Rowley	1	*M. pensilis.* Southern Baja California (may belong in Echinocereeae)
Nopalxochia Britt. & Rose	4	Mexico
Nyctocereus (Berger) Britt. & Rose	ca. 6	Southern Mexico, Guatemala, and Nicaragua
Peniocereus (Berger) Britt. & Rose	ca. 15	Mexico, Central America (includes *Cullmannia* and *Neoevansia*)
Selenicereus (Berger) Britt. & Rose	ca. 15	Mexico, the West Indies, and northern South America (includes *Deamia*)
Strophocactus Britt. & Rose	1	*S. wittii.* Amazonian Brazil
Weberocereus Britt. & Rose	4	Costa Rica and Panama

Table 1.2 *(continued)*

Taxonomic grouping and authority[a]	Estimated number of species	Distribution
Werckleocereus Britt. & Rose	2	Chiapas, Mexico, Guatemala, and Costa Rica
Wilmattea Britt. & Rose	1–2	Guatemala, Belize, and Honduras
TRIBE LEPTOCEREEAE		
Armatocereus Backeb.	ca. 10	Ecuador and Peru
Calymmanthium Ritter	1–2	Northern Peru
Jasminocereus Britt. & Rose	1	*J. galapagensis*. Galápagos Islands
Lemaireocereus Britt. & Rose	1–4	Includes *L. hollianus* Puebla; possibly several species from Ecuador and Peru (inclusion here uncertain; otherwise classified in Pachycereeae)
Leptocereus Britt. & Rose	8–10	West Indies
Neoabbottia Britt. & Rose	2	Hispaniola
Neoraimondia Britt. & Rose	2	Western Peru (includes *Neocardenasia*)
Samaipaticereus Card.	1	*S. corroanus*. Bolivia
TRIBE NOTOCACTEAE		
Austrocactus Britt. & Rose	1 to several	Patagonia
Blossfeldia Werderm.	1	*B. liliputana*. Northern Argentina and Bolivia (probably belongs in *Frailea*)
Copiapoa Britt. & Rose	16–20	Northern Chile
Corryocactus Britt. & Rose	ca. 15	Peru, Bolivia, and northern Chile (includes *Erdisia*)
Discocactus Pfeiff.	5–6	Northern Paraguay, eastern Bolivia, and adjacent Brazil
Eriosyce Phil.	1	*E. ceratistes*. Chile
Eulychnia Phil.	6 or more	Chile
Frailea Britt. & Rose	ca. 10	South America, east side of the Andes (probably should include *Blossfeldia*)
Gymnocalycium Pfeiff.	35 or more	Argentina to southern Brazil and Bolivia, east of the Andes
Lymanbensonia Kimn.	1	*L. micrantha*. Peru
Neoporteria Britt. & Rose	ca. 70	Argentina, Chile, and Peru (includes *Horridocactus, Islaya, Neochilenia*, and *Pyrrhocactus*)
Neowerdermannia Frič	1–3	Northern Bolivia to northern Argentina
Notocactus (Schum.) Frič	20 or more	Southern South America, east side of the Andes (includes *Eriocactus, Malacocarpus*, and *Wigginsia*)
Parodia Speg.	30–50	Southern South America, east side of the Andes
Rebutia Schum.	20 or more	Bolivia and northern Argentina
Rhipsalis Gaertn.	40–50	Northern Mexico and Florida to southern South America, Old World Tropics (includes *Hatiora, Lepismium, Pfeiffera, Pseudozygocactus*, and *Rhipsalidopsis*)
Schlumbergera Lem.	5–6	Brazil (includes *Epiphyllanthus* and *Zygocactus*)
Sulcorebutia Backeb.	10–20	Northeastern Bolivia
Uebelmannia Buin.	2–4	Eastern Brazil
Weingartia Werderm.	5–6	Northern Argentina and eastern Bolivia
TRIBE PACHYCEREEAE		
Backebergia Bravo-H.	1	*B. militaris*. West central Mexico
Carnegiea Britt. & Rose	1	*C. gigantea*. Sonoran Desert of Arizona, Sonora, Mexico, and California (barely)
Cephalocereus Pfeiff.	ca. 20	Mexico, Central America, northernmost South America, and the West Indies (includes *Haseltonia, Neodawsonia*, and *Pilocereus*)
Escontria Rose	1	*E. chiotilla*. West central to southern Mexico
Lemaireocereus Britt. & Rose See Leptocereeae		
Lophocereus Britt. & Rose	2	Sonoran Desert in Mexico

Table 1.2 *(continued)*

Taxonomic grouping and authority[a]	Estimated number of species	Distribution
Mitrocereus Backeb.	1	*M. fulviceps.* Puebla, Mexico
Myrtillocactus Cons.	4	Mexico and Guatemala
Neobuxbaumia Backeb.	6–7	Central to southern Mexico (includes *Rooksbya*)
Pachycereus (Berger) Britt. & Rose	5	Northwestern Mexico to Guatemala (includes *Marginatocereus*)
Polaskia Backeb.	2	Puebla and Oaxaca, Mexico (includes *Heliabravoa*)
Pterocereus Macdoug. & Mir.	2	Southernmost Mexico and Guatemala (inclusion here uncertain)
Stenocereus (Berger) Riccob.	23–24	Arizona, Mexico, Guatemala, the West Indies, and coastal Venezuela (includes *Hertrichocereus, Isolatocereus, Machaerocereus, Rathbunia,* and *Ritterocereus,* but not the type of *Marshallocereus*)
TRIBE TRICHOCEREEAE		
Acanthocalycium Backeb.	6–10	Argentina
Borzicactus Riccob.	20–25	Andes (includes *Arequipa, Bolivicereus, Loxanthocereus, Matucana,* and *Oreocereus,* among others)
Cleistocactus Lem.	25 or more	Central Peru to Argentina
Denmoza Britt. & Rose	1	*D. rhodacantha.* Northwestern Argentina
Echinopsis Zucc.	30 or more	Southern South America east of the Andes (includes *Setiechinopsis*)
Espostoa Britt. & Rose	8–10	Andes (includes *Pseudoespostoa* and *Thrixanthocereus*)
Haageocereus Backeb.	25–30	Andes (includes *Weberbauerocereus*)
Leocereus Britt. & Rose	3–5	Eastern Brazil
Lobivia Britt. & Rose	25 or more	Argentina to central Peru
Mila Britt. & Rose	1	*M. caespitosa.* Western Peru
Oroya Britt. & Rose	3–5	Peru
Trichocereus (Berger) Riccob.	20 or more	Western and southern South America (should be combined with *Echinopsis*)
Zehntnerella Britt. & Rose	2	Northeastern Brazil

a. Some of the genera recognized here may eventually have to be combined with other genera because they are not distinctive enough at the generic level.

appeared in completely unrelated evolutionary branches (lineages, or clades).

For these reasons the naming and classification of cacti has been extremely difficult; and cactus taxonomy has been complicated even more by the publication of a plethora of binomials, including those for minor variants that were described from plants grown under artificial conditions. Indeed, many professional taxonomists have purposely avoided investigating Cactaceae. Yet, despite all this, the species of Cactaceae are now as well known as those in other families of comparable size, thanks especially to the individual efforts of many amateur collectors, who not only have assembled large collections of the variants in a species or genus but also have studied these organisms throughout their geographic range and have published some of their important observations, often in cactus journals, such as the *Cactus and Succulent Journal* of the United States and that of Great Britain.

Whereas the description and naming of cacti has developed very rapidly, particularly in the last 70 years, there is still a great need to evaluate carefully the composition of each genus and to determine how the genera are interrelated. Some of these issues are discussed later in the book. Nevertheless, the reader needs to have some general understanding of cactus classification that can be used with this book and as a guide in reading the cactus literature. For these purposes and on the basis of our current systematic knowledge (which admittedly is woefully incomplete), Table 1.2 provides a classification of the genera, giving the correct spelling and the authority for the names, the number of species, and the geographic range.

Like others before, the classification presented in Table 1.2 recognizes three major subdivisions of the family, which we call the subfamilies Pereskioideae, Opuntioideae, and Cactoideae. Subfamily Pereskioideae consists of two genera: *Pereskia,* which has, as mentioned earlier, the primitive features of all Cactaceae; and the highly specialized, low, caespitose, leaf-bearing cushion plants of *Maihuenia,* which live in the Andean foothills of Chile and Argentina. Opuntioideae are classified into seven genera, although the subfamily now needs a critical study to determine how the many evolutionary branches should be grouped into genera. Species of opuntioids have two derived characters that set them apart from the rest of the family: glochids on the areoles and an aril (typically a white, bony covering that encases the seed). The remaining species—for example, the columnars, the epiphytes, and many species with small growth habits—constitute the very large subfamily Cactoideae (80% of all cactus species).

The subfamily Cactoideae was formerly known as Cereoideae. The official rules of botanical nomenclature require this to be called Cactoideae because included in this subfamily is the type species (that is the namesake of the family), specifically, a species of *Cactus* Linnaeus. But the genus *Cactus* is no longer recognized. The first species Linnaeus listed was *Cactus mammillaris,* and this would have become the type for the genus and the family. Instead, this same species became the type for a new genus, *Mammillaria* of Haworth (1812). Fortunately or not, 100 years later the generic name *Mammillaria* was officially conserved—which means that no one can ever change it—so that the name *Cactus,* which was not conserved, had to become a synonym of *Mammillaria.* All this is just as well because the genus *Cactus* was very artificial and was already a much-confused name. Of course, this is not the only case where a family has a Latin name that stems from an unrecognized genus. Other examples include teas (Theaceae), hollies (Aquifoliaceae), and carnations (Caryophyllaceae), as well as the remarkable case of the Himantandraceae, which consists of one Australasian species, *Galbulimima belgraveana.*

Within the subfamily Cactoideae we have recognized nine tribes. These are modifications of categories invented by the famous Austrian cactologist Franz Buxbaum. Although some of Buxbaum's genera and tribes have since been redefined and species have been moved from one genus to another, by and large the composition of his tribes (as defined in the 1970s) are quite useful in describing the lineages of cacti.

SELECTED BIBLIOGRAPHY

Backeberg, C. 1958–62. *Die Cactaceae.* 6 vols. Gustav Fischer-Verlag, Jena.

——— 1981. *Cactus lexicon,* 2nd ed., trans. L. Glass. Royal Horticultural Society Enterprises, London.

Barthlott, W. 1979. *Cacti: botanical aspects, descriptions, and cultivation,* trans. L. Glass. Thornes, London.

Benson, L. 1969a. *The cacti of Arizona,* 3rd ed. University of Arizona Press, Tucson.

——— 1969b. *The native cacti of California.* Stanford University Press, Stanford.

——— 1982. *The cacti of the United States and Canada.* Stanford University Press, Stanford.

Boke, N. H. 1980. Developmental morphology and anatomy in Cactaceae. *Bioscience* 30:605–610.

Bravo, H. H. 1979. *Las Cactáceas de México,* 2nd ed., vol. 1. Universidad Nacional de México, Mexico.

Britton, N. L., and J. N. Rose. 1919–23. *The Cactaceae.* 4 vols. Carnegie Institute Washington Publication 248, Washington, D.C.

Buxbaum, F. 1950. *Morphology of cacti.* Section I. *Roots and stems.* Abbey Garden Press, Pasadena.

——— 1953. *Morphology of cacti.* Section II. *The flower.* Abbey Garden Press, Pasadena.

——— 1955. *Morphology of cacti.* Section III. *Fruits and seeds.* Abbey Garden Press, Pasadena.

——— 1958. The phylogenetic division of the subfamily Cereoideae, Cactaceae. *Madroño* 14:177–206.

Gibson, A. C., R. Bajaj, J. L. McLaughlin, and K. Spencer. 1986. The ever-changing landscape of cactus systematics. *Annals of the Missouri Botanical Garden.* In press.

Gibson, A. C., and K. E. Horak. 1978. Systematic anatomy and phylogeny of Mexican columnar cacti. *Annals of the Missouri Botanical Garden* 65:999–1057.

Hunt, D. R. 1967. Cactaceae. Pp. 427–467 in *The genera of flowering plants,* vol. 2, ed. J. Hutchinson. Oxford University Press, Oxford.

Kluge, M., and I. P. Ting. 1978. *Crassulacean acid metabolism: analysis of an ecological adaptation.* Springer-Verlag, Berlin–Heidelberg–New York.

Leuenberger, B. E. 1976. Die Pollenmorphologie der Cactaceae und ihre Bedeutung für die Systematik. *Dissertationes Botanicae* 31:1–321.

Marshall, W. T., and T. M. Bock. 1941. *Cactaceae.* Abbey Garden Press, Pasadena.

Nobel, P. S. 1977. Water relations and photosynthesis of a barrel cactus, *Ferocactus acanthodes,* in the Colorado Desert. *Oecologia* 27:117–133.

——— 1983. *Biophysical plant physiology and ecology.* W. H. Freeman, New York.

Nobel, P. S., and J. Sanderson. 1984. Rectifier-like activities of roots of two desert succulents. *Journal of Experimental Botany* 35:727–737.

Rauh, W. 1958. *Beitrag zur Kenntnis der peruanischen Kakteenvegetation.* Sitzungsberichte Heidelberger Akademie der Wissenschaften, Mathematisch-Naturwissenschaftlichen Klasse. Springer-Verlag, Heidelberg.

Shaw, E. A. 1976. The genus *Cactus* Linn. *Cactus and Succulent Journal* (Los Angeles) 48:21–24.

Steenbergh, W. F., and C. H. Lowe. 1977. *Ecology of the saguaro.* II. *Reproduction, germination, establishment, growth, and survival of the young plant.* National Part Service Scientific Monograph Series no. 8. U. S. Government Printing Office, Washington, D.C.

2

Cacti with Primitive Features

To elucidate the evolution of form and physiology within a particular group, it is imperative to determine the ancestral condition — what was the starting point? Sometimes an investigator studies fossils to identify the primitive features. For example, paleontologists studying the origin of mammals used extinct forms to document the structural transitions leading from reptiles to mammals. For most taxa, however, there is no direct fossil evidence to pinpoint the ancestral condition, so workers must infer the primitive characteristics by comparisons of living forms.

In Cactaceae there are no known early fossils. In fact, the oldest fragments date back a mere 22,000 years.* Compare this to the fossil records of sycamores and beeches, remains of which have been found from as long ago as 60,000,000 years. Therefore, the entire evolutionary history of cacti must be inferred from extant (surviving) species. In this chapter we shall look at several cacti that have many primitive features. By studying these cacti, we can identify many evolutionary trends within the family. Assuming that evolution has proceeded in a logical manner, we might reason that such a group of stem succulents originated either from another family of succulents or as a novel evolutionary line from a nonsucculent dicotyledon. Fortunately, there are some cacti — called leaf-bearing cacti — that have prominent, thin leaves and relatively nonsucculent, unspecialized stems. Even though the taxonomy of the cactus family has caused much controversy and debate, cactologists have consistently agreed that the features considered to be primitive for the cactus family can all be found in certain species of *Pereskia.*

* For a review of known cactus fossils, see N. F. McCarten, "Fossil Cacti and Other Succulents from the Late Pleistocene," *Cactus and Succulent Journal* (Los Angeles) 53(1981):122–123.

Vegetative Structure of a Nonsucculent *Pereskia*

Certainly no living species of *Pereskia* is identical to the one from which all other cacti evolved — evolution just does not work that way. The ancestor of the cactus family probably became extinct many millions of years ago. About 15 species of *Pereskia* are extant and are the products generated by uncountable episodes of geographic and genetic subdivisions. Each speciation event has been accompanied by the origin of new characteristics, the loss of others, and, most important, the extinction of any intermediate forms. Some of these extant derivatives have diverged relatively little from the initial condition, but some are now quite highly specialized in structure. For example, one obvious specialization appears in the viney (scandent) *P. aculeata,* a plant naturalized widely throughout the Caribbean region. It has a pair of strong, downward-pointing, curved spines that are located just above the leaf and aid in climbing (Fig. 2.1).

Despite its highly specialized growth habit, *P. aculeata,* which is the species most commonly cultivated, was frequently used as the prototype of Cactaceae. In the early 1960s, Irving W. Bailey of Harvard University and his co-worker, Lalit M. Srivastava, analyzed the internal structure of leaf-bearing cacti and concluded that *P. sacharosa* has vegetative features that more closely resemble those of "typical" dicotyledons than do those of any other cactus. *Pereskia sacharosa* from Argentina and Bolivia is a highly branched, small tree that grows up to 8 m in height and has nonsucculent stems and large, fairly thin leaves. In other words, if you saw this plant for the first time in the wild, you would identify it as an ordinary, spiny, nonsucculent, tropical plant with

Figure 2.1. *Pereskia aculeata;* a vigorously growing shoot with thin, widely spaced leaves and pairs of strong, downward-pointing, curved spines produced on the axillary buds. The spines enable the shoot to become fastened to and entangled with neighboring plants.

drought-deciduous leaves, not as a cactus. In concert with this, Norman H. Boke of the University of Oklahoma convincingly demonstrated that the very "primitive-looking" flowers of the pereskias indeed are representative of the ancestral condition of cacti.

General Structure of the Leaf

A typical leaf blade of *Pereskia sacharosa* is 7 to 10 cm long, obovate in outline, about 0.6 mm thick, and attached to the stem by a very short and poorly defined petiole (Fig. 2.2*A*). The margin of the leaf is smooth, lacking lobes or teeth (that is, the leaf is entire); the venation is pinnate, having a conspicuous midvein from which diverge the smaller lateral (secondary) veins. When a leaf is examined under a microscope, the midvein appears as a composite of numerous vascular bundles arranged in a U-shaped arc (Fig. 2.2*B*). The bundles on the right and left progressively diverge into the two respective halves of the leaf blade until only two small vascular bundles remain in the midvein near the leaf tip (Fig. 2.2*C* and *D*).

As in typical dicotyledons, the leaf of *P. sacharosa* is the chief photosynthetic organ, and the manufacture of sugars and other photosynthetic products takes place in an internal tissue rich in chloroplasts (chlorenchyma). Chloroplasts absorb certain wavelengths of sunlight and convert this energy into chemical energy. In this step of photosynthesis, water is supplied by the xylem in the vascular bundle and is split; oxygen is thereby released into the atmosphere through the stomates. Carbon dioxide moves in the other direction and diffuses into the leaf through the stomates. In this species the stomates occur on both the lower, or abaxial, and the upper, or adaxial, side of the leaf.

Leaves of *Pereskia* (Figs. 1.23 and 2.2*E*) resemble those of typical dicotyledons structurally. They have an upper and a lower epidermis, each tissue consisting of a single cell layer coated with a waxy material (the cuticle); the upper epidermis has a relatively thick cuticle. The region of chlorenchyma is relatively thick (over 0.5 mm) and is formed by about 10 cell layers. Chlorenchyma cells are rectangular and vertically elongate; but they are not palisadelike as in most plants. Leaves of *Pereskia* also lack the mechanical supporting cells with thick, strong walls (sclerenchyma) that make some leaves tough and hard. The absence of thick-walled cells in the petiole facilitates leaf shedding during drought.

General Structure of the Stem

The abundance of stomates on the leaves (over 100 mm^{-2}) is in sharp contrast to their almost complete absence on the stem (Fig. 2.3*A–C*). The bulging outer cell walls of the one-layered stem epidermis are covered by a prominent cuticle (Fig. 2.3*B*), and the cells are produced in well-defined vertical (axial) rows (Fig. 2.3*C*). When the stem is 6 to 8 mm in diameter, the cells of the epidermis divide in a plane parallel to the surface of the organ (a periclinal division) and form a new layer of cells, the cork cambium (phellogen). Cork cells, which seal the stem from the environment and thus largely prevent water loss but also prevent carbon dioxide uptake, are produced to the outside of this cambium; thin-walled cells are produced to the inside and add a small amount of succulence to the stem. Because the young stem lacks stomates and is soon covered by cork, stem photosynthesis in this species is apparently minimal.

The young stem has no internal features that would uniquely distinguish this plant from a nonsuc-

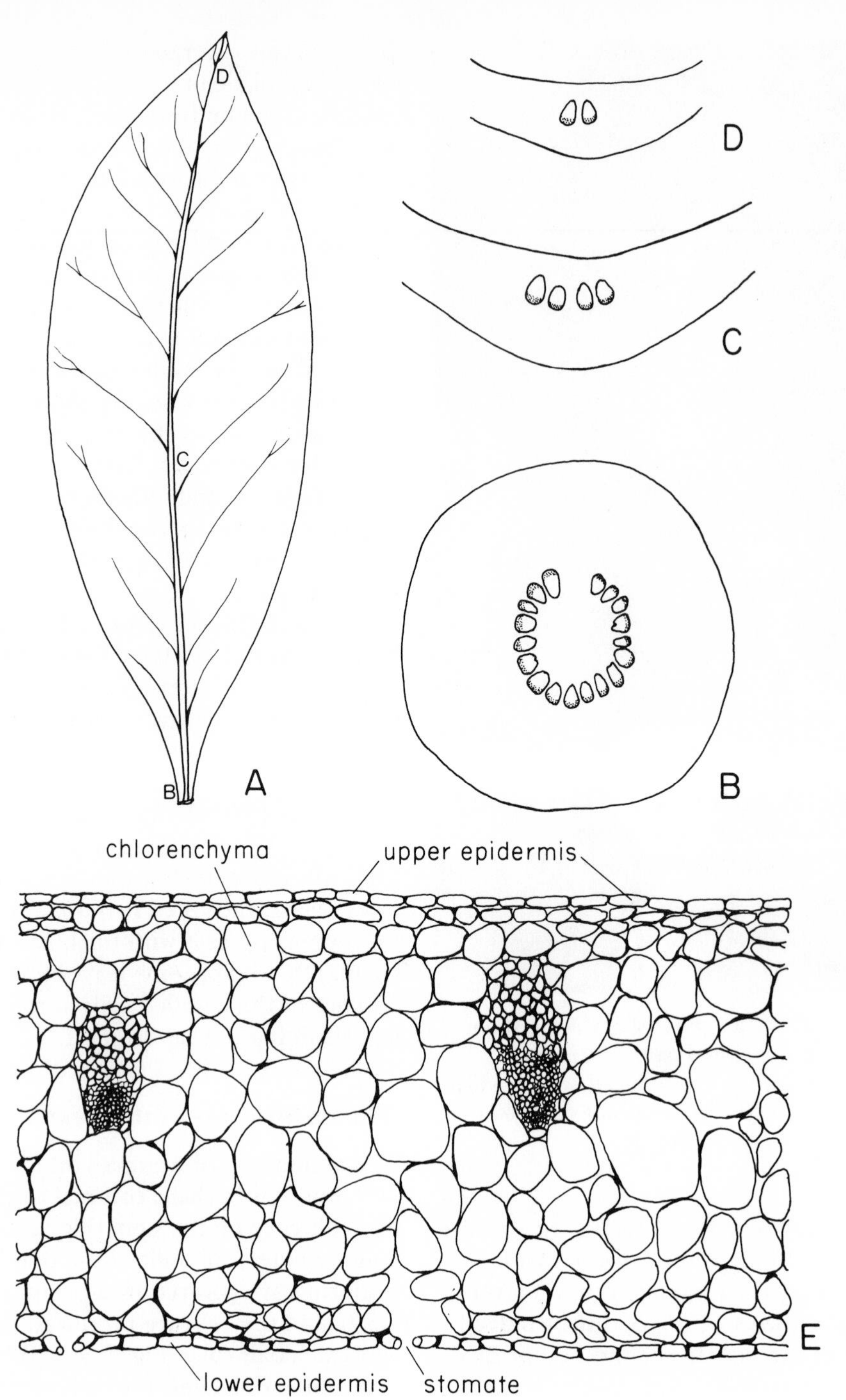

Figure 2.2. Leaf structure of *Pereskia sacharosa.*
(A) Surface view of a typical leaf, which has a major midvein, pinnate venation in the blade, and a very short petiole. Capital letters B, C, and D, on panel *A* identify the positions of sections shown in panels *B*, *C*, and *D*.
(B) Transection through the petiole, showing the U-shaped arrangement of numerous vascular bundles. Xylem is located on the inner side of each vascular bundle and phloem (*stipples*) on the outer side.
(C) Transection through the midvein at the midpoint of the blade; only four vascular bundles are present.
(D) Transection through the tip of the blade; only two vascular bundles are present.
(E) Transection of the blade, showing the cellular structure: upper epidermis, chlorenchyma (mesophyll), two vascular bundles with phloem *(lower)* and xylem *(upper)*, and lower epidermis with a stomate.

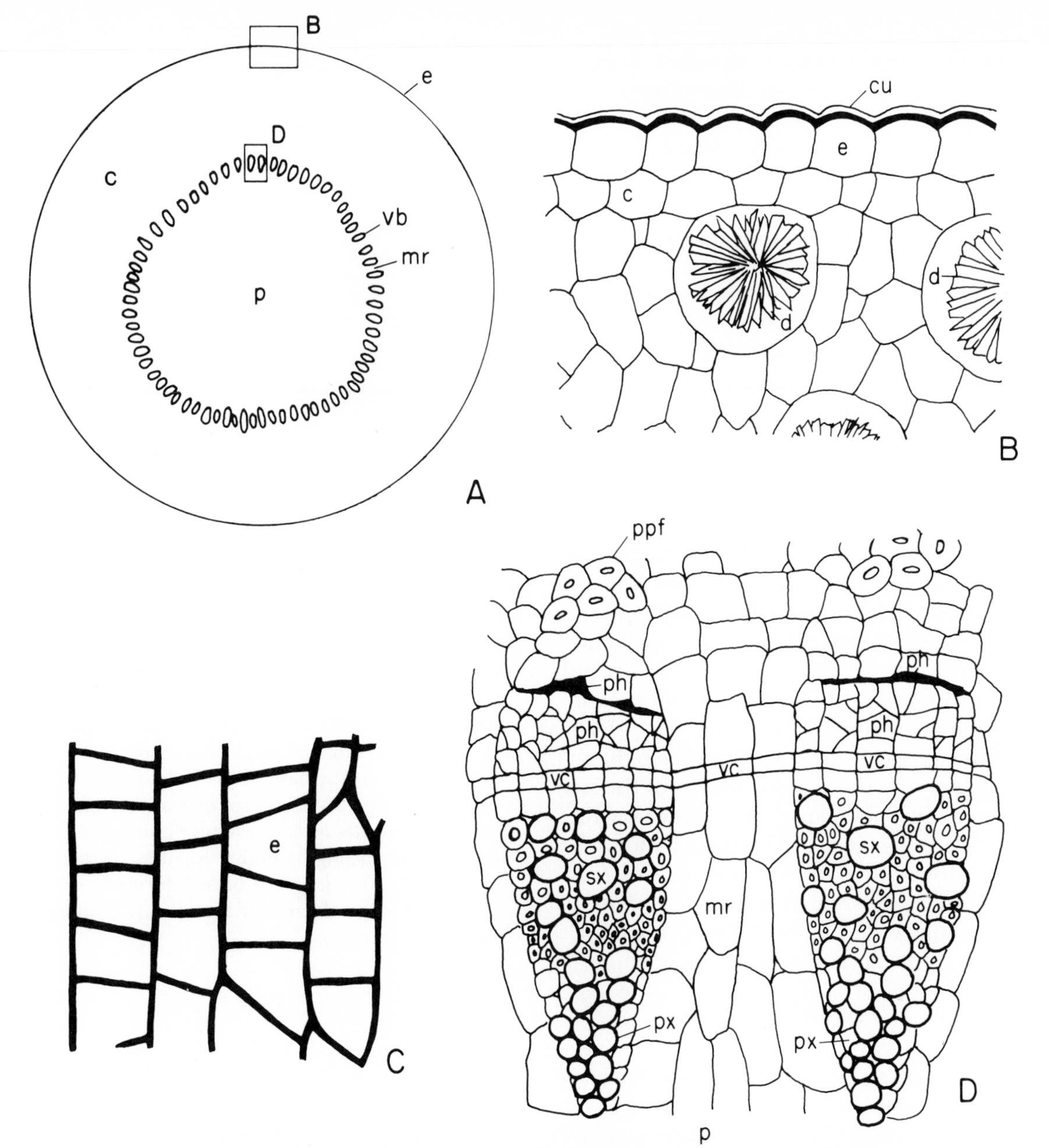

Figure 2.3. Structure of young stems of *Pereskia sacharosa.* ***(A)*** Diagram of a stem, showing the general location of the epidermal layer (e), cortex (c), vascular bundles (vb), medullary rays (mr), and pith (p).
(B) Close-up of a region at the surface of the stem, where the epidermis lacks stomates and has a very prominent cuticle (cu). Star-shaped crystal aggregates composed of calcium oxalate, called druses (d), occur in the cortex.
(C) View of the surface of the epidermal cells, which are formed in short vertical files; stomates are not present.
(D) Close-up of the vascular tissues are viewed in transection; outside to inside, each vascular bundle has primary phloem fibers (ppf), the conducting cells of the phloem (ph), the vascular cambium (vc), and the newly formed (secondary, sx) and the old (primary, px) xylem. The soft-walled parenchyma cells located in the center of the stem are collectively called the pith, and the soft tissue between two vascular bundles in the cylinder is called a medullary ray.

culent dicotyledon. Beneath the epidermis is the cortex, which consists of many, thin-walled, living cells (parenchyma) that are small adjacent to the epidermis and larger in the center. Cells of the outer cortex are closely packed; but in the inner cortex, cell enlargement is greater, and prominent intercellular air spaces develop. Scattered within the cortex, especially just inside the epidermis, are large cells (idioblasts) specialized for the formation of a crystal aggregate known as a druse, which is composed of prismatic units of calcium oxalate (Fig. 2.3*B*). The cortex lacks mucilage cells, which are commonly present in the advanced Cactaceae.

Inward from the cortex lies the vascular tissue (Fig. 2.3*D*), which consists of a ring of vascular bundles with phloem (involved in sugar transport) on the outside and xylem (for water transport) on the inside. Each vascular bundle is separated from adjacent vascular bundles by thin-walled parenchyma cells that form a tissue called the medullary ray; and the individual bundles are capped by a group of elongate, thick-walled cells called primary phloem fibers. A ring of vascular bundles surrounds the pith, which consists of spherical, thin-walled cells similar to those of the inner cortex; some pith cells contain a druse. No mucilage cells are formed in the pith.

A small segment of the vascular system of the stem and leaf is portrayed in Figure 2.4 to show how the leaf receives its U-shaped arc of vascular bundles. In particular, a group of vascular strands are diverted from the central cylinder into the leaf. The entire three-dimensional organization of this cylinder is quite complex and will be discussed later (Chapter 3).

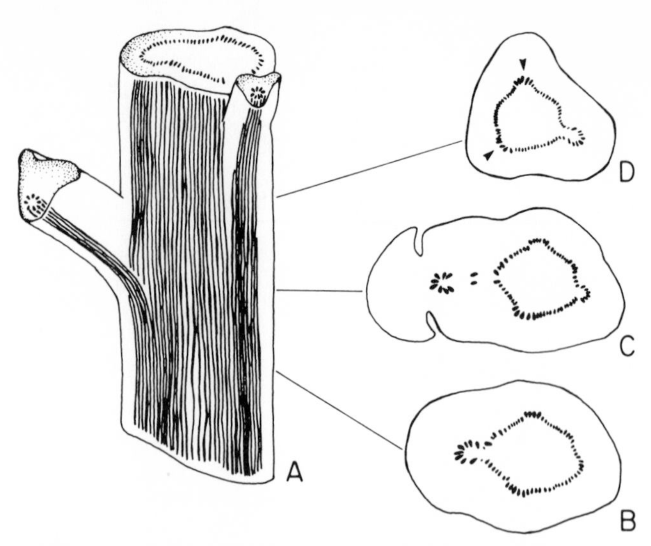

Figure 2.4. Reconstruction of a small portion of the first-formed vascular system of a leaf-bearing pereskia, *Pereskia grandifolia*.
(A) Three-dimensional view showing how a number of vascular bundles in the stem diverge into the petiole of a leaf. ***(B–D)*** Cross sections through *A*, showing how the shape of the vascular cylinder and the size and number of vascular bundles change along the stem. In *B*, the loop of vascular bundles is beginning to diverge toward the leaf; in *C*, the vascular bundles for the left leaf are close to the point where they enter the base of the leaf, and another loop of vascular bundles has begun to form at the right for the next leaf; in *D*, the loop of vascular bundles for the leaf at the right is diverging from the vascular cylinder, whereas the site of the first loop has been reformed. Also in *D*, one can see that loops of vascular bundles are forming for two more leaves *(arrows)*.

Wood Anatomy

The vascular tissue in the stem increases in thickness by a new phase of cell division called secondary growth, which is carried out by the vascular cambium (Fig. 2.3*D*). The vascular cambium is a single layer of cells that encircles the stem and is formed mainly from cells located between the early xylem and phloem of the vascular bundles. The products of the vascular cambium are wood (secondary xylem) and inner bark (secondary phloem). In *P. sacharosa,* the vascular cambium produces a solid cylinder of wood with only a small, thin-walled (parenchymatous) primary ray opposite where a leaf is attached (Fig. 2.5). On the phloem (outer) side, some cells produced by the vascular cambium develop into clusters of fiberlike sclereids, which are hard because they are heavily impregnated with a complex chemical called lignin.

Although we shall return to wood structure in Chapter 8, it is exceedingly important to understand the composition of the wood in *P. sacharosa* (Fig. 2.6). Much research, especially by Bailey, has shown that the degree of specialization of a group can be determined by examining the cellular characteristics of the wood. Dicotyledonous woods are composed of three cell types: (1) dead, empty, water-conducting cells with a hole at each end, called vessel elements, which form a vertical pipe (a vessel); (2) living or dead, long, supporting cells called wood fibers; and (3) small, living cells called axial wood parenchyma and ray parenchyma.

Bailey and Srivastava determined that *P. sacharosa* has a very specialized dicotyledonous wood. First, the vessel elements (Fig. 2.7*A*) are moderately short (0.15 to 0.25 mm in length) and relatively narrow; they have a single round hole (perforation plate) at each end; and the lateral walls are covered by numerous small, bordered pits (multiseriate pitting). Second, the wood fibers are libriform fibers (Fig. 2.7*B*), having small diameters, tapered fine ends, and relatively few, small pits on the lateral walls. Moreover, these fibers are living at maturity. The rays of

Figure 2.5. Photograph of a solid cylinder of branch wood of *Pereskia sacharosa.* An arrow indicates the location of an unlignified primary ray, which is located directly beneath (that is, opposite) the areole and here is very small.

the wood do not have primitive features and store abundant starch grains. Finally, the wood parenchyma (Fig. 2.7*C*) touches the vessels (paratracheal arrangement). In addition, the woods of *Pereskia* are partially storied, meaning that some elements are produced in lateral tiers (Fig. 2.7*D*). The occurrence of these highly specialized features in a "primitive" cactus species clearly indicates that the stock leading to cacti had achieved a high level of cellular specialization in its wood structure, both in the stem and in the root, prior to the development of the family and the origin of stem succulence (see Chapter 11).

Organogenesis of the Shoot System

The shoot of *P. sacharosa* has two distinctive structures that are present in all cacti but absent in typical dicotyledons: the areole and a special pattern of cellular zonation in the growing point at the shoot tip (apical meristem). These characters will be described together because the formation of the areole occurs near the apical meristem.

Shoot Apical Meristem

Located at the tip of each shoot is a small group of cells called the shoot apical meristem; these cells divide repeatedly to produce the tissues of the shoot. (A similar region occurs at the other end of the plant, where each root is terminated by an apical meristem that is protected by a rootcap.) The shoot apical meristem of a plant is usually dome-shaped, although in cacti it tends to be a dome that is broad and gentle or even nearly flat. In *Pereskia,* the apical dome is up to 0.32 mm in diameter, but some cactus apices are up to 1.5 mm across—apparently the largest apical meristems in the plant kingdom. Apices of *Pereskia* and *Opuntia* are exposed at the tip of the shoot, but in many cacti the dome is actually sunken in a pit and further protected by hairs and spines.

Cells in the apical meristem are classified either as initials or as recent derivatives. An initial divides to form two identical (daughter) cells: one cell, the

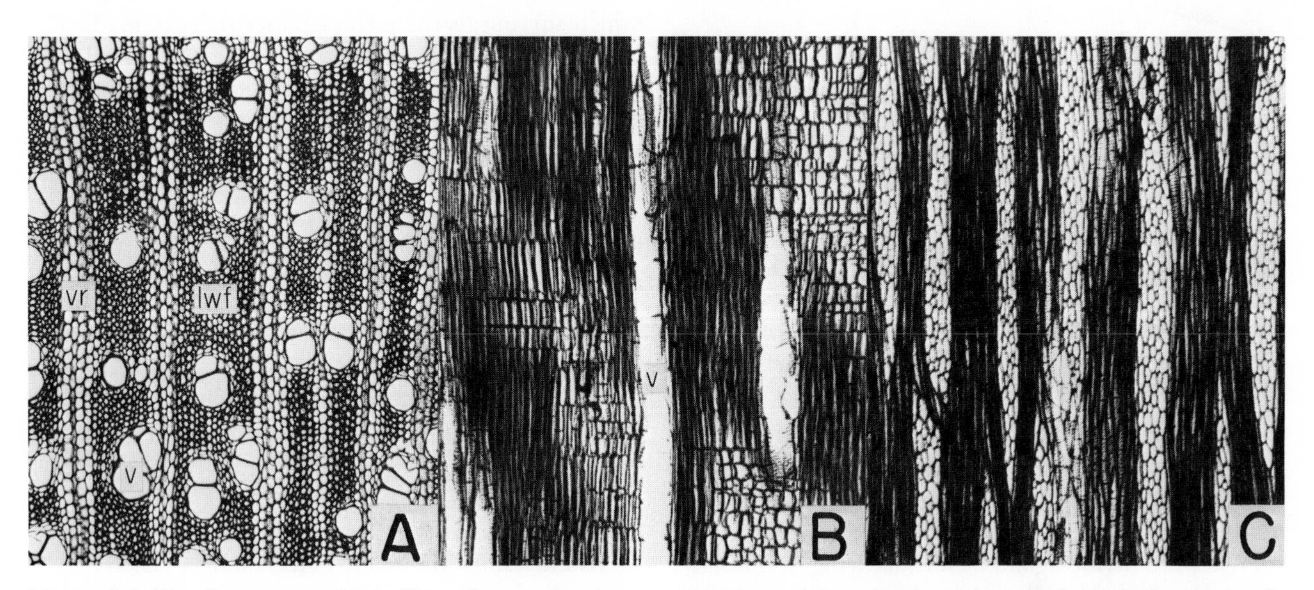

Figure 2.6. Wood structure of *Pereskia sacharosa,* showing a fully lignified cactus wood. ×50.
(A) Transection.
(B) Radial section (a section cut along the long axis of the wood through the pith).
(C) Tangential section (a section cut along the long axis of the wood at a tangent to the outside of the cylinder of wood). v, Vessel; lwf, libriform wood fibers; vr, vascular ray.

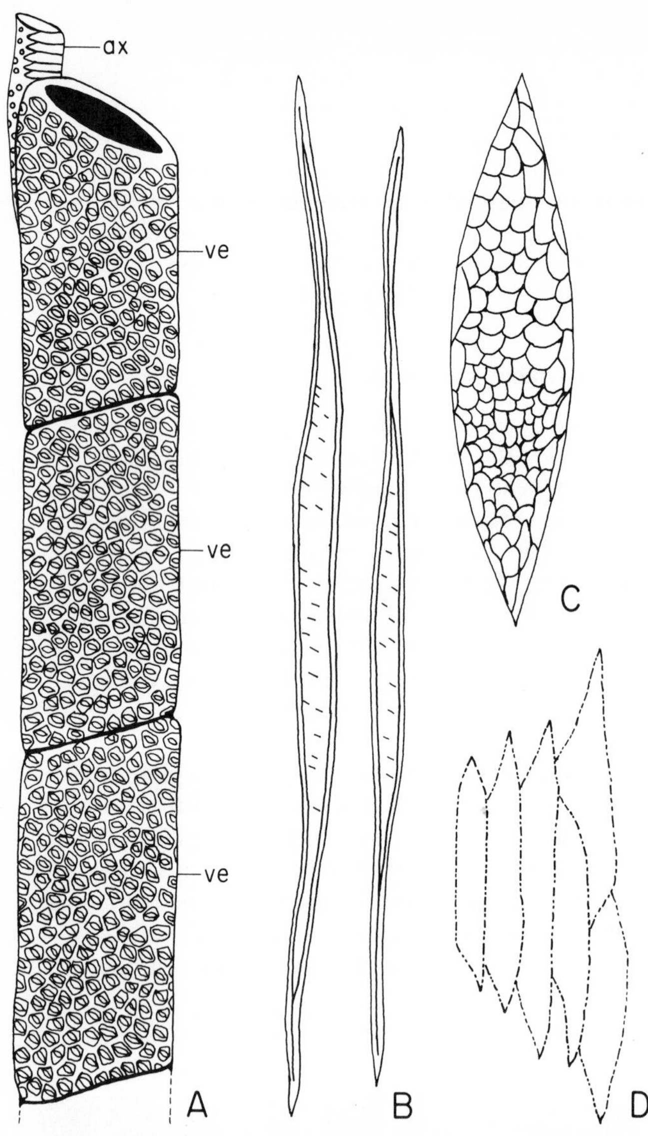

Figure 2.7. Cellular components of stem wood of *Pereskia sacharosa.* (Redrawn from Gibson, 1975.)
(A) A vessel element (ve) and an attached axial parenchyma cell (ax).
(B) Libriform wood fibers; both fibers have very narrow, diagonal pits on their lateral walls.
(C) Multiseriate (many-layered) vascular rays as viewed in tangential section (see Fig. 2.6*C*), which shows the shapes of the cells that are present.
(D) Surface (tangential) view of the cells of the vascular cambium; the cells are arranged in lateral tiers.

derivative, becomes incorporated into the body of the plant (for example, as part of the pith, cortex, or epidermis) and the other, an initial, remains in the meristem to divide again, never maturing or specializing. Hence, each cell of a shoot can in theory be traced back to the apical meristem and indeed to some particular initial.

The typical shoot apex of an angiosperm has a tunica–corpus organization (Fig. 2.8*A*). In general, the apical dome consists of one or two (rarely, three) discrete surface layers, the tunica (Latin, cap), and a central mass of initials and derivatives called the corpus (Latin, body). Cell divisions in the outer tunica layer are always perpendicular to the surface of the organ (anticlinal divisions), and this single layer becomes the epidermis. Divisions of the second tunica layer produce the cells that form the outermost cortex, and divisions of the corpus produce the cells for the central plant tissues: the inner cortex, vascular tissues, and pith.

In the 1940s and 1950s, Boke described the unique organization in the apical meristem of cacti known as apical zonation (Fig. 2.8*B* and *C*). On the surface of the shoot apex is a single tunica layer that is called the dermatogen (producer of the skin), and that eventually differentiates into the one-layered epidermis, as in other angiosperms. In contrast, the structure of the corpus is complex, consisting of three very distinct zones. At the top of the corpus is a block of cells, the central mother cell zone (or initial zone). These initials have conspicuous vacuoles (cellular water sacs) and divide very slowly. Directly beneath the central zone is the pith–rib meristem zone, a solid block of cells in the shape of a truncated cone (Fig. 2.8*C*). Here the cells are large, highly vacuolated, and arranged in long files radiating downward toward the older part of the plant (Fig. 2.8*B*). These vertical files are produced by cell divisions perpendicular to the axis of the shoot (anticlinal divisions). Cells produced by the pith–rib meristem later differentiate into the pith. The remaining corpus region is the peripheral zone, which surrounds the pith–rib meristem zone like a collar (Fig. 2.8*C*); its cells are very small, have dense cytoplasm (no large vacuoles are present), and divide rapidly to produce the cells that later form the vascular cylinder, cortex, and lateral organs such as leaves.

Recent developmental and stereological studies of zonation in cactus apices by James D. Mauseth of the University of Texas have shown that throughout the family—from the primitive pereskias to the most highly derived growth forms—the relative volume occupied by each of the meristematic zones in the corpus is remarkably constant, regardless of the total size and volume of the apex. Yet, although we like to think that cacti are unique, it appears that many other plants have regions within the corpus that correspond to the central, peripheral, and pith–rib meristem zones so visually conspicuous in cacti.

Origin of Leaf Primordia

The origins of leaves and areoles has not been ascertained for *P. sacharosa* (our sample specimen up

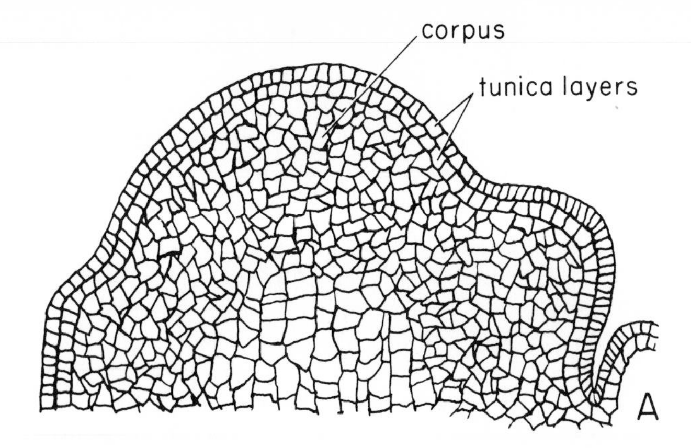

Figure 2.8. The organization of shoot apices.
(A) Drawing of a median longitudinal section of a typical dicotyledonous shoot apex with a tunica–corpus organization. This drawing shows two discrete tunica layers covering the central corpus.
(B) Drawing of a median longitudinal section of a cactus shoot apex with apical zonation (redrawn from work by Boke), showing the single-layered dermatogen (d), the central mother cell zone (cmz), the pith–rib meristem zone (prmz), and the peripheral zone (pz).
(C) Three-dimensional drawing of a cactus shoot apex, showing the four zones identified in *B*. (*C* redrawn from Mauseth and Niklas, 1979.)

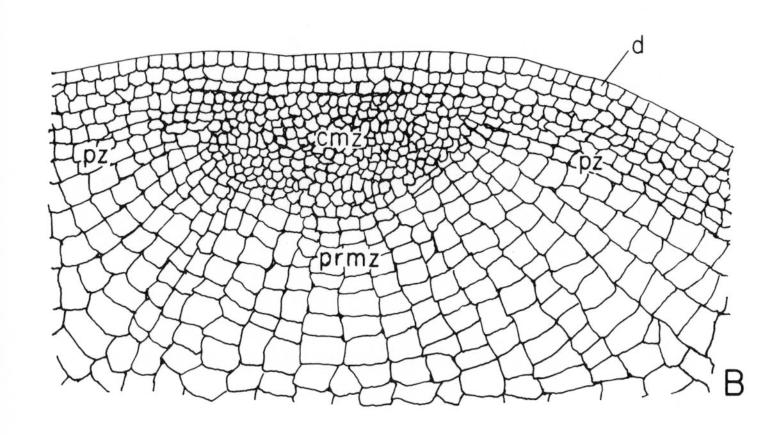

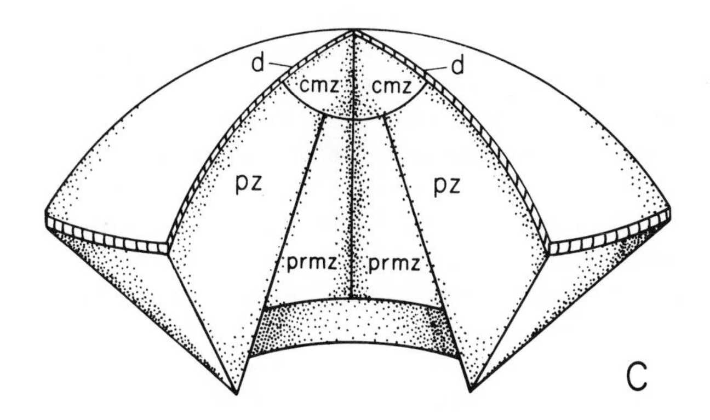

to now), but they have been studied in great detail by Boke for a very close relative, *P. grandifolia,* and for two other species of the genus.

The origin and growth of a leaf in *Pereskia grandifolia* is like that found in any dicotyledon with a simple, entire leaf. The arrangement of leaves is helically alternate, which means that only one leaf is initiated at a time from the apical meristem. The second leaf is produced at an average of 137.5° around the stem from the first leaf, the third one another 137.5° around, and so forth; thus, an ascending helix could be drawn through the successive leaf bases. This is the most common leaf arrangement found in dicotyledons and will be described in more detail in Chapter 6.

For the formation of a new leaf, cell divisions begin on the flank of the apical meristem and produce a small surface mound or dome that enlarges by further cell division at its tip (apical growth) into a peglike projection (Fig. 2.9*A*). This small projection is the leaf primordium, and its growth results from cell divisions along its axis as well as at its tip. When the primordium is about 1 mm long, (Fig. 2.9*B*), apical growth decreases. By this time "wings" have begun to form on opposite sides along the axis of the leaf primordium; these structures will

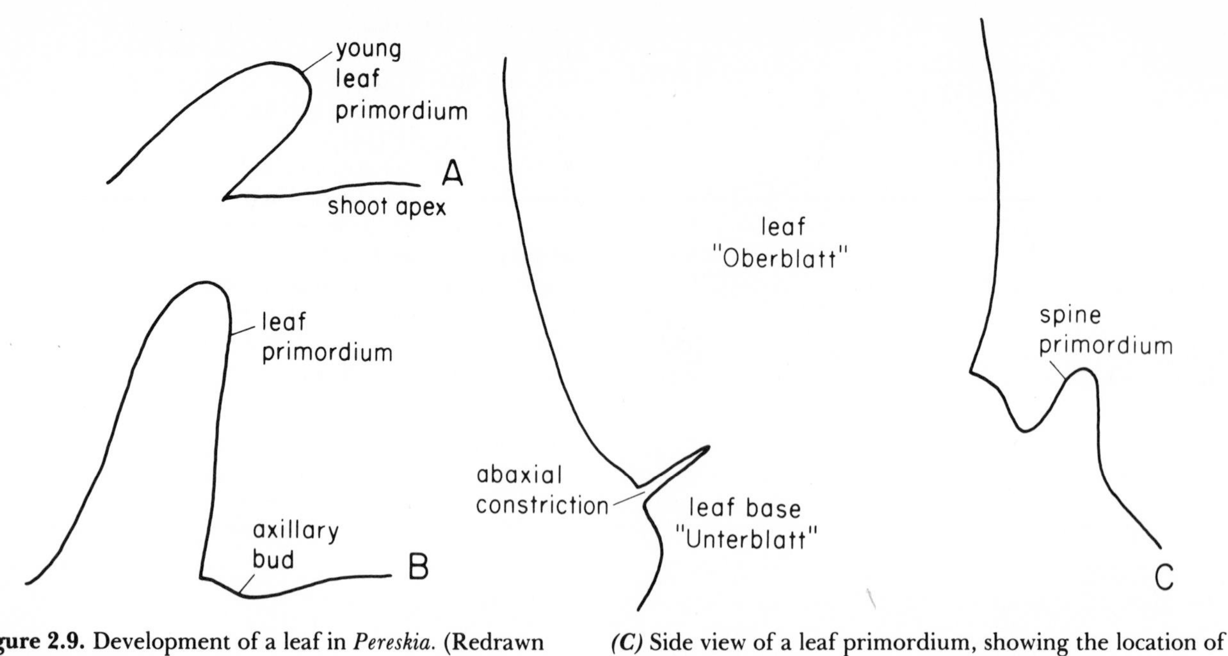

Figure 2.9. Development of a leaf in *Pereskia.* (Redrawn from Boke, 1954.)
(A) Formation of a small leaf primordium on the shoot tip.
(B) The leaf blade has begun to develop by cell divisions along the original cylindrical axis.
(C) Side view of a leaf primordium, showing the location of the constriction between the petiole of the leaf and the leaf base (*Unterblatt*).

become the two halves of the leaf blade. Along the edge of each wing is a strip of actively dividing cells, called a marginal meristem; and the divisions of this meristem determine not only the size and shape of the leaf blade but also the various cell layers of the leaf.

Areole and Tubercle

No aspect of cactus morphology and anatomy is more important to understand than the origins and growth of the areole and the tubercle. The leaf proper (that is, the blade and petiole), which in the old German literature is called the *Oberblatt* (literally, upper leaf), is delimited by a small constriction on the lower (abaxial) side (Fig. 2.9*C*). This abaxial constriction forms after the leaf primordium is only about 2 mm long. Beneath this is the leaf base, called the *Unterblatt* (literally, lower leaf). Enlargement of the leaf base results in the formation of the cactus tubercle.

In the angle between the leaf and the stem, a typical dicotyledon forms a bud (called an axillary bud, or a lateral bud). This bud is actually a dormant apical meristem with or without small leaf primordia that can, if stimulated, develop into an entire, new shoot. This is precisely what happens when a shrub is pruned—lateral shoots are initiated. Generally the formation of an axillary meristem lags behind the growth and development of the adjoining leaf. But in cacti in general and in these leafy pereskias in particular, development of the axillary meristem proceeds immediately and the bud becomes observable before the leaf primordium is 0.1 mm long (Fig. 2.9*B*).

In cacti the axillary bud forms the areole. Initially the leaf base and the axillary bud are adjacent to and facing the apical meristem. But as the shoot tip expands in diameter, the leaf base and accompanying axillary meristem are turned outward until the bud is reoriented perpendicular to the axis of the shoot and above the projecting leaf base. Unlike the axillary buds in other plants, these meristems are not dormant in cacti but instead begin to form primordia, which become the spines. Spine primordia are the structural equivalents of leaf primordia (described in the preceding section) and arise in the same manner. In a spine primordium, however, cell divisions in the apical meristem cease very early; the cells at the tip of the primordium become elongated, with thick, lignified cell walls, which make them extremely hard. Further cell division in the spine primordium is concentrated in a disk of cells at the base of the spine—a region called a basal meristematic zone or an intercalary meristem (Chapter 5). The derivatives of these cell divisions are produced on the adaxial side; therefore, the spine elongates from the base. (These types of cell division and intercalary meristem also are found at the base of petioles in dicotyledons or in leaves of monocotyledons, such as grasses.) A spine can grow for an extended period; but eventually cell division stops, and the spine is hardened completely to the base.

In the first year the areolar meristem produces

spines but no leaves. Trichomes (plant hairs) are also formed from the epidermal cells of the areole, and they grow to form a mat that covers the areolar meristem and that also can hide the apical meristem of the shoot. These trichomes are characteristically multicellular and are produced as a long, single chain (Chapter 5).

In *P. sacharosa* and its closest relatives each areole becomes a spur shoot, or short shoot. A regular shoot (long shoot) has nodes (places where a leaf is attached) separated by long internodes (smooth areas where no leaves occur) (Fig. 2.10*A*); however, the areole is so highly condensed that there are no internodes. At the end of the first growing season, the adjoining leaf produced by the shoot apical meristem is shed, and the remaining areole looks like a mound of grayish hair with some long, smooth, projecting spines. During the next growing season the areolar meristem once again becomes active. One to three leaves are produced on the spur shoot meristem; and the leaves appear to arise out of the center of the hair and above the spines (Fig. 2.10*B*). Leaf production may continue for several seasons, and spines may also be produced. This continued activity of the areolar meristem is the reason why the stems of *Pereskia* appear more knobby with increasing age. The spur shoot is a "cheap" way for these species to produce leaves year after year without investing more energy in the manufacture of new stems.

To summarize the events that produce the external features of the leaf-bearing cactus, we can say that leaf initiation and development are typical of those for a dicotyledon except that leaf production is accompanied by precocious and rapid development of the axillary meristem, which becomes the areolar meristem. The primordia produced by the areolar meristem in the first year develop not as leaves but rather as spines, with a different developmental sequence.* The adjacent leaf base, later to be called the tubercle, becomes reoriented by an increase in stem diameter, and the spine-bearing areole is thus shifted into a position such that the spines point outward from the stem. In the second and subsequent growing seasons, leaves and some spines are produced from areoles on the old stems (Fig. 2.10*B*), now called spur shoots.

The careful developmental studies of Boke confirmed the models of numerous early workers, especially that of the German morphologist Karl Goebel, who had used form and position to conclude

* Mauseth has shown that many cellular characteristics of leaf and spine primordia are different. J. D. Mauseth, "A Morphometric Study of the Ultrastructure of *Echinocereus engelmannii* (Cactaceae). IV. Leaf and Spine Primordia," *American Journal of Botany* 69(1982):546–550.

Figure 2.10. Photographs of shoots of *Pereskia sacharosa.* ***(A)*** Newly formed shoot, which has a leaf that underlies each spine-bearing areole, just as a typical dicotyledon has a leaf adjoining an axillary bud. ***(B)*** Two-year-old shoot, which has leaves that emerge *above* the spines from the apical meristem of the spur shoot. Note that the leaves on this shoot have short petioles and pinnate venation.

Figure 2.11. Seedlings of *Pereskia sacharosa.*
(A) A seedling in which the cotyledons have just expanded (the seed coat is still hanging on the left cotyledon).
(B) A 1-month-old seedling that has four well-developed foliage leaves and a pair of fully expanded, leafy cotyledons.

that spines are modified leaves and not modified hairs. Cactologists had been tempted to speculate that cactus spines arose by the modification of leaves, that is, a gradual loss of the blade and concomitant hardening of the axis. But no such transitions can be observed in the leaf-bearing cacti, and it is likely that spines arose from a leaf primordium without any structural transition from a leaf.

Near the end of his study on the leaf-bearing cacti, Bailey examined the seedling structure of *P. sacharosa* and confirmed that the design of this species is what one might expect for a primitive-looking cactus. The young seedling bears two thin, elliptical or slightly ovate cotyledons (Fig. 2.11) that could easily pass for the cotyledons of any typical, nonsucculent dicotyledon. These cotyledons, which have pinnate venation and remain green and photosynthetic for one to several months, aid the development of the young plant. Young seedlings also have the characteristic areoles and spines of the Cactaceae even in the first month. The occurrence of areoles in the seedling has forced cactologists to conclude, as did Bailey, that the ancestors of Cactaceae (the protocacti) must have possessed spines.

Reproductive Structures of *Pereskia*

A unique feature of *Pereskia*—unique both within the family and within the angiosperms—is the flower. Cactologists have long recognized that some pereskias have an ovary that is superior, the condition in which the ovary is situated above the attachment of the perianth parts and the stamens. Conversely, the ovary in a typical cactus flower is inferior (Fig. 1.33*A*), buried beneath the base of the perianth and stamens. In this section we shall try to explain how the reproductive structures arose in *Pereskia.*

Flower

Figure 2.12*A* shows the major features of the pereskia-type flower. As in most cacti, the pereskia flower is large, colorful, and open. The showy perianth, which is white, lavender, or violet, consists of 12 or more large "petals" that surround a mass of over 200 stamens, whose anthers split open (dehisce) longitudinally. Rising through the stamens is the pistil (ovary, style, and stigma), and in the open flower only the twelve or more stigma lobes can be seen.

Turning the flower over (Fig. 2.12*B*), we can see that it was produced at the end of a stalk (the peduncle), which may be up to 4 cm in length. Leaves are present at the top of the peduncle and on the wide receptacle. These receptacular leaves (bracts) are similar to vegetative leaves in that each has an enlarged base, a basal constriction, and an areole. From the upper rim of the receptacle arise the perianth parts. In *P. sacharosa,* the receptacular bracts and the perianth parts cannot be sharply distinguished, although the perianth parts are colored white or lavender rather than green and lack typical leaf characteristics such as the tubercle and the areole.

When this same flower is cut lengthwise through the center, other important features are revealed (Fig. 2.13). The lining of the ovary chamber (locule) has a series of longitudinal ridges. These ridges formed when twelve or more separate units (called carpels) were fused at their bases to form the ovary chamber. The fused carpel margins extend to the top of the locule, and the opening within the locule extends upward into the style as a central canal. The fused carpel margins are absent from the floor of the chamber, which is smooth. The ovules are produced at the bases of the fused carpel margins.

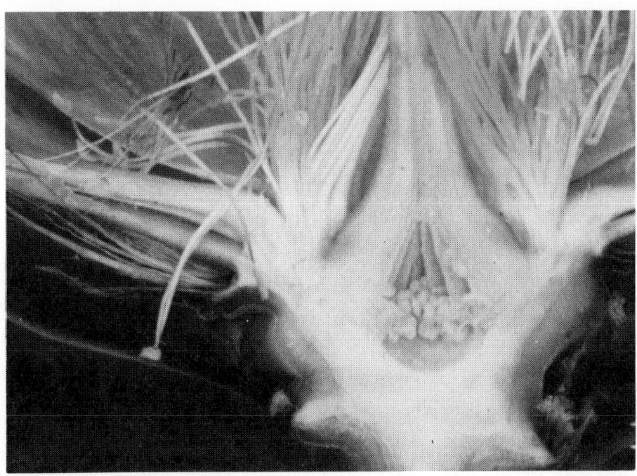

Figure 2.13. Photograph of the inside of the flower of *Pereskia sacharosa,* showing the receptacle, the stem tissue with areoles present outside the ovary, the position of perianth parts and stamens, and the central position of the ovary and style. Inside the ovary in the locule are the longitudinal ridges (septa) that bear the ovules on the lower half.

Figure 2.12. Flower of *Pereskia sacharosa.*
(A) Top view of an open flower, showing the numerous stamens and perianth parts and the central group of stigma lobes (in this flower, there are about 15 separate, rose-lavender "petals").
(B) Bottom view of an open flower, showing the short peduncle and the gradual transition from green leaves on the receptacle to the lighter perianth parts.

The pereskia flower develops into a stalked, fleshy fruit. On the outside of this fruit are leaflike bracts, and their widely spaced, hair-bearing areoles lack spines. The dried, withered perianth parts may also persist. In species with a superior ovary (*P. sacharosa, P. diaz-romeroana,* and *P. aculeata*), the small fruits are green and obovate (Fig. 2.14), although those of the more specialized *P. grandifolia* are often red, turbinlike, almost polyhedral, and up to 6 cm across.

Figure 2.14. Fruits of pereskias. *Pereskia grandifolia* has a large fruit *(left)* with a short stalk (not shown); *P. sacharosa* has a medium-sized fruit *(right)* with a long stalk; and *P. aculeata* has small, round fruits *(two fruits in center)* with prominent leafy bracts and a long, slender stalk.

These same species differ greatly in other ways. For example, the fruits of *P. sacharosa* are commonly solitary, but in *P. grandifolia* the fruits occur in groups (Fig. 1.23).

The fruit wall of *Pereskia* is often thick (usually over 5 mm) and mucilaginous, even though the stem may have only a small amount of mucilage. The fruit wall is also filled with malic acid, which has a sour taste and a strong odor. Projecting from the wall into the locule are the thick and juicy funiculi, each of which bears a seed. The seed (Fig. 2.15) has a smooth, black coat (testa); and as in other cactus seeds, the embryo is strongly curved around the nutritive perisperm tissue (Chapter 1). There are usually fewer than 40 seeds in a pereskia fruit; and the species with large seeds, such as *P. grandifolia* (6 mm long), have an even smaller maximum number of seeds. *Pereskia grandifolia* is atypical because it has an open, relatively dry locule. In species with small fruits, the locule is filled by the seeds and the fleshy funiculi.

Formation of the Inferior Ovary

The development of an inferior ovary is a process that has intrigued botanists for more than a century. Because floral organs are formed sequentially at the floral apex, with the sepals formed first and the carpels formed last, it was difficult to understand how the part of the carpel that becomes the ovary could "sink" below organs formed before it. Now there is wide agreement among botanists that this sinking occurs in two ways.

The most common way is termed the appendicular model, so called because the outer structures of the flower are appended to (that is, fused with or adnate to) the ovary. Figure 2.16 presents the likely stages in the evolution of a flower with a superior ovary to one with an inferior ovary. Initially, the perianth and the stamens arise from a disc-shaped or cup-shaped structure called the hypanthium (Fig. 2.16*A*). Next, the hypanthium elongates and becomes collarlike; thus, it forms a deep cup that is close to and encloses the ovary (Fig. 2.16*B*). In this example the lower half of the hypanthium has fused with the lower half of the ovary. In the final stage (Fig. 2.16*C*), the hypanthium becomes totally fused to the ovary; consequently the perianth and stamens appear to arise from the top of the ovary, an arrangement producing an inferior ovary (epigyny).

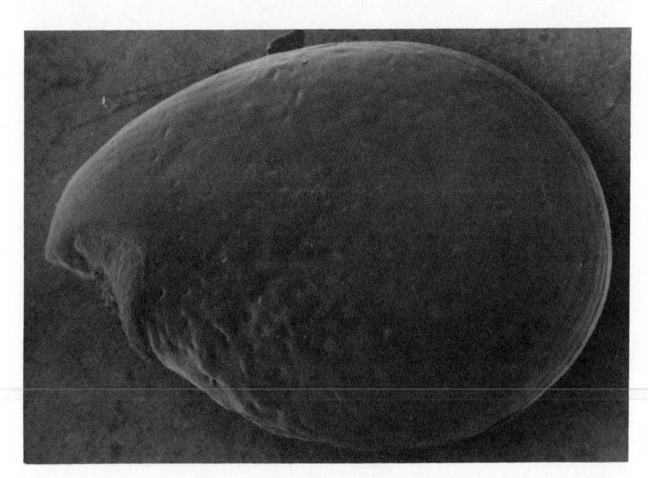

Figure 2.15. A scanning electron photomicrograph of the relatively smooth seed of *Pereskia sacharosa.* The seed is about 5 mm long and was attached to the funiculus at the hilum, seen at the left corner of the seed.

An inferior ovary can also be formed when it becomes "sunken" in the receptacle. In the receptacular model, the ovary is covered by stem (receptacular) tissue and is not engulfed by the outer floral appendages. The receptacular model (Fig. 2.17) explains why the ovary of the cactus flower is covered by bracts (leaves) and areoles (axillary buds).

Many authors recognized that the cactus ovary fits the receptacular model, but the details of how this sinking occurred were unknown until the 1960s. Boke used two lines of anatomical evidence to differentiate between the two models: (1) how the ovary is vascularized and (2) how the ovary develops and is ultimately overtopped by other tissues.

Figure 2.16*A*–*C* shows how the vascular bundles, which carry water and nutrients to the floral parts, enter a dicotyledonous flower directly from the stem, progressing upward into the ovary and style and also passing through the hypanthium to enter the outer floral parts (perianth and stamens). Even though the hypanthium becomes fused to the ovary, these same vascular pathways are retained, that is, the vasculature of the ovary proper remains separate from the vasculature of the hypanthium.

Vascularization of a pereskia flower is quite different (Fig. 2.17). Longitudinal sections of a pereskia flower show that the principal vascular bundles do not pass directly into the base of the ovary. Instead, the 12 to 30 major bundles from the stem enter the receptacle, spread toward the outside, and then travel upward to the level of the outermost stamens. Collectively, these bundles are called the ascending receptacular system. At the highest point, just below the perianth, the ascending receptacular bundles fuse to produce a network, from which emerge the vascular bundles (traces) for the perianth segments and the upper bracts on the receptacle. At that point the receptacular bundles turn abruptly inward and downward, passing just underneath the bases of the stamens en route to the ovary; from this recurrent system arise the vascular bundles for the stamens. After the vascular trace for the innermost

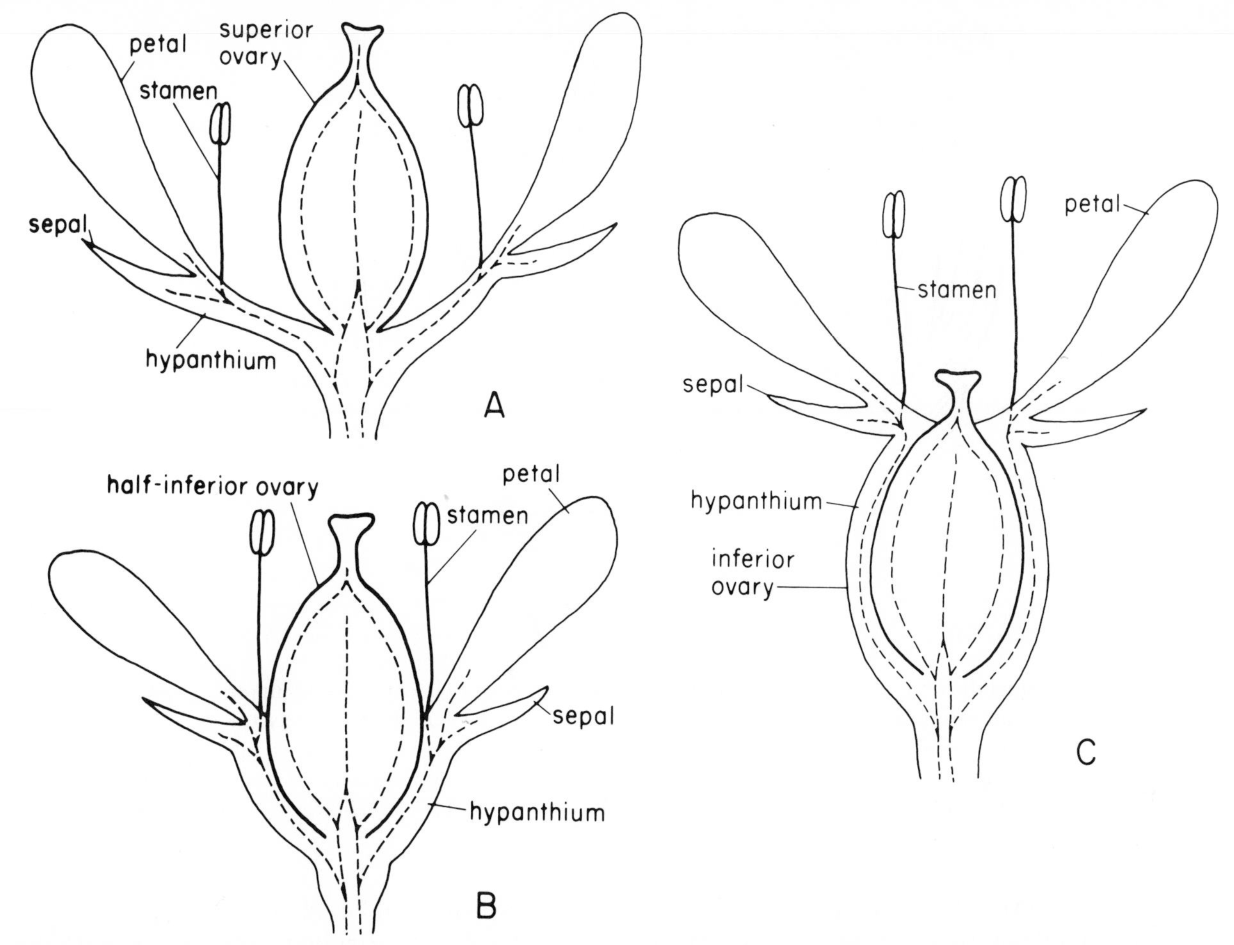

Figure 2.16. Appendicular model for the origin of the inferior ovary.
(A) Longitudinal drawing of a hypothetical flower with a hypanthium and a superior ovary.
(B) The hypanthium begins to enclose the ovary and fuses with its lower half.
(C) The hypanthium has fused with the ovary, and the perianth parts and stamens now arise from the top of the ovary. Major vascular bundles of the ovary and hypanthium are indicated throughout.

stamen has diverged, the recurrent bundle splits. The upper bundle, or dorsal bundle, enters the roof of the ovary and continues into the style and stigma lobe of that particular segment of the ovary (one per carpel), whereas the lower bundle, or ventral bundle, enters the floor of the carpel and gives rise to the vascular trace to each ovule. All of the ventral bundles converge toward the lowest point of the ovary, but they do not reach it and therefore never fuse. Consequently, the very center of the ovary floor is devoid of vascular tissue.

From this vascular pattern we can see how the receptacular model fits *Pereskia* because it predicts that the ovary vasculature should be "bent" downward with the sinking of the ovary. Therefore, the original base of the ovary would not be at the lowest point of the ovary near the center of the flower but rather would be higher up, where the dorsal and ventral bundles diverge as they enter the carpel. But the piece of evidence that clinches this theory is the arrangement of the xylem and the phloem. The phloem remains on the same side of the bundle, as if it had been bent; the arrangement is very different from what it would be if the bundle had developed directly from below (see Fig. 2.16*C*).

Boke's studies also showed exactly how the female tissues are formed and developmentally relocated (Fig. 2.18). Carpels arise in a ring around the broad floral apex (floral meristem, Fig. 2.18*A*). In longitudinal view, each carpel primordium appears as a small mound (Fig. 2.18*A*), which forms an upper and a lower portion (Fig. 2.18*B–D*). The upper portion of the carpel becomes the style and stigma and the lower portion forms the ovary, where the ovules and eventually the seeds are produced (Fig. 2.18*C* and *D*). While the ovary is forming, however, the stamens and the tissues beneath the stamens develop in such a way that the ovary is relocated below the level of the

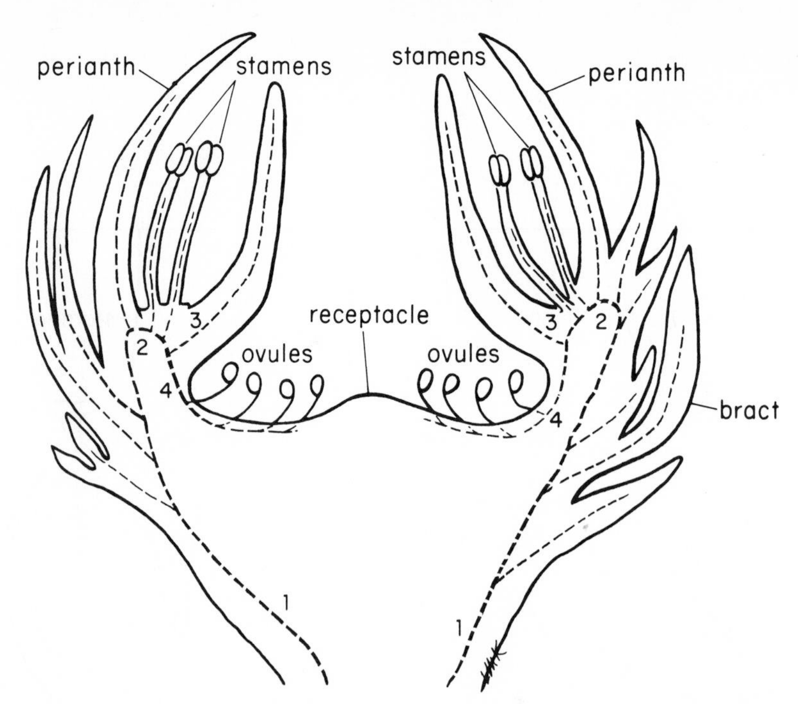

Figure 2.17. Longitudinal drawing of a flower of *Pereskia sacharosa,* showing the vascularization of the various parts. The ascending receptacular bundle (1) travels upward and bears the traces for the bracts and perianth parts. After the innermost perianth part is vascularized, the vascular bundle bends to become the recurrent system (2) and gives rise to vascular bundles into the stamens. Upon entering the ovary, the system splits into a dorsal bundle (3), which enters the style and stigma, and a ventral bundle (4), which develops toward the receptacle and vascularizes the ovules.

stamen bases. Moreover, tissue enlargement in the vicinity of the floral meristem also causes the ovary to be relocated to the side. Consequently, when the female portion of the flower is ready to receive pollen (Fig. 2.18*D* and *E*), the carpel—in particular, the ovary—has been rotated downward; hence, its vascular system is also bent downward.

While each carpel is being relocated downward, other events are happening. The carpels arise in a ring around the broad floral apex (Fig. 2.18*E* and *F*). Several of the pereskias have 12 to 14 carpels, but the number of carpels varies from 4 or 5 in *P. aculeata* to 18 in *P. pititache.* Each carpel is initiated as a separate unit that is folded; the "point" (dorsal midpoint) of the carpel primordium faces outward and the two "wings" (ventral margins) project toward the center. As the carpels grow, they form a solid ring, and the margins fuse together (Fig. 2.18*E* and *F*). These suture lines are the vertical ridges observed on the inner ovary wall (Fig. 2.12*A*). Although the bases fuse to form an ovary with one large locule, the tips of the carpels remain free (Fig. 2.18*E*). As a consequence, the locule is not completely closed but is open through the top. Even later when the style forms from the lateral fusion of the upper portions of the carpels, the extreme tips are free (the stigma lobes) and the style becomes a cylinder that is lined with abundant trichomes.

Developmental studies of pereskias reveal that at anthesis the carpels are tilted downward only slightly, so that most of the sinking of the carpels occurs during fruit development. During this period the region between the base of the stamens and the floor of the ovary is deepened by cell divisions as the receptacle enlarges. In addition to this, the carpels can also be reoriented to a fully "inferior" position by the resumed growth of the floral meristem, which produces a mound of cells in the floor of the ovary.

In classical systems of classification, cacti were associated with families that have parietal placentation; in this arrangement the placentas that bear the ovules are located on the outer (dorsal) ovary wall in vertical rows or on the partitions that may intrude into the locule (chamber). Another common type of placentation in dicotyledons—often disguised as parietal placentation—is axile placentation. In axile placentation, the placentas are located around a central axis so that the ovules radiate out from the inner (ventral) wall like spokes on a wheel. Boke was able to show that the ovules of pereskias are, in fact, produced on the fused ventral margins of the carpels and not on the dorsal walls. In a nutshell, the placentation of pereskias is fundamentally axile because, if the carpels were pulled together into a fused structure, the ovules would radiate from a central axis.

Our knowledge of the flowers and fruits of *Pereskia*

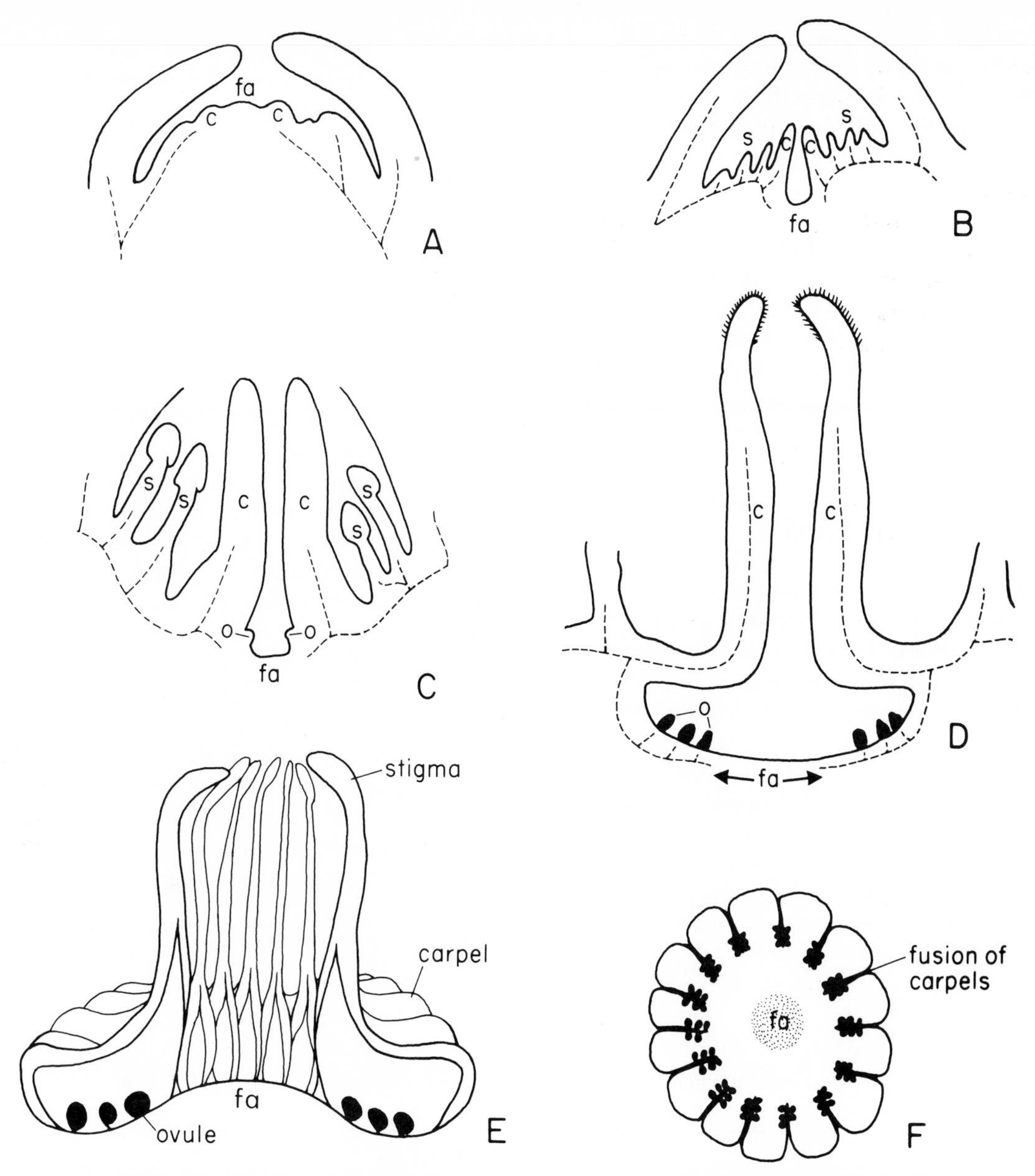

Figure 2.18. Development of the female portion of a pereskia flower. (After Boke, 1963, 1964, 1966; and Kaplan, 1967.)
(A) Longitudinal view through a very young flower bud, showing that the carpels (c) arise as small "bumps" on the edge of the floral apex (fa) before the stamens begin to develop. At this point, the vascular tissues follow a "typical," upward course.
(B) Longitudinal view of a slightly older flower bud, in which the stamens (s) have begun to form and the carpel (one on each side of the center) have produced an upper portion (style and stigma) and a lower portion (ovary). Growth of the stamens and associated tissues has caused the ovary to be relocated beneath the stamens so that the vascular bundles entering the ovary (ventral bundle) are also headed downward.
(C) Longitudinal view through a flower bud that has begun to form ovules (o) in the ovaries and anthers on the stamens.
(D) Longitudinal view through the female portion of the flower at the time that the flower opens (anthesis); the stigma has become receptive and the ovules are ready to be fertilized. The ovary has been rotated downward by the expansion of the region beneath the floral apex *(arrows);* because the styles have not fused, a canal remains in the center. The trichomes that line this canal have not been drawn.
(E) Three-dimensional drawing of half of the carpels in a ring around the floral apex. The ovaries of the once-separate carpels have fused, but the styles and stigmas are still distinct.
(F) Cross section through the lowest part of the ovaries, showing the 14 carpels arranged in a ring and their fusion laterally along the ventral margins, where the ovules are attached. The fused margins of the carpels are the septa observed in the cactus ovary (Fig. 2.13); the stippled region is the floral apex.

supports the notion that pereskias are "primitive" cacti because they occupy an intermediate position between typical dicotyledons and the highly derived cacti. If we were asked to choose the one species with the fewest specialized reproductive structures, we would choose *P. sacharosa, P. diaz-romeroana,* or *P. aculeata. Pereskia aculeata* and *P. diaz-romeroana* have small flowers, a low carpel number, few seeds (only one ovule per carpel), and a sharply delimited perianth; and these features could be interpreted as either primitive or specialized. Of course, there is no objective way to determine how close the structure of *P. sacharosa* is to that of the very first cactus plant.

Types of Leaf-Bearing Cacti

Even if we are wrong about our candidates for the species with the most primitive features, at least we can compare these plants with other leaf-bearing cacti to identify the major trends of specialization leading toward the more specialized "leafless" forms. A discussion of leaf-bearing cacti should include three groups: *Pereskia,* which includes arborescent and shrubby, nonsucculent and succulent species; *Maihuenia,* which is the other genus in the subfamily Pereskioideae and includes highly specialized succulent plants; and *Pereskiopsis* and *Quiabentia,* the two genera of the subfamily Opuntioideae that have flattened, succulent leaves.

Taxonomic Subdivisions of *Pereskia*

Very little detailed taxonomic research has been published on leaf-bearing cacti; consequently, there is no widely accepted list of the species and no stable classification of the groups within these genera. Some workers have attempted to subdivide *Pereskia* into subgenera and even to remove species and erect a segregate genus. In 1926 the subgenus *Eupereskia* was described by Alwin Berger, a famous German cactologist, for *P. sacharosa* and *P. aculeata,* the species with the most primitive ovary; he created the subgenus *Rhodocactus* to accommodate the other species. Nine years later Knuth reshuffled the species into three groups. The genus *Pereskia* included two subgenera: subgenus *Pereskia,* consisting of some species with superior ovaries; and subgenus *Neopereskia,* which included the small-flowered species of Peru and Bolivia. Most of the pereskias were then reclassified in a new genus, *Rhodocactus.* But most cactologists accepted only one genus—*Pereskia.*

The first detailed contribution to the classification of pereskias was made by Bailey, who in the 1960s published a series of 18 papers on cacti.* Bailey discovered that all species of *Pereskia* have thick-walled cells (called sclereids) in the secondary phloem. He identified three distinct sclereid types and found that they correlate with other vegetative features and can be used to subdivide the genus into

* This series was published from 1960 to 1968 in the *Journal of the Arnold Arboretum.*

Table 2.1. Tentative groups of species in the genus *Pereskia,* based on the anatomical research of Bailey (after Gibson, 1975).

Species	Distribution
Group I	
Pereskia sacharosa	Argentina and Paraguay
P. bleo (including *P. bahiensis* and *P. moorei*)	Colombia and Brazil
P. grandifolia (including *P. tampicana*)	Brazil and Mexico
P. corrugata	Uncertain
Group II	
Pereskia aculeata	Tropical America
P. pititache (=*P. conzattii*)	Southern Mexico, possibly Guatemala
P. autumnalis	Guatemala and El Salvador
P. nicoyana	Costa Rica
P. diaz-romeroana	Eastern Bolivia
P. humboldtii	Northern Peru
P. weberiana	Bolivia
Group III	
Pereskia colombiana	Colombia
P. guamacho	Venezuela
P. cubensis	Cuba
P. portulacifolia	Haiti

Figures 2.19 and 2.20. Genera of Opuntioideae with prominent leaves.
2.19. *Pereskiopsis porteri* from western Mexico; this species has leaves up to 3 cm long and approximately that wide. Many trichomes are produced on the areoles; the areoles also bear glochids, which can be observed on the areole in the center of the photograph. The name *Pereskiopsis* means that this plant resembles *Pereskia* in vegetative features.
2.20. *Quiabentia chacoensis* from northern Argentina; the leaves, which are up to 5 cm long, are succulent; the spines are conspicuous.

three groups (Table 2.1). The first group consists of species with elongate sclereids that are aggregated in longitudinally oriented clusters. Classified here are some species of Backeberg's subgenus *Pereskia* (but not *P. aculeata*) and some species of *Rhodocactus.* Species in the second group have diffusely distributed, fiberlike sclereids. In this group Bailey included *P. aculeata,* all species of the subgenus *Neopereskia,* and a group of species from *Rhodocactus* that occur from southern Mexico to Costa Rica. In the third group are species of the Greater Antilles and northern South America that have ordinary, roundish sclereids (brachysclereids) aggregated in massive irregular clusters. Generally speaking, Bailey's rearrangement not only explained the morphological observations better but also showed that closely related species have adjacent geographic ranges.

Bailey's classification, tentative as it is, can be used as a starting point to study how structure evolved in pereskias. Leaf venation is an example. Leaves of the least specialized species in the genus and in each group have pinnate venation: the lateral veins of a leaf diverge from a single, broad, prominent midvein. The most specialized, thickest leaves in the genus, for example, those of *P. pititache* and *P. autumnalis,* have pseudopalmate venation: the central midvein is less conspicuous than several prominent lateral veins that arc independently into the upper leaf. To confirm this, Bailey also showed that pinnate venation is a primitive characteristic for the cotyledons of *Pereskia;* the most highly derived species have a modified type of pinnate venation. Bailey concluded that the origin of succulence in the cactus leaf was coupled with a shift in leaf venation from pinnate to pseudopalmate and a decrease in leaf size.

While leaf structure was evolving in pereskias, succulence in the stem was increasing. According to Bailey, the stem wood of *P. sacharosa* is similar to that of a nonsucculent dicotyledon (Fig. 2.6); this species has a small pith, relatively dense wood with fully lignified cells and relatively little storage of starch, and considerable accumulation of wood. Other species, for example, *P. pititache,* have much more succulence; the stem of *P. pititache* has a wide pith, relatively succulent wood with large zones of unlignified storage cells (and, consequently, fewer and weaker cells), and a great amount of starch storage in the wood. A special case of succulence occurs in the roots of the Andean species, in which soft tissue forms in the center of the root before a cylinder of lignified wood is formed.

Specialization in the Opuntioideae

The traditional reasons for classifying *Pereskiopsis* (Fig. 2.19) and *Quiabentia* (Fig. 2.20) in subfamily Opuntioideae and not in subfamily Pereskioideae are well established. Opuntioids, including the large-leaved species, have glochids on the areole and a hard, bony aril that surrounds the seed. Moreover, opuntioid flowers do not have a peduncle. Bailey found two additional features that support this taxonomic distinction. First, species of *Pereskiopsis* and *Quiabentia* have no sclereids in the secondary phloem; and second, they have a very conspicuous layer of calcium oxalate crystal aggregates (druses) in the outermost layer of the stem cortex (Fig. 2.21). This notwithstanding, Bailey considered these genera to be the next stage in the evolution of succulence in cacti; they have small, succulent leaves with palmate venation (no midvein develops), and the stems are very succulent. Wood production in *Pereskiopsis* and *Quiabentia* is very scanty, and most of the wood cells lack thick walls and lignin.

Maihuenia, a Specialized Pereskioid

It may seem inappropriate to discuss the highly specialized *Maihuenia* at the end of a chapter on the "primitive," leaf-bearing cacti; nevertheless, this peculiar genus, the second genus of Pereskioideae, illustrates how cacti with broad leaves ultimately gave rise to bizarre evolutionary novelties. Species of *Maihuenia* are profusely branched, low-caespitose plants, 30 to 40 cm in height and often 1 m across (Figs. 2.22 and 2.23). Few cactologists have ever seen these magnificent mounds growing in their out-of-the-way cool habitats in the plains and Andean foothills of Argentina and Chile. Each plant has a large main root; and the center of this root, as in *P. humboldtii,* consists of mostly parenchyma rather than xylem vessels and fibers. Numerous underground, creeping stems radiate from the root, and from these arise the vertical shoots (Fig. 2.22*B*). A vertical shoot grows apically, as do the shoots of the pereskias; but, in contrast, the nodes of *Maihuenia* are very condensed. When the stem is young and still bears its original leaves, the axillary buds begin to develop as lateral shoots. Initially, a bud produces one to three spines; and soon after this, small leaves form on these spur shoots (Figs. 2.22*C* and 2.23*C*). Hence, the surface of each vertical shoot is densely covered with many short, lateral shoots (Fig. 2.23*A* and *B*).

Leaves of *Maihuenia* are approximately cylindrical (terete) and grow up to 10 mm long. Figures 2.24 and 2.25 show a cross section of a young leaf primordium of *P. humboldtii* from Peru next to one of *Maihuenia patagonica* from Argentina, both at a similar stage of development. This representative peres-

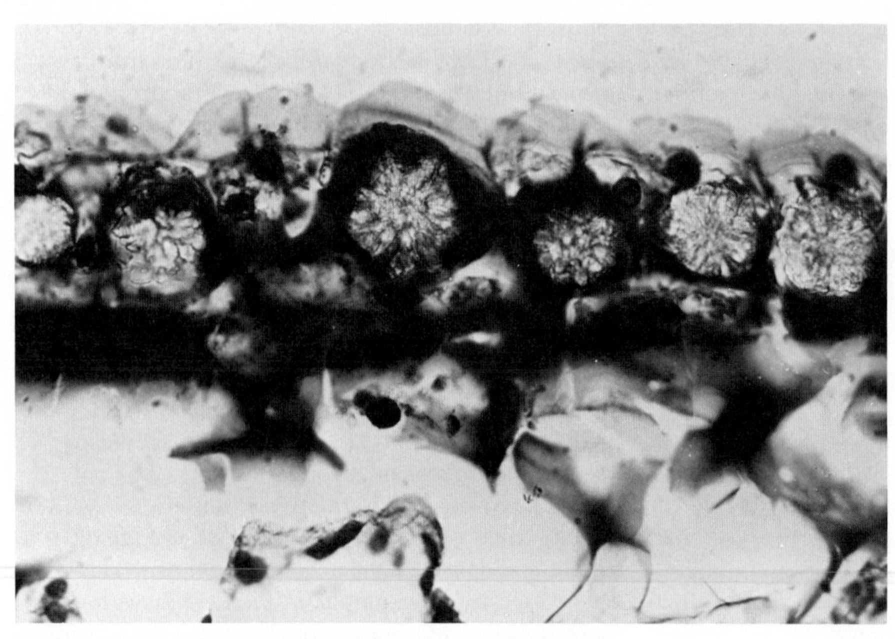

Figure 2.21. *Quiabentia zehtneri,* light microscopic view of the outer stem. Each cell of the outer layer of the thick-walled hypodermis has a large druse of calcium oxalate, a feature occurring in all genera and nearly all species of the subfamily Opuntioideae.

Figures 2.22 and 2.23. *Maihuenia* from southern South America.
2.22. *M. patagonica,* from west central Argentina. This plant grows as a dense, caespitose cushion plant over 1 m in diameter *(A);* the decumbent stems *(B)* give rise to hundreds of vertical shoots. *(C)* Close-up showing the small terete leaves and the areoles with prominent spines and precocious development of leafy shoots. (*C* from Gibson, 1977.)
2.23. *M. poeppigii,* from eastern Chile, illustrating the terete leaves and the spines (*A* to *C*), the white, areolar trichomes *(C)*, and a mature fruit with persistent leafy bracts on the outside of the fruit *(B)*.

kia has two broad halves of the leaf blade that are produced by cell divisions on the two leaf edges from cells of the marginal meristems. A leaf widens as new cells are added to the inner side of the meristem. *Maihuenia* leaves do not have marginal meristems and therefore do not have true leaf blades. These genera also differ in mucilage production; *Pereskia* generally has few, if any, mucilage cells in the leaf, whereas the small leaf of *Maihuenia* has abundant and large mucilage cells, composing over half of the leaf volume. Consequently, a *Maihuenia* leaf has relatively little chlorenchyma per unit volume; it is small and produces no blade. *Maihuenia* has some additional curiosities, including a thick-walled epidermis on the leaf, precocious development of bark (periderm) on the stem, and an extremely large mucilage reservoir in the center of the stem.

The wood of *Maihuenia* generally has few, if any, wood fibers; thin walls with remarkably little lignin; and, hence, little mechanical strength. The cellular architecture of this wood parallels that of other low-caespitose cacti in the other subfamilies and is

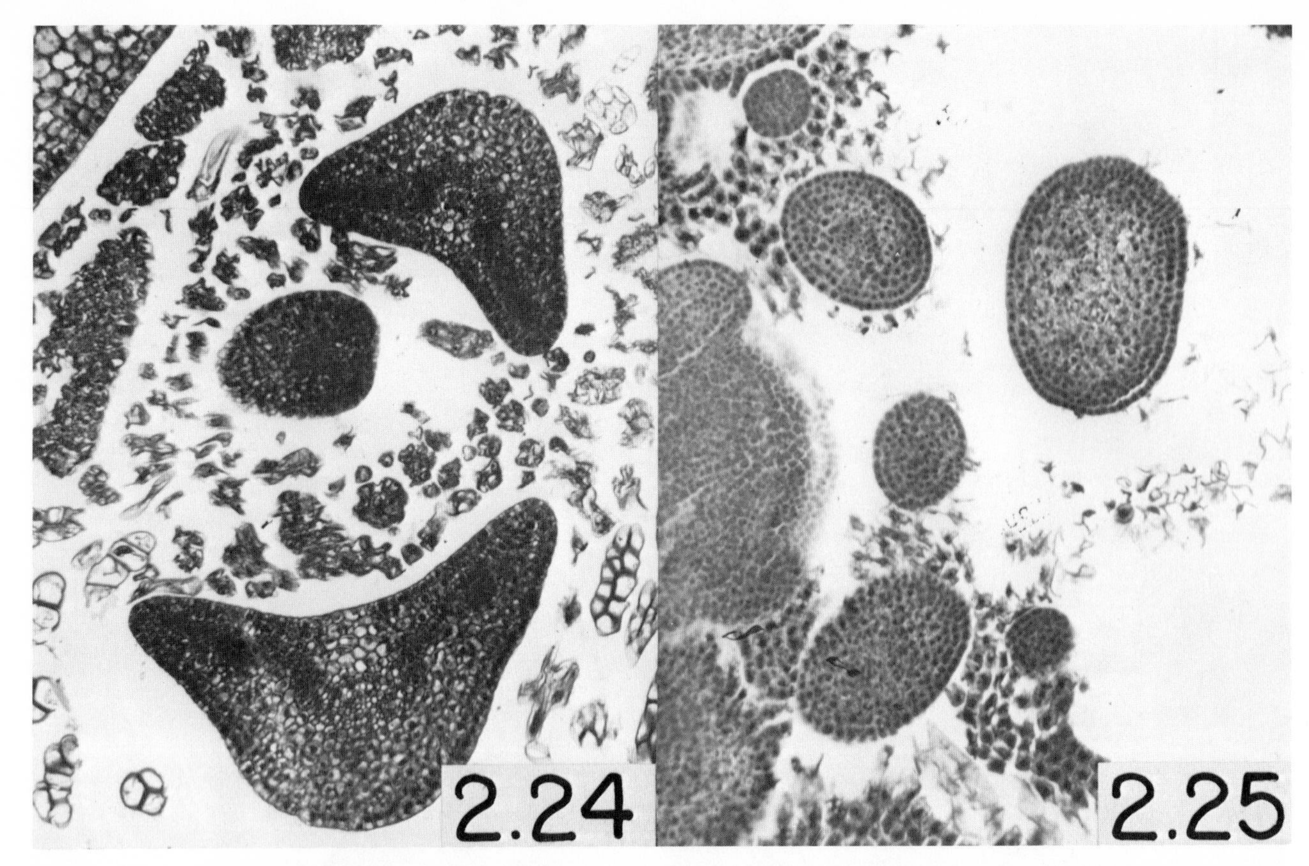

Figures 2.24 and 2.25. Photomicrographs of young leaf primordia of cacti forming on the shoot tip. Both pictures are printed at the same magnification to show the relative differences in the size and shape of the leaf primordia. ×90.
2.24. *Pereskia humboldtii,* a species with broad leaves; the two halves of the leaf blade form as wings on the leaf primordia.

2.25. *Maihuenia patagonica,* a species with small terete (cylindrical) leaves. The two halves of the blade never form on the leaf axis; thus, the leaf primordium is circular to oval in shape and does not have the "wings" seen in Figure 2.24.

not at all similar to the woods we have just discussed for *Pereskia,* its closest relative (see Chapter 8).

The distinctiveness of *Maihuenia* is also apparent in its reproductive features, which unfortunately are seldom seen and therefore poorly described. *Maihuenia* differs from *Pereskia* in several major respects: the flowers are sessile and borne terminally on the stem; the receptacle sometimes has bristly spines; and the perianth is yellow. Although the seeds of *Maihuenia* are black and glossy, like those of *Pereskia,* the seedling has very narrow, long-lived cotyledons.

In short, *Maihuenia* is quite extraordinary in most vegetative and reproductive features. There are no "missing links" to indicate how this taxon actually evolved from a pereskia, if indeed it did. The same problem faces scientists who want to demonstrate a link between *Pereskia* and subfamily Opuntioideae on the one hand and with subfamily Cactoideae on the other.

The lesson to be gained from this survey of leaf-bearing cacti is fairly simple. A few species of *Pereskia* are relatively nonsucculent organisms that have primitive, thin leaves but also several specializations in external morphology: areoles, a shoot apex with apical zonation, and a flower with an ovary that has started to sink into the receptacle. These and other characters show that the ancestors of the cactus family were already highly derived dicotyledons. What happened in the evolution of cacti, even in the evolution of leaf-bearing cacti, was a reduction in leaf size, and a shifting of the location of photosynthesis from the leaves to succulent stems.

SELECTED BIBLIOGRAPHY

Bailey, I. W. 1960. Comparative anatomy of the leaf-bearing Cactaceae, I. Foliar vasculature of *Pereskia, Pereskiopsis* and *Quiabentia. Journal of the Arnold Arboretum* 41:341–356.

——— 1961a. Comparative anatomy of the leaf-bearing Cactaceae, II. Structure and distribution of sclerenchyma in the phloem of *Pereskia, Pereskiopsis* and *Quiabentia. Journal of the Arnold Arboretum* 42:144–156.

——— 1961b. Comparative anatomy of the leaf-bearing Cactaceae, III. Form and distribution of crystals in *Pereskia, Pereskiopsis* and *Quiabentia*. *Journal of the Arnold Arboretum* 42:334–346.

——— 1962. Comparative anatomy of the leaf-bearing Cactaceae, VI. The xylem of *Pereskia sacharosa* and *Pereskia aculeata*. *Journal of the Arnold Arboretum* 43:376–388.

——— 1963a. Comparative anatomy of the leaf-bearing Cactaceae, VII. The xylem of pereskias from Peru and Bolivia. *Journal of the Arnold Arboretum* 44:127–137.

——— 1963b. Comparative anatomy of the leaf-bearing Cactaceae, VIII. The xylem of pereskias from southern Mexico and Central America. *Journal of the Arnold Arboretum* 44:211–221.

——— 1963c. Comparative anatomy of the leaf-bearing Cactaceae, IX. The xylem of *Pereskia grandifolia* and *Pereskia bleo*. *Journal of the Arnold Arboretum* 44:222–231.

——— 1963d. Comparative anatomy of the leaf-bearing Cactaceae, X. The xylem of *Pereskia colombiana, Pereskia guamacho, Pereskia cubensis,* and *Pereskia portulacifolia*. *Journal of the Arnold Arboretum* 44:390–401.

——— 1964a. Comparative anatomy of the leaf-bearing Cactaceae, XI. The xylem of *Pereskiopsis* and *Quiabentia*. *Journal of the Arnold Arboretum* 45:140–157.

——— 1964b. Comparative anatomy of the leaf-bearing Cactaceae, XII. Preliminary observations upon the structure of the epidermis, stomata, and cuticle. *Journal of the Arnold Arboretum* 45:374–389.

——— 1965. Comparative anatomy of the leaf-bearing Cactaceae, XIV. Preliminary observations on the vasculature of cotyledons. *Journal of the Arnold Arboretum* 46:445–452.

——— 1967. Comparative anatomy of the leaf-bearing Cactaceae, XVII. Preliminary observations on the problem of transitions from broad to terete leaves. *Journal of the Arnold Arboretum* 49:370–376.

Boke, N. H. 1941. Zonation in the shoot apices of *Trichocereus spachianus* and *Opuntia cylindrica*. *American Journal of Botany* 28:656–664.

——— 1944. Histogenesis of the leaf and areole in *Opuntia cylindrica*. *American Journal of Botany* 31:299–316.

——— 1954. Organogenesis of the vegetative shoot in *Pereskia*. *American Journal of Botany* 41:619–637.

——— 1963. Anatomy and development of the flower and fruit of *Pereskia pititache*. *American Journal of Botany* 50:843–858.

——— 1964. The cactus gynoecium: a new interpretation. *American Journal of Botany* 51:598–610.

——— 1966. Ontogeny and structure of the flower and fruit of *Pereskia aculeata*. *American Journal of Botany* 53:534–542.

——— 1968. Structure and development of the flower and fruit of *Pereskia diaz-romeroana*. *American Journal of Botany* 55:1254–1260.

——— 1980. Developmental morphology and anatomy in Cactaceae. *Bioscience* 30:605–610.

Gibson, A. C. 1975. Another look at the cactus research of Irving Widmer Bailey. Pp. 76–85 in *1975 Yearbook, Supplement, Cactus and Succulent Journal* (Los Angeles), ed. C. Glass and R. Foster.

——— 1976. Vascular organization in shoots of Cactaceae. I. Development and morphology of primary vasculature in Pereskioideae and Opuntioideae. *American Journal of Botany* 63:414–426.

——— 1977. Vegetative anatomy of *Maihuenia* (Cactaceae), with some theoretical discussions of ontogenetic changes in xylem cell types. *Bulletin of the Torrey Botanical Club* 104:35–48.

Kaplan, D. R. 1967. Floral morphology, organogenesis and interpretation of the inferior ovary in *Downingia bacigalupii*. *American Journal of Botany* 54:1274–1290.

Mauseth, J. D. 1978a. An investigation on the morphogenetic mechanisms which control the development of zonation in seedling shoot apical meristems. *American Journal of Botany* 65:158–167.

——— 1978b. An investigation of the phylogenetic and ontogenetic variability of shoot apical meristems in the Cactaceae. *American Journal of Botany* 65:326–333.

Mauseth, J. D., and K. J. Niklas. 1979. Constancy of relative volumes of zones in shoot apical meristems in Cactaceae: implications concerning meristem size, shape, and metabolism. *American Journal of Botany* 66:933–939.

Ross, R. 1982. Initiation of stamens, carpels, and receptacle in the Cactaceae. *American Journal of Botany* 69:369–379.

3

Succulence

We have until now repeatedly described cacti as succulents without defining this term. *Succulence* and *succulent* are words that appear in standard dictionaries in reference to a plant or plant part that is greatly thickened, juicy, or fleshy, and thereby modified to store large quantities of water. Unfortunately, an exact scientific definition for succulence has never been achieved because some plants with fleshy leaves or stems that have water contents greater than 90%—certain terrestrial and aquatic herbs—will die very quickly when they are deprived of water. The subtle but important distinction between a plant that is juicy or fleshy and one that is succulent rests in the ability of the succulent to tolerate drought for extremely long periods by drawing upon water stored in tissues. Succulents can be removed from the soil and left to dry for days or even years without fatal consequences; in fact, even in a rootless condition they can remain in good health and be capable of renewed growth. We shall also see that cacti additionally have a form of succulence at the cellular level (Chapter 4).

Anatomy of Stem Succulence in Cacti

To develop an understanding of succulence and cactus physiology, the reader must be familiar with the structure of the tissues and cells of a "typical" cactus stem. For this purpose, a cactus must be "drawn and quartered" so that the internal tissues can be viewed longitudinally and in cross section (Figs. 3.1 and 3.2). Any succulent stem can be used for this exercise because all species have the same general organization even though plant shape and therefore the dimensions of each tissue may vary.

The cactus has a tough but flexible outer covering that is popularly referred to as the skin. Holding a peel of the skin in front of a lamp shows that this structure is translucent and therefore permits light to pass through to the internal tissues of the cactus. Beneath the skin is the cortex, which is composed of water-filled parenchyma cells. Parenchyma is the general name for the living cells of a plant that can undergo cell division and that store most of the water in the plant (even one that is not succulent). In the cortex, the outer 2 to 3 mm is composed of dark green chlorenchyma, a special type of parenchyma in which the chloroplasts (the sites for photosynthesis) are localized; and because chlorophyll is absent, the color of the deeper cell layers in the cortex is nongreen—usually white, nearly clear, pink, or yellow. Throughout the cortex we can observe the white, threadlike vascular bundles that pass through the parenchyma; and in many species we can also see large, clear cells that are the mucilage-bearing cells. Inside the cortex is a thin band of phloem—mostly secondary phloem—that is produced by divisions of the elongate initials of the vascular cambium. The vascular cambium cannot be located exactly because it is a single layer of cells, less than 0.02 mm thick. This ring of dividing cells is very important, however, because to the inside it produces the cylinder of wood (secondary xylem) that surrounds the parenchymatous, succulent central pith. The woody cylinder is generally not solid but has sections of soft parenchyma cells called primary rays, which provide succulent connections between the cortex and the pith. Vascular bundles may be observed in the pith of many species of the subfamily Cactoideae (medullary bundles).

When we make thin sections of the various tissues and study each region under a microscope, we observe different sets of structural characteristics; by this technique we can discover how the fine structure of the plant part relates to its function.

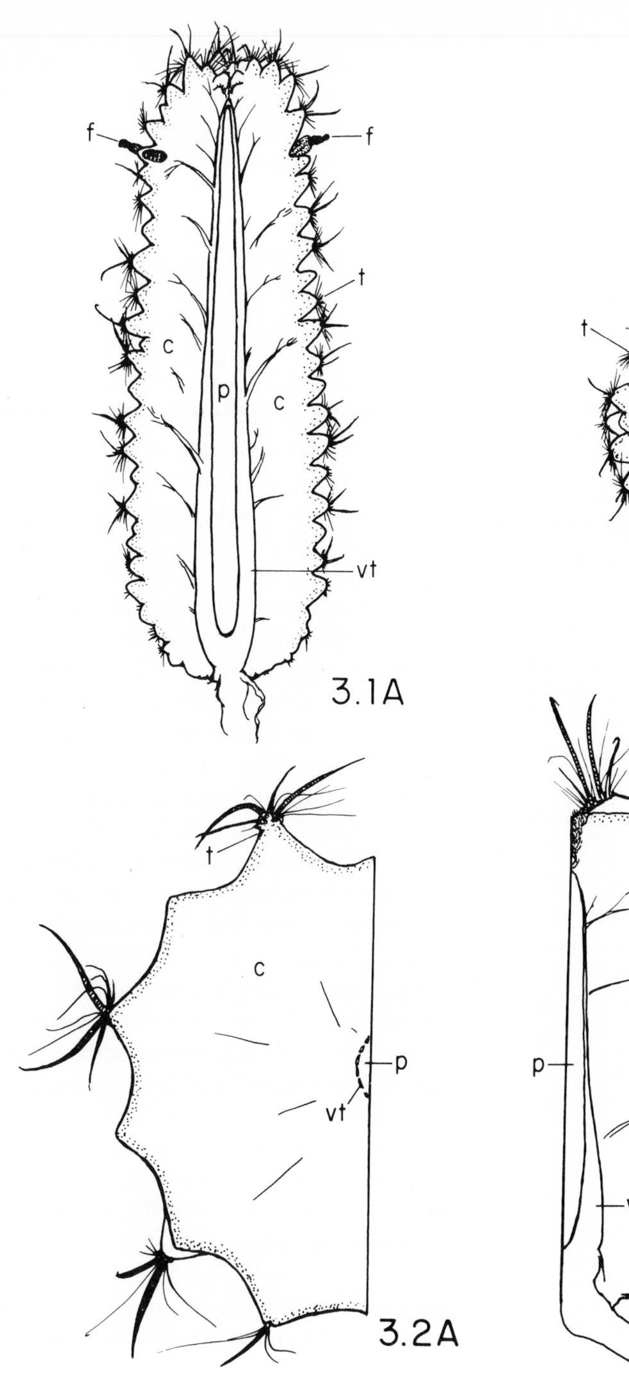

Figures 3.1 and 3.2. General structure of a cactus plant.
3.1. *Mammillaria dioica,* from Arizona.
(A) Longitudinal section through the stem and uppermost portion of the root, showing the spine-bearing tubercles (t), the photosynthetic chlorenchyma *(stippling),* cortex (c) and pith (p) (the major succulent tissues), and the vascular tissue (vt), which is thick at the base of the plant because wood has formed. Two fruits (f) are also illustrated.
(B) Cross section of the same plant.
3.2. *Ferocactus* sp., a fairly young barrel cactus that is still globular.
(A) Cross section, showing the tubercles (which are arranged into ribs), the cortex, the vascular tissues (mostly xylem), and the pith.
(B) Longitudinal section of the same plant.

Skin

Skin is a convenient term for a complex and very important structure. This plant covering consists of the epidermis, which is coated with a thick, waxy cuticle, and the hypodermis, which is the outermost region of the cortex and has cells with thick walls (Figs. 3.3–3.5). The skin restricts the loss of water from the stem and yet permits carbon dioxide to enter the plant tissues from the atmosphere. It provides mechanical integrity to the stem and thus acts as the first line of defense against fungi, bacteria, and foraging animals. It also influences the amount of solar radiation that is absorbed and reflected by the stem.

Epidermis. In the vast majority of cacti, the stem epidermis is one cell layer thick (uniseriate; Figs. 3.3 and 3.4). The outer cell walls are generally flat or convex; but in some of the highly specialized forms, such as *Opuntia basilaris* and *Pterocactus kuntzii,* the cells have bulging outer walls, called papillae, that look like small nipples.* Cacti with a multiple epidermis (two or more cell layers) are uncommon, but there are some excellent examples in certain columnar cacti of the North American tribe Pachycereeae (Fig. 3.5). The outermost wall of the epidermis is impregnated with and covered by a wax called cutin. This wax is a fatty substance that prevents water vapor from escaping from the plant and repels surface water. Cutin is also indigestible by small organisms that might try to enter a cactus plant and to grow and reproduce in its watery tissues. Compared with the cuticles of most plants, the cuticle of a cactus is relatively thick; and there are some species with a cuticle so thick that pure wax can be scraped off the plant with a knife or a fingernail. Stems that are very gray, bluish, silvery, or white in appearance (glaucous), for example, those of *Stenocereus beneckei* and *Myrtillocactus,* have a powdery surface layer called epicuticular wax, which can be removed easily by touching or gently rubbing the stem.

Because the stem is the principal photosynthetic organ of most cacti, the stem epidermis has numerous stomates. Each stomate or pore is formed as an opening between two crescent-shaped or kidney-shaped cells, called guard cells (Figs. 1.29–1.31, 3.6). When these guard cells are relatively flaccid, they appress each other and thereby close the pore (Figs. 1.30, 3.6*A*); but when the guard cells become turgid as a result of the uptake of water, the cells expand in such a way as to open the pore, which is shaped like a biconvex lens (Figs. 1.31, 3.6*B*). The fine structure of a cactus guard cell is like that of the guard cells of other flowering plants.*

In many species of cacti stomates are evenly distributed over the entire stem surface, but on a ribbed plant stomates may occur only on the sides of the ribs and not on the ridges and in the creases. In cacti with ribs the long axes of the stomates are most frequently oriented along the long axis of the stem; whereas in cacti without ribs the pore axes are randomly oriented. Cactus stems usually have 15 to 70 stomates per square millimeter, a lower value than that found for the leaves of dicotyledons, which generally have more than 100 stomates per square millimeter. In the anatomical studies that have been done on cacti, most authors have found that the pair of guard cells is flanked by one or two cells (called subsidiary cells) outside of each guard cell and parallel with it (Figs. 1.30 and 3.6). Guard cells are usually positioned right at the surface of the epidermis, although some of the highly specialized species—especially those with a thick cuticle—have sunken stomates and the pore is somewhat hidden from the stem surface (Fig. 1.15).

Hypodermis. The hypodermis is composed of a type of cells called collenchyma, which are used by plants for mechanical support. The living cells of the collenchymatous hypodermis differ from parenchyma cells because their walls are unevenly thickened and contain high concentrations of pectin (a water-holding substance) and hemicellulose (celluloselike molecules) but contain no lignin (hardening substance). The pectin attracts water, which fills these walls and makes them hard but flexible. Because the stem has this cylinder of tough, flexible collenchyma, it can expand and contract, as well as bend back and forth, without experiencing damage to the cells or breakage of the skin. In fact, collenchyma can shrink markedly as water is lost without experiencing major harmful effects. Although some of the small, highly specialized cacti lack a collenchymatous hypodermis, the majority of species have this tissue; and in the gigantic columnar cacti, the hypodermis may be over 1 mm thick (more than 10 cell layers thick). The tough hypodermis also functions as a deterrent to feeding by small animals, especially insects. As we shall discuss in Chapters 6, 9, and 10, the hypodermis as well as the epidermis often contains crystals, which may also discourage insects with chewing mouthparts.

The hypodermis is not a solid structure because, if it were, the stem could not function for gas ex-

* Epidermal papillae in cacti were first illustrated in a paper by A. F. Hememway and M. J. Allen, "A Study of the Pubescence of Cacti," *American Journal of Botany* 23(1936):139–144.

* C. D. Faraday, W. W. Thomson, and K. A. Platt-Aloia, "Comparative Ultrastructure of Guard Cells of C_3, C_4, and CAM Plants," in *Crassulacean Acid Metabolism,* ed. I. P. Ting and M. Gibbs, pp. 18–30 (Rockville, Maryland; American Society of Plant Physiologists, 1982).

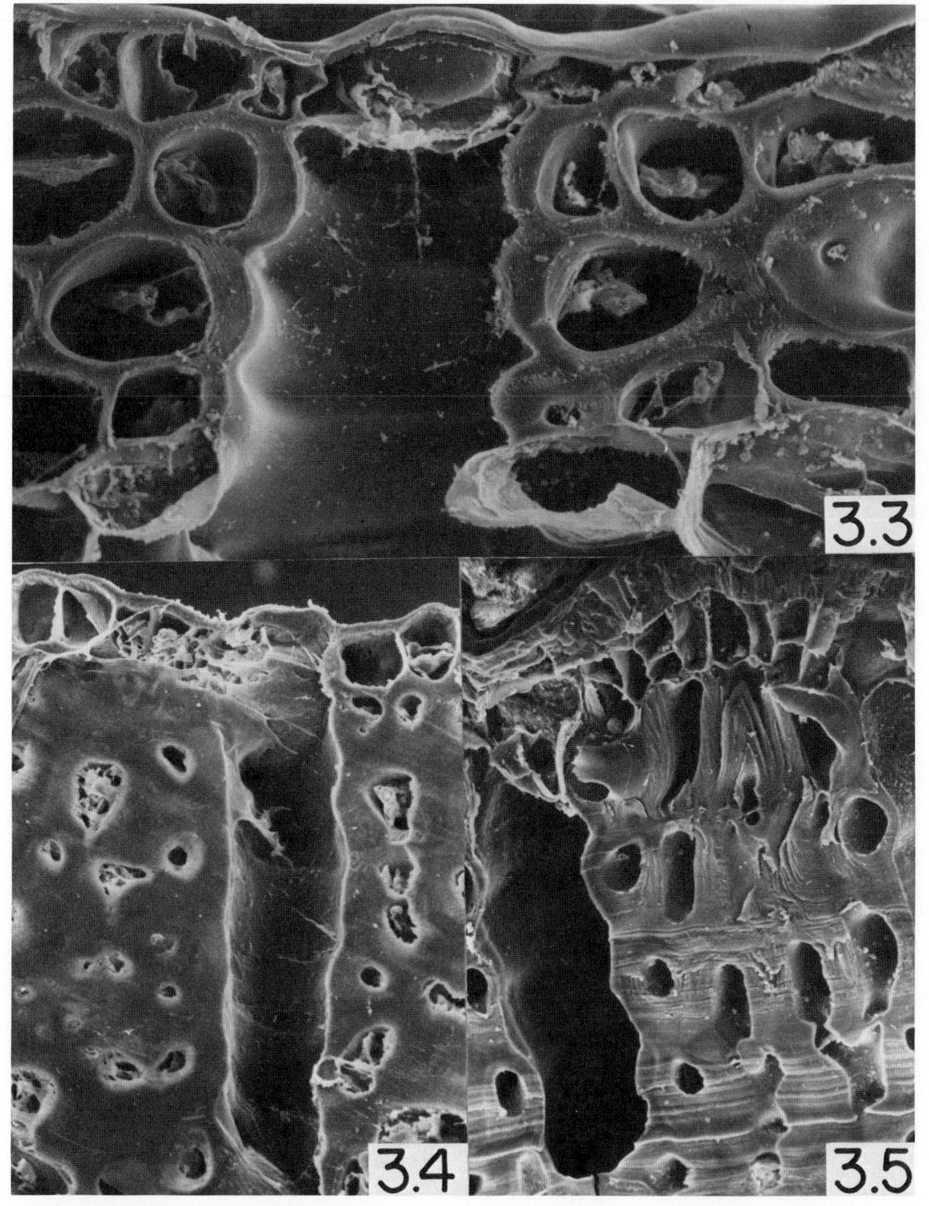

Figures 3.3–3.5. Scanning electron photomicrographs of cactus skins.
3.3. *Werckleocereus tonduzii,* an epiphyte with a fairly weak skin (0.12 mm thick) that consists of a one-layer epidermis and a three-layer, collenchymatous hypodermis with relatively thin-walled cells. A prominent substomatal canal (center; beneath a guard cell) forms a passageway for gas exchange between the atmosphere and the photosynthetic tissue (chlorenchyma) below the hypodermis.
3.4. *Stenocereus gummosus,* or agria, a columnar cactus of the Sonoran Desert; the skin (0.25 mm thick) is composed of a one-layer epidermis and a fairly thick hypodermis, which has fairly thick-walled cells. A substomatal canal is visible in this section.
3.5. *Pachycereus pringlei,* or cardón, from the Sonoran Desert. In this section (0.25 mm thick) a thick cuticle covers an epidermis that is two or three cells thick; the photomicrograph shows the very tough, thick-walled hypodermal cells.

change. Instead, a passageway, called the substomatal canal (Figs. 3.3–3.5), develops between the epidermis and the inner tissues of the plant through the hypodermis.

Hypodermis covers the plant from near the apical meristem down to ground level, but it does not occur in cactus roots. Even though very old branches —50 years old or more—can have intact skin, the skin will eventually be replaced by bark (periderm). Periderm develops typically from the epidermis as a natural consequence of aging or in deeper tissues of the cortex following an injury that breaks the skin (Chapter 6). When bark forms, the thick cell walls of the hypodermis thin down; and ultimately the cells

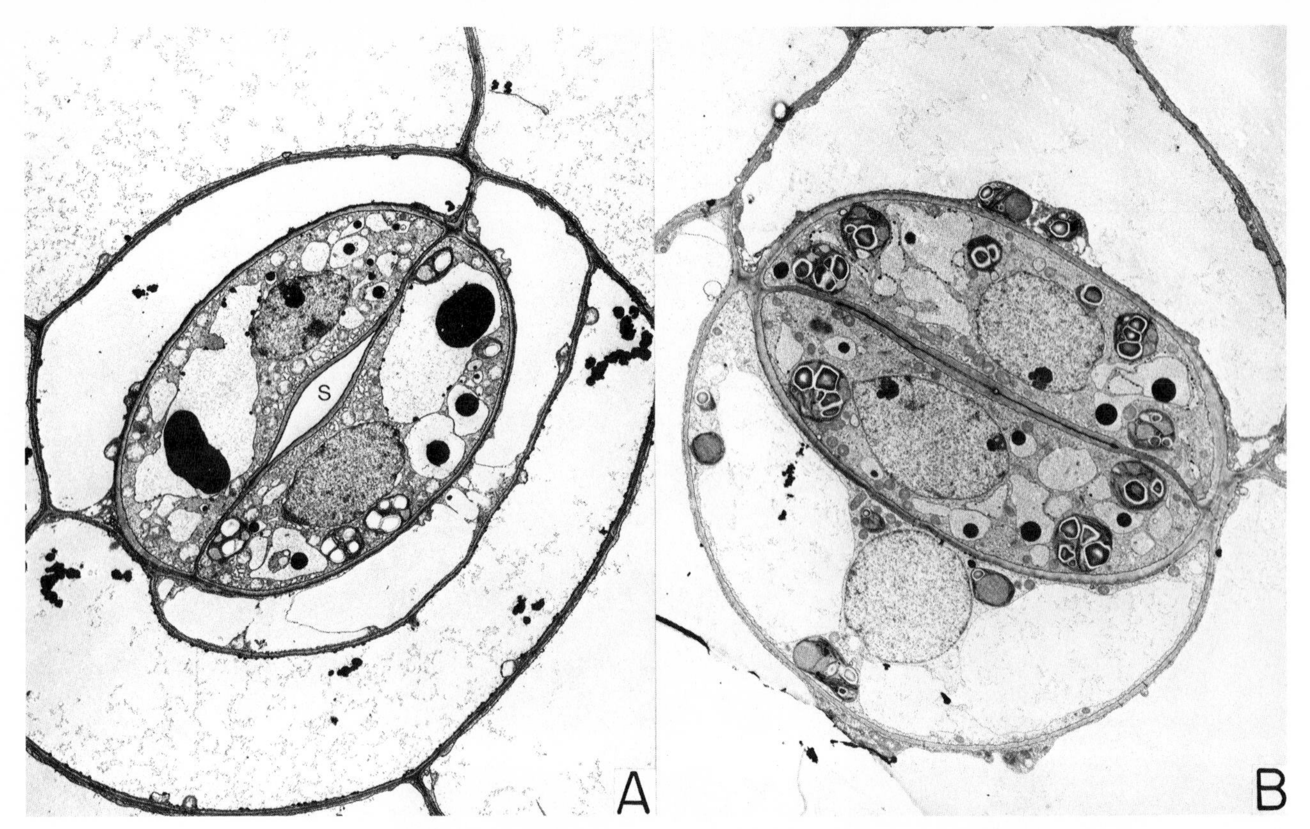

Figure 3.6. Transmission electron photomicrographs (×2500) of the lower leaf of *Pereskia aculeata*. These photomicrographs were generously provided by Christopher D. Faraday, University of California, Riverside. ***(A)*** Open stomate (s) on a leaf that was killed and fixed at midday. The two crescent-shaped guard cells are enclosed by crescent-shaped subsidiary cells. ***(B)*** Closed stomate of a leaf killed and fixed at night.

of the hypodermis may be indistinguishable from parenchyma cells of the cortex.

Temperature effects on skin. The effects of particular skin features on cactus stem temperature are largely unknown. For example, if the skin is gray, silvery, or white—as caused by the presence of much wax or a very rough surface—much of the sun's radiation will be reflected, a condition that theoretically could result in lowered stem temperatures. Remarkably, the surface temperature of a cactus growing in the field can rise to over 50°C. For most nonsucculent plants there is extensive damage to cell proteins at temperatures above 45°C, and as a consequence the tissue dies. Researchers have been able to show, however, that cactus skin can tolerate temperatures as high as nearly 70°C without fatal consequences (Chapter 7). The exact biochemical adaptations that permit cacti to withstand such high temperatures have not been identified.

Chlorenchyma

Chlorenchyma (Fig. 3.7) are the green cells of the cactus stem in which carbon dioxide (CO_2), water, and sunlight are used to make sugars. As we shall describe in Chapter 4, CO_2 from the atmosphere diffuses as a gas into the stem through the stomates and the substomatal canals of the hypodermis to the cell walls of the chlorenchyma cells, where CO_2 must become dissolved in water (the liquid phase) so that it can pass into the living, watery cell. Water is stored mostly in large storage structures (vacuoles) in the parenchyma cells. Typically each cell has a single, central vacuole, which occupies up to 95% of the cell volume. Light passes through the skin to the chlorenchyma, and certain wavelengths of this light are absorbed by the pigments (chlorophyll and carotenoids) present in the chloroplasts of the chlorenchyma cells. The absorbed light energy is then converted into chemical energy that can be used by the cell.

The structures of chlorenchyma and of the individual cells are designed to maximize photosynthesis. Air spaces between the cells (intercellular air spaces; Fig. 3.7*B*) allow CO_2 to diffuse between the cells of the tissue. The cell walls are very thin, a characteristic that facilitates the uptake of CO_2 by the cells; and the chloroplasts of the cells are arranged evenly around the periphery of the cells (because a large vacuole occupies the center of each cell) and thus are

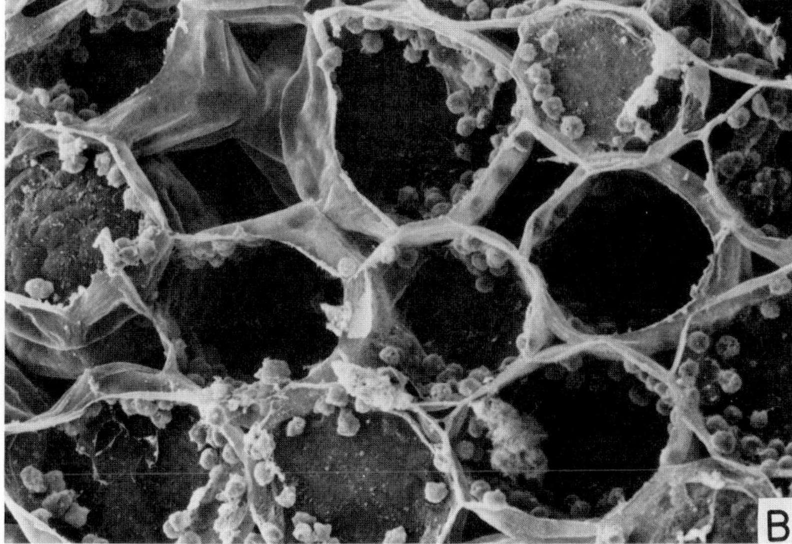

Figure 3.7. *Stenocereus gummosus,* or agria, from Baja California, Mexico.
(A) Cross section through the stem, showing a single-layered epidermis; a seven-layer hypodermis, which consists of cells with thick, collenchymatous walls; and outer cortex, which is the photosynthetic tissue (chlorenchyma). The skin is interrupted by canals *(upper left)* through which the gases diffuse between the chlorenchyma and the atmosphere. A large, oval mucilage cell lies in the cortex. ×120. (From Gibson, 1982.)
(B) Scanning electron photomicrograph of chlorenchyma cells. Although some cells have experienced partial collapse in the preparation, most cells still appear round in cross section. The chloroplasts (small, irregular balls) are arranged around the periphery of cells because the vacuole occupied the central region. Intercellular air spaces are large and permit gases to diffuse easily between the cells. ×285. (From Gibson, 1982.)

directly adjacent to the cell membrane, through which the CO_2 must enter. The central vacuoles temporarily store CO_2 by incorporating it into organic acids (Chapter 4).

Cortex and Pith

From an anatomist's point of view, the water-storing tissues of cacti are monotonous. Succulent tissues of the cortex and pith (collectively termed the ground tissue) are composed of large, spheroidal, parenchyma cells with thin cell walls. We say they are spheroidal rather than spherical because each cell is flattened where it touches an adjacent cell (Figs. 3.7 and 3.8). These cells are formed from three zones of cell divisions (Chapter 6). The cells of the pith result from cell divisions of the pith–rib meristem, which is the central cone of cells located in the apical meristem (Chapter 2). The cortex develops either from the peripheral zone of the apical meristem or from another meristem called the peripheral meristem, or the subprotodermal meristem, which produces the cells of the hypodermis and the outer succulent cortex of the ribs and tubercles, including chlorenchyma. During cell enlargement water uptake by the cell vacuole causes the cells to round up and thus to achieve their mature spheroidal shapes.

Because the vacuole occupies the center of the cell, the living protoplast of the cell is, as in chlorenchyma, positioned next to the cell membrane and the cell wall. Such succulent tissues rarely contain cells with thick, lignified cell walls, such as fibers or stone cells; rather it is the responsibility of the thin-walled cells of the ground tissue, swollen (turgid) with water, to provide much of the support that holds the stem erect. There are a few species of arborescent columnar cacti, for example, *Stetsonia coryne* of Argentina and *Pachycereus pringlei* of northwestern Mexico, that have some lignified cell walls in the cortex, especially in the chlorenchyma; and in a few columnar cacti, the pith can be firm because its cells have slightly thickened (but still unlignified) walls. Some epiphytes, for example, *Strophocactus wittii* and *Disocactus amazonicus,* have parenchyma cells with relatively thick, lignified walls. Also, some epiphytes and columnar cacti have primary phloem fibers present in the major vascular bundles of the ribs.

In cacti, the succulent ground tissue may be very soft and "spongy" as a result of the presence of abundant, soft, mucilaginous structures (Figs. 3.9 and 3.10). Mucilage is a very slippery, complex, and indigestible carbohydrate (Chapter 9). Many common cactus species are loaded with mucilage, from the outer cortex to the center of the pith. But this is not always the case: hundreds of species lack mucilage cells in the stem and have them only in the flowers and fruits; some of the least derived columnar cacti fall into this latter category. Perhaps surprisingly, many highly specialized species of North American cacti in the tribe Cacteae, such as the barrel cactus genus *Ferocactus,* also lack mucilage cells in the stem.

Mucilage cells in the stem seem to have evolved first in the cortex and much later in the pith. This statement is based on the evidence that there are many cacti with mucilage cells in the cortex but not in the pith and no known examples of the reverse condition. Moreover, the most highly derived species tend to have the most numerous and often the largest mucilage cells in a lineage.

Crystals (Figs. 3.11 and 3.12), such as the calcium oxalate crystals that occur in the majority of cacti, may impart a grittiness to succulent tissues. Generally speaking, these crystals are considered to be waste products of metabolism — excess calcium ions from the water and probably excess oxalate ions produced by the plant are combined into an insoluble salt. Calcium salts are collected inside the vacuole of a special cell called an idioblast, so called because it has a function different from that of its neighbors. There can be one crystal per cell, but in most species a number of crystals "grow" in the same vacuole and form an aggregate of crystals referred to as a druse (Fig. 3.12; Chapter 9). These whitish crystal aggregates can be clearly seen with a hand lens and even with the naked eye.

Starch and a variety of poisonous chemicals (Chapter 9) are stored in ground tissue. In species with few or no stem mucilage cells and a narrow pith, most of the starch is stored in the inner cortex, phloem, primary rays, and outer pith, that is, in close proximity to the vascular cylinder. But as the pith increased in diameter during the evolution of Cactaceae, much of the long-term storage of starch in stems was shifted to the pith, where the starch bodies (called starch grains) became very large (Fig. 3.13). Of course, there are some exceptions to this generalization, and some advanced cacti store abundant lipids (oil or fat molecules) rather than starch (Chapter 9).

Vascular System of the Succulent Shoot

Researchers on cacti have been concerned primarily with the water-storing cells of the cortex and pith and have paid little attention to the cylinder of vascular tissues that is situated between the cortex and the pith or to the vascular bundles that appear within the succulent regions. But cactologists should appreciate that the anatomy of succulence must include a description of the vascular tissues, which supply the ground tissues with water and minerals

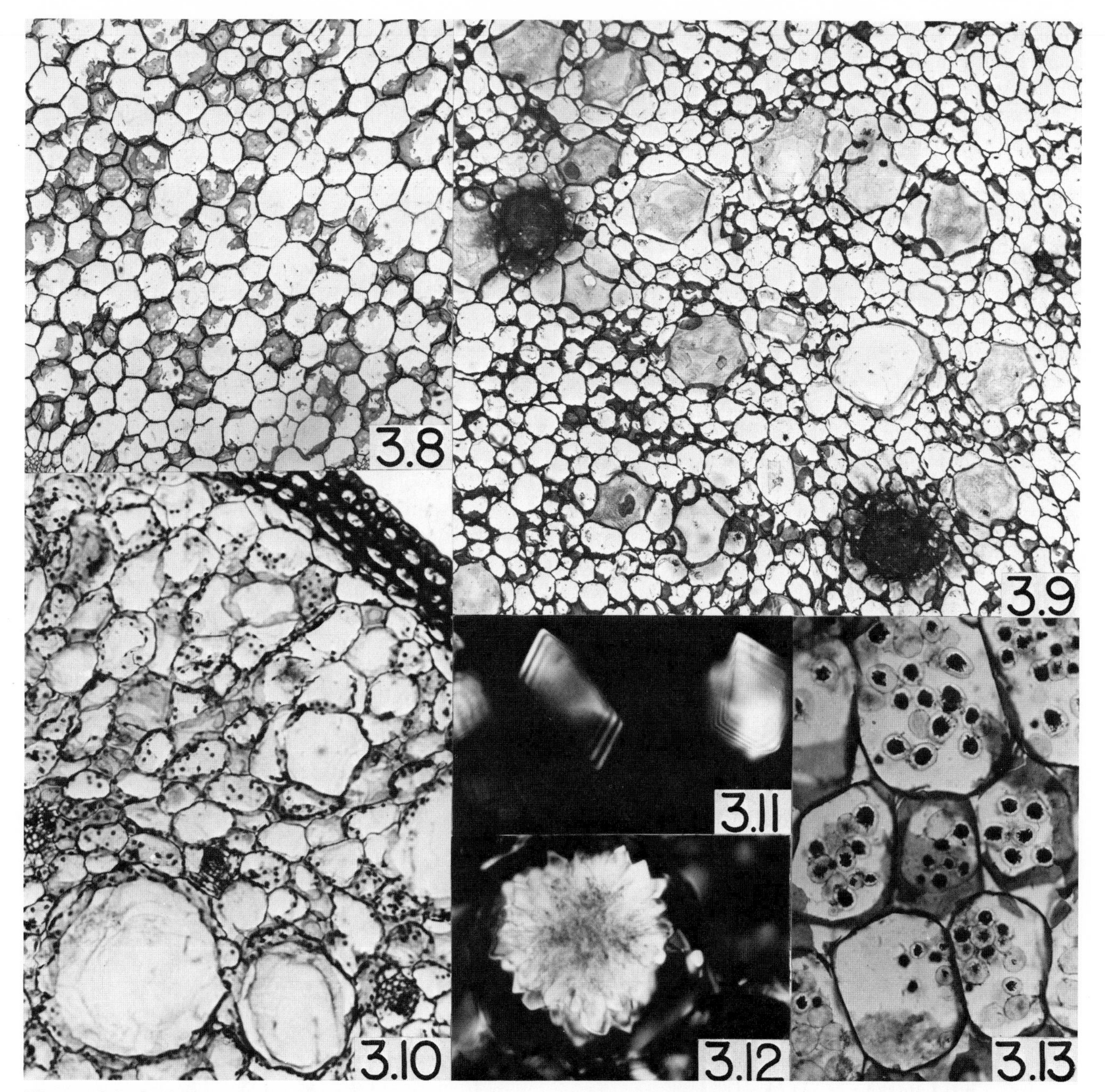

Figures 3.8–3.13. Structures of cactus stems.
3.8. *Peniocereus marianus.* Cross section through the pith, which consists of spheroidal parenchyma cells but not mucilage cells (section width about 1 mm).
3.9. *Selenicereus spinulosus.* Cross section of the pith at the same magnification as Figure 3.8; this pith has large mucilage cells.
3.10. *Werckleocereus* sp. Cross section through the outer stem, which has a skin (upper right) that has three layers of collenchyma and covers the chloroplast-bearing cells of the chlorenchyma. The chlorenchyma has scattered mucilage cells that are smaller than those found in the pith of *Selenicereus spinulosum* (they appear larger because the magnification of Fig. 3.10 is over twice that of Fig. 3.9).
3.11. *Mitrocereus fulviceps.* Large, solitary crystals of calcium oxalate (15 μm across), as viewed under polarized light.
3.12. *Leptocereus wrightii.* A druse (crystal aggregate) of calcium oxalate in a wood parenchyma cell, as viewed under polarized light (aggregate about 30 μm in diameter). (From Gibson, 1973.)
3.13. *Eccremocactus bradei.* Spheroidal starch grains are abundant in the wood parenchyma cells (each starch grain is about 5 μm in diameter).

from the soil via the roots and which transport the sugars made in the chlorenchyma to the rest of the plant. Apparently, the great enlargement of succulent tissues in cacti has been matched by increases in the number of vascular bundles in the stem.

Vascularization of the pereskias and opuntias. Before we describe the vascular systems of more specialized cacti, we want to describe the organization of vascular bundles in *Pereskia,* which represents the ancestral, nonsucculent condition. Figure 2.4 shows that an arborescent pereskia has a vascular cylinder that consists of many bundles and that the bundles directly beneath a leaf enter that leaf base to carry water into and carbohydrates out of the leaf. But this is only a small segment of the vascular system, and a fairly long section of the shoot must be analyzed to determine how the bundles of the stem and leaf are actually connected. To make this reconstruction, an anatomist must study a series of transverse (cross) sections that progress from the shoot apex downward and are cut at intervals of about 10 μm (0.00001 m) or less; these serial sections show where the bundles branch and where they fuse. This tedious procedure eventually provides enough information to produce a diagram of the entire system.

Although vascular systems are most easily visualized in three dimensions, the complex vascular system of a cactus is most easily illustrated with a two-dimensional diagram that is made by splitting the system lengthwise and then flattening it to show all of the bundles (Fig. 3.14). In *Pereskia humboldtii,* there are five major vertical bundles (called axial bundles) that continuously develop upward in the growing shoot. The many vascular bundles (traces) that diverge into a leaf arise from just one of these five axial bundles; for example, the vasculature for leaves 2, 7, and 12 arises from the central axial bundle in the figure, and that for leaves 1, 6, 11, and 16 arises from the axial bundle shown on the far right. This species has a vascular system slightly simpler than *P. sacharosa,* which has more bundles entering each leaf. In Figure 3.14, the bundles that approach the areole (above each leaf) are formed from two axial bundles, one on each side of the areole.

The system just illustrated is called an open vascular system because the axial bundles are independent of each other. Another type, called a closed vascular system, has axial bundles that are fused into a network or reticulum because bundles (bridges) are formed between them. In a closed system, phloem contents can move freely to any part, from one side of the plant to the other. Conversely, in an open

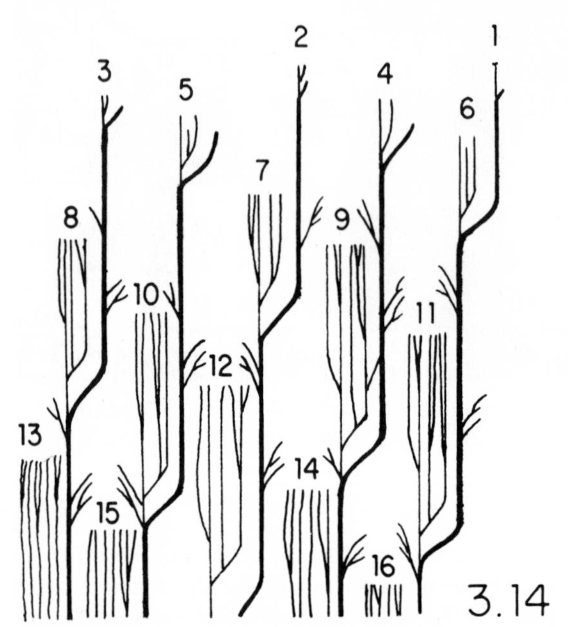

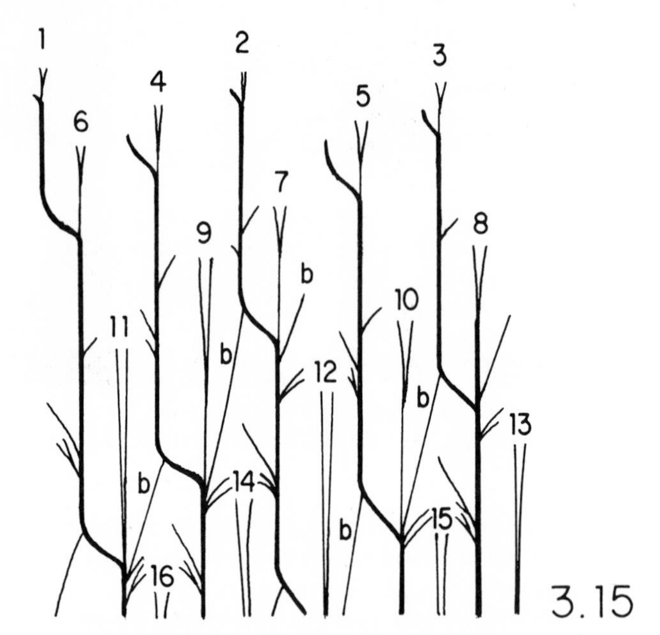

Figures 3.14 and 3.15. Diagrams of the primary vascular systems of pereskias, determined by tracing the longitudinal courses of vascular bundles in the shoot and then drawing them to show the vascular system as if it were split along one side, opened out, and then flattened. The heavy lines are called axial bundles, from which the traces for the leaves arise; the vascular bundles leading to each leaf are numbered. (Redrawn from Gibson, 1976.)

3.14. *Pereskia humboldtii,* a leaf-bearing species with numerous vascular bundles per leaf and an open vascular system. This arrangement is called open because the axial bundles are not interconnected.

3.15. *Pereskia aculeata,* another leaf-bearing species with thin stems and a simple vascular system, consisting of five axial bundles and a single trace leading to each leaf, which bifurcates before both bundles enter the leaf. This arrangement is called a closed system because the axial bundles are interconnected by strands (b, bridges).

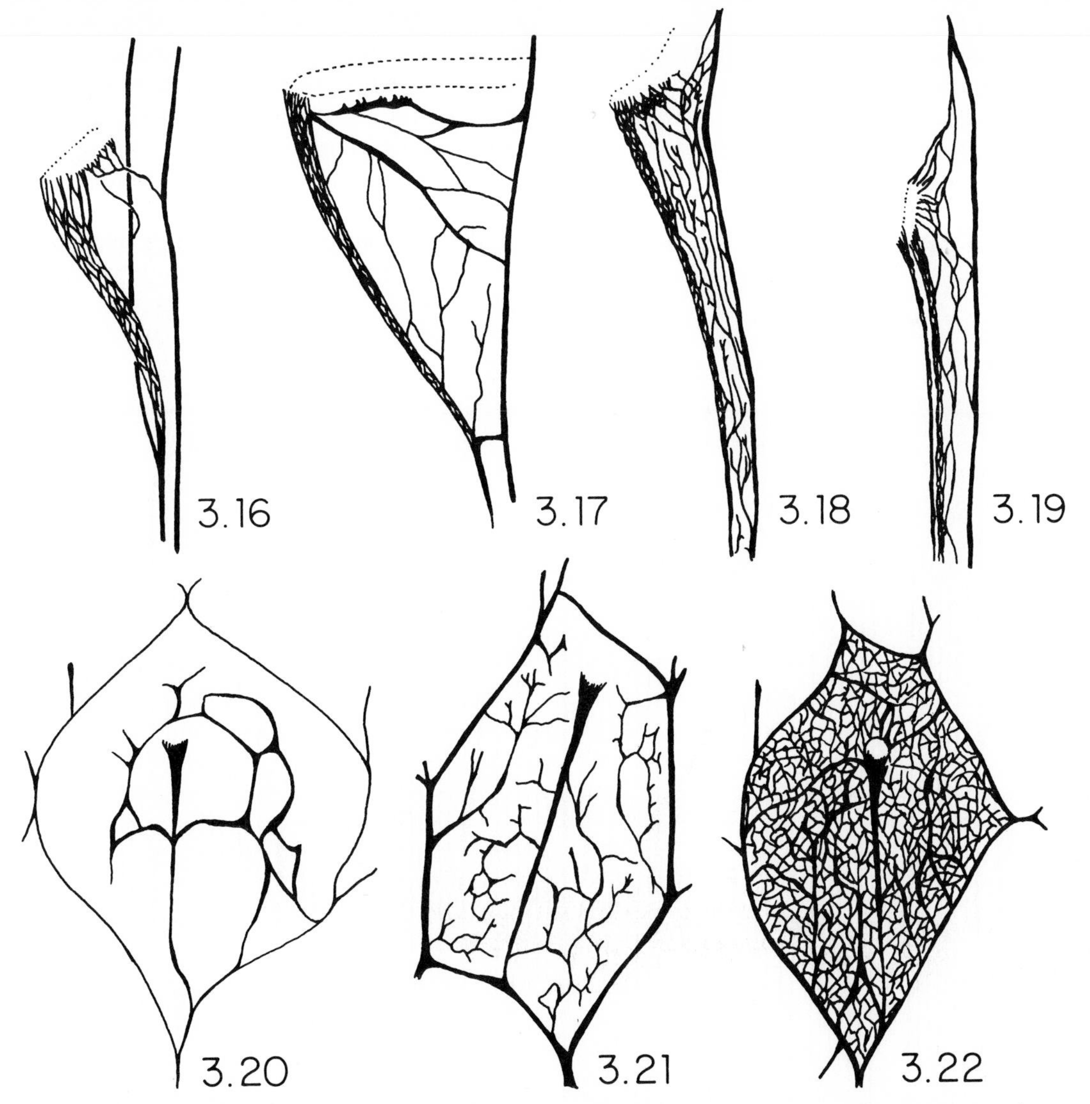

Figures 3.16–3.22. Vascular systems in stems of opuntias, showing the vasculature of the leaf (thick, central strand) and other vascular bundles. (From Gibson, 1976.)
3.16. *Opuntia ramosissima,* or diamond cholla; a cholla with very low tubercles.
3.17. *Opuntia spinosior,* or cane cholla; a cholla with high, wide tubercles.
3.18. *Opuntia versicolor,* or staghorn cholla; a cholla with fairly long tubercles.
3.19. *Opuntia arbuscula,* or pencil cholla; a cholla with relatively low tubercles.
3.20. *Opuntia microdasys,* or bunny ears; a platyopuntia with relatively few vascular bundles associated with each leaf and areole.
3.21. *Opuntia phaeacantha* var. *discata;* a platyopuntia with an intermediate amount of tubercle vascularization.
3.22. *Opuntia tomentosa;* a tree platyopuntia with a dense and highly reticulate vascular system associated with each tubercle.

system the axial bundles are not interconnected; thus sugars can move easily up and down but cannot move laterally.

The fairly simple vascular system of *P. aculeata* (Fig. 3.15) is a closed vascular system. This system also has five axial bundles and a very simple way of forming vascular bundles for the leaf. The vasculature for every fifth leaf arises from a single axial bundle, for example, 2, 7, and 12 or 1, 6, 11, and 16. In this system, however, bridges form between adjacent axial bundles and join the axial bundles into a single network. In *P. aculeata,* only two vascular bundles enter the leaf base, and as before the areolar traces arise from the two adjacent axial bundles.

The closed vascular system found in *P. aculeata* is similar to the vascular systems of all Opuntioideae, which appear at first glance to be very complicated. In opuntioids, numerous minor (accessory) vascular bundles have been added to the system to supply the enlarged succulent tissues of the tubercles (in cylindropuntias) and to the broader succulent tissues of the vascular cylinder of the cladode (in platyopuntias). Notice in Figures 3.16–3.22 how a major vascular supply still diverges toward (and then into) the small, ephemeral leaf of an opuntioid.

In Pereskioideae and Opuntioideae, there are no vascular bundles in the pith.

Vascularization in the subfamily Cactoideae. Whereas the vascular systems of Pereskioideae and Opuntioideae are very tightly linked to the formation of the leaves, the complex vascular systems of Cactoideae cannot be linked to leaf formation because the leaf blade is vestigial and inconspicuous (Fig. 1.18). As a consequence, in the succulent Cactoideae the single bundle or pair of bundles that diverges from the vascular cylinder into the tubercle or rib does not

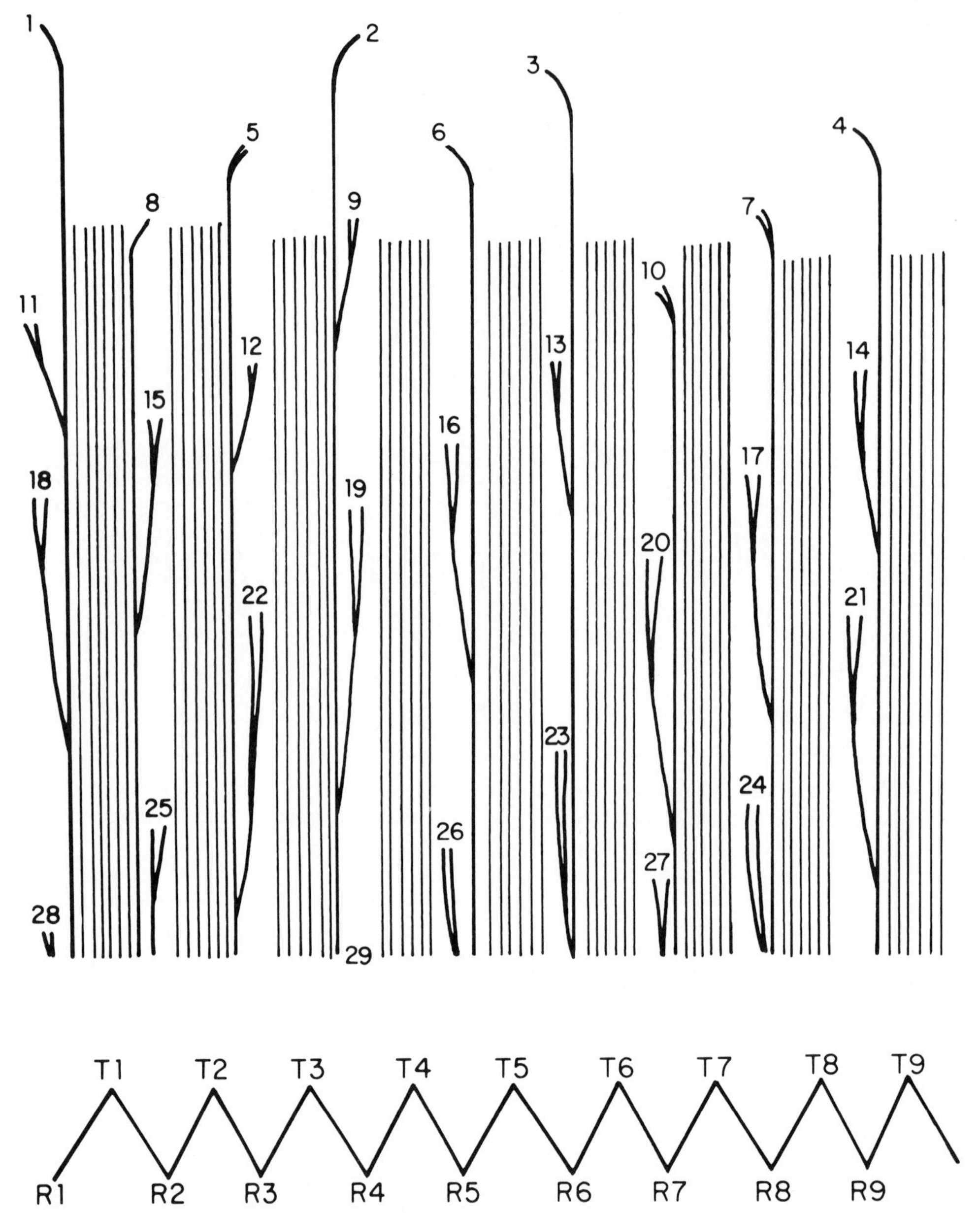

Figure 3.23. *Polaskia chende,* a Mexican columnar cactus with a fairly simple primary vascular system. The vascular bundles of the vascular cylinder are presented as if the system were slit down one side and opened up to make this two-dimensional diagram. The heavy vertical lines represent the vascular bundles that give rise to the vascular tissues leading to the leaf and the areole, and the finer vertical lines are extra bundles that fill the space in the vascular cylinder. A wide space occurs where the vascular bundles diverge outward toward the areoles. The diagram at the bottom of the figure shows the position of the nine ribs (R1–R9) and nine troughs (T1–T9); note how the wide vertical gap in the vascular system and the origins of the areolar traces are positioned directly beneath the points of the ribs. The areolar traces are numbered in reverse order of formation: 1 is the newest and 29 the oldest.

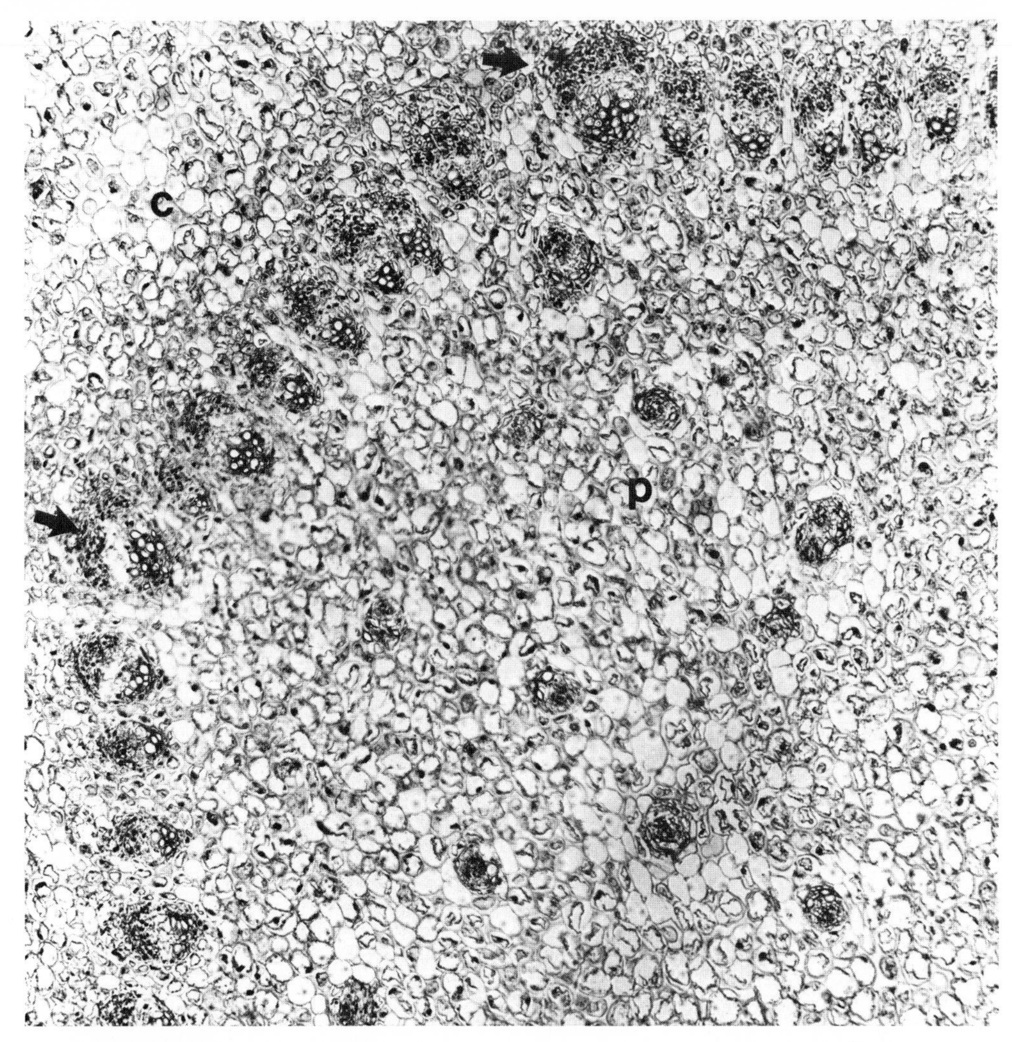

Figure 3.24. *Polaskia chende;* cross section of the stem vascular cylinder showing pith (p) and cortex (c). The large vascular bundles (arrows) are diverging toward the areoles, as diagrammed in Figure 3.23; between the large bundles are additional, parallel stem bundles (cauline bundles). Vascular bundles (medullary bundles) are present in the pith.

develop rapidly toward the leaf but instead develops toward the areole, which is the prominent structure. Vascular tissues generally do not enter the spines (Chapter 2).

No species of the subfamily Cactoideae can be used as typical, or representative, because both open and closed vascular systems are found in these species. Therefore, to make a simple point about shoot vasculature in these forms, we shall concentrate on one system—an open pattern in a Mexican cactus, *Polaskia chende.* Figure 3.23 illustrates the vascular system of this columnar species, which is a relatively unspecialized member of its tribe.

As in earlier examples, Figure 3.23 is a two-dimensional diagram of a shoot and was compiled from a very long series of stem transverse sections. This plant has nine ribs (R1 – R9), and the tubercles on those ribs are numbered sequentially from the youngest to the oldest. The vascular cylinder is composed of many vascular bundles, which can for convenience be referred to either as axial bundles or as cauline bundles. As before, the axial bundles give rise to traces that diverge from the vascular cylinder toward the outer stem, but in this case the ultimate destinations for the bundles are the areoles. The figure shows that a bundle diverges from the vascular cylinder from a position just opposite the spine-bearing rib, which is, of course, the most direct course from the vascular cylinder to the areole. Shown with finer lines are the cauline bundles of the vascular cylinder. These are parallel, unbranched, and slightly thinner bundles that fill in the space within the vascular cylinder (Fig. 3.24). Therefore, unlike the systems that we have already discussed, in which only trace-producing axial bundles are present, *P. chende* has both trace-producing axial bundles and unbranched cauline bundles.

In the nine-ribbed specimen illustrated in Figure

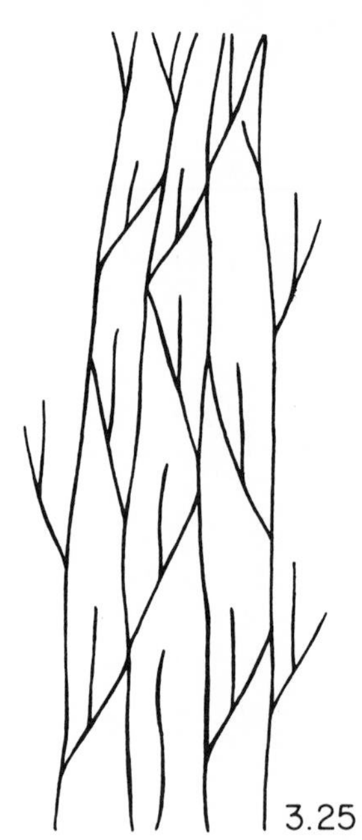

Figures 3.25 and 3.26. Examples of closed vascular systems of cactus stems; the bundles are fused into a netlike structure.
3.25. *Notocactus haselbergii;* a portion of the vascular system is shown with four of the principal vertical bundles that are fused into a network.

3.26. *Mammillaria fragilis;* three-dimensional drawing of a closed vascular system with only five vertical bundles that are interconnected, in which one bundle diverges toward each node (leaf and areole).

3.23, we can observe the pattern of trace origin for each rib. For example, the axial bundle that vascularizes R1 includes traces for primordia (nodes) 1, 11, 18, and 28, and all traces diverge to the left; whereas the axial bundle for areoles on R4 produces traces for 2, 9, 19, and 29, and all traces diverge to the right. If this were a specimen with eight ribs—a number in the Fibonacci sequence (Chapter 6)—then R1 would include traces for primordia 1, 9, 17, and 25, that is, a regular interval of 8, and the traces for all ribs would always diverge in one direction, either all to the left or all to the right (this is the pattern in common nonsucculent dicotyledons with eight axial bundles). Instead, the nine-ribbed specimen has intervals of 7 and 10 between traces, and selected axial bundles must be left-handed or right-handed. Hence, here we can see how the evolution of ribbing in cacti has caused important changes within the internal vascular system of the plant, namely, by altering the way that traces arise.

Vascular cylinders similar to the one just described in *Polaskia* also occur in other species of the tribes Pachycereeae, Leptocereeae, and Hylocereeae, and this type of vasculature may be the primitive condition for the subfamily. In most of these species the cauline bundles are interconnected by some bridges. So little is known about the vascular systems in this subfamily that at this time it would be unwise to speculate much on the significance of any design, except to say that the addition of cauline bundles to the vascular cylinder was probably necessitated by the enlargement of the pith and, consequently, the increase in the circumference of the vascular cylinder. Indeed, wherever the pith has become very wide, the vascular cylinder has been increased by the addition of many cauline bundles, so that the vascular cylinder of saguaro has about 200 bundles in each cross section. Whenever the pith has become narrow (that is, narrower than in *Polaskia*), the number of bundles in the vascular cylinder has decreased and often equals the number of ribs or vertical rows of tubercles; thus, the system has only axial bundles. The closed vascular systems of many small, globular cacti have only axial bundles (Figs. 3.25 and 3.26).

Enlargement of the pith has also influenced the evolution of vascular bundles in the pith; as expected, wherever the pith is large, a medullary vascular system occurs and supplies the cells with sugars and water. It is possible that the development of medullary vascular bundles helped cacti to use the pith for starch storage. But having pith vasculature is not a requirement for storing starch in the pith because many narrow-stemmed epiphytes that lack medullary bundles store abundant starch grains (Fig. 3.13). Boke has observed that medullary bundles seem to arise from the vascular cylinder; but an understanding of the evolution of pith vasculature will have to await an analysis of closely related species (Chapter 10) and detailed developmental studies.

Many species of Cactoideae also have a third, fairly discrete vascular system in the cortex, which is present outside the vascular cylinder and which increases the vascular supply to the succulent tissues of the tubercles and ribs. Once again we can argue that this design makes good physiological sense, because thick tissues need more vascular tissues; but researchers still have not studied the origin and design of cortical vascular bundles and certainly have not determined how this system is related to the major areolar traces. In future studies, radioactively labeled tracers will be required to follow the flow of carbohydrates from the site of their manufacture at the surface of the stem to other portions of the plant to determine how important cortical and medullary vascular tissues are for the succulent cacti.

Succulence in Woody Cylinders

The nature of secondary growth. In typical dicotyledons, as well as in the arborescent pereskias, wood (secondary xylem) and inner bark (secondary phloem) are produced by a cylinder of dividing cells called the vascular cambium. This cylinder forms from thin-walled cells located between the primary (original) xylem and the primary phloem of the vascular bundles and from thin-walled parenchyma cells located between the vascular bundles in the medullary rays (Fig. 3.27). Each cell in this meristematic cylinder is termed an initial, which is a cell that

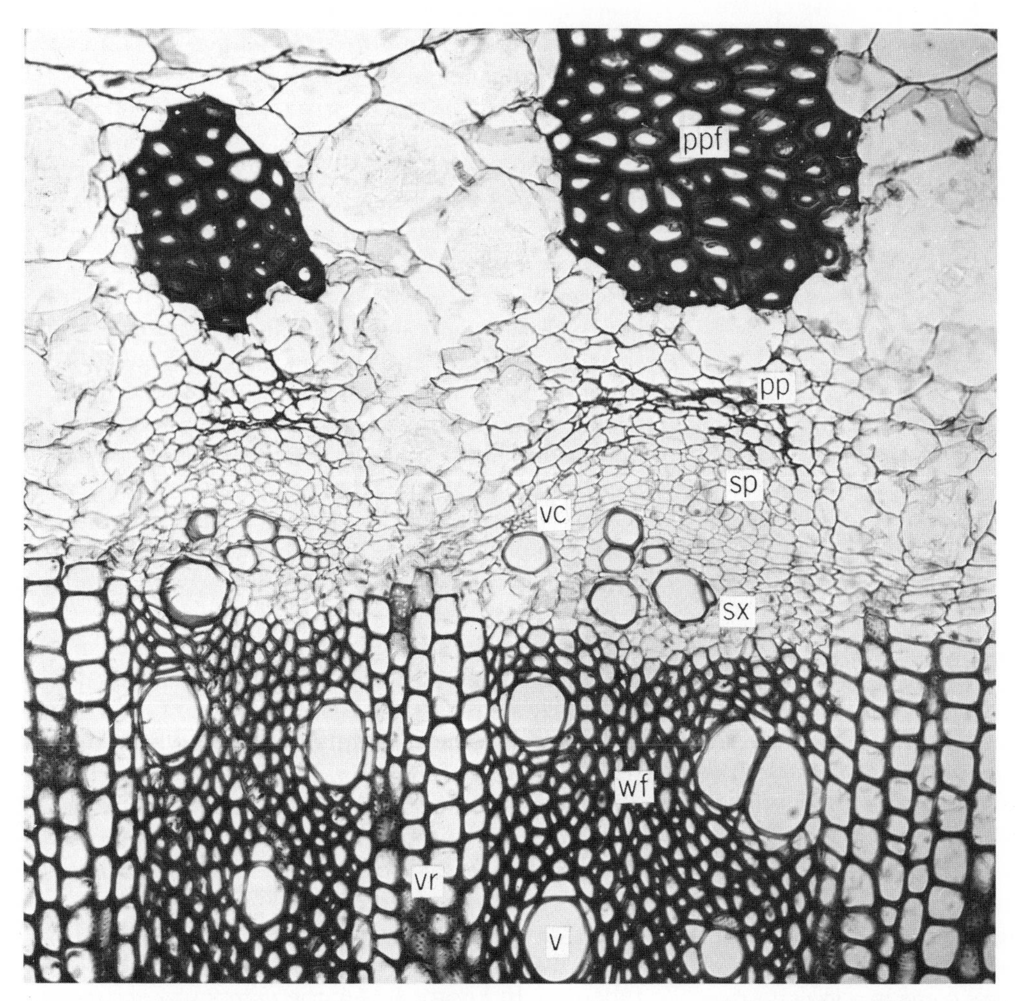

Figure 3.27. *Hylocereus purpusii*, an epiphyte; cross section of the stem, showing the vascular tissues: thick-walled, primary phloem fibers (ppf); crushed primary phloem (pp); functional secondary phloem (sp); the vascular cambium (vc); recently formed secondary xylem (sx) containing vessels but not wood fibers; and last year's secondary xylem, having vascular rays (vr), wood fibers (wf), and vessels (v). Some of the large cells adjacent to the primary phloem fibers contain large druses of calcium oxalate. The width of the original section was about 1 mm.

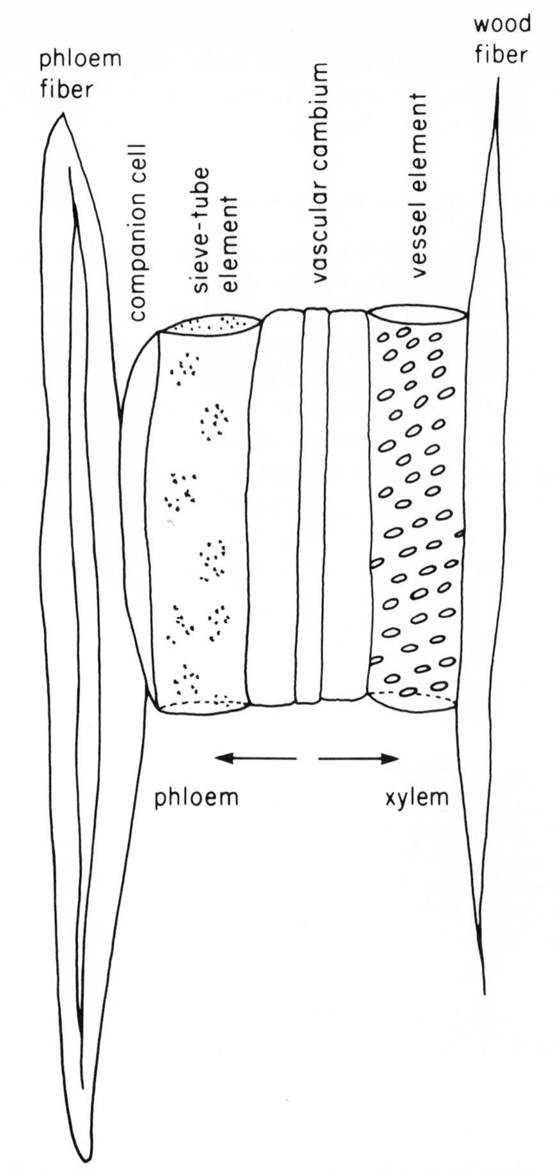

Figure 3.28. Cells near the vascular cambium, as viewed from a longitudinal section cut from the pith to the epidermis. The initial of the vascular cambium divides to produce a cell on either the xylem or the phloem side. The xylary derivative may become a vessel element, a parenchyma cell, or a wood fiber; the phloem derivative may become a sieve-tube element, a phloem fiber, or a parenchyma cell (for example, a companion cell). Both the vessel element and the sieve-tube element are approximately the same length as the cambial initial, but the wood or phloem fiber can become very long after they are initiated from the initial.

repeatedly divides throughout the life of the plant and never matures to become a cell of the xylem or the phloem. Cambial initials come in a variety of shapes, but they are generally classified into two categories: fusiform initials, which are several to many times longer than they are wide and which produce the vertically elongate cells of the xylem (Fig. 2.7) and phloem; and ray initials, which are squarish or rectangular in outline and form parenchyma cells that are squarish or rectangular in outline.

The initials divide in a plane parallel to the surface of the stem, and the daughter cells that are produced become part of either the xylem or the phloem (Fig. 3.28). Cells that are produced to the inside of the vascular cambium lay down lignified cell walls after their shapes have been determined; and therefore the woody cylinder that is made is solid. In contrast, only some of the cells produced to the outside for the phloem—generally the phloem fibers (Figs. 3.27 and 3.28)—form thick, lignified cell walls. The sieve-tube elements (the conducting cells of the phloem) and the parenchyma cells of the phloem typically lack thick walls.

Even though *Pereskia sacharosa* appears to have a solid cylinder of wood (Fig. 2.5), close examination of the wood reveals that there is a small, often inconspicuous unlignified region in the cylinder opposite every spur shoot. Because the spur shoots are widely spaced along the stem, these unlignified rays do not strongly influence wood design.

Patterns of succulence in woody cylinders. Visitors to the deserts of the New World have often seen and marveled at the wood skeletons of the succulent cacti (Figs. 3.29–3.33). When the stem of a cactus dies and dries, the skin falls off and the succulent tissues dry up and crumble (Figs. 3.29 and 3.31). This exposes the central woody cylinder, which has holes in it—places where lignified cells were never present (Figs. 3.30–3.33). Each species has a characteristic wood skeleton, and some of the sturdiest ones have been used by artisans to make craft objects, souvenirs, and even furniture or shelters. More important from our point of view, the architectural designs of these wood skeletons can also be used to study the evolution of succulence in Cactaceae. We have already laid the groundwork for interpreting the design of the woody cylinder of a cactus by our discussion of the structure of the original vascular systems of these plants.

A fairly simple example of the origin of the holes in the cactus wood skeleton can be observed in the stem of the cylindropuntia, *Opuntia leptocaulis* (Fig. 3.34). Figure 3.34*A* shows the external structure of this young cholla shoot; and Figure 3.38*B* shows how the vascular system is organized—it is similar to that described for *Pereskia aculeata* (Fig. 3.15) and is a closed system with five axial bundles. But in this cholla many accessory bundles vascularize the tubercles. In Figure 3.34*B*, the empty spaces surrounding the vascular tissues are occupied by soft, unlignified parenchyma cells; the region of soft cells opposite the tubercles are the primary rays.

As a stem gets older, wood is made. The vascular

Figures 3.29–3.33. Wood skeletons of cacti.
3.29. *Stenocereus thurberi,* or organ pipe cactus; the succulent tissues have dried up and fallen off most of the stems of this dead plant, leaving the erect cylinders of wood.
3.30. The wood skeletons of four species of chollas from Arizona, which have different sizes and shapes of tubercles and therefore different sizes and shapes of the holes in the wood skeleton.
3.31. *Opuntia fulgida,* or jumping cholla; the succulent tissues have dried and disappeared, but some of the spine-bearing areoles are still in their original location above holes in the wood skeleton.
3.32. *Mammillaria dioica;* tubercles and holes in the wood skeleton are both helically arranged.
3.33. *Ferocactus townsendianus;* in this barrel cactus, which has well-developed ribs, the wood skeleton has large vertical cables of wood and less-pronounced diagonal bridges.

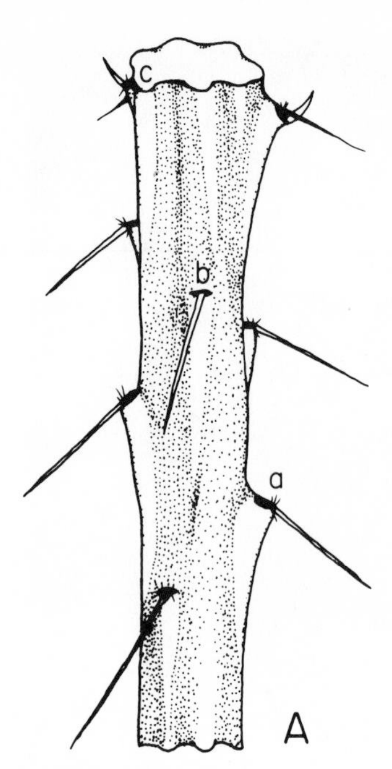

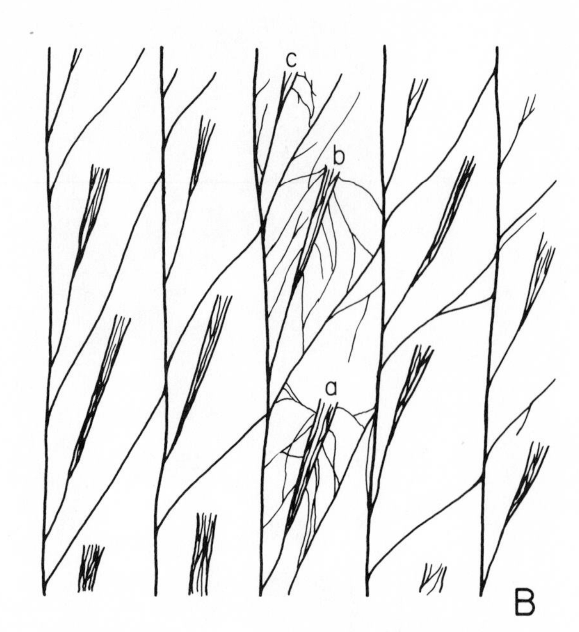

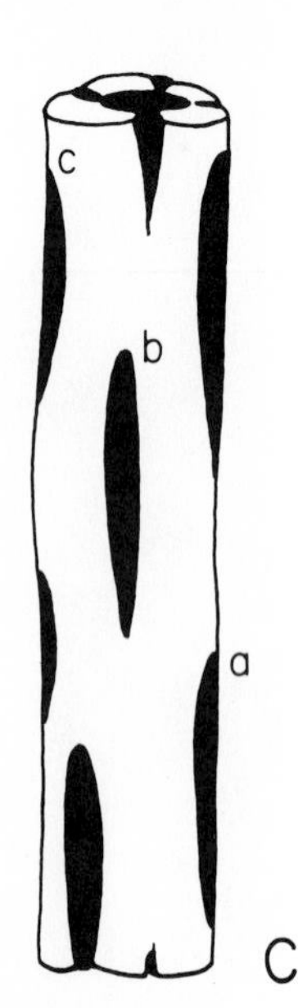

Figure 3.34. Vascular tissues in a narrow-stemmed cholla, *Opuntia leptocaulis.*
(***A***) Side view of a shoot, on which three areoles are labeled (a to c).
(***B***) View of the first-formed vascular system of the same stem, showing that the system is a network (a closed system) in which the principal bundles of the network are not present beneath the tubercles but instead are located beneath the indented portions of the stem. A vascular bundle diverges from the vascular network at approximately the same level at which the tubercle begins to show on the stem surface. Additional vascular bundles are present (only shown for tubercles a through c) to vascularize the tubercles. All of the blank area under tubercles a through c is parenchyma.
(***C***) Three-dimensional drawing of the wood skeleton of this same stem projected ahead 5 years. The areas that were mostly soft parenchyma form the holes, whereas the original lattice of vascular bundles becomes cables of wood.

cambium forms over the netlike vascular cylinder; thus, the vertical vascular bundles and the bridges connecting them begin to accumulate wood and become thicker and harder. At the same time, the primary rays of the vascular cylinder remain soft and parenchymatous, apparently because they are not stimulated to form a cambium and wood. Consequently, the woody cylinder resembles the latticelike pattern of the original vascular bundles, and the holes or primary rays assume the basic shape of the tubercle (Fig. 3.34*C*). In *Opuntia leptocaulis,* the holes are long and narrow; but in species that have broad tubercles, the holes are elliptical or circular (Fig. 3.30).

In many cacti, the vascular cambium does not develop across a primary ray for a long time; and in species in which the primary rays are very wide, the vascular cambium may develop gradually across the primary ray but never fully form a complete sheath of cells. This is why some holes may persist even in very old cactus stems. Hence, the persistence of large primary rays is one avenue available for the development of stem succulence and for maintaining a parenchymatous connection between the pith and the cortex.

Another way of developing succulence in the woody cylinder is shown by the platyopuntias (Figs. 3.35 and 3.36). In contrast to the opuntioids just described, the platyopuntias do not have large primary rays; instead they have a very fine, fenestrate network of vascular tissue that is formed by the elaborate system of accessory bundles (Figs. 3.20–3.22). A complete vascular cylinder forms in the cladode of a platyopuntia, but lignified cells are produced only from initials that are located opposite the original vascular bundles of the system. Collectively, these initials are called the fascicular cambium. Unlignified parenchyma cells (rays) are produced from ray initials of the interfascicular cambium. The resultant woody cylinder therefore has the same design as the original vascular cylinder, and water can diffuse between cortex and pith at any point in the wood through these parenchymatous rays.

In subfamily Cactoideae the diversity of architectural designs of wood skeletons is quite splendid. One of the most fascinating and now easily understood patterns is the wood skeleton that characterizes the tribe Pachycereeae (Figs. 3.37–3.39). In all species of this North American tribe the young woody cylinder is subdivided into a ring of discrete, parallel (fastigiate) rods. Each rod of wood forms

Figures 3.35 and 3.36. Architecture of wood skeletons in platyopuntias.
3.35. *Opuntia pilifera,* a small tree from southern Mexico, which has a netlike, young wood skeleton that mimics the initial structure of the cladode vascular system (for example, Figs. 3.20–3.22).
3.36. *Opuntia quimilo,* a tree platyopuntia from Argentina; the skeletonized portion of a large branch, showing the places where unlignified parenchyma was present in the woody stem.

Figures 3.37 and 3.38. Wood skeletons of arborescent species in the tribe Pachycereeae, which have a ring of parallel, discrete rods arranged around a very wide pith.
3.37. *Carnegiea gigantea,* or saguaro, from southern Arizona. (*A*) Skeletonized plant, showing the discrete rods of wood. (*B*) Close-up of the stem about 1.5 m from the ground where the rods are connected because the plant shifted to a higher number of ribs and, consequently, a higher number of rods (one per stem valley).
3.38. *Neobuxbaumia mezcalaensis,* from Puebla, Mexico; skeleton of a very tall, solitary columnar cactus, which has discrete rods of wood.

Figures 3.39–3.41. Changes in the structure of wood skeletons from the inside *(A)* to the outside *(B)*.
3.39. *Stenocereus* sp. (tribe Pachycereeae), a species with discrete, parallel rods of wood *(A)* that are separated by vertical rows of primary rays (arrows), which tend to become obscured by subsequent growth of the woody cylinder *(B)*.
3.40. *Cleistocactus areolatus* (tribe Trichocereeae), a species with a very delicate, netlike (reticulate) vascular system when wood is initiated *(A)*. During secondary growth, the large primary rays disappear and the outer portion of the woody cylinder is solid, with only small holes *(B)*.
3.41. *Cereus dayamii* (tribe Cereeae); the inner wood skeleton is an irregular network of strands *(A)*, which result in a wood skeleton with no geometric pattern *(B)* but has a large hole beneath the areole.

opposite a valley of the stem, where the cauline bundles are clustered (Figs. 3.23 and 3.24); and each vertical panel of succulent tissue, which is a vertically continuous series of unlignified primary rays, forms opposite the point of a rib. Because the axial bundles are not fused together at an early stage into a tight, interconnected system, the sectors of the vascular system can be spread apart when the pith increases in circumference, a change that also can cause the primary rays to increase in width. This clever design of the vascular system permits some of these cacti to develop a pith as large as 20 cm in diameter — for example, *Carnegiea* (Fig. 3.37), *Neobuxbaumia* (Fig. 3.38), and *Mitrocereus* — and even to enlarge the pith in old stems without having deleterious effects on the vascular system. In other species of the tribe that have a relatively narrow pith, the rods of wood are not widely spaced; and after the pith has stopped enlarging, the rods can fuse and then form a solid cylinder (Fig. 3.39*B*).

Other tribes of the subfamily have basic designs different from the rodlike pattern of Pachycereeae.

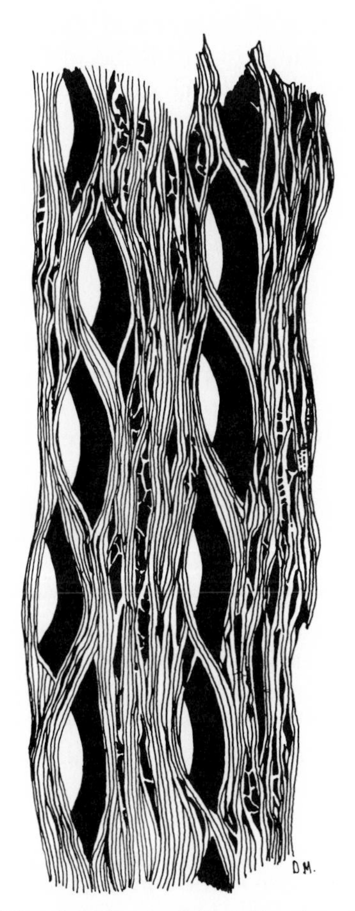

Figure 3.42. Wood skeleton of the tall, arborescent cactus, *Trichocereus terscheckii,* from Argentina; large, lens-shaped holes are arranged in vertical series. (From Gibson, 1978b.)

Some of these are elaborate and reticulate (Figs. 3.31 – 3.33, 3.36, 3.37, and 3.40*A*) and are similar in some respects to the vascular systems of chollas (Fig. 3.30), whereas in others the wood architecture is very irregular (Fig. 3.41*A*). The way to see the fundamental nature of the woody cylinder is to remove the pith and see the first-formed pattern of the wood (Figs. 3.39*A*, 3.40*A*, and 3.41*A*) because the basic pattern can be obscured after many years of wood are produced (Figs. 3.39*B*, 3.40*B*, and 3.41*B*).

Tribe Trichocereeae can be used to demonstrate how wood skeletons have evolved. All species have one large, unlignified primary ray per tubercle, and the first-formed system is a reticulate, closed vascular system (Fig. 3.40*A*). If the ribs are broad and high, as in *Trichocereus terscheckii,* the holes in the wood skeleton will appear in vertical series (Fig. 3.42); but if the ribs are low, narrow, and very numerous, the holes will appear in their original helical arrangement and the vertical series will be more difficult to discern. Note how the wood skeleton of *T. terscheckii* of Argentina is completely different from that of *Carnegiea gigantea* of Arizona (Fig. 3.37), even though they have identical growth forms and growth habits.

Although Gibson has previously described many of the patterns of woody cylinders in cacti, there still remains a need for more detailed research in all tribes to determine whether the evolutionary lineages of cacti can be identified on the basis of the architectural design of the wood skeleton.

Roots and Root Succulence

Whereas the stems of most cacti contain much succulent tissue, the roots of most cacti are relatively nonsucculent. A typical cactus root has no pith and a fairly thin cylinder of cortex. The root is covered at a very early stage of growth by orange and brown plates of corky bark; and even on the thinnest roots, these plates are easily dislodged when the roots are handled.

Types of Root Systems

No one has ever made an extensive study of cactus root systems, but there appear to be five common designs. Many taxa, including most of the opuntioids, barrel cacti, and columnar cacti, have a set of shallow, horizontal roots that radiate from the plant in all directions. These roots are generally unbranched and are positioned no more than 5 to 15 cm below the soil surface. Because some roots are close to the surface, they can take advantage of rains that moisten only the uppermost soil layer. A second design is found in some large columnar cacti (for example, senita of the Sonoran Desert, *Lophocereus schottii*), which have long, shallow, horizontal, lateral roots and a central, main root that penetrates to deeper soil layers and anchors the plant. A third design occurs in hundreds of small cacti, which have a compact root system consisting of many short laterals that develop directly beneath the plant. These roots take advantage of the water that drips off the plant and collects in the soil at its base. A fourth design occurs in some of the small, globular cacti, such as peyote (*Lophophora williamsii*, Fig. 1.14) and many species of *Lobivia*. These cacti have a single, fleshy taproot with no or only a few small lateral roots. The fifth design may escape notice, but almost 200 species of cacti have exclusively adventitious root systems. These species can be found among the epiphytic cacti (Fig. 3.43), in creeping or prostrate terrestrial plants such as creeping devil (*Stenocereus eruca*, Fig. 1.25), and in many small platyopuntias that form roots on the lower side of the stem.

There are some cacti that have enlarged, underground structures (Figs. 3.44 and 3.45). We have already mentioned the Andean species of *Pereskia* and the closely related species of *Maihuenia* of southern South America, which have converted the center of the root wood into a soft region. Development of succulence in the roots of Opuntioideae is also generally restricted to the secondary xylem. Fibers and cells with lignified cell walls have been replaced by vacuolate, thin-walled parenchyma cells. Some examples of this are found in *Opuntia macrorhiza*, *O. tunicata*, and *Tephrocactus russellii*. The most dramatic example of this substitution occurs in the storage root of *Pterocactus* (Fig. 3.44). This cactus geophyte, which lives hidden in sand for most of the year, produces one or several shoots at ground level; each shoot can produce a flower and fruit before it dies back to ground level. The enlarged root portion of this plant is formed by secondary xylem, which has no fibers.

Figure 3.43. Adventitious roots have broken through the skin in this epiphytic species of *Hylocereus*.

In addition to the many small cacti of subfamily Cactoideae that possess a fleshy taproot and several species that have a root similar to that of *Pterocactus*, for example, some of the Chilean species of *Neoporteria* (Fig. 3.45), there are two other outstanding enlarged root forms. The most massive of these occurs in *Peniocereus*. In this genus of geophytes, succulence occurs in the cortex and is already well developed in young plants. Roots of *Peniocereus* can weigh 4 kg or more. Much different in structure are the swollen, white "tubers" of *Wilcoxia*, which are about the size of a small sausage. In these structures, most of the succulence develops in the cortex, but some thin-walled cells are also present in the center.

Rain Roots

Because the cactus plant and its roots are very moist even though the soil surrounding them is exceedingly dry, we might expect that the cactus would lose water to the soil. There is no hard evidence that this occurs or, if it occurs, that water loss to the soil is a serious problem for the plant. One reason why the root systems do not lose much water is that they seem to "seal themselves up" during periods of soil water stress. In the dry seasons, the roots are long and have few, if any, thin lateral branches, and the entire root is insulated by layers of bark. Upon wetting a soil that has been dry for a considerable period—from weeks to many months—the existing roots become rehydrated; and within a couple of days one can observe the growth of new roots, called rain roots.

Rain roots develop from root buds (root primordia) that are hidden in the cortex of the old roots. When water is supplied, the vacuoles of the root cells take up large quantities of water and the cells thereby elongate rapidly. These white roots become

Figures 3.44 and 3.45. Cactus geophytes.
3.44. *Pterocactus kuntzei,* formerly *P. tuberosus,* unearthed from its native habitat in central Argentina and displayed next to a 30-cm-long digging tool.
3.45. *Neoporteria glabrescens,* which normally produces one to several small photosynthetic shoots. The shoots normally project only a centimeter or so from the ground; the tubercles are colored and textured to mimic rocks in their habitat (Chapter 8).

visible within a matter of hours after wetting the soil because only a little cell division is required for their elongation; they continue to grow rapidly after they have been formed by cell division from the root apical meristem. After a few days, water uptake by the newly formed rain roots can exceed the amount of water taken up by the rehydrated old roots.

When the soil begins to dry out, the rain roots shrivel; and by the time the soil is dry, the rain roots have been shed and the main roots have been sealed for the drought.

Significance of Succulence for Growth and Survival

Germination and Seedling Succulence

Anyone can successfully grow cacti from seeds because cactus embryos need little encouragement to germinate when placed in a humid container such as a petri dish or a pan with a glass cover, on moist sand, or on moist filter paper or a paper towel. Experiments in the early 1960s by Stanley Alcorn and Edwin Kurtz of the University of Arizona demonstrated that germination of cactus seeds is extremely high when they are placed under red light (660 nanometers, nm) or constant white light but is strongly suppressed when they are placed in the dark or under far-red light (730 nm). When placed under the right conditions of light, moisture, and temperature, the seeds imbibe water, the embryos inside the seeds swell, and the young roots of the embryos (the radicles) start to elongate. As the embryos expand, the seeds open like boxes with hinged lids (operculate germination) and the radicles emerge to penetrate the soil.

Seedlings of pereskias (Fig. 2.11) and some of the epiphytes and opuntioids are not very succulent and have thin, leafy cotyledons. But the typical cactus seedling has a succulent hypocotyl, which is the structure between the root and the cotyledons, and two fleshy and often very small cotyledons (Figs. 3.46–3.49). In fact, many of the smaller cacti produce a seedling with a hypocotyl that is greatly enlarged and cotyledons that are greatly reduced in size (Figs. 3.48 and 3.49*A*). The radicle is thin, and some research indicates that periderm begins to form on this radicle before it is 1 week old.

The epicotyl, which is the portion of the seedling above the two cotyledons, is the site of the shoot apical meristem. This meristem produces the aboveground part of the plant (Chapter 2). In the seed, the shoot tip has no areoles; but soon after the cells of a seedling are fully expanded—a period of only a few days—the first leaves and areoles have formed and spines appear on the shoot (Figs. 3.46–3.49).

In the pereskias, which have thin, leafy cotyledons,

Figures 3.46–3.49. Seedlings of cacti.
3.46. *Opuntia echinocarpa,* which has two large, succulent cotyledons. Between the cotyledons can be seen the spines of the areoles on the shoot.
3.47. *Ferocactus wislizenii,* a barrel cactus from Arizona. The long axis of the seedling, called the hypocotyl–root axis, is succulent and bears two small, pointed cotyledons; the first areole of the shoot has erect, straight spines.
3.48. *Ferocactus acanthodes,* another barrel cactus, which has a short, fat hypocotyl–root axis topped by two very tiny cotyledons.
3.49. *Ariocarpus trigonus;* side view *(A)* and top view *(B).* The seedling has a short, fat hypocotyl–root axis and no clearly defined cotyledons: the young areoles produce small, clasping spines that have prominent cellular projections. Spines are absent on the adult plant, which has very elongate tubercles.

new shoot growth can be rapid and can resemble the rapid shoot growth of typical woody dicotyledons. In the succulent cactus seedling, however, no stem axis is produced for the first weeks of growth; and growth of the shoot tip therefore must depend on foods translocated from the succulent hypocotyl and cotyledons, which are green.

Changes in Volume-to-Surface Relationships

Great patience is required to watch a freshly germinated seedling with two tiny cotyledons and a short, fleshy hypocotyl develop into a mature, flowering plant because the growth of a cactus is very slow. After the initial stage of formation, with the accompanying development of an apical meristem and several weakly spined areoles, the shoot experiences great changes in form and diameter. The apical meristem, which gradually increases in diameter, produces new leaves with their associated tubercles and areoles (Chapter 2); but simultaneously the cortex and the pith develop from the peripheral zone and pith-rib meristem zone, respectively, of the shoot apex (Fig. 3.50), thereby producing the succulence in the stem portion. The first-formed, helically arranged tubercles are small and flat, but later tubercles are larger because the leaf bases enlarge by new cell divisions (Chapter 6).

When the upward growth of the plant closely matches the increase in stem diameter, the plant shape remains globular; but when the stem elongates more rapidly, a club-shaped or even a cylindrical form occurs. The transition from globular to cylindrical may take many years, as it does in a barrel cactus. When the transition from globular to cylindrical begins, the new tubercles may become arranged as vertical ribs, as in the columnar cacti and the epiphytes, or may remain as separate, helically arranged projections, as in the cylindropuntias and cylindrical species of *Mammillaria.*

Shifts in growth form from small and globose with flat tubercles to globose with well-developed tubercles and then to cylindrical with ribs have great relevance to our discussion on succulence. One of the chief benefits of a globular growth form is that there is a great volume of water stored per unit of transpiring surface area. We can calculate the volume-to-surface ratio of a sphere:

$$\frac{\text{volume}}{\text{surface}} = \frac{\frac{4}{3}(\pi r^3)}{4\,\pi r^2} = \frac{r}{3} \qquad (3.1)$$

where r is the radius. For a cylinder, neglecting the area of the ends, we note that

$$\frac{\text{volume}}{\text{surface}} \cong \frac{\pi r^2 l}{2\pi r l} = \frac{r}{2} \qquad (3.2)$$

where l is the length of the cylinder. These two equations indicate that the wider the sphere or the cylinder, the more water-storing volume per unit surface area; when the radius is doubled, the volume-to-surface ratio is also doubled. Additionally, for a given total surface area (including the area of the ends of the cylinder), a sphere can store a greater volume of water than can a cylinder; consequently, a sphere is the ideal shape for water retention. On the other hand, as we shall see in Chapter 6, formation of tubercles and ribs decreases the volume-to-surface ratio by increasing the surface area. This increase in the photosynthetic area of the stem can thereby increase the growth rate; thus, the demand for greater volume for water storage can conflict with the area demand of photosynthesis.

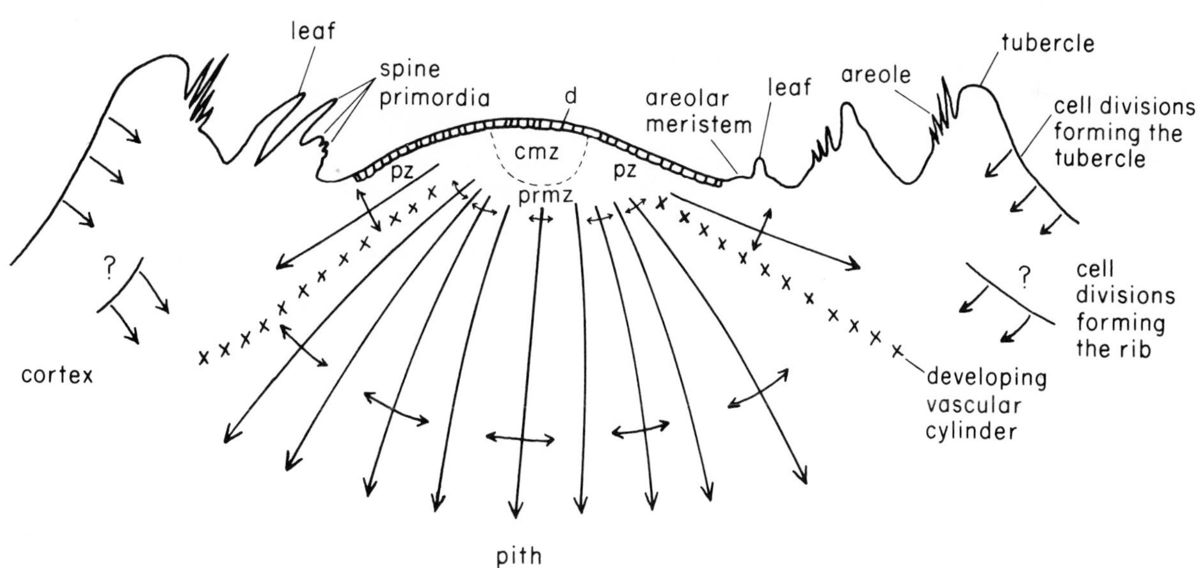

Figure 3.50. Diagram of a cactus shoot apex, drawn as a median longitudinal section to show the regions where succulent tissues are formed. As illustrated in Fig. 2.8*B* and *C*, the apical meristem has four meristematic zones: dermatogen (d), for the formation of the epidermis and outermost layer of the cortex; central mother cells (cmz); peripheral zone (pz), for formation of the outer cortex; and pith–rib meristematic zone (prmz), which produces the pith. Arrows are placed to show the degree to which the tissues are expanding. The peripheral meristem, beneath the epidermis, produces the outer cortex of the tubercles, where the presumed "primary thickening meristem" produces cells in the formation of vertical ribs (see Chapter 6). (Redrawn and modified from Boke, 1941, 1944.)

To relate all this material to cacti, let us briefly consider how the volume-to-surface ratio of a barrel cactus changes as it grows. For a 7-day-old seedling of *Ferocactus acanthodes* about 2 mm tall and growing in the laboratory, the ratio is 0.05 cm^3/cm^2. This ratio increases to 0.11 cm at 2 months and 0.56 cm after 1 year, when the plant is about 1.5 cm tall. The greater stem volume per unit surface area is crucial for seedling survival because the greater water storage per unit area available for water loss of the older seedlings allows them to survive longer droughts. For a mature plant 50 years old and 75 cm tall, the volume-to-surface ratio can increase to 9.1 cm, which means that there is an average depth of 9.1 cm of water-bearing tissue behind every transpiring point, or nearly 9 cm^3 of water stored for each square centimeter of surface area. Such water storage is crucial for the tolerance of long drought periods by barrel cacti in the field.

Water Relations of Succulent Tissues

Most people who grow cacti have probably never heard of the expression *water relations,* and none of the common dictionaries give a definition for it. When scientists describe the water relations of a plant, they talk about the ways in which the plant gains and loses water; these processes include water uptake from the soil by the roots, water storage in succulent tissues, water flow through plant tissues and plant organs, and water escape from the plant to the atmosphere through the stomates and the cuticle. Whereas cacti take up most of their water through their roots, some plants, for example, bromeliads (Bromeliaceae), take up water through the leaves; a few workers have described some water uptake through the spines and areoles of cacti.

One simple law of nature must be understood before we can write anything about the water relations of cacti: water tends to flow spontaneously toward regions of lower energy, that is, downhill. Three effects control this water flow. First, water tends to flow downhill because of gravity. Second, water under pressure (called hydrostatic pressure) tends to move toward lower pressures. This is what happens when a water faucet is opened; the pressure at the faucet end of the pipe, which was very high, drops suddenly to air pressure, and water flows toward the low pressure. But for our discussions here on water movement in plants—especially in cacti—it is important to concentrate on the third effect, that is, water flows from regions of low salt or solute concentration to regions of higher concentration. If you place a bag of salt water in fresh water—assuming that water but not the dissolved salt can pass freely through the walls of the bag—fresh water will flow into the bag and eventually cause it to burst. A similar observation can be made by placing intact but wrinkled fruit in water; the fruit will tend to swell as water enters.

When water diffuses from one solution into another because of a difference in salt concentration between the two solutions, the process is called osmosis. In the preceding example, the salty region receives water because it has a higher osmotic pressure. Osmotic pressure is proportional to solute concentration; a 1 molar solution (1 mole of solute molecules per liter of solution) has an osmotic pressure of 24 atmospheres (atm) at 20°C, which is approximately the osmotic pressure of seawater. For expressing osmotic pressure, scientists now use a pressure unit called a pascal (Pa), which is a pressure of 1 newton per square meter. Because the size of the pascal is not particularly convenient for expressing osmotic pressures in plants, the megapascal (abbreviated MPa) has been accepted for use in plant research. Thus, 1 MPa equals 10^6 pascals or 10.13 atm. A typical osmotic pressure for the stem of a cactus, which has many salts and other solutes, is 0.5 MPa, which corresponds to approximately 5 atm.

The water potential, which is a measure of the energy of water, is dependent on the three factors that we have just mentioned:

$$\text{water potential} = \text{hydrostatic pressure} - \text{osmotic pressure} + \text{gravitational effect} \quad (3.3a)$$

which in symbols can be represented by

$$\Psi = P - \pi + \rho_w g h \quad (3.3b)$$

where the constant $\rho_w g$ equals 0.010 MPa m^{-1} and h is height in meters (ρ_w is the water density and g is the gravitational acceleration). When Ψ, the water potential, equals zero, all of the factors on the right side of the equation sum to zero, although each component may be different from zero. In plants, tissues generally do not have $\Psi = 0$; rather, the values of Ψ are generally negative. The more negative the value of the water potential, the greater the capacity for water uptake.

We are now in a position to analyze the movement of water from the soil into a cactus. As stated earlier, the osmotic pressure in a healthy, well-watered cactus stem is generally about 0.5 MPa. The lowest possible value for the hydrostatic pressure in a succulent cell is zero; this condition exists when there is no internal pressure. Cells with zero hydrostatic pressure are flaccid (they have no turgidity); such a condition occurs in a wilted leaf. If water is made available to these cells, it will flow into them and the cells will become turgid as they develop an internal hydrostatic pressure. Thus, for soil water to be able

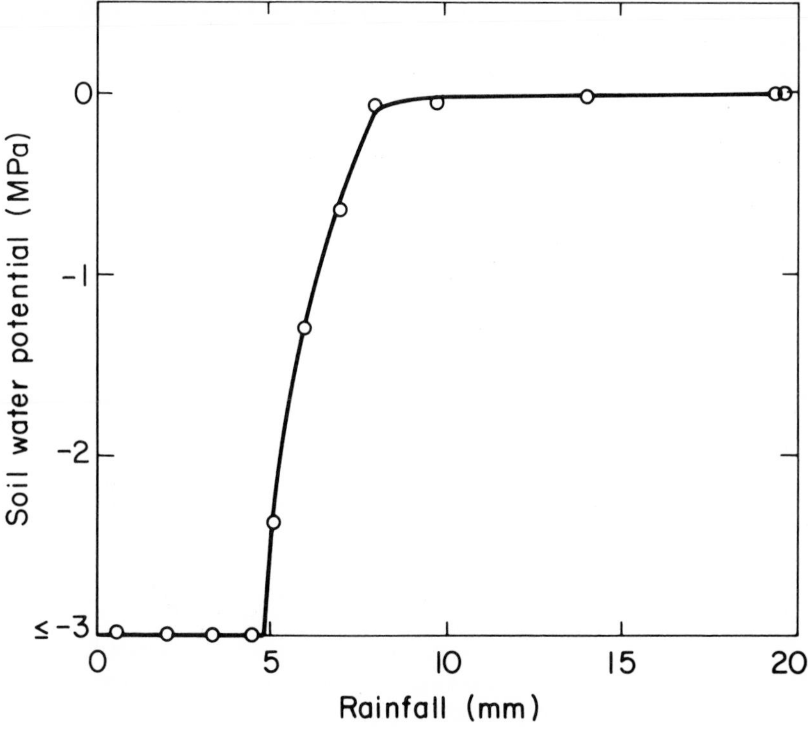

Figure 3.51. Relationship between depth of rainfall on dry soil and the resulting soil water potential in the root zone at 10 cm below the surface. (Adapted from Nobel, 1976.)

to enter flaccid cells, the water potential of the soil must be greater (higher, that is, less negative) than the water potential of the cells. Gravitational effects are generally small; therefore, for the water to enter a cactus, the soil water potential must only be greater than the negative osmotic pressure of the stem, for example, above −0.5 MPa (see Eq. 3.3).

Under natural conditions, the soil water potential is raised by rainfall. However, rainfalls less than about 6 mm (about ¼ inch) do not raise the water potential of an initially dry soil to even −1.0 MPa at a depth of 10 cm (Fig. 3.51), which is a shallow depth but one representative of the roots of many species of cacti. Therefore, very light rainfalls have little effect on cacti because nearly all of the water is rapidly evaporated from the uppermost part of the soil before it has a chance to get to the roots—unless several light rainfalls occur in close succession or the soil is initially wet. Rainfalls of 10 mm or more can generally raise the soil water potential in the rooting zone above −0.1 MPa (Fig. 3.51) and thus lead to water uptake by the plant.

Depending on the amount and intensity of the rainfall, the initial wetness of the soil, and the water retention properties of the soil, the soil water potential will stay above the plant water potential for various time periods. But even after the soil dries so that water can no longer spontaneously enter the plant via the roots, cacti can store enough water in their succulent organs for stomatal opening and other manifestations of metabolic activity to continue unabated for considerable periods. For instance, in the barrel cactus *Ferocactus acanthodes*, stomatal opening can continue for 1.5 months after the soil becomes drier than the plant (Fig. 1.28). This use of water stored in the succulent tissues is one of the most remarkable features of cacti and greatly improves their ability to survive in regions with low and sporadic amounts of rainfall. Thus, by having a high water content, a high volume-to-surface ratio, and a low stomatal frequency with its associated low water vapor conductance (Chapter 1), cacti can maintain a relatively high stem water potential and thereby prolong nocturnal CO_2 uptake long after the soil water potential falls below the stem water potential.

Reaction of Succulent Tissues to Dehydration

Because stomates continue to open long after water cannot be extracted from the soil, the plant naturally decreases in volume and gradually changes shape. Wherever careful observations have been made on the shrinkage of cactus stems, it has been noted that the tubercles and ribs in particular show effects: these structures that were flat or bowed out slightly when fully hydrated lose water and become narrower and often closer together. One way to observe this change as hydration changes is to watch what happens in a species with prominent spine clusters, for example, in a species like teddy bear cholla, *Opuntia bigelovii,* of North America. The stems of teddy bear cholla are exceedingly light in color when dry and appear greener right after an effective rain because when the stems are dry the spine clusters are more closely packed than they are when the stems are turgid. Therefore, as the spines become separated under hydrated conditions, more of the green stem can be seen.

Recently Wayne Barcikowski and Park Nobel at UCLA quantified how much water is lost from the various tissues within the stem. In 3-year-old stems of saguaro *(Carnegiea gigantea)* and a barrel cactus *(Ferocactus acanthodes),* most water is stored in the cortex internal to the chlorenchyma (Fig. 3.1). The cells of the succulent cortex are very large, so it is not surprising that four times more water is lost from these cells with huge vacuoles than from the smaller cells of the chlorenchyma. What is noteworthy, however, is that the percentage of water loss from the inner cortex and the pith (80 to 82%) was significantly greater than the percentage of water lost from the chlorenchyma (73%). Data obtained in this study suggest that water was preferentially lost from the storage tissues (succulent cortex and pith, which are connected by primary rays) because the number of solutes in these cells generally decreases, thereby lowering their osmotic pressure. Water then diffuses from this storage region into regions of higher solute osmotic pressure, such as the chlorenchyma. By maintaining higher water content in the chlorenchyma, nocturnal opening of stomates and the CAM metabolism discussed in the next chapter are allowed to continue for a longer period than they would if all the tissues of the plant were to dry at the same rate. Similar results were obtained for the platyopuntia *O. basilaris,* except in this species most water storage occurs in the pith. How the plant accomplishes this innovative adaptation for survival is not fully known.

SELECTED BIBLIOGRAPHY

Alcorn, S. M., and E. B. Kurtz, Jr. 1959. Some factors affecting the germination of seeds of the saguaro cactus *(Carnegiea gigantea). American Journal of Botany* 46:526–529.

Bailey, I. W., and L. M. Srivastava. 1962. Comparative anatomy of the leaf-bearing Cactaceae, IV. The fusiform initials of the vascular cambium and the form and structure of their derivatives. *Journal of the Arnold Arboretum* 43:187–202.

Barcikowski, W., and P. S. Nobel. 1984. Water relations of cacti during desiccation: distribution of water in tissues. *Botanical Gazette* 145:110–115.

Benson, L. 1982. *The cacti of the United States and Canada.* Stanford University Press, Stanford.

Boke, N. H. 1941. Zonation in the shoot apices of *Trichocereus spachianus* and *Opuntia cylindrica. American Journal of Botany* 28:656–664.

——— 1944. Histogenesis of the leaf and areole in *Opuntia cylindrica. American Journal of Botany* 31:299–316.

Bregman, R., and F. Bouman. 1983. Seed germination in Cactaceae. *Botanical Journal of the Linnean Society* (London) 86:357–374.

Buxbaum, F. 1950. *Morphology of cacti.* Section I. *Roots and stems.* Abbey Garden Press, Pasadena.

Esau, K. 1977. *Anatomy of seed plants,* 2nd ed. John Wiley & Sons, New York.

Fearn, B. 1981. Seed germination: the modern approach. *Cactus and Succulent Journal* (London) 43:13–16.

Freeman, T. P. 1969. The developmental anatomy of *Opuntia basilaris.* I. Embryo, root, transition zone. *American Journal of Botany* 56:1067–1074.

Gasson, P. 1981. Epidermal anatomy of some North American globular cacti. *Cactus and Succulent Journal* (London) 43:101–108.

Gibson, A. C. 1973. Comparative anatomy of secondary xylem in Cactoideae (Cactaceae). *Biotropica* 5:29–65.

——— 1976. Vascular organization in shoots of Cactaceae. I. Development and morphology of primary vasculature in Pereskioideae and Opuntioideae. *American Journal of Botany* 63:414–426.

——— 1978a. Structure of *Pterocactus tuberosus,* a cactus geophyte. *Cactus and Succulent Journal* (Los Angeles) 50:41–43.

——— 1978b. Architectural designs of wood skeletons in cacti. *Cactus and Succulent Journal* (London) 40:73–80.

——— 1982. The anatomy of succulence. Pp. 1–17 in *Crassulacean acid metabolism,* ed. I. P. Ting and M. Gibbs. American Society of Plant Physiologists, Rockville, Maryland.

Gibson, A. C., and K. E. Horak. 1978. Systematic anatomy and phylogeny of Mexican columnar cacti. *Annals of the Missouri Botanical Garden* 65:999–1057.

Hamilton, M. W. 1970a. Seedling development in *Opuntia bradtiana. American Journal of Botany* 57:599–603.

——— 1970b.The comparative morphology of three cylindropuntias. *American Journal of Botany* 57:1255–1263.

Jordan, P. W., and P. S. Nobel. 1981. Seedling establishment of *Ferocactus acanthodes* in relation to drought. *Ecology* 62:901–906.

Kurtz, E. B., Jr., and S. M. Alcorn. 1960. Some germination requirements of saguaro seeds. *Cactus and Succulent Journal* (Los Angeles) 32:72–74.

MacDougal, D. T., and E. S. Spalding. 1910. *The water-balance of succulent plants.* Carnegie Institute of Washington Publication 141, Washington, D.C.

Mauseth, J. D. 1983a. Introduction to cactus anatomy. Part 4. Mature cells. *Cactus and Succulent Journal* (Los Angeles) 55:113–118.

——— 1983b. Introduction to cactus anatomy. Part 5.

Secretory cells. *Cactus and Succulent Journal* (Los Angeles) 55:171–175.
——— 1984a. Introduction to cactus anatomy. Part 7. Epidermis. *Cactus and Succulent Journal* (Los Angeles) 56:33–37.
——— 1984b. Introduction to cactus anatomy. Part 8. Inner body. *Cactus and Succulent Journal* (Los Angeles) 56:131–135.
——— 1984c. Introduction to cactus anatomy. Part 9. Primary and secondary growth. *Cactus and Succulent Journal* (Los Angeles) 56:181–184.
Nobel, P. S. 1976. Water relations and photosynthesis of a desert CAM plant, *Agave deserti. Plant Physiology* 58:576–582.
——— 1983. *Biophysical plant physiology and ecology.* W. H. Freeman, New York.
Nobel, P. S., and J. Sanderson. 1984. Rectifier-like activities of roots of two desert succulents. *Journal of Experimental Botany* 35:727–737.
Preston, C. E. 1900. Observations on the root system of certain Cactaceae. *Botanical Gazette* 30:348–351.
——— 1901. Structural studies in southwestern Cactaceae. *Botanical Gazette* 32:35–55.
Reichert, E. T. 1913. *The differentiation and specificity of starches in relation to genera, species, etc.* Carnegie Institute of Washington Publication 173, Washington, D.C.
Soule, O. H., and C. H. Lowe. 1970. Osmotic characteristics of tissue fluids in the saguaro giant cactus *(Cereus giganteus). Annals of the Missouri Botanical Garden* 57:265–351.
Walter, H., and E. Stadelmann. 1974. A new approach to the water relations of desert plants. Pp. 213–310 in *Desert biology,* vol. II, ed. G. W. Brown, Jr. Academic Press, New York.

4

Gas Exchange and Crassulacean Acid Metabolism

Stem succulence is certainly one of the preeminent adaptations of a cactus plant for living in a desert environment. By having large quantities of stored water in the stem, a cactus can remain green and metabolically active during long periods of water shortage—when no water can be obtained from the soil. Other woody plants of the desert are more directly dependent on the amount of moisture in the soil.

Cacti differ from other desert plants in another way. They tend to open their stomates at night and close them during the day. This behavior is a characteristic of plants with Crassulacean acid metabolism (CAM). The name for this type of metabolism derives from the observation reported in 1815 by Benjamin Heyne that during the night large quantities of organic acid accumulated in the succulent leaves of *Kalanchoe calycina (=Bryophyllum calycinum)*, a species of the Crassulaceae (the stonecrop family). Heyne did not perform a highly sophisticated scientific experiment—he simply took bites out of his succulent plants at various times. He found that the leaves of *Kalanchoe* were very acid in the morning and then lost their acid taste during the daytime, only to regain acidity by the next morning. Nighttime CO_2 uptake, another salient feature of CAM, was first reported in 1804 by the French chemist Th. de Saussure from observations made on darkened pads of *Opuntia.* But the causal relationship between nocturnal uptake of CO_2 and the manufacture and accumulation of organic acids was not realized until the middle of the twentieth century. Moreover, only very recently have scientists begun to understand the ecological importance of CAM, which is now known to occur in over 20 families of land plants. This metabolic trait is a very special adaptation, because most plants, including the leaf-bearing pereskias, fix carbon dioxide only during the daytime.

General Characteristics of Gas Exchange

The release and uptake of gases such as water vapor and carbon dioxide is called gas exchange. We must begin our discussion of this process by considering the physical pathway for CO_2 entry into a plant and its cells and for water vapor loss from these cells to the atmosphere. A few central equations will be presented so that we can put all subsequent discussions of cactus biology on a quantitative footing. Most of the features of the pathways and how gas exchange is controlled actually fit all types of plants. Of course, we shall show how the gas exchange parameters that we have introduced can be used to understand the behavior of a typical plant—represented in this case by *Pereskia* (Fig. 4.1)—before we explain the more highly specialized features exhibited by a CAM plant.

Gas Exchange Pathway

Like all gases, carbon dioxide obeys a set of natural gas laws that were discovered by physical scientists. One property of gases is that they can move toward regions of lower concentration by spontaneous, random, thermal motions, a process called diffusion. The CO_2 that enters a plant diffuses from a site of high concentration in the atmosphere to sites of low concentration in the chloroplasts or wherever CO_2 will be used. If the difference in CO_2 concentration between the atmosphere and the chloroplast is small, then CO_2 uptake by the plant must also be small; but if the difference is great, then CO_2 uptake can be substantial.

To reach the chloroplasts, CO_2 entering a plant must diffuse across a number of more or less discrete regions or structures. First it must cross the so-called

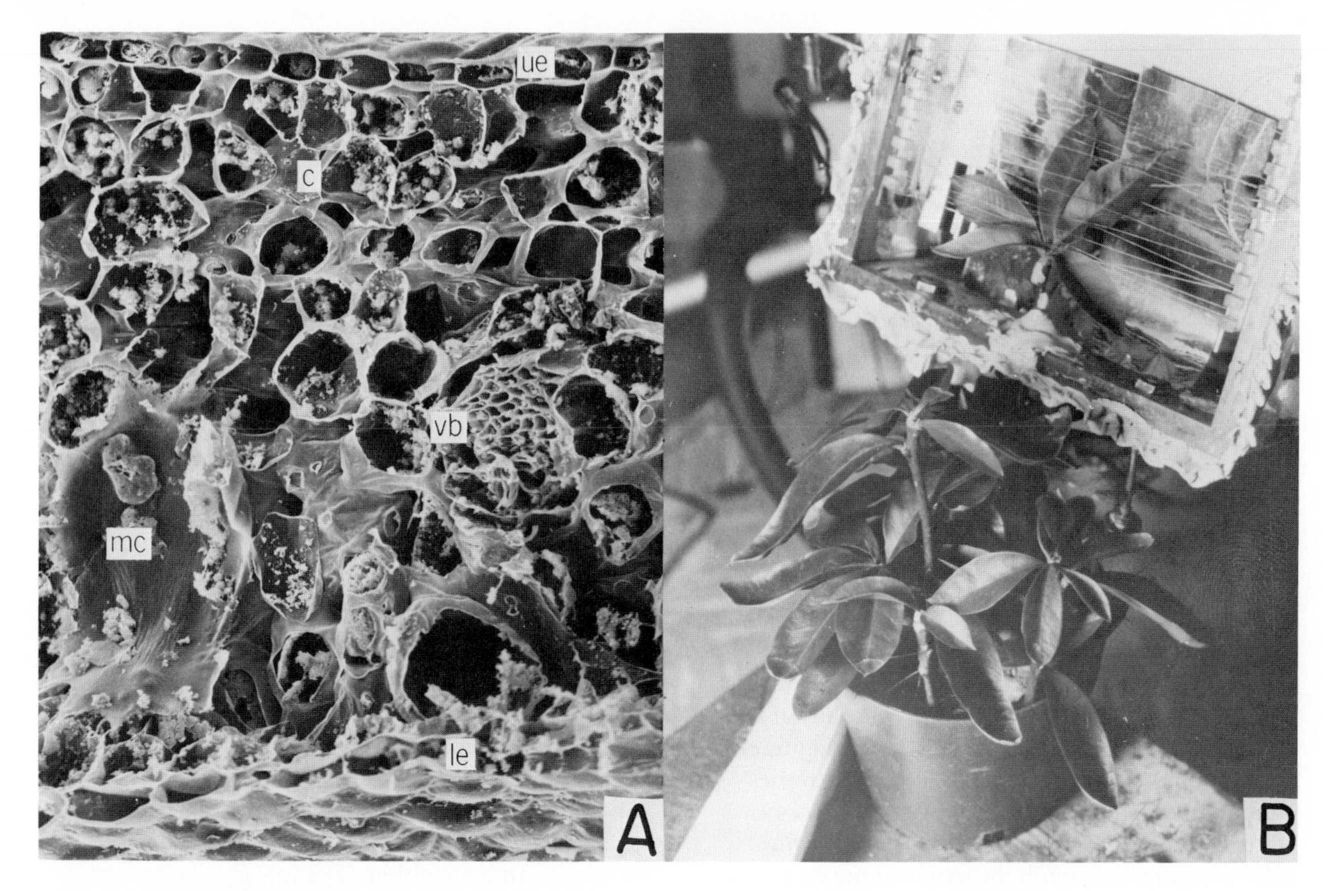

Figure 4.1. *Pereskia grandifolia.*
(A) Scanning electron photomicrograph of a leaf, showing the upper epidermis (ue), lower epidermis (le), chloroplast-bearing cells of the chlorenchyma (c) that have large intercellular air spaces, a mucilage cell (mc), and a vascular bundle (vb).
(B) A plant, used for measuring gas exchange, with several leaves sealed into a chamber.

boundary layer—a layer of essentially still air that surrounds all plant parts. Indeed, all exchanges between a plant and its environment involve movement across this boundary layer. (Although boundary layers will not be discussed again in this chapter, they become particularly important when we discuss heat exchange and temperature stresses for cacti in Chapters 5 and 7.)

Once CO_2 has diffused across the boundary layer, it enters the plant proper by diffusing either across the cuticle or through the stomates (Figs. 4.1*A* and 4.2*A*). Essentially no gas exchange occurs across the cuticle, because wax is not an easy pathway for either CO_2 entry or water vapor loss. Almost all CO_2 diffuses into the plant through the stomates, which offer the least resistance to diffusion, and closure of stomates can therefore effectively stop gas exchange. Hence, the guard cells serve as the main regulator of gas exchange by plants.

The CO_2 that enters the leaf diffuses through the intercellular air spaces to reach the surfaces of the chlorenchyma cells (mesophyll cells; Figs. 4.1*A* and 4.2). At this point a major change occurs in the pathway. Carbon dioxide can no longer diffuse in a gas phase but must be dissolved in the water that is contained in the minute pores in the cell wall (Fig. 4.2*B*); that is, it must enter a liquid phase. Diffusion of CO_2 is much more difficult in water than in air because water is so much more dense than air. But the cell walls of the chlorenchyma cells are relatively thin, and consequently the distance that CO_2 must diffuse is small. To enter the inside of the cell, CO_2 must now diffuse across a membrane, the plasmalemma (Fig. 4.2*B*), a process that is fairly easy because the plasmalemma is very thin and permits the very small, uncharged CO_2 molecules to pass freely.

Once across the plasmalemma, CO_2 must cross some cytosol (Fig. 4.2*B*; also called cytoplasm). Because the chloroplasts of most plants occur as a layer just inside the plasmalemma (Fig. 4.2*B*), the distance across the cytosol is very short and is not a substantial barrier to diffusion. Also, it is relatively easy for CO_2 to enter a chloroplast across the two thin membranes that surround the organelle.

The opening of the stomates to permit CO_2 to enter the leaf has the inevitable consequence of

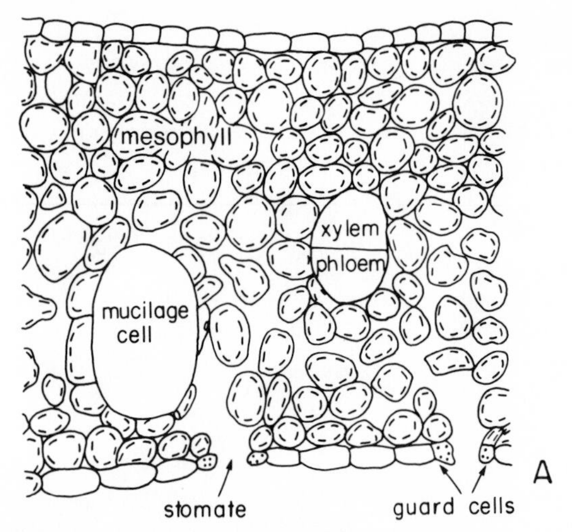

Figure 4.2. Leaf structures involved in gas exchange. *(A)* Section of a leaf of *P. grandifolia;* note the chloroplasts in the mesophyll cells (see Fig. 4.1*A*).

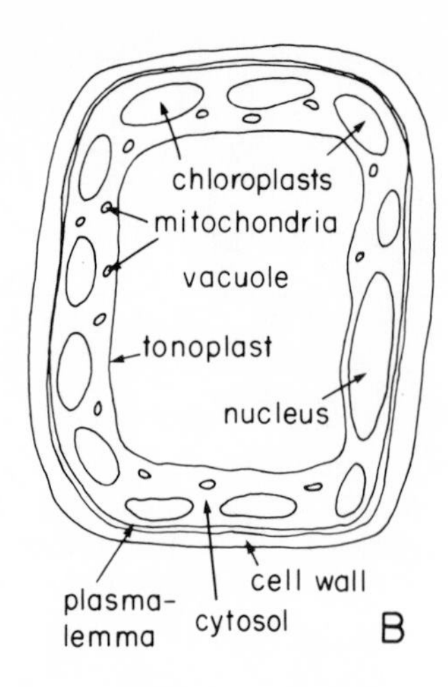

(B) Mesophyll cell; organelles and other cellular structures are labeled.

letting water vapor diffuse out. Water lost during transpiration evaporates from the cell walls of the chlorenchyma cells and especially from the inner cell walls of epidermal cells that are adjacent to the stomatal pores. This water vapor, which is a gas, diffuses easily across a small distance in the intercellular air spaces, through the stomatal pores, across the boundary layer of air, and finally into the surrounding atmosphere. Thus, net diffusion of water vapor occurs from the inside of the plant, where the water vapor concentration is extremely high (essentially 100% saturation), to the atmosphere, which typically has a much lower concentration of water vapor.

Control of Gas Exchange

In Chapter 1 we introduced an equation for transpiration, $J_{wv} = g_{wv} \Delta c_{wv}$ (Eq. 1.1), which describes the amount of water leaving unit area of a leaf per unit time. Let us now use this equation in a more formal consideration of gas exchange. In the preceding paragraph we mentioned the drop in water vapor concentration from the inside of the stem or leaf to the air, Δc_{wv}. Now we need to discuss g_{wv}, or what we called stomatal conductance in Chapter 1. Actually, g_{wv} is the overall gas-phase conductance and is referred to as the water vapor conductance; it includes effects of the intercellular air spaces, the stomates, the cuticle, and the boundary layer. But practically speaking, the stomates control g_{wv}. When the stomates are closed, the water vapor conductance is near zero; when they are open, it is high. Indeed, g_{wv} is approximately proportional to the stomatal pore area per unit area of plant surface, so that the highest water vapor conductances are obtained from a plant part with many large stomates that are wide open.

We have already indicated that cacti generally are adapted to conserve water, which means that transpiration (J_{wv}) is kept low. Equation 1.1 immediately leads us to the two possible choices for reducing J_{wv}: the plant can have either (1) a low g_{wv}, meaning reduced stomatal opening and thereby reduced CO_2 uptake, or (2) a low Δc_{wv}, meaning a small drop in water vapor concentration from the chlorenchyma to the ambient air.

Let us first explore the consequences of reducing transpiration by reducing g_{wv}. To analyze this properly, we need an expression indicating how partial closure of the stomates affects CO_2 uptake. For the gas-phase part of the pathway, CO_2 diffuses in and water vapor diffuses out, but the pathway is basically the same—boundary layer, stomatal pores, and intercellular air spaces. But the CO_2 molecule is just over twice as heavy as the H_2O molecule; and hence CO_2 diffuses more slowly. The rate of diffusion is related to molecular weight; therefore, the CO_2 gas-phase conductance ($g_{CO_2}{}^{gas}$, where the superscript means that we are referring to the gas phase) is lower than the H_2O gas-phse conductance (g_{wv}); in fact, $g_{CO_2}{}^{gas}$ is equal to 0.63 g_{wv}. We can relate the net CO_2 uptake (J_{CO_2}) to the drop in CO_2 concentration across the gas phase ($\Delta c_{CO_2}{}^{gas}$) as follows:

$$
\begin{aligned}
J_{CO_2} &= g_{CO_2}^{gas} \Delta c_{CO_2}^{gas} \\
&= g_{CO_2}^{gas} (c_{CO_2}^{air} - c_{CO_2}^{ias})
\end{aligned}
\tag{4.1}
$$

where $c_{CO_2}{}^{air}$ is the CO_2 concentration in the atmospheric air (for example, 350 ppm, which corresponds to 14.3 mmol CO_2 m^{-3} at 25°C and an air pressure of 1 atm) and $c_{CO_2}{}^{ias}$ is the CO_2 concentration in the intracellular air spaces.

Next we must consider the liquid (cellular) phase of the pathway, where the CO_2 conductance can be represented by $g_{CO_2}{}^{liquid}$. Because the same amount of CO_2 that entered the leaf continues on to the chloroplasts, J_{CO_2} is the same as before (Eq. 4.1) and we can write

$$J_{CO_2} = g_{CO_2}^{liquid}(c_{CO_2}^{ias} - c_{CO_2}^{chl}) \qquad (4.2)$$

where $c_{CO_2}{}^{chl}$ is the CO_2 concentration in the chloroplasts, which unfortunately has never been measured for cacti — or for any other plant, for that matter. It is often assumed that CO_2 fixation processes reduce the chloroplast CO_2 concentration to a very low — essentially zero — concentration; in which case we can write

$$J_{CO_2} = g_{CO_2}^{liquid} c_{CO_2}^{ias} \qquad (4.3)$$

We can then substitute $c_{CO_2}{}^{ias}$ as defined by Equation 4.3 into Equation 4.1 and rearrange; this manipulation leads to a very useful relation describing the movement of CO_2 from the atmosphere all the way to the sites of photosynthesis:

$$J_{CO_2} = \frac{g_{CO_2}^{gas} g_{CO_2}^{liquid}}{g_{CO_2}^{gas} + g_{CO_2}^{liquid}} c_{CO_2}^{air} \qquad (4.4)$$

We are now in a position to evaluate the options for water conservation by cacti from the point of view of gas exchange. As previously stated, when the stomates begin to close, transpiration is proportionally reduced; a 10-fold reduction in stomatal opening reduces water loss 10-fold. The situation for CO_2 uptake is more complicated and depends on the value of the CO_2 gas-phase conductance relative to the value of the CO_2 liquid-phase conductance. If $g_{CO_2}{}^{liquid}$ were much greater than $g_{CO_2}{}^{gas}$, then the gas phase will limit photosynthesis; Equation 4.4 then essentially reduces to $J_{CO_2} \cong g_{CO_2}{}^{gas} c_{CO_2}{}^{air}$, and hence any reduction in stomatal opening would proportionally reduce net CO_2 uptake — not a very attractive option. If $g_{CO_2}{}^{liquid}$ were much less than $g_{CO_2}{}^{gas}$, then the liquid phase will limit maximum photosynthesis. Equation 4.4 would become $J_{CO_2} \cong g_{CO_2}{}^{liquid} c_{CO_2}{}^{air}$, which is independent of stomatal opening; but the hypothesis of a low $g_{CO_2}{}^{liquid}$ means a low net CO_2 uptake — also an unattractive option. Thus, we might expect that $g_{CO_2}{}^{gas}$ and $g_{CO_2}{}^{liquid}$ are comparable in magnitude, which is indeed true for the cacti that have been studied as well as for other plants.

Let us consider the case where the gas-phase and liquid-phase conductances are initially the same. A 10-fold reduction in $g_{CO_2}{}^{gas}$ (that is, a 10-fold reduction in g_{wv}) would then reduce net CO_2 uptake by 82%. Thus, a 10-fold decrease in water loss would require less than a 5-fold reduction in net CO_2 uptake. Thus, stomates can exert a greater control on transpiration than on photosynthesis. When saving water becomes of greater consequence to the plant than the accompanying loss in CO_2 uptake, partial stomatal closure is a very effective means of balancing the need for CO_2 against the inevitable loss of water. But we can also reduce transpiration by reducing Δc_{wv}, namely, by opening stomates only during periods when Δc_{wv} is less, for example, at night. When we discuss cacti exhibiting CAM, we shall see that they follow this alternative strategy for conserving water.

Fate of CO_2

Let us now consider the fate of the CO_2 that has diffused into a plant. For net CO_2 uptake to occur, the CO_2 must be taken out of solution by being incorporated into some chemical compounds, that is, the CO_2 must be "fixed." Otherwise, the CO_2 concentration would not be lowered inside the plant, and hence inward CO_2 diffusion would stop. In particular, photosynthesis uses CO_2 from the atmosphere to make simple and then more complex organic compounds in the chloroplasts. The process has been studied in great detail, but for our purposes the important step that uses CO_2 can be summarized as a single chemical reaction. (For convenience in balancing the chemical reaction, all the molecules in Equation 4.5a are shown in their uncharged form, whereas in actuality hydrogen ions (H^+) dissociate from the phosphate ($-OPO_3H_2$) and carboxyl ($-COOH$) groups, leaving these groups with various negative charges. The term 3-phosphoglycerate refers to such a charged form. By convention, H_2O is often omitted, especially when focusing on the compounds of interest.)

$$^{*}CO_2 + \begin{array}{c} CH_2OPO_3H_2 \\ | \\ C{=}O \\ | \\ HCOH \\ | \\ HCOH \\ | \\ CH_2OPO_3H_2 \end{array} + H_2O \rightarrow \begin{array}{c} CH_2OPO_3H_2 \\ | \\ HCOH \\ | \\ COOH \\ + \\ ^{*}COOH \\ | \\ HCOH \\ | \\ CH_2OPO_3H_2 \end{array} \qquad (4.5a)$$

where the asterisk traces the carbon of CO_2. For the compunds of interest and in their appropriate charge state, we can rewrite Equation 4.5a in the following form:

carbon dioxide + ribulose 1,5-bisphosphate
→ two 3-phosphoglycerate (4.5b)

This chemical reaction occurs in the light, and the 3-phosphoglycerate that is formed subsequently enters a special chemical cycle leading to photosynthetic products such as the sugar glucose. This particular series of chemical reactions leading from carbon dioxide to glucose is called the C_3 pathway or C_3 cycle, because the initial product after CO_2 incorporation — 3-phosphoglycerate — contains three carbon atoms. Plants that possess this pathway as the sole means of processing CO_2 from the atmosphere are referred to as C_3 plants. The C_3 pathway is the dominant type of photosynthesis on earth, and, as we shall soon see, species of *Pereskia* are C_3 plants.

If you placed the initial chemicals of Equation 4.5b into a test tube, you would have to wait a long, long time for much 3-phosphoglycerate to form. What has to be present for this reaction to proceed rapidly is an enzyme, which is a protein that speeds up (catalyzes) the chemical reaction. In living systems, enzymes allow photosynthesis, or any other complicated metabolic process, to occur in a controlled fashion at a reasonable rate. For photosynthesis, the enzyme required to catalyze the reaction represented by Equation 4.1 is known by the rather unwieldy name of ribulose-1,5-bisphosphate carboxylase/oxygenase, or Rubisco for short. The name reflects the fact that this enzyme can do two things: under certain conditions it can catalyze the carboxylation (addition of CO_2 in Eq. 4.5b) of ribulose 1,5-bisphosphate and under different conditions it can oxidize ribulose 1,5-bisphosphate (addition of oxygen). It is interesting to note that Rubisco is the most prevalent protein on earth.

For healthy plants in the light and at a moderate temperature, Rubisco acts to lower the CO_2 concentration in the chloroplasts. The CO_2 level in the chloroplast thereby becomes less than the CO_2 level in the atmosphere, where the average CO_2 level in 1986 was about 350 parts per million (ppm) by volume. (CO_2 concentration is increasing at nearly 2 ppm per year, mainly as a consequence of the burning of fossil fuels.) More CO_2 then diffuses in from the atmosphere to replace the CO_2 being used in the chloroplast, and photosynthesis keeps going.

Gas Exchange of *Pereskia*

Little information is available on transpiration and photosynthesis by leaf-bearing cacti — in particular, by *Pereskia*. Because such information is important for understanding the physiological and evolutionary attributes of Crassulacean acid metabolism, we rooted some cuttings of *Pereskia aculeata* from our cactus patch (Fig. 1.1) and gathered some specimens of *P. grandifolia* (Fig. 1.23) from our greenhouse. With the excellent technical assistance of Terry L. Hartsock, we then proceeded to see how the gas exchange of these two species responded to temperature and light (see Fig. 4.1*B*). We also wanted to learn whether they exclusively used the C_3 pathway of photosynthesis (the initial reaction of which is indicated in Eq. 4.5) or whether these primitive cacti relied upon CAM.

Temperature Dependence of Transpiration

Figure 4.3 shows how transpiration was influenced over a wide range of temperatures that are representative of the temperatures these plants experience in their native, dry, tropical habitats. From about 15°C (59°F) to 35°C (95°F), the water vapor loss averaged about 20 mg m^{-2} s^{-1} for *P. aculeata* and 25 mg m^{-2} s^{-1} for *P. grandifolia* (Fig. 4.3*A*). These

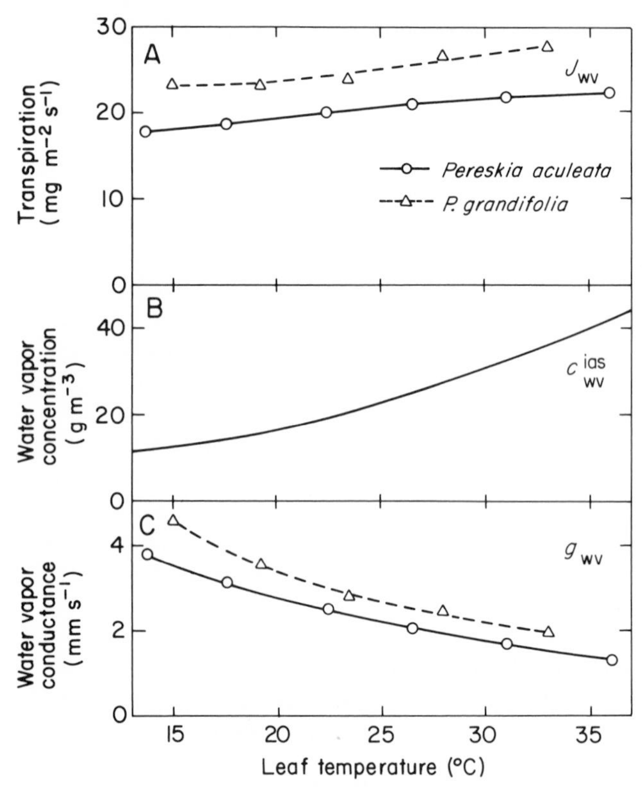

Figure 4.3. Temperature dependence of transpiration by two species of *Pereskia.* Plants were grown under well-watered conditions in environmental chambers with 12-hour days at an air temperature of 30°C and 12-hour nights at 20°C.

(A) Transpiration was measured for attached leaves enclosed within a transparent chamber (Fig. 4.1*B*; see Nobel and Hartsock, 1978).

(B) Saturation water vapor concentration; air in the environmental chambers and gas exchange system was maintained at 60% relative humidity, that is, at 60% of the value indicated.

(C) Water vapor conductance, calculated from Equation 1.1 ($J_{wv} = g_{wv}\Delta c_{wv}$).

values represent moderately low rates of water loss, being surpassed by nearly all crop plants under the conditions used; but they are similar to those for ferns, trees, and many other plants in their native environment (crop plants tend to have high rates of transpiration and the associated high rates of photosynthesis). Another interesting aspect of the measured transpiration rates is their relative constancy—indeed, the variation about the mean was only about ±10% from 15°C to 35°C. Such constancy occurred even though the water vapor concentration difference, to which transpiration is proportional, varied considerably.

In Chapter 1 we indicated that the saturation water vapor concentration is very dependent on temperature. This is illustrated in Figure 4.3*B*, where the water vapor concentration in the intercellular air spaces of the chlorenchyma is plotted versus temperature. Because water can readily evaporate from all the cell walls bordering the intercellular air spaces within a leaf (Figs. 4.1*A* and 4.2) or a stem, the air surrounding the chlorenchyma cells is essentially saturated with water vapor; therefore, the saturation water vapor concentration is plotted. Under the experimental conditions employed, the air outside the leaves was maintained at 60% relative humidity, as might be appropriate for the average value in the native habitats of these two pereskias. Hence, recalling Equation 1.1—$J_{wv} = g_{wv}\Delta c_{wv} = g_{wv}(c_{wv}^{ias} - c_{wv}^{air})$—we note that transpiration J_{wv} is here equal to $g_{wv}(c_{wv}^{ias} - 0.6c_{wv}^{ias})$, or $g_{wv} \times 0.4c_{wv}^{ias}$ (leaves and air were at essentially the same temperature, and so $c_{wv}^{air} \cong c_{wv}^{ias}$). In going from 15°C to 35°C, c_{wv}^{ias} increases just over threefold (Fig. 4.3*B*). Because transpiration is fairly constant, the water vapor conductance g_{wv} must decrease. Indeed, over this temperature range g_{wv} decreases essentially threefold (Fig. 4.3*C*). This is a remarkable adjustment; and it is done by the stomates. Specifically, the stomates partially close, thereby causing the area available for water vapor diffusion out of a leaf at 15°C to be reduced to only one third of the area available for water vapor diffusion at a leaf temperature of 35°C. This change prevents excessive water loss at the higher temperatures. Thus, we see that *Pereskia* can make adjustments that conserve water. We shall soon see that advanced cacti have an even more potent way of conserving water.

Temperature Dependence of Photosynthesis

What happens to photosynthesis as the leaf temperature of these two pereskias is varied? For both *P. aculeata* and *P. grandifolia*, net CO_2 uptake was maximal near 24°C (Fig. 4.4*A*). Both species also had approximately the same maximal rate, 9 μmol m^{-2}

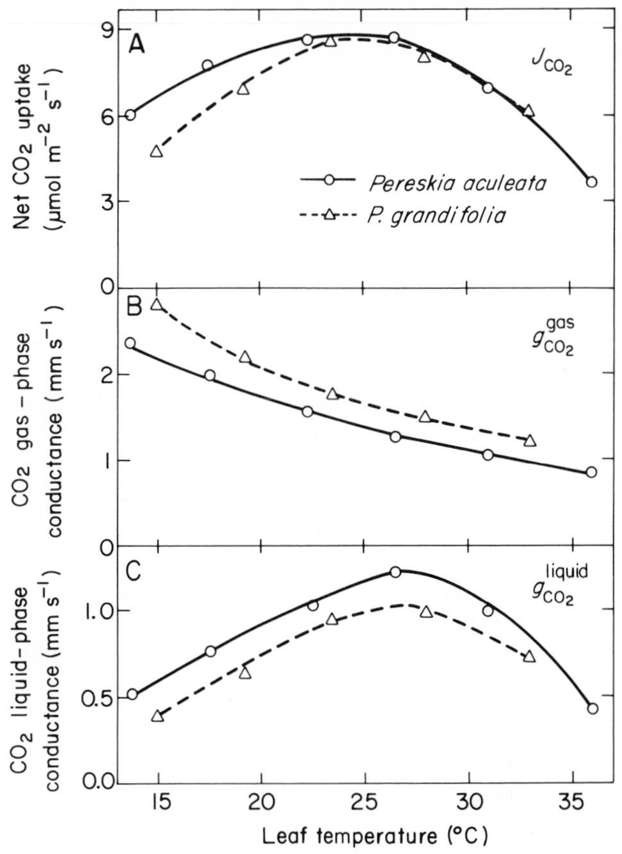

Figure 4.4. Temperature dependence of carbon dioxide uptake by two species of *Pereskia*.
(A) Net CO_2 uptake was measured at the same time and hence under the same conditions as the water vapor loss presented in Figure 4.3. The CO_2 concentration in the gas exchange apparatus was 340 ppm by volume, and the light level was saturating for net CO_2 uptake.
(B) The CO_2 gas-phase conductance was calculated from the water vapor conductance in Figure 4.3*C* ($g_{CO_2}^{gas} = 0.63\,g_{wv}$).
(C) Once $g_{CO_2}^{gas}$ is known, $g_{CO_2}^{liquid}$ can be calculated from Equations 4.2 and 4.4.

s^{-1}. This is a much lower rate than that observed for cultivated crops such as maize, sorghum, and sugar cane under ideal conditions, for which the net photosynthetic rate can be 40 μmol m^{-2} s^{-1}. The rate for the pereskias, however, is similar to the maximal rates of ferns, trees, and many other wild plants, that is, net CO_2 uptake by these two species of *Pereskia* is typical of that for healthy plants under favorable field conditions.

Figure 4.4 also contains information on how the CO_2 gas-phase conductance and the CO_2 liquid-phase conductance vary with temperature. Equation 4.4 shows that these two conductances ($g_{CO_2}^{gas}$ and $g_{CO_2}^{liquid}$) determine the net CO_2 uptake rate (J_{CO_2}), because the CO_2 concentration in the ambient air is essentially constant. As the leaf temperature increases, the accompanying stomatal closure, discussed earlier, causes the CO_2 gas-phase conductance to decrease steadily (Fig. 4.4*B*; compare with Fig. 4.3*C*). The CO_2 liquid-phase conductance, which describes the effect of the chlorenchyma cells on net CO_2 uptake, is greatest near 27°C. This represents the temperature where Rubisco and the other enzymes that process CO_2 work the fastest. Indeed, 27°C is a rather moderate temperature, and moderate temperatures are optimal for many plant enzymes.

Another salient feature of the two types of conductances can be gleaned from Figure 4.4. Namely, at temperatures leading to maximum CO_2 uptake, the CO_2 gas-phase conductance is similar in magnitude to the CO_2 liquid-phase conductance (about 1.2 mm s^{-1}). Using Equation 4.2, we can calculate the CO_2 concentration in the intercellular air spaces ($c_{CO_2}{}^{ias}$). For photosynthesis proceeding as in Figure 4.4 at 25°C, $c_{CO_2}{}^{ias}$ averages 200 ppm for these two species of *Pereskia*. Thus, a drop in CO_2 concentration from 340 ppm in the atmosphere to 200 ppm in the intercellular air spaces is necessary to allow enough CO_2 to diffuse in to replace that used in photosynthesis. This means that the chlorenchyma cells are surrounded by only 200 ppm CO_2 as the carbon source for all growth and productivity of these cacti.

Dependence of Photosynthesis on Photosynthetically Active Radiation

Everyone knows that photosynthesis requires light—the very word means that light is used to cause chemical changes leading to the creation of new molecules. Plant biologists, however, use a slightly different concept to characterize light: they focus on that part of the electromagnetic spectrum that can be absorbed by photosynthetic pigments like chlorophyll because only these wavelengths can be used in photosynthesis. This usually means wavelengths from 400 nm (1 nanometer = 10^{-9} meter), which we perceive as violet, to 700 nm, which we perceive as red. Perhaps more important, light is treated as discrete, countable particles referred to as photons. Each photon that is absorbed can excite a photosynthetic pigment, which is the initial event of photosynthesis. Thus, the word *light,* which refers to wavelengths perceived by the human eye, is replaced by the term photosynthetically active radiation (PAR) when we discuss the response of plants to sunlight or artificial light. Because photons can be counted in the same way molecules can, units for PAR are the same as those for CO_2 uptake or any other flow of particles, for example, μmol m^{-2} s^{-1}. When the sun is directly overhead at noon on a cloudless day, the PAR level on a horizontal surface is just over 2000 μmol m^{-2} s^{-1}. We will next see how net CO_2 uptake by our two species of *Pereskia* responds to PAR.

As PAR increases, net CO_2 uptake by *P. aculeata* and *P. grandifolia* increases (Fig. 4.5). Eventually a PAR level is reached above which net CO_2 uptake no longer increases, that is, uptake is then PAR saturated. Saturation for both *Pereskia* species occurs near a PAR level of 500 μmol m^{-2} s^{-1} (Fig. 4.5), which is just under one fourth of full sunlight. Ninety percent saturation, a criterion that is often easier to determine, is reached at about 300 μmol m^{-2} s^{-1}. This is a relatively low PAR level for 90% saturation and is characteristic of C_3 plants from moderately shady habitats. Not surprisingly, both species of *Pereskia* used in the study live in tropical woodland or tropical scrub, where they are shaded much of the time by neighboring plants.

Figure 4.5 indicates that a net loss of CO_2 occurs at a PAR level of 0 μmol m^{-2} s^{-1}, that is, in the dark. Of course, no photosynthesis can take place in total darkness because the energy carried by photons is necessary to drive the chemical reactions of photosynthesis. The negative portion of the curve indicates that CO_2 is lost, that is, more CO_2 is released inside the leaves by respiration in the mitochondria (Fig. 4.2) than is fixed from the atmosphere. As the PAR level increases, some photosynthesis occurs. At the PAR compensation point, CO_2 uptake by photosynthesis balances CO_2 released by mitochondrial respiration and another CO_2-releasing process re-

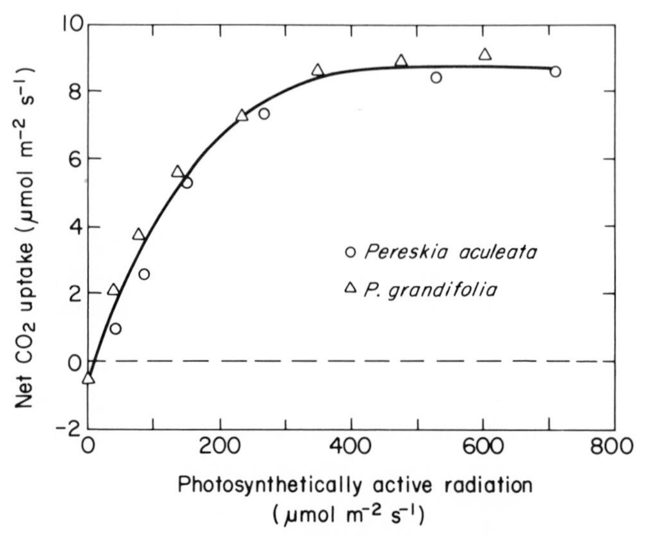

Figure 4.5. Response of carbon dioxide uptake by two species of *Pereskia* to photosynthetically active radiation (PAR). PAR for wavelengths from 400 to 700 nm was provided by a metal-halide arc lamp and refers to the value on the upper side of the leaves (the lower side received about 20% as much). Data were obtained as for Figures 4.3 and 4.4 (where the PAR level was 500 μmol m^{-2} s^{-1}), except the temperature was held at 24°C.

ferred to as photorespiration. A plant has neither a net gain nor a net loss of CO_2 at the PAR compensation point and hence cannot grow. For the two species of *Pereskia* studied, the PAR compensation point occurs at a PAR level of 10 μmol m^{-2} s^{-1} (Fig. 4.5). This physiologically important concept will be particularly important when we relate PAR to plant form in Chapter 8.

Daily Patterns

Because C_3 plants lose CO_2 in the dark and take it up when the PAR level is sufficiently high, it becomes important to study the net CO_2 exchange over an entire day. In this way, we can see whether the two pereskias completely follow the C_3 pattern. We can also examine how much CO_2 is taken up during an entire day, which will eventually lead us to a consideration of seasonal and annual productivity of cacti.

Figure 4.6*A* shows the course of CO_2 exchange for *P. grandifolia* over an entire 24-hour period. It is quite typical of a C_3 plant. At night a CO_2 loss occurs. When we add up the total loss for the entire night, it amounts to 8 mmol m^{-2}. Such a loss is much less than the gain during the daytime for a healthy leaf under favorable conditions. Specifically, the leaf considered here had a net CO_2 gain of 229 mmol m^{-2} for the daytime period, 6 a.m. to 6 p.m. (Fig. 4.6*A*; 6 p.m. corresponds to 18 hours). Figure 4.6 also presents information on the speed of response of the stomates to changes in the PAR level. Under the experimental conditions employed, the PAR level was abruptly increased to its maximum value (500 μmol m^{-2} s^{-1}) at the beginning of the day and similarly reduced at the end. The stomates did not respond instantly, but at "dawn" the stomates took about 15 minutes to reach half of their maximum opening, and at "dusk" half of their closing required nearly 20 minutes—indicated by the time course of the change in the water vapor conductance (Fig. 4.6*C*). This pattern is fairly typical of stomatal changes of C_3 plants under natural conditions. Also, the stomates were open only during the daytime—the typical pattern for C_3 plants.

Water-Use Efficiency

We can calculate a performance index indicating how well *P. grandifolia* has performed with respect to carbon gain and water loss. This index is called the water-use efficiency, which has a number of definitions depending on the data available and the intended application. A definition that is satisfactory for our present purpose is

$$\text{water-use efficiency} = \frac{\text{mass } CO_2 \text{ fixed}}{\text{mass } H_2O \text{ transpired}} \qquad (4.6)$$

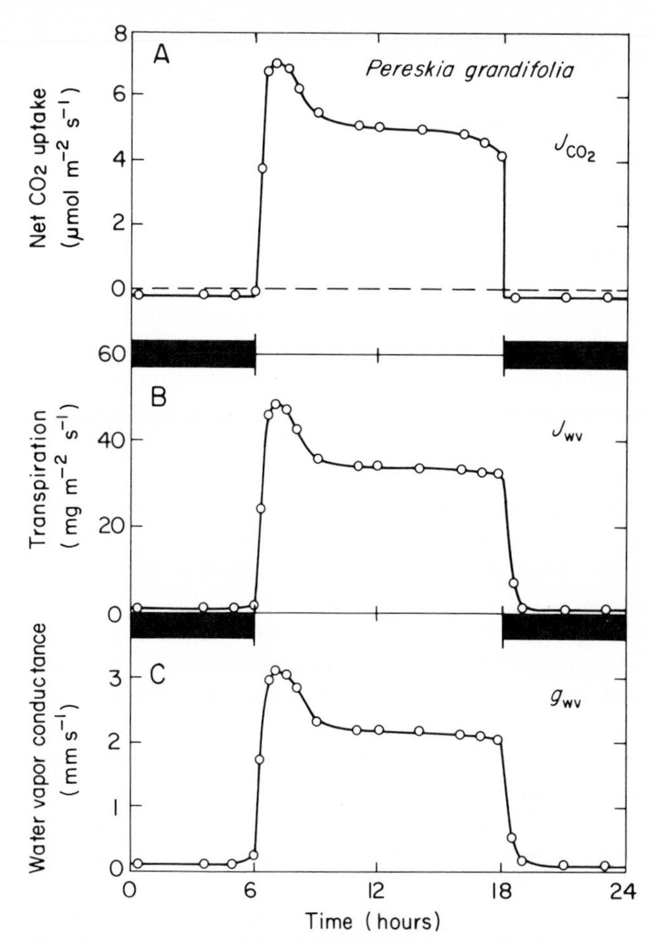

Figure 4.6. Gas exchange for *Pereskia grandifolia* over a 24-hour period. Conditions were basically the same as those described for Figures 4.3 to 4.5, with a daytime air temperature of 30°C and leaf temperature of 32°C at a PAR level of 500 μmol m^{-2} s^{-1} and nighttime leaf and air temperatures of 20°C. The solid areas indicate nighttime intervals.

For example, we can express the water-use efficiency in grams of CO_2 fixed per kilogram of H_2O transpired (we note that 1 mol of CO_2 corresponds to 44 g, so we can readily convert our mole units to mass units, and that 1 kg = 1000 g).

For *P. grandifolia* (Fig. 4.6), the water-use efficiency was 8 g CO_2/kg H_2O during the daytime and 7 g CO_2/kg H_2O over the entire 24-hour period. Thus, it gains 8 g of CO_2, which it can use for growth and maintenance for every kilogram of water transpired to the atmosphere. This is a slightly better use of water than for most crop plants, in which the water-use efficiency during the daytime is generally 2 to 6 g CO_2 fixed/kg H_2O transpired. However, it is substantially below the performance in terms of carbon gained per unit of water lost for cacti exhibiting CAM (see following section).

Gas Exchange in Stem-Succulent Cacti

By now readers should be familiar with gas exchange parameters of certain leafy cacti. To recapitulate, nonsucculent pereskias are interesting because they are *not* special, that is, they act like ordinary C_3 plants that have conspicuous leaves. Because these least-derived species of cacti of the subfamily Pereskioideae have C_3 photosynthesis, we can conclude that special physiological adaptations of cacti leading to CAM arose after they diverged from a nonsucculent dicotyledonous ancestor. This permits investigators to ask questions such as, "What is the adaptive significance of the new physiological features?"

By now readers are also aware that the stem succulent cacti are special because they are CAM plants: they open their stomates at night and close them during the day; they experience great diurnal changes in stem acids; and they take up and store CO_2 in a manner different from that of C_3 plants. We shall now consider the gas exchange characteristics of two species: *Opuntia ficus-indica,* a platyopuntia (Fig. 4.7) and a representative of subfamily Opuntioideae, and *Ferocactus acanthodes,* a barrel cactus (Fig. 4.8) and a representative of the subfamily Cactoideae. Even though these two species are distinctly different morphologically (as are the two subfamilies they represent), they are amazingly similar in terms of their gas exchange characteristics. Because platyopuntias and *Ferocactus* have been separated in an evolutionary sense for tens of millions of years, the strong physiological similarities of these two distinctive forms suggest that CAM in cacti is pretty much the same in a wide variety of species.

Daily Patterns

Figure 4.9*A* shows that *Opuntia ficus-indica* has no net CO_2 uptake during the entire daytime—in fact, there is a slight loss of CO_2 then. However, at night, when we would expect a net loss of CO_2 from C_3 plants, *O. ficus-indica* has a net CO_2 uptake. The maximum rate of net CO_2 uptake is just over 10 μmol m^{-2} s^{-1}, which is about the same as the uptake rate that occurs during the daytime for many native plants under favorable conditions. Thus, the rate is not as unusual as the fact that CO_2 uptake is occurring at night, when no photosynthesis can take place because there is no photosynthetically active radiation then (other than a very small amount from moonlight and starlight).

Because the ambient air always has a water vapor concentration lower than that in the chlorenchyma (except under the conditions that lead to dew formation, which can be important for some cacti), water always diffuses from the stem to the surrounding air. Thus, transpiration is always positive (Fig. 4.9*B*); but during the daytime it is much lower than it is at night. Figure 4.9*C* shows that the water vapor conductance is very low during the daytime, an observation indicating that stomates are closed then. At dusk, which occurs abruptly at 18 hours for these laboratory plants, the stomates begin to open. The water vapor conductance is about half-maximal about 1 hour into the dark period, and increases in

Figure 4.7. *Opuntia ficus-indica,* a platyopuntia cultivated on a limited basis worldwide.
(A) Five-year-old plants in Til Til, Chile (50 km north-northwest of Santiago), which are harvested for their fruits. *(B)* Thirteen-year-old plants near Fillmore, California (70 km northwest of Los Angeles), which are harvested for their young cladodes.

Figure 4.8. *Ferocactus acanthodes,* a barrel cactus common in the Sonoran Desert. Plants shown are at the University of California Philip L. Boyd Deep Canyon Desert Research Center near Palm Desert, California.

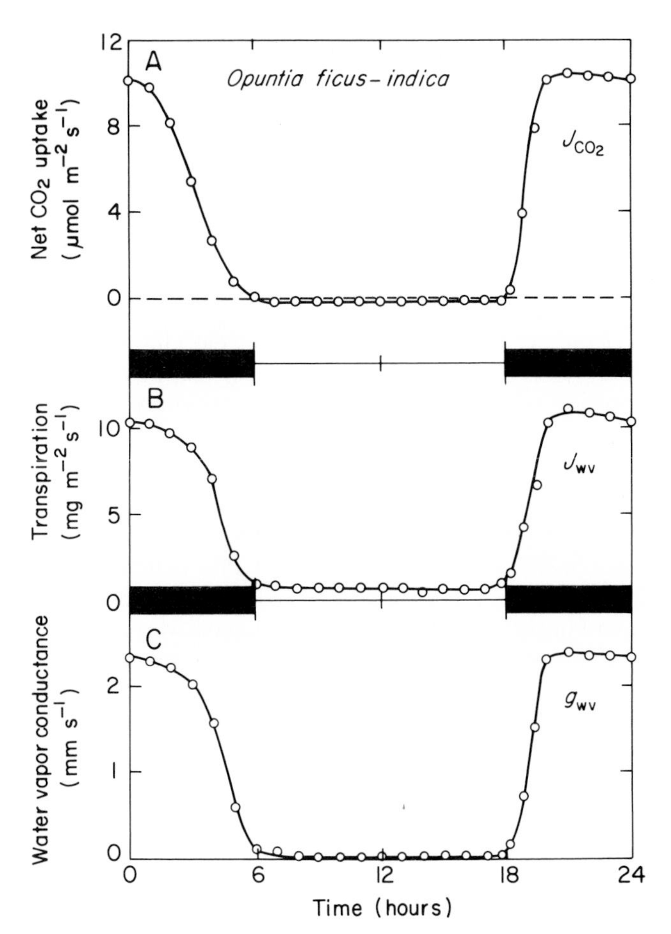

Figure 4.9. Gas exchange for *Opuntia ficus-indica* over a 24-hour period. The plant was maintained in a laboratory growth chamber with a daytime air temperature of 25°C, a PAR level of 560 μmol m^{-2} s^{-1} in the planes of the cladodes provided for 12 hours per day, a relative humidity of 40%, and 345 ppm CO_2. At night the air temperature was 15°C and the relative humidity was 60%. Chlorenchyma temperatures reached 28°C during the day and fell to 14°C at night. (Curves are adapted from Nobel and Hartsock, 1984, and from previously unpublished data.)

transpiration accompany the increases in stomatal opening. Stomates begin closing considerably before dawn (Fig. 4.9*C*); this action leads to decreases in transpiration (Fig. 4.9*B*) and parallels the decrease in net CO_2 uptake over the last part of the night (Fig. 4.9*A*).

A rather similar daily gas exchange pattern occurs for *Ferocactus acanthodes* in the field (Fig. 4.10), but some new features deserve mention. Again, a net CO_2 flow out of the stem can occur during the daytime (in this case, from solar times of 8 to 17 hours). CO_2 uptake occurs for a few hours at the beginning of the daytime (Fig. 4.10*A*), and probably largely represents typical C_3 photosynthesis. Associated with this is a slight reopening of the closing stomates (Fig. 4.10*C*), an action causing transpiration to increase slightly (Fig. 4.10*B*). At the end of the daytime the stomates tend to anticipate the upcoming nocturnal events by beginning to open just before dusk, which, of course, comes more gradually under field conditions than in the laboratory where the lights are switched off abruptly. Transpiration rates are about the same for *F. acanthodes* (Fig. 4.10*B*) and *O. ficus-indica* (Fig. 4.9*B*) because the generally lower water vapor conductance of *F. acanthodes* is largely compensated for by an air humidity lower than that for *O. ficus-indica. Ferocactus acanthodes* also has a considerably lower water vapor conductance (Fig. 4.10*C*) than does *O. ficus-indica* (Fig. 4.9*C*), 2.3 mm s^{-1} versus 1.2 mm s^{-1}, a result that further accounts for its lower maximal rate of net CO_2 uptake.

Biochemistry of CAM

We have already discussed the movement of CO_2 from the atmosphere to the site of photosynthesis in the chloroplasts, including the biochemical reaction involved with CO_2 fixation in C_3 plants (Eq. 4.1). Because CO_2 uptake by the stem succulents *O. ficus-indica* and *F. acanthodes* occurs predominantly in the dark (Figs. 4.9 and 4.10), we might expect that the enzyme responsible for fixing CO_2 might be different from that in C_3 plants. Indeed, this is true. The following reaction takes place in the dark and is the step whereby CO_2 is initially fixed in CAM plants (compare with Eq. 4.1, which is the reaction responsible for CO_2 fixation in C_3 plants; as in Eq. 4.5a, the molecules in Eq. 4.7a are represented in their uncharged forms):

$$
{}^*CO_2 + \begin{array}{c} COOH \\ | \\ COPO_3H_2 \\ \| \\ CH_2 \end{array} + H_2O \quad \xrightarrow[H_3PO_4]{} \quad \begin{array}{c} COOH \\ | \\ C{=}O \\ | \\ CH_2 \\ | \\ {}^*COOH \end{array} \quad \xrightarrow{2H} \quad \begin{array}{c} COOH \\ | \\ HCOH \\ | \\ CH_2 \\ | \\ {}^*COOH \end{array} \tag{4.7a}
$$

which for the compounds of interest and in their appropriate charge state can be written as

$$
CO_2 + \text{phosphoenolpyruvate} \rightarrow \text{oxaloacetate} \rightarrow \text{malate} \tag{4.7b}
$$

The enzyme responsible is phosphoenolpyruvate carboxylase (PEP carboxylase), which occurs in the cytosol. Thus, the initial fixation of CO_2 during CAM occurs in the cytosol, not in chloroplasts, as it does in C_3 plants. The initial product, oxaloacetate, is rapidly converted to malate and some other organic acids—in most CAM plants, the principal compound accumulating upon CO_2 fixation at night is malate.

The malate produced at night cannot immediately be converted into photosynthetic products because photosynthesis requires light as an energy source and there is no light at night. This circumstance underscores one of the dilemmas of CAM and also helps us bring together the cell succulence we discussed in Chapter 3 with the early observations of Heyne on the change in acidity of a CAM plant during a daily cycle. As PEP carboxylase converts the CO_2 to organic acids such as malate in the cytosol, the CO_2 concentration is thereby lowered so that more CO_2 can diffuse in from the atmosphere. However, the malate level begins to build up, and the acidity of the cell increases because the hydrogens on either end of the malate in Equation 4.7a are actually dissociated (technically, the term malate refers to the charged form, and the uncharged form is called malic acid):

$$
\begin{array}{c} COOH \\ | \\ HCOH \\ | \\ CH_2 \\ | \\ COOH \end{array} \rightarrow \begin{array}{c} COO^- \\ | \\ HCOH \\ | \\ CH_2 \\ | \\ COO^- \end{array} + 2H^+ \tag{4.8}
$$

malic acid → malate + hydrogen ions

As more malate is produced, more hydrogen ions accumulate, a process causing the chlorenchyma to become more and more acidic as the night progresses. Indeed, cows and other animals that feed on cacti and other CAM plants often develop intestinal problems as a result of the acidity, especially in the morning. We can now better appreciate the significance of the word *acid* in the phrase "Crassulacean acid metabolism."

The buildup of acidity in CAM plants during the night could be disruptive of other metabolic processes in the cells. Cacti and other CAM plants have solved this possible problem by sequestering the malate in the central vacuoles of the cells (Figs. 3.8, 4.2*B*, and 4.11), where it is thus separated from the rest of the metabolic machinery. The vacuoles need to be fairly large if the cells are to take up much CO_2 and form much malate. This is true for CAM plants ranging from the epiphytic *Tillandsia usneoides* (Spanish moss) to the Crassulaceae (where CAM was

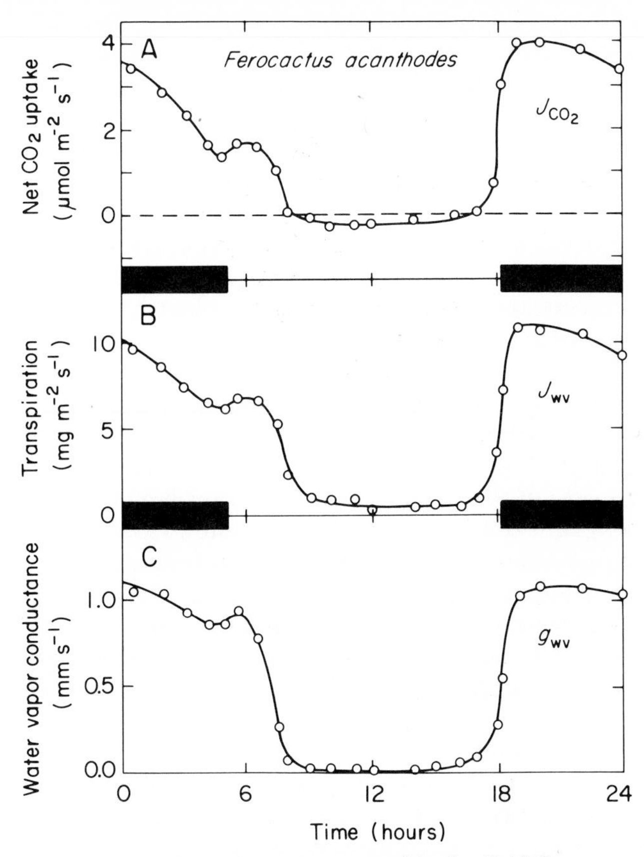

Figure 4.10. Gas exchange measured in the field for *Ferocactus acanthodes* over a 24-hour period. A plant near the site pictured in Figure 4.8 was studied on a spring day when the daytime air temperature reached 27°C and midstem chlorenchyma temperatures reached 40°C; the minimum nighttime air temperature was 14°C and the minimum midstem chlorenchyma temperature was 11°C. The minimum daytime relative humidity was 12% and the maximum nighttime relative humidity was 32%. (Curves are adapted from Nobel, 1977, and previously unpublished data.)

first observed) to our present concern, the Cactaceae. Thus, we can begin to understand why the special metabolism of CAM plants bestows certain requirements on their internal structure. Also, to keep taking in CO_2 and producing malate, the cells must keep making the "acceptor" for CO_2, phosphoenolpyruvate (PEP; Fig. 4.11). This three-carbon compound is made from starch or another glucan (large molecules composed of many six-carbon sugar units). When the supply of glucan runs out, chlorenchyma cells can no longer make phosphoenolpyruvate and net CO_2 uptake must cease.

Other biochemical events take place in CAM plants during the daytime, events that have so far been monitored in about a dozen species of the subfamily Opuntioideae and another dozen species of the subfamily Cactoideae. At dawn the large central vacuole is full of malate. Malate begins to move out of the vacuole into the cytosol, where an enzyme helps break it down (decarboxylation) into CO_2 and pyruvate (Fig. 4.11*B*). We are thus in the familiar position of having CO_2 in the cytosol during the light period; and the consequences are the same as for C_3 plants. Namely, the CO_2 diffuses from the cytosol into the chloroplasts where our old friend Rubisco is waiting to fix the CO_2 (Eq. 4.5), just as for C_3 plants. The immediate source of CO_2 for CAM plants is malate, not atmospheric CO_2. In fact, the decarboxylation of malate raises the CO_2 level in the cytosol above the atmospheric level; and, hence, CO_2 tends to diffuse out of cacti during the daytime (Figs. 4.9*A* and 4.10*A*). Indeed, the CO_2 level in the cytosol can exceed 1% (10,000 ppm), about 30 times the level in the atmosphere. It is a good thing that the stomates are closed during the daytime or even more CO_2 would diffuse out along with water vapor.

The pyruvate formed in the cytosol (produced by decarboxylation of malate) and the sugars (produced as photosynthetic products in the chloroplasts) lead to the formation of glucan, which seems to be stored in the chloroplasts (Fig. 4.11) but may also accumulate in the cytosol. Hence, the glucan level in the chlorenchyma cells gradually increases during the daytime. This starch, or some other glucan, is the

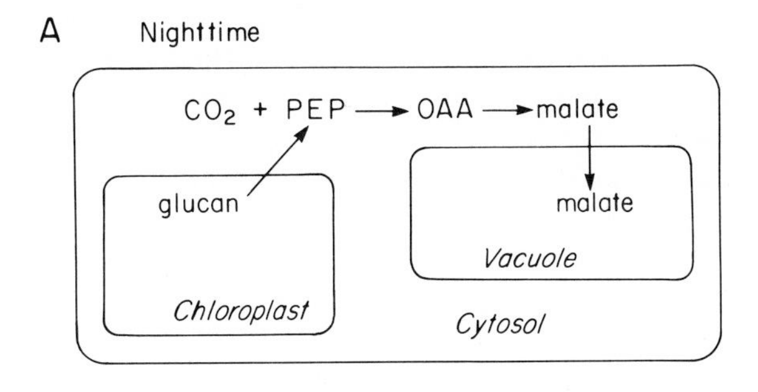

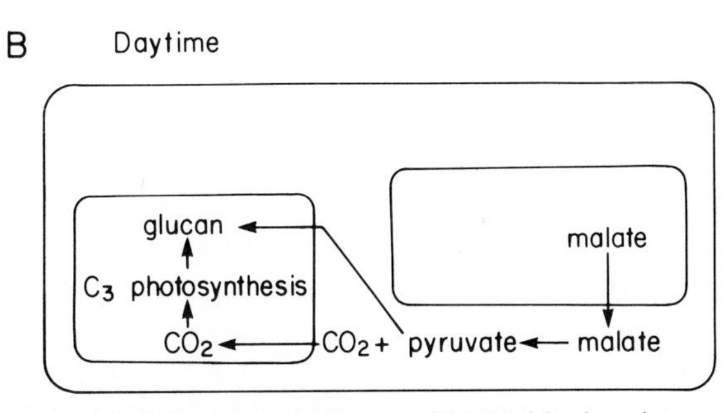

Figure 4.11. Schematic summary of CAM biochemistry during *(A)* nighttime and *(B)* daytime, indicating the cellular components and overall reactions involved for chlorenchyma cells. PEP, Phosphoenolpyruvate; OAA, oxaloacetate.

precursor for the PEP needed for the next night to keep the cycle going day by day.

Although glucan is an uncharged molecule, it gets converted to a three-carbon compound (PEP; Fig. 4.11) and eventually leads to acid as a product. We can write the following equation to indicate such nocturnal acid production in CAM plants:

$$\begin{array}{l}\text{3 carbons from}\\ \quad\text{starch or}\\ \text{another glucan}\end{array} + CO_2 \rightarrow \text{malate} + 2H^+ \qquad (4.9)$$

Thus, by measuring the change in acidity (that is, the change in concentration of hydrogen ions), we can determine how much CO_2 is taken up.

All the CO_2 used to make malate is not taken up directly from the atmosphere. Carbon dioxide can also be produced by respiration in mitochondria, as mentioned earlier. This internally released CO_2 can then be fixed at night by PEP carboxylase and can result in an increase in acidity. Indeed, under severe water stress the stomates of cacti may remain closed throughout both the day and the night, thereby essentially eliminating transpirational water loss as well as net CO_2 exchange with the atmosphere. In this case, the respiratory release of CO_2 at night causes the buildup of malate and acidity during the night; then during the daytime this malate is decarboxylated and the released CO_2 is fixed in the chloroplasts, a process leading to the resynthesis of glucan. This pattern of no net daily CO_2 uptake but day–night oscillations in acidity has been named "idling" by Irwin P. Ting of the University of California at Riverside, who has pioneered the biochemical studies of CAM plants. Such idling with its daily oscillations in malate can even occur in the pereskias when they are subjected to water stress. Even though during water stress cacti may simply be idling their time away, when a chance desert shower occurs, the plants can rapidly take up water (see Fig. 1.28); and then within a few days nocturnal stomatal opening and nocturnal CO_2 uptake is resumed.

Stomatal Functioning

We have talked about stomatal opening in many places in this book because it is crucial for the exchange of gases between a plant and its environment. We have also indicated that stomates generally open during the daytime for C_3 plants and during the night for CAM plants and that for CAM plants CO_2 uptake at night ceases when the chlorenchyma cells run out of glucan. As CO_2 fixation slows down toward the end of the night (Figs. 4.9*A* and 4.10*A*), the level of CO_2 increases in the intercellular air spaces. High levels of CO_2 induce partial stomatal closure (Figs. 4.9*C* and 4.10*C*), which in turn reduces transpiration toward the end of the night (Figs. 4.9*B* and 4.10*B*; transpiration for *F. acanthodes* decreases faster than the water vapor conductance because in the field the stem temperature is also decreasing during the night). Stomatal closure at high internal CO_2 levels makes good sense from the plant's point of view because high internal CO_2 levels mean that there is then adequate CO_2 to keep up with the fixation processes. Thus, the stomatal closure prevents unnecessary loss of water without greatly impeding CO_2 fixation.

Besides depending on CO_2, stomatal opening also depends on temperature, PAR, and water status because metabolic activity and physical factors in the guard cells as well as in the chlorenchyma affect the opening of the stomatal pores. For opening to occur, the guard cells must accumulate solutes. In particular, potassium ions are transported in from the surrounding epidermal cells (Fig. 4.12). This energy-requiring process causes the osmotic pressure to increase in the guard cells, a change that lowers the water potential (recall Eq. 3.3b, $\Psi = P - \pi + \rho_w gh$). As the water potential of the guard cells decreases, water flows into the guard cells and thereby increases their internal hydrostatic pressure, that is, they become more turgid. This increase in turgidity causes the stomatal pore to widen (Fig. 4.12), just as increasing the air pressure in a flat innertube causes the hole in the center to become larger. Stomatal opening occurs for the two *Pereskia* species in the

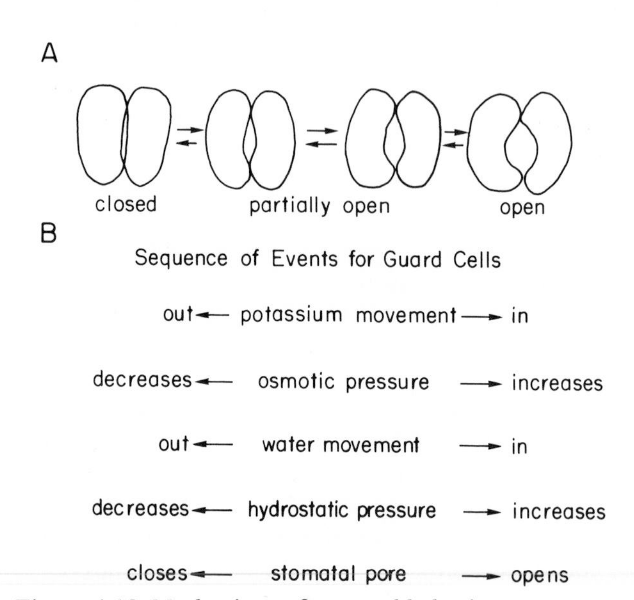

Figure 4.12. Mechanisms of stomatal behavior. *(A)* Degrees of stomatal opening. *(B)* Sequence of events in the opening or closing of stomatal pores. The quantities refer to those conditions inside the pair of guard cells. Arrows to the right indicate stomatal opening, and those to the left, closing.

light (see Fig. 4.6 for *P. grandiflora*) and for the two stem succulents in the dark (Figs. 4.9 and 4.10), and so light can act in different ways. The effect of light can be mediated through CO_2, as we just indicated, even though the enzyme initially fixing CO_2 in C_3 plants is different from that in CAM plants.

Stomatal closure is initiated by potassium movement from the guard cells to the surrounding epidermal cells, an activity that lowers the osmotic pressure in the guard cells. The water potential in the guard cells is thus raised and so water flows out, lowering the turgor of guard cells and inducing closure of the stomatal pore (Fig. 4.12). The time during the day for stomatal closure of C_3 plants is obviously different from that of CAM plants, but the underlying cellular events are actually the same, as far as is known (Fig. 4.12). Water status can have an overriding effect on stomatal opening, because under severe stress stomates for both C_3 and CAM species remain closed, regardless of the light or CO_2 levels. During periods when water is available, lowering the relative humidity of the air, a change that tends to increase transpiration, leads to partial stomatal closure, a response that can prevent the potentially excessive transpiration, at least for some species. Thus, regulation of stomatal opening is fairly complicated, and many of the molecular details have yet to be clarified for cacti.

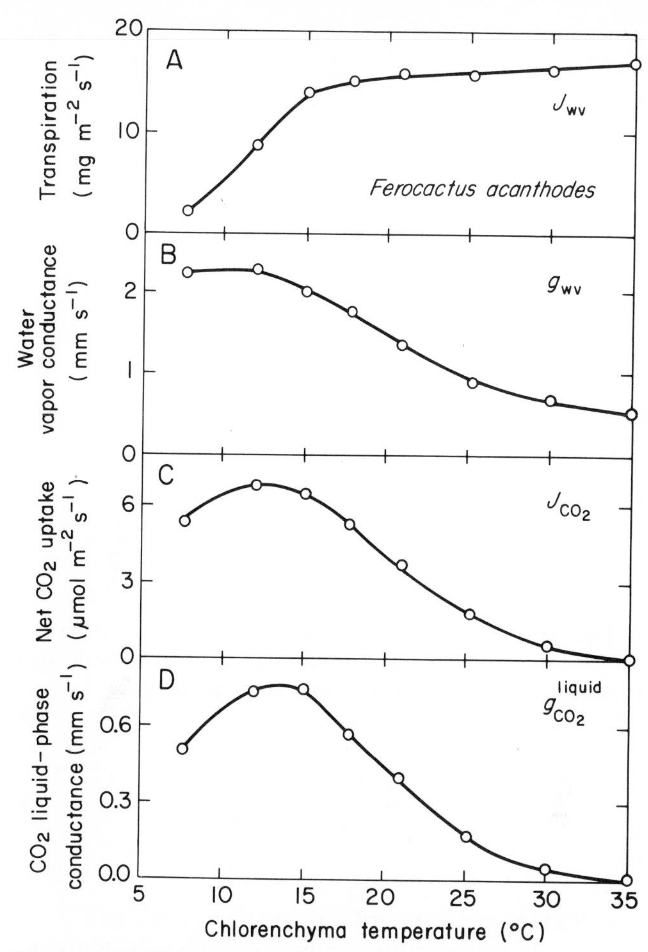

Figure 4.13. Temperature dependence of gas exchange for *Ferocactus acanthodes.* Transpiration *(A)* and CO_2 uptake *(C)* were determined in the laboratory for conditions similar to those described for Figure 4.3 during that part of the night when CO_2 uptake is normally maximal and fairly constant (1 to 5 hours after the beginning of the dark period). The air water vapor concentration was 6.8 g m^{-3} (39% relative humidity at 20°C).

Temperature Dependence of Gas Exchange

As the nighttime temperature is increased above about 15°C, transpiration by *Ferocactus acanthodes* is relatively constant (Fig. 4.13*A*). The progressive closure of stomates with increasing temperature reduces the water vapor conductance (Fig. 4.13*B*); transpiration is kept constant even though the water vapor concentration drop from chlorenchyma to the ambient air increases nearly fivefold from 15°C to 35°C, a response similar to the control of daytime water loss exerted by the stomates of the two species of *Pereskia* considered previously (Fig. 4.3*A*). The optimal temperature for net CO_2 uptake in *Pereskia* is near 24°C, which is relatively low compared to the optima for most C_3 plants; but it is still far above the optimal temperature of only 12°C for *F. acanthodes* (Fig. 4.13*C*). Indeed, in most CAM plants net nocturnal CO_2 uptake is maximal at a nighttime temperature from 10°C to 15°C. Such high activity at relatively low temperatures is a considerable departure from most biochemical processes, in which biochemical activity usually increases along with temperature. The high CO_2 uptake at relatively low temperatures is in large measure a consequence of the relatively low temperature at which maximal CO_2 liquid-phase conductance occurs (Fig. 4.13*D*), a characteristic also found in other CAM plants.

The low temperature for maximal CO_2 uptake is a very important part of the water-conserving attributes of CAM. In Chapter 1 we showed that for a given degree of stomatal opening and a representative air water vapor content of 8.1 g m^{-3}, transpiration is 3-fold less at 25°C and 30-fold less at 10°C than it would be at 40°C. We all know how much harder it is to dry clothes outside at night rather than during the daytime! This is mainly a consequence of the great increase with temperature in the concentration of water vapor that can be held in saturated air, such water vapor saturation occurring in the pores of clothing as well as in the intercellular air spaces of the chlorenchyma (Fig. 4.3*B*). Thus, opening stomates during the night when air and

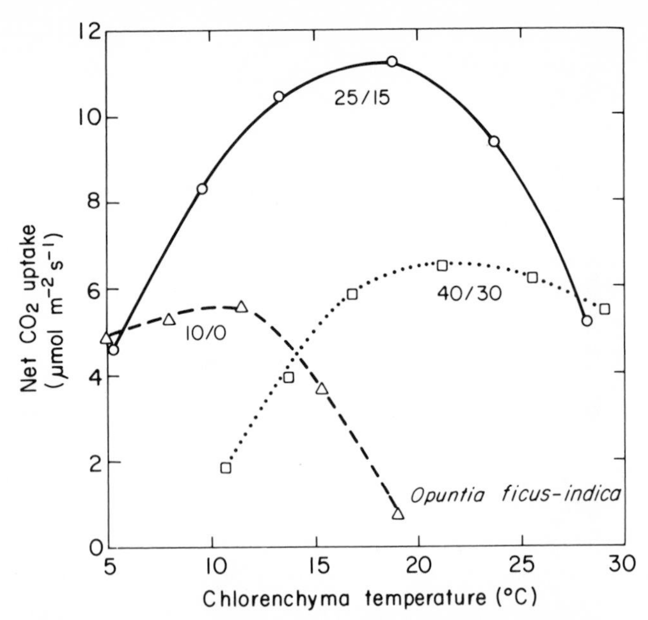

Figure 4.14. Shift in the temperature response for net nocturnal CO_2 uptake by *Opuntia ficus-indica* as the growth temperature is changed. Data were obtained (under conditions similar to those described for Fig. 4.9) at 2 to 6 hours after the beginning of the night. The day/night air temperatures during growth are indicated in °C (25/15 corresponds to the conditions for Fig. 4.9).

stem temperatures are relatively low leads to a reduced transpiratory water loss. The biochemistry of stem-succulent cacti is also advantageously adapted to such low nighttime temperatures.

Cacti typically live in habitats where there are seasonal as well as daily changes in temperature, so it has interested scientists to determine how CAM plants respond to different day/night temperature regimes. Some of the best data exist for *Opuntia ficus-indica*. For *O. ficus-indica* growing in environmental chambers with low temperatures of 10°C during the daytime and 0°C at night, the optimal temperature for net CO_2 uptake is near 10°C, which shifts up to about 20°C when the same plants are transferred to high day/night temperatures of 40°C/30°C (Fig. 4.14). This acclimation, which apparently occurs at the enzyme level, allows the plants to be opportunistic and to take advantage of the prevailing (seasonal) environmental conditions. Nonetheless, the maximal CO_2 uptake rates for plants grown at the low or at the high temperatures are only about half as great as the maximal rate for plants grown at moderate day/night air temperatures of 25°C/15°C (Fig. 4.14). Thus, although cacti can tolerate quite high desert temperatures, as we shall see later, CO_2 uptake by cacti is favored at moderate temperatures, especially if accompanied by cool nights. For *O. ficus-indica*, total nocturnal CO_2 uptake over the course of the whole night is greatest for nighttime temperatures of 10°C to 15°C; and nocturnal CO_2 uptake decreases 50% at a low nighttime temperature of 0°C or a relatively high nighttime temperature of 25°C. Other cacti, including *F. acanthodes*, also have shifts in the temperature optimum for net nocturnal CO_2 uptake from winter to summer, such shifts occurring in response to changes in nighttime temperature. Ability to perform well at higher environmental temperatures is also important for cacti in the tropics, where the minimum nighttime temperatures are relatively high and fairly constant throughout the year. Indeed, the CO_2 uptake reactions of cacti may have adapted to seasonal shifts in temperature as they radiated from tropical regions to regions with lower nighttime temperatures.

A large daily range in tissue temperatures, generally greater than 15°C, results in maximum malate oscillations in many CAM plants. Some *Opuntia* species have an optimal nighttime temperature for CO_2 uptake of 15°C and an optimal daytime temperature near 40°C for their C_3 photosynthesis. Indeed, high daytime temperatures favor the decarboxylation of malate. Such high daytime temperatures and relatively low nighttime temperatures occur for many cacti in the deserts of North and South America.

PAR Dependence of Net CO_2 Uptake

Because very little CO_2 uptake occurs during the daytime for *F. acanthodes, O. ficus-indica,* and other stem-succulent cacti, at first sight it might appear that CO_2 uptake does not depend on the amount of photosynthetically active radiation. But the light-requiring reactions of C_3 photosynthesis taking place during the daytime are necessary for the net generation of glucan; and glucan is needed for forming phosphoenolpyruvate, the acceptor for CO_2 at night (Fig. 4.11). Thus, there must be some dependence of CO_2 uptake on the PAR received.

For most plants, researchers measure how much CO_2 is taken up by the plant at a particular PAR level—this procedure was used for *Pereskia* (Fig. 4.5). Looking back at Figure 4.5, we can see that at a PAR level of 300 μmol m^{-2} s^{-1} the instantaneous value of CO_2 uptake was 8 μmol m^{-2} s^{-1}. However, in CAM plants, where the event of CO_2 uptake is temporally separated from the use of CO_2 in making carbohydrates, a more convenient method must be used to relate CO_2 uptake to PAR. This is why researchers use the total amount of PAR received during the entire daytime to compare with the total CO_2 uptake or the overall change in malate level at night.

When expressed on a relative basis, the response of nocturnal CO_2 uptake to total daily PAR is nearly identical for *F. acanthodes* in the field and *O. ficus-indica* in the laboratory (Fig. 4.15). Net CO_2 uptake was negative up to a total daily PAR of nearly 4 mol m^{-2}. In other words, below about 3 or 4 mol m^{-2} these CAM plants would have a negative carbon balance, meaning that they would respire away their stored sugars and glucan. This has ecological implications as well as implications on how best to treat cacti maintained as indoor plants. We shall discuss the influence of PAR on cactus distribution and growth form in Chapter 8. Here we can at least note that cacti will slowly degenerate and eventually die if they are kept in a low PAR environment year round. For example, a total daily PAR of 3 mol m^{-2} corresponds to an instantaneous PAR level of 70 μmol m^{-2} s^{-1} for a 12-hour day. Although to the human eye this would be a brightly lit room, it is only 3% of full sunlight. It is surprising, therefore, that some relatively expensive cacti are placed in dimly lit rooms where they never see the light of day; we hope that the owners of these plants will heed our warning about the PAR requirements of desert cacti.

As the daytime PAR level is increased, the nocturnal CO_2 uptake increases and eventually reaches saturation (Fig. 4.15). Ninety percent of saturation can occur for a total daily PAR of 20 to 25 mol m^{-2}. When the cacti are exposed to suboptimal temperatures or other unfavorable conditions, 90% saturation occurs at a somewhat lower total daily PAR. In any case, daily PAR levels above about 25 mol m^{-2} do not increase net CO_2 uptake very much. In Chapter 8 we shall consider PAR levels available at various latitudes and for various seasons in some detail, but let us note here that 20 mol m^{-2} corresponds to the average total daily PAR available on vertical surfaces on clear days. Thus, shading of cacti by other plants or cloudy days can easily reduce the available total daily PAR below 20 mol m^{-2} and hence reduce the nocturnal CO_2 uptake (Fig. 4.15). As a result, stem-succulent cacti can be PAR limited, even in the high radiation environment of a desert.

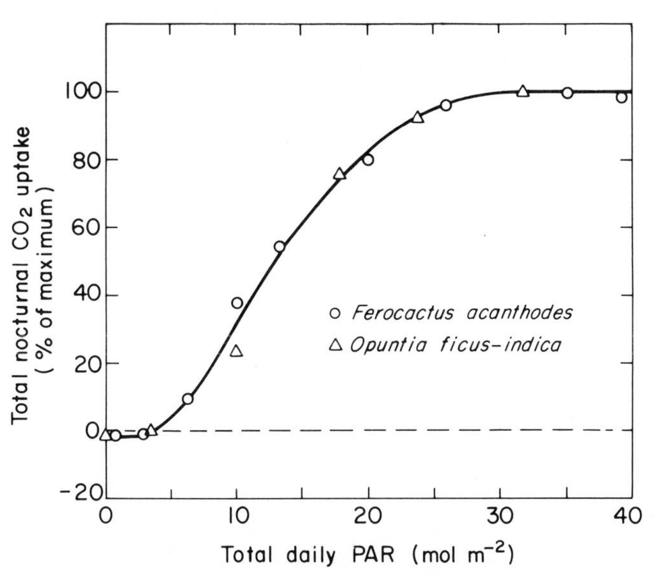

Figure 4.15. Response of total nocturnal CO_2 uptake to total daily PAR by *Ferocactus acanthodes* and *Opuntia ficus-indica*. The maximum CO_2 uptake was 130 mmol m^{-2} for *F. acanthodes* (Nobel, 1977) and 340 mmol m^{-2} for *O. ficus-indica* (Nobel and Hartsock, 1983, and previously unpublished data).

Water Relations

The key to desert existence is water. In Chapter 1 we introduced water vapor conductance and considered the relation between rainfall, drought periods, stomatal opening, and the water content of the stem (Fig. 1.28). In Chapter 3 we considered the relation between rainfall and soil water potential (Fig. 3.51) and also discussed water storage in succulent stems. Now that we have considered stomatal opening in a little more detail, let us bring these various themes together and examine some field water relations data for *F. acanthodes*.

Rainfall (Fig. 4.16*A*) raises the soil water potential (Fig. 4.16*B*) at the Sonoran Desert site considered, especially in the winter and early spring. The stomates of *F. acanthodes* respond accordingly, with the water vapor conductance exceeding 1 mm s^{-1} sometime during the night, a value indicating appreciable stomatal opening for the winter–spring period. Drought usually occurs at the field site in the northwestern Sonoran Desert in early summer and indeed can last all summer. As drought progresses, stomatal opening is initiated later and later during the night and the maximum opening also progressively decreases. In 1975 and 1976 appreciable nocturnal stomatal opening ceased in July, and in 1977 it ceased in May (Fig. 4.16*C*). If rainfall sufficient to wet the soil in the root zone occurs, for example, over 10 mm of rainfall (Fig. 3.51), appreciable nocturnal stomatal opening will resume, even at the hottest times of the year. But the higher temperatures in the summer cause the maximal stomatal opening under wet conditions to be less than the maximum in the winter, a consequence of the temperature dependence of stomatal opening discussed earlier (see Fig. 4.13).

Besides the water-conserving strategy of nocturnal stomatal opening, cacti also can maintain an extremely low water vapor conductance during periods of water stress. For instance, g_{wv} can be below 0.005 to 0.010 mm s^{-1} throughout the daily cycle for *F.*

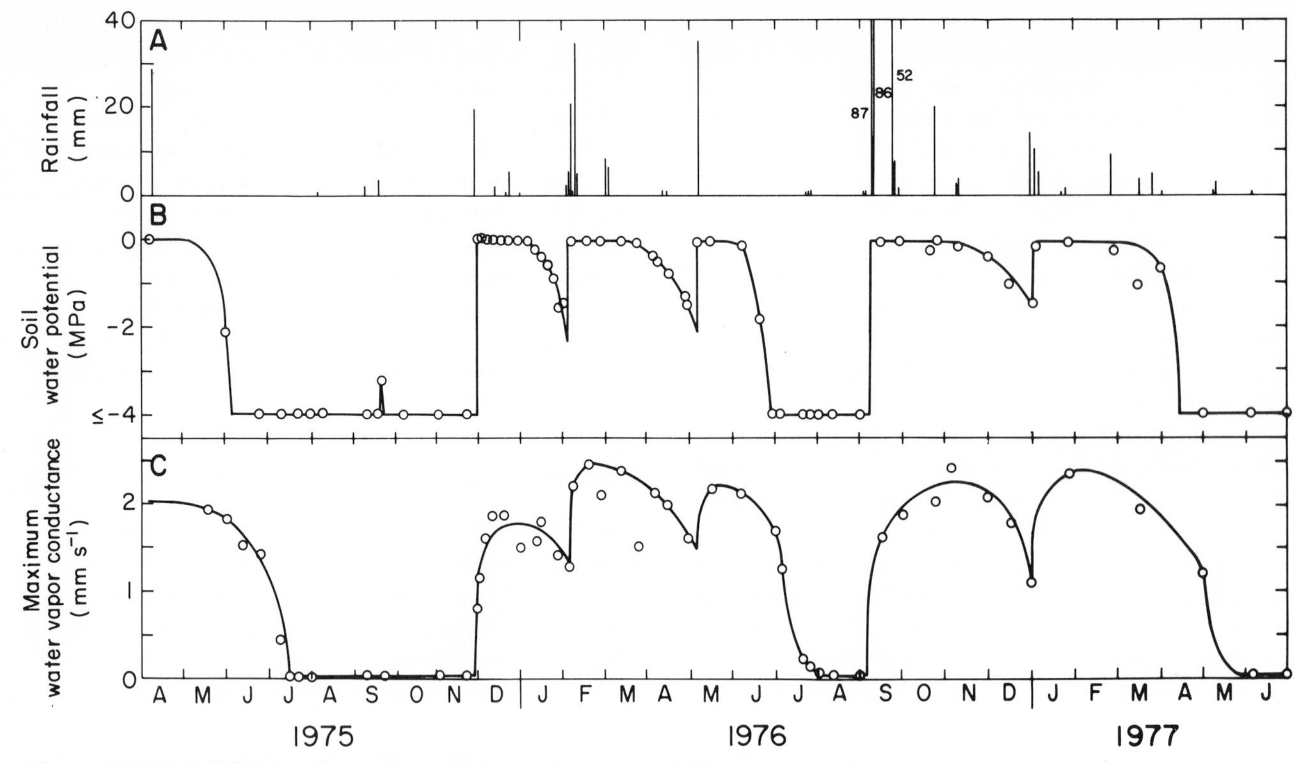

Figure 4.16. Rainfall *(A)*, soil water potential *(B)*, and maximum water vapor conductance *(C)* observed for *Ferocactus acanthodes* in the northwestern Sonoran Desert. The maximum water vapor conductance was obtained from daily patterns similar to that shown in Figure 4.10*C*. (Data are from Nobel, 1977, and previously unpublished observations.)

acanthodes (Fig. 4.16) and other cacti. This is a consequence of the thick, conspicuous waxy cuticle and is in contrast to crop plants, for which the minimum g_{wv} is often 0.1 to 0.2 mm s^{-1}. Thus, under drought conditions the rate of water loss from the stems may be a factor of 10 to 400 lower than comparable water loss from agronomic plants. This low water vapor conductance under water stress conserves water and thus helps increase the water-use efficiency of cacti when considered for an extended period such as a year. On the other hand, the maximal water vapor conductance for stem-succulent cacti under wet conditions is generally below 3 mm s^{-1}. This value is considerably lower than maximal values for non-CAM plants and reflects the fact that cacti have fewer stomates per unit area than do most other plants (about 50 stomates/mm^2 for the two species considered here versus 200/mm^2 for most C_3 plants). The low stomatal frequency would tend to limit the maximal rate of CO_2 uptake; but because lowering the gas-phase conductance has a greater proportional effect on water loss than on CO_2 uptake (as discussed earlier), the lower maximal gas-phase conductance also causes the water-use efficiency of cacti during periods of water availability to be higher than for non-CAM plants.

Let us next consider the values of water-use efficiency for stem-succulent cacti by reexamining Figures 4.9 and 4.10. The total net CO_2 uptake over the 24-hour period for *O. ficus-indica* (Fig. 4.9) was 298 mmol m^{-2}, and the transpiratory water loss was 385 g m^{-2}. Hence, its water-use efficiency (Eq. 4.6) for this day was 34 g CO_2/kg H_2O. For *F. acanthodes* (Fig. 4.10), the comparable values were 137 mmol CO_2 m^{-2} and 452 g H_2O m^{-2}, values corresponding to a water-use efficiency of 13 g CO_2/kg H_2O. The water-use efficiency during the growing season for agronomic plants under similar environmental conditions is usually only 2 to 6 g CO_2/kg H_2O. Thus, the CAM pathway of the stem succulents leads to high carbon gains per unit of water lost.

The higher water-use efficiency of stem-succulent cacti is even more apparent when considered on a seasonal or an annual basis. For instance, for a 1-year period from mid-July 1975 to mid-July 1976, the water-use efficiency for *F. acanthodes* in the field (Fig. 4.16) was 14 g CO_2 fixed/kg H_2O transpired. Only a few crop plants can fix even 3 g CO_2/kg H_2O lost on a seasonal or annual basis. During this same period, the total rainfall was 154 mm (Fig. 4.16*A*). On the basis of the ground area explored by the roots of the particular 30-cm-tall *F. acanthodes* considered and the measured transpiration, we calculate that approximately 22% of the incident rainfall was transpired by the plant. Water transpired times the water-use efficiency (CO_2 uptake/water lost) indicates the productivity, our final topic in this consideration of CAM metabolism.

Productivity

Physical factors such as water status, temperature, and PAR affect productivity, which we can define as the dry weight produced per unit land area per unit time (dry weight refers to the weight after drying to remove water and is the conventional basis for expressing plant productivity). Often we consider just the above-ground productivity, which for the entire land area of the earth averages about 0.7 kg m^{-2} yr^{-1}. Such a figure is not ecologically very meaningful because it averages in highly fertilized agricultural lands and deserts and even suburban shopping malls. However, relatively few measurements of productivity have been made for cacti and almost none under natural conditions. Yet it is still worthwhile for us to consider the principles involved and to make some preliminary estimates of the productivity that might be expected for cacti.

The obvious starting place is with water because deserts are most reasonably defined as regions where water availability is the major factor influencing growth. Figure 4.16 shows that the only periods where the soil water potential was high enough for cacti to take up water was during and immediately after periods of rainfall. We also note that stems of cacti can store water and thus can extend the periods of stomatal opening and CO_2 uptake beyond the point where the soil is too dry. Yet we can readily appreciate that average productivity of, for example, *Copiapoa cinerea* in the Atacama Desert of Chile, where it may not rain for several years, would be much lower than for irrigated *Opuntia ficus-indica* near Santiago, Chile, where its fruits are harvested and sold in the local supermarkets.

Next let us consider temperature—we shall avoid considering the extreme conditions of wintertime freezing and summertime overheating. Indeed, in regions of moderate temperature, temperature is not a limiting factor for cacti, especially if temperature is low during the night, a condition that favors stomatal opening and net CO_2 uptake (Figs. 4.13 and 4.14). Also, cacti can acclimate to changing environmental temperatures. Moreover, if temperature is too high during the early part of the night, a condition that can occur in some places where cacti are common, major stomatal opening may be delayed until a later, cooler part of the night.

We finally turn to PAR and ask what might be optimal for productivity. If the cacti are far apart so that they do not shade each other and if other plants do not shade them, then the PAR is obviously maximized. However, productivity per unit area of the ground is not. On the other hand, if the cacti are jammed too close together, there would be much shading; therefore, the total daily PAR might drop below the compensation level (negative CO_2 uptake in Fig. 4.15) and there would be a consequent net loss of dry matter over the year. Is there a more reasonable compromise? Let us assume that the total surface area of the stems is the same as the ground area, a condition that is approximately the case for many cultivated platyopuntias, and that the cacti are so arranged that there is very little shading. For a plant of *O. ficus-indica* with all of its cladodes facing east–west at Santiago, Chile, PAR limitations could cause net CO_2 uptake to average 55% of maximal over a year, taking into consideration an annual average PAR reduction of 25% due to clouds. Because maximum net CO_2 uptake by this species can be 340 mmol m^{-2} day^{-1}, the annual net uptake of carbon would be 68 mol m^{-2} stem area yr^{-1}, a value that translates into 2.0 kg of carbohydrates (the principal photosynthetic product) per square meter of ground area per year (assuming ground and stem areas are equal). These values represent the productivity predicted for such a cactus under conditions where temperature and water are not limiting.

Opuntia ficus-indica is cultivated on a limited basis in tropical and subtropical areas as well as in hot regions of the temperate zone; therefore, its productivity has been studied (Table 4.1). Stem and especially fruit productivities tend to increase with plant age. In central Chile maximum fruit productivity occurred for plants 16 to 20 years old (Table 4.1). Ignacio Badilla and Edmundo Acevedo of the Universidad de Chile, working with Nobel, found that the productivity of 5.4-year-old *O. ficus-indica* in commercial plantations was nearly 1.4 kg above-ground dry matter m^{-2} yr^{-1} (Table 4.1). Because the irrigation occurs only three times per year (with a rainfall equivalent of 80 mm each time) and does not keep the soil wet year-round and because the temperature can be suboptimal for net CO_2 uptake, this is really quite good agreement between prediction (2.0 kg m^{-2} yr^{-1}) and measurement. But a few words of caution are in order. First, irrigation is an artificial condition; thus, productivity under natural conditions would be expected to be less than that occurring when rainfall is not limiting, as will be illustrated for *F. acanthodes* in the following paragraph. Second, *O. ficus-indica* has the highest nocturnal acid accumulation and net CO_2 uptake rate so far reported for any cactus; thus, other species could well have lower productivities. Finally, shading by

Table 4.1. Annual productivity of cultivated *Opuntia ficus-indica.*

Location	Conditions: Annual rainfall (mm)	Irrigated	Plant age (yr)	Plant part	Annual productivity[a]: Wet weight ($kg\ m^{-2}yr^{-1}$)	Dry weight ($kg\ m^{-2}yr^{-1}$)	Source[b]
California	450	+	—	Fruit	1.0	—	3
Argentina	300	−	—	Stem	—	0.25	2
		−	—	Fruit	—	0.02	5
Brazil	~700	−	—	Stem	9.3	—	5
Chile	300	+	3	Fruit	0.3	—	6
	300	+	4–15	Fruit	0.6	—	6
	300	+	16–20	Fruit	1.6	—	6
	300	+	21–35	Fruit	0.8	—	6
	300	+	1.2	Stem	—	0.38	1
	300	+	1.2	Fruit	—	0.01	1
	300	+	2.2	Stem	—	0.53	1
	300	+	2.2	Fruit	—	0.03	1
	300	+	5.4	Stem	—	1.05	1
	300	+	5.4	Fruit	—	0.32	1
Tunisia	280	−	—	Stem	≤5.0	—	4

a. 1 kg wet weight equals approximately 0.12 kg dry weight (depending on the relative water content), and 1 kg $m^{-2}yr^{-1}$ equals 10 metric tons $hectare^{-1}yr^{-1}$. All areas refer to ground area.

b. 1; Avecedo, Badilla, and Nobel (1983); 2, Braun, Cordero, and Ramacciotti (1979); 3, Curtis (1977); 4, Le Houerou (1970); 5, Metral (1965); 6, Reñasco (1976).

spines or by other vegetation would reduce the total daily PAR on the stem and thereby reduce net CO_2 uptake. Nonetheless, 2.0 and even 1.4 kg m^{-2} yr^{-1} represent substantial productivities.

Let us now return to our discussion of *F. acanthodes* in the field—in particular, from mid-July 1975 to mid-July 1976 (Fig. 4.16). During this 1-year period, the 30-cm-tall *F. acanthodes* took up about 221 g CO_2 and increased in height by 1.5 cm. Such relatively slow growth is characteristic of barrel cacti. The net CO_2 uptake of 221 g would correspond to 151 g of dry weight if all were converted to carbohydrates such as sugars and starch. On the basis of the ground area explored by the roots (0.45 m^2; corresponding to a circle 76 cm in diameter), the net CO_2 uptake represents an annual dry weight gain of 0.34 kg/m^2 ground area. During the 1-year period considered, the rainfall was 154 mm (Fig. 4.16*A*). This lower water input and the lower inherent maximal CO_2 uptake rate account for the fact that the productivity of *F. acanthodes* in the Sonoran Desert is lower than that of irrigated *O. ficus-indica* in central Chile. However, a productivity of 0.34 kg m^{-2} yr^{-1} is considerably greater than the average productivity of desert ecosystems—0.1 kg m^{-2} yr^{-1}—and is similar to that of some desert agaves that have been studied recently.

The stem plus fruit productivity approaching 1.4 kg dry weight m^{-2} yr^{-1} for *O. ficus-indica* (Table 4.1) is one of the highest productivities ever reported for a CAM plant. Other high values for cultivated CAM plants occur for pineapple and for agaves harvested for fiber. Indeed, the productivity of *O. ficus-indica* is similar to the productivity of corn or wheat at the turn of the century and still compares favorably with some conventional agricultural crops. Such high productivities for cacti can have major implications for the use of arid and semiarid regions in the future. For the present, we can assuredly say that we are beginning to understand how these remarkable plants respond to environmental factors and that they have great potential for productivity in inhospitable environments where conventional crops would not be suitable.

SELECTED BIBLIOGRAPHY

Acevedo, E., I. Badilla, and P. S. Nobel. 1983. Water relations, diurnal acidity changes, and productivity of a cultivated cactus, *Opuntia ficus-indica. Plant Physiology* 72:775–780.

Braun, W. R. H., A. Cordero, and J. Ramacciotti. 1979. Productividad, ecología y valor forrajero de tunales *(Opuntia ficus-indica)* de los Llanos, provincia de la Rioja. *Cuaderno Técnico, Iadiza* 1–79:29–37.

Curtis, J. R. 1977. Prickly pear farming in the Santa Clara Valley, California. *Economic Botany* 31:175–179.

García de Cortázar, V., E. Acevedo, and P. S. Nobel. 1985. Modeling of PAR interception and productivity by *Opuntia ficus-indica. Agricultural and Forest Meteorology* 34:145–162.

Gerwick, B. C., G. J. Williams III, M. H. Spalding, and G. E. Edwards. 1978. Temperature response of CO_2 fixation in isolated *Opuntia* cells. *Plant Science Letters* 13:389–396.
Hanscom, Z., III, and I. P. Ting. 1977. Physiological responses to irrigation in *Opuntia basilaris* Engelm. & Bigel. *Botanical Gazette* 138:159–167.
Heyne, B. 1815. On the deoxidation of the leaves of *Cotyledon calycina. Transactions of the Linnean Society of London* 11, part 2:213–215.
Kluge, M. 1979. The flow of carbon in Crassulacean acid metabolism (CAM). Pp. 113–125 in *Encyclopedia of plant physiology,* new series, vol. 6: *Photosynthesis* II, ed. M. Gibbs and E. Latzko. Springer-Verlag, Berlin–Heidelberg–New York.
Kluge, M., and I. P. Ting. 1978. Crassulacean acid metabolism: analysis of an ecological adaptation. Springer-Verlag, Berlin–Heidelberg–New York.
Körner, C. H., J. A. Scheel, and H. Bauer. 1979. Maximum leaf diffusive conductance in vascular plants. *Photosynthetica* 13:45–82.
Le Houerou, H. N. 1970. North Africa: past, present, future. Pp. 227–278 in *Arid Lands in Transition,* ed. H. E. Dregne. American Association for the Advancement of Science, Publication no. 90, Washington, D.C.
Longstreth, D. J., T. L. Hartsock, and P. S. Nobel. 1980. Mesophyll cell properties for some C_3 and C_4 species with high photosynthetic rates. *Physiologia Plantarum* 48:494–498.
Metral, J. J. 1965. Les cactes fourragères dans le Nord–Est du Brasil, plus particulierment dans l'état du Ceara. *Agronomía Tropical* 20:248–261.
Nisbet, R. A., and D. T. Patten. 1974. Seasonal temperature acclimation of a prickly-pear cactus in southcentral Arizona. *Oecologia* 15:345–352.
Nobel, P. S. 1977. Water relations and photosynthesis of a barrel cactus, *Ferocactus acanthodes,* in the Colorado Desert. *Oecologia* 27:117–133.
——— 1983. *Biophysical plant physiology and ecology.* W. H. Freeman, New York.
——— 1984. Productivity of *Agave deserti:* measurement by dry weight and monthly prediction using physiological responses to environmental parameters. *Oecologia* 64:1–7.
——— 1985. Photosynthesis and productivity of desert plants. In *Progress in desert research,* ed. L. Berkovsky and M. G. Wurtele. In press.
Nobel, P. S., H. W. Calkin, and A. C. Gibson. 1984. Influences of PAR, temperature, and water vapor concentration on gas exchange by ferns. *Physiologia Plantarum* 62:527–534.
Nobel, P. S., and T. L. Hartsock. 1978. Resistance analysis of nocturnal carbon dioxide uptake by a Crassulacean acid metabolism succulent, *Agave deserti. Plant Physiology* 61:510–514.
——— 1981. Shifts in the optimal temperature for nocturnal CO_2 uptake caused by changes in growth temperature for cacti and agaves. *Physiologia Plantarum* 53:523–527.
——— 1983. Relationships between photosynthetically active radiation, nocturnal acid accumulation, and CO_2 uptake for a Crassulacean acid metabolism plant, *Opuntia ficus-indica. Plant Physiology* 71:71–75.
——— 1984. Physiological responses of *Opuntia ficus-indica* to growth temperatures. *Physiologia Plantarum* 60:98–105.
Osmond, C. B. 1978. Crassulacean acid metabolism: a curiosity in context. *Annual Review of Plant Physiology* 29:379–414.
Osmond, C. B., K. Winter, and H. Ziegler. 1982. Functional significance of different pathways of CO_2 fixation in photosynthesis. Pp. 479–547 in *Encyclopedia of plant physiology,* new series, vol. 12B: *Physiological plant ecology II,* ed. O. L. Lange, P. S. Nobel, C. B. Osmond, and H. Ziegler. Springer-Verlag, Berlin–Heidelberg–New York.
Patten, D. T., and B. E. Dinger. 1969. Carbon dioxide exchange patterns of cacti from different environments. *Ecology* 50:686–688.
Rayder, L., and I. P. Ting. 1981. Carbon metabolism in two species of *Pereskia* (Cactaceae). *Plant Physiology* 68:139–142.
Reñasco, G. 1976. *Cultivo de tunales.* Boletín Divulgativo no. 44, Servicio agrícola y ganadero, Santiago, Chile.
Samish, Y. B., and S. J. Ellern. 1975. Titratable acids in *Opuntia ficus-indica* L. *Journal of Range Management* 28:365–369.
Smith, S. D., and P. S. Nobel. 1986. Deserts. In *Topics in photosynthesis,* ed. N. R. Baker and S. P. Long, vol. 7: *Photosynthesis in specific environments.* Chapman and Hall, London.
Szarek, S. R., and I. P. Ting. 1975. Photosynthetic efficiency of CAM plants in relation to C_3 and C_4 plants. Pp. 289–297 in *Environmental and biological control of photosynthesis,* ed. R. Marcelle. W. Junk, The Hague.
Ting, I. P., and M. Gibbs, eds. 1982. *Crassulacean acid metabolism.* American Society of Plant Physiologists, Rockville, Maryland.

5

Areoles and Spines

The areole is an axillary bud that is situated on an enlarged and persistent leaf base (a tubercle) and produces spines instead of normal leaves. These terms and the origin of the areole were introduced in Chapter 1; but such an important feature of cacti needs to be fully understood, so we shall describe it here in detail. The areole, with its associated spines and trichomes, not only provides great appeal to the collector, who treasures species with the most unusual areoles, but also is credited with much of the evolutionary success of the cactus family. In this chapter we shall explore the range of areolar features and attempt to ferret out the developmental and functional reasons for this great variation.

The Nature of Spines

Some Generalizations

All spines that have been studied arise on the edge of the areolar meristem—also called the spine meristem—in the same manner as a first-formed leaf primordium. Cell divisions produce a small dome of cells, which appear like a cone with a rounded tip (Fig. 5.1*A*). For a very short period the spine grows at the tip, but this type of apical growth ceases when the spine primordium is about 0.1 mm in length. Subsequently, the spine primordium grows by cell divisions at its base, a region that is called the basal meristematic zone or an intercalary meristem (Fig. 5.1*B*). The basal meristematic zone can be envisioned as a plate of cells, several cells thick, in which all cells of the plate are dividing to produce cells only toward the spine tip. Cells that are produced by the basal meristematic zone become elongated and stretched as they mature. Meanwhile, beginning from the tip down, the elongate cells become hardened (sclerified) because they produce thick cell walls that are impregnated with lignin. A thick cuticle is generally deposited on the outer walls of the spine epidermal cells. Hence, in the typical spine the upper part of the spine is not only hard but also waterproof, even while the cells at the bottom of the spine are still undergoing elongation.

When you make a section of a cactus spine for microscopic examination—and this is not easy because a spine is very hard—or when you break a spine and examine it with a hand lens or, better yet, with a scanning electron microscope (Fig. 5.2), you find no vascular tissues (xylem and phloem). In leaves, vascular tissues enter the base and differentiate to the tip; but for cactus spines, the vascular tissues leading to each developing spine typically terminate in the areolar tissue beneath the spine base and do not penetrate the basal meristematic zone. To a developmental botanist, this makes perfect sense. First, vascular tissues for a primordium are usually guided in their development by a message received from the growing tip of the primordium; but for a cactus spine, the tip is dead and sclerified. Second, vascular tissues do not penetrate the basal meristematic zone while cells are actively dividing. When cell division stops, sclerification of the existing cells follows immediately and rapidly; thus, there is little opportunity for vascular tissues to form in the lower portion of the spine.

When a spine is actively growing from the base, you can pull a spine away from the areole easily because the spine will separate from the plant through the basal spine meristem, where the cell walls are very thin. In contrast, a mature spine usually cannot be removed easily. Spine attachment is strengthened by two structural events. First, cells of the basal meristematic zone finally mature and form hard walls like those of the sclerified cells in the upper part of the spine. Second, each spine becomes fastened to the areole and to other spines on the

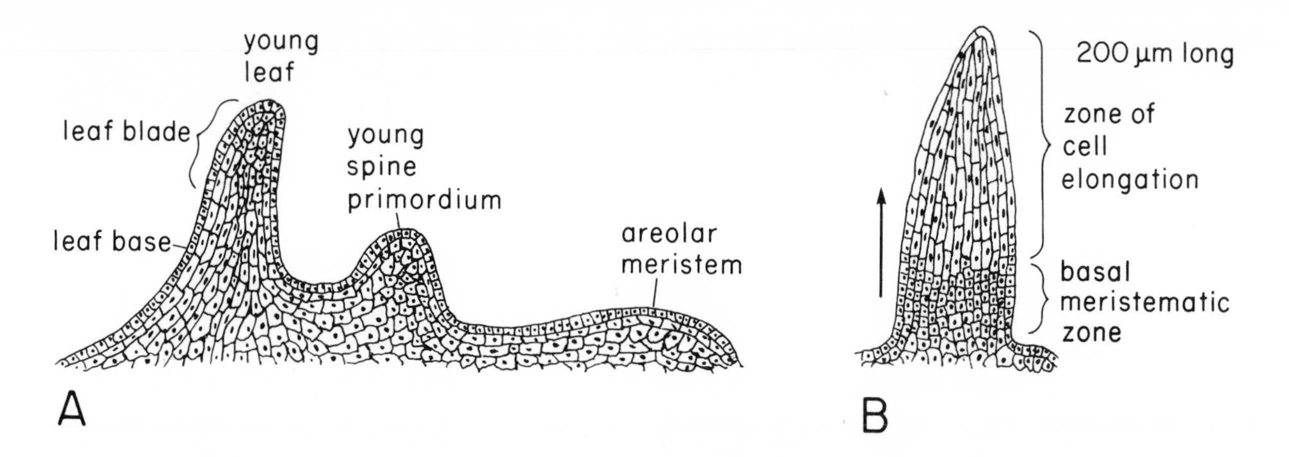

Figure 5.1. Origin and development of a spine on an areole. *(A)* Vertical, longitudinal section through a very young areole, showing the young leaf with a very short blade and an enlarged leaf base. The young spine primordium arises as a dome of cells to the inside of the leaf on the flanks of the areolar meristem. *(B)* Longitudinal section through a growing spine 200 μm long. The original cells of the spine tip have undergone elongation; the cells at the base of the spine primordium give rise to all future cells. The arrow indicates that the spine grows upward from the base.

areole by cork cells, which are produced by a phellogen (cork cambium) that arises in the outer cortical (hypodermal) region around the base of each spine. Because of this arrangement spines are often shed as clusters, either when a branch decays or when a fruit ripens. The bases of the individual spines become fused by cork formation between them. A spine cluster makes a more effective defense than separate spines; and the development of a corky covering on part of the areole may also decrease water loss.

Whereas typical spines are firmly fastened to the areole, the glochids of Opuntioideae (Figs. 5.3 and 5.4) are deciduous and easily dislodged when they are either touched by something or blown by a strong wind. Glochids are tiny spines that are formed as primordia in the same manner as persistent spines;

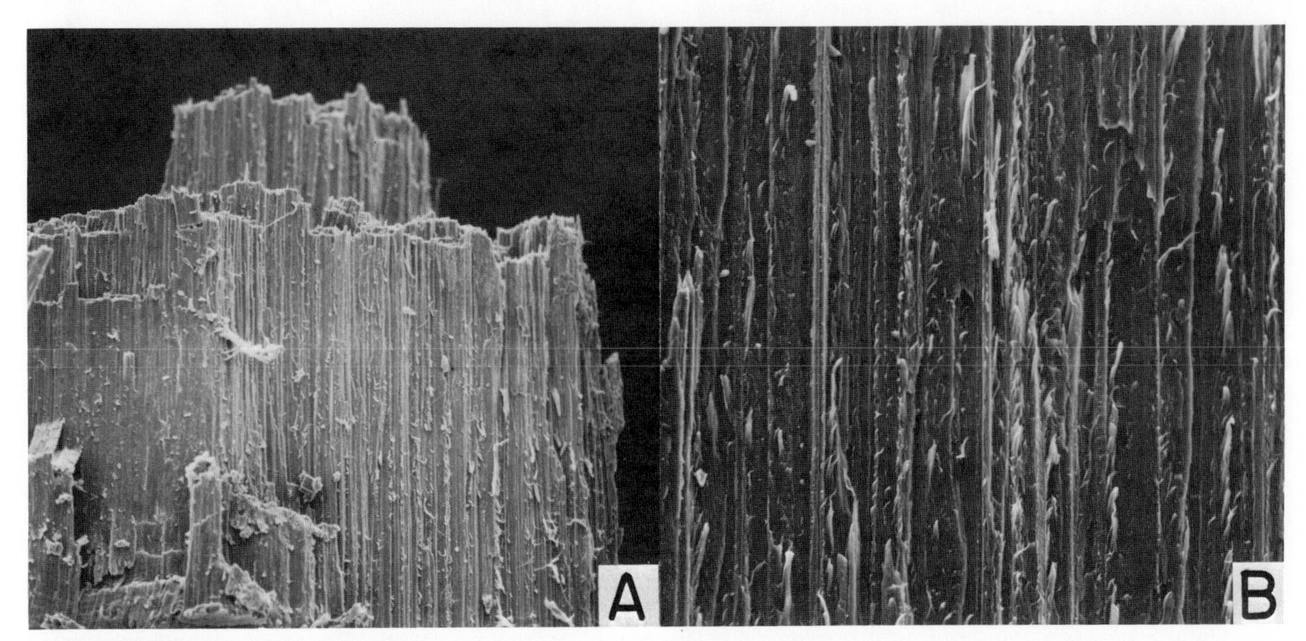

Figure 5.2. Spine structure of *Pereskia sacharosa,* as viewed with the scanning electron microscope.
(A) Low-magnification view of the interior of a spine, showing many long, narrow cells.
(B) Portion of the spine enlarged about 10 more times to show the elongate cells with their lignified cell walls.

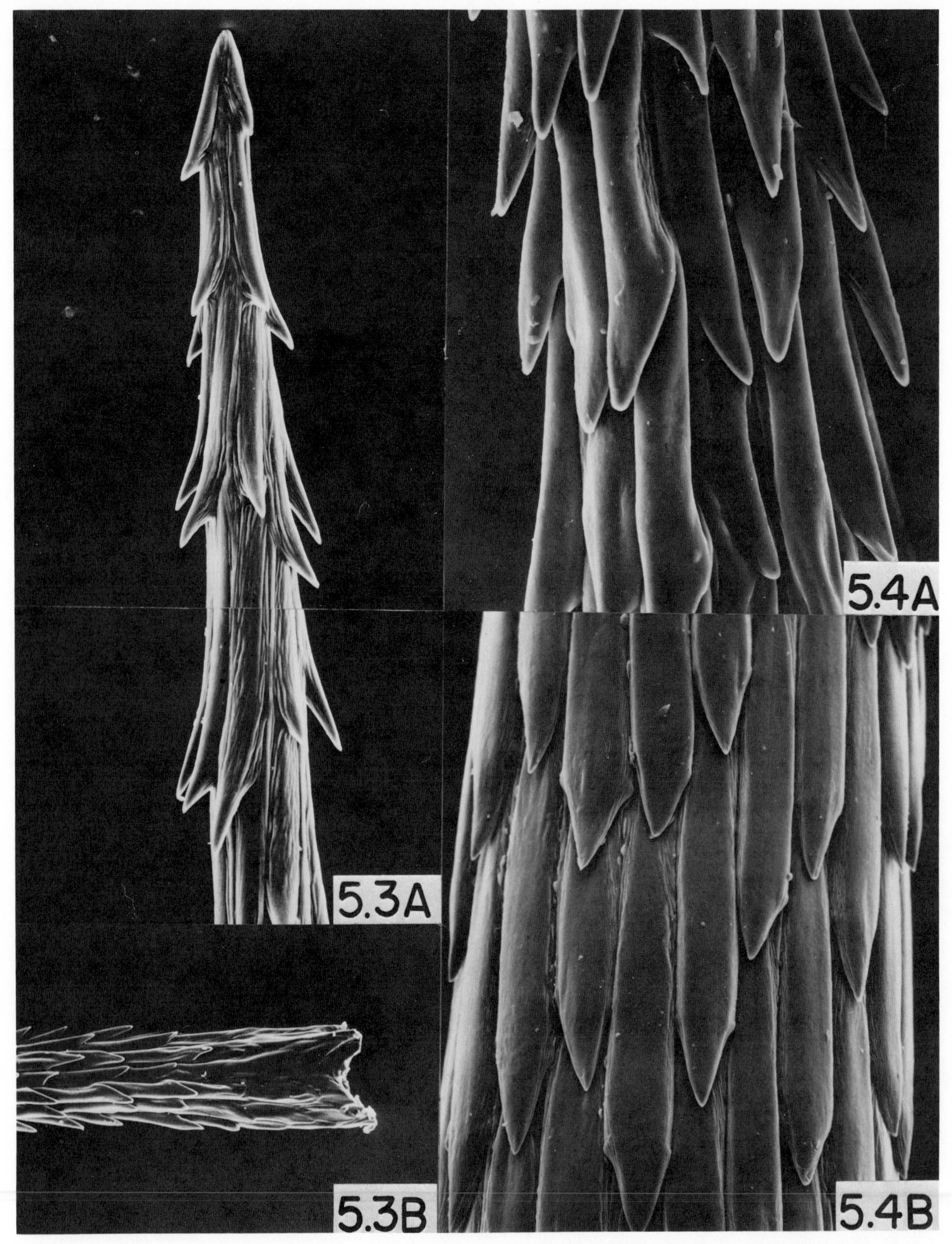

Figures 5.3 and 5.4. Scanning electron photomicrographs of spines of opuntias.
5.3. *Opuntia microdasys,* bunny ears, a Mexican platyopuntia. ×750.
(A) The tip of a glochid, showing the barbs (epidermal cells) that permit the spine to lodge in the skin.
(B) The base of a glochid, which is not strongly attached to the areole and so can easily be broken off.

5.4. *O. echinocarpa,* a cholla from southern California. ×510.
(A) The very prominent, downward-pointing barbed epidermal cells of a glochid.
(B) The surface of a spine, showing that the epidermal cells are more closely appressed to the spine and therefore do not act as barbs.

but they are shorter and thinner than persistent spines and the base of the glochid does not become sclerified. Therefore, a weak zone remains where the basal meristematic zone was once located.

Because a spine grows from a basal plate of cells, it can stop growing at any time. Understanding this simple fact helps us interpret some of the variability observed in the spine structure on a single plant. For example, on a single areole we often find one or more long, central spines and a number of shorter, lateral spines. The long, central spines merely grow for a longer period than the others; and they are thickest because they are produced by the most robust primordia that have the widest basal meristems. Spines produced on the areole of a seedling are characteristically shorter and much thinner than spines produced on the mature plant because spines of the seedling have fairly short growth periods and develop from very narrow primordia. Cacti can also show the opposite condition; for example, in *Rhipsalis baccifera,* the seedling has noticeable spines whereas the mature plant usually has none (Fig. 1.24). Actually, one and rarely two or three spine primordia can occur on the adult areoles of *R. baccifera,* but they stop growing when they are about 1 mm in length. The length of all spines is related in part to plant nutrition and climate, so it is not surprising to find that specimens of the same species — even cuttings of the same plant — can show different spine lengths when they are grown under different environmental conditions. Unfortunately, some cactus collectors of the past did not understand how environment can modify spine length, and this lack of knowledge led some of them to describe many

Figure 5.5. Variations in spination within a single species, *Opuntia violacea.*
(A) *O. violacea* var. *violacea,* having stiff spines on the upper half of the cladode.
(B) *O. violacea* var. *macrocentra,* having very long spines, mostly restricted to the upper margin of the cladode and only glochids on the lower areoles.
(C) *O. violacea* var. *gosseliniana,* having bristlelike (flexible) spines arising from all areoles on a cladode.
(D) *O. violacea* var. *santa-rita,* having only glochids arising from the areoles. Actually, each variety can have consideration variability in spination.

aberrant varieties and forms on the basis of spine properties of cultivated materials.

Variability in spine development is nicely illustrated by one highly variable species, *Opuntia violacea,* which is a widespread species of platyopuntia in the southwestern United States and adjacent Mexico. Four varieties are commonly recognized: *Opuntia violacea* var. *violacea, O. violacea* var. *macrocentra, O. violacea* var. *gosseliniana,* and *O. violacea* var. *santa-rita.* Variety *violacea* has spines that are mostly restricted to the upper half of the pad (Fig. 5.5*A*), and a typical areole has one long central spine up to 7 cm long, a shorter central spine, several very short lateral spines about 1 cm long, and several dozen sharp glochids. Variety *macrocentra* (Fig. 5.5*B*) lacks conspicuous persistent spines over most of the pad surface and instead has them concentrated on and just below the upper margin of the pad. In this variety the central spine may be up to 17 cm long, and all of the areoles bear numerous glochids. In variety *gosseliniana* (Fig. 5.5*C*), long, thin, persistent spines that bend easily are produced both on the upper and the flat surfaces of the pad; thus, an old plant is covered with long gray spines. In contrast, variety *santa-rita* (Fig. 5.5*D*) has either a few short spines along the upper margin of the pad or, more typically, no visible, persistent spines.

Arrangement and Initiation of Spines

Great variability in spine arrangement (Fig. 5.6) can be observed by viewing any cactus garden or by thumbing through any cactus book. Nobody has ever attempted to classify every variation in cactus spine clusters, but two basic types are generally recognized: (1) the radial areole and (2) the asymmetrical, or unilateral, areole. In the radial areole, the spines radiate from a central point and diverge either more

Figure 5.6. Arrays of spines in the genus *Gymnocalycium.* ***Top row:*** *G. cardenasianum (left),* 5 thick, curved, radial spines; *G. leeanum (middle),* 7 or more thin, straight, radial spines and 1 thin, straight, projecting central spine; *G. chiquitanum (right),* 7 radial spines and 1 central spine, which all project outward and are similar in curvature and thickness. ***Bottom row:*** *G. bruchii (left),* 12 or more very thin, short, radials; *G. denudatum (middle),* asymmetrical areole with 5 thin, bent radials that are closely appressed to the tubercle; and *G. stellatum* var. *paucispinum (right),* 3 relatively straight radials that point sideways (2) and downward.

Figures 5.7 and 5.8. Pectinate spine arrangement.
5.7. *Echinocereus reichenbachii,* one of several species of hedgehog cacti that have no central spines; all spines are of the same thickness and project to the sides like the teeth of a comb.

5.8. *Pelecyphora pseudopectinata,* a small globose cactus of the Chihuahuan Desert in Mexico, which has hatchet-shaped areoles with soft spines. Each spine is highly modified, having prominent hairs and little, if any, lignin in the cell walls.

or less evenly in all directions (hemispherical) or, when no centrals are present, like the spokes of a wheel (flattened) (Fig. 5.6). In the asymmetrical areole, the arrangement of spines is biased toward one side, commonly the lower or outer (abaxial) side.

Radial areoles are most highly developed in species with pectinate spine clusters (Figs. 1.7, 5.7, and 5.8); central spines are not present, so the cluster consists of a series of 20 or more fine, lateral spines that are evenly spaced around the areole. Pectinate areoles are generally elongate; and the spines form a nearly solid shield for the areole, thus concealing the stem surface near the top of the plant. Pectinate areoles are present only in highly advanced taxa.

An understanding of the origins of the different types of spine arrangements can be achieved by studying the patterns of spine initiation on the areoles. These studies involve two steps. Initially the plant is dissected at various ages, thus allowing the close observation and photography of the sequences of organ formation. Then the plant parts, such as young shoot tips at various ages, are killed, sectioned with a microtome, and stained so that the minute details of cell division and enlargement can be observed.

Following up on some observations by several European workers, Boke studied the sequence of spine initiation in selected species of cacti. In these studies, he found the following general pattern:

1. When the leaf blade (that portion of the leaf above the tubercle) is approximately 0.1 mm in length, the areolar meristem is already apparent. The areolar meristem develops at approximately the same site at which an axillary bud would develop, that is, near the inner or upper (adaxial) side of the leaf base, which faces the shoot apex (Fig. 5.9*A*).

2. Two spine primordia (sometimes only one) arise immediately next to and inside of the rudimentary leaf (Fig. 5.9*B*). This is termed the abaxial (lower) side of the areolar meristem. These primordia typically become lateral spines.

3. Several more spine primordia are formed on the abaxial side of the areolar meristem, and they arise to the inside of the older pair of spine primordia (Fig. 5.9*C*). The third spine primordium may be formed as a central spine, as shown in Figure 5.9*C*, or primordia may be initiated to the sides and become lateral spines.

4. Additional spine primordia may be formed, proceeding in a linear sequence from the oldest primordium situated next to the tubercle to the newest ones produced next to the areolar meristem (Fig. 5.9*D*). When more than one central spine is present, the primordium is initiated slightly out of sequence, that is, further away from the areolar meristem than the most recent pair of lateral spines. In most cases all spine primordia arise on the abaxial side of the areolar meristem; but in some instances several of the last-formed spines are initiated on the adaxial (upper) side and the meristem is encircled.

Figure 5.9*E* is the same areole of *Echinocactus horizonthalonius* that was just described, as viewed

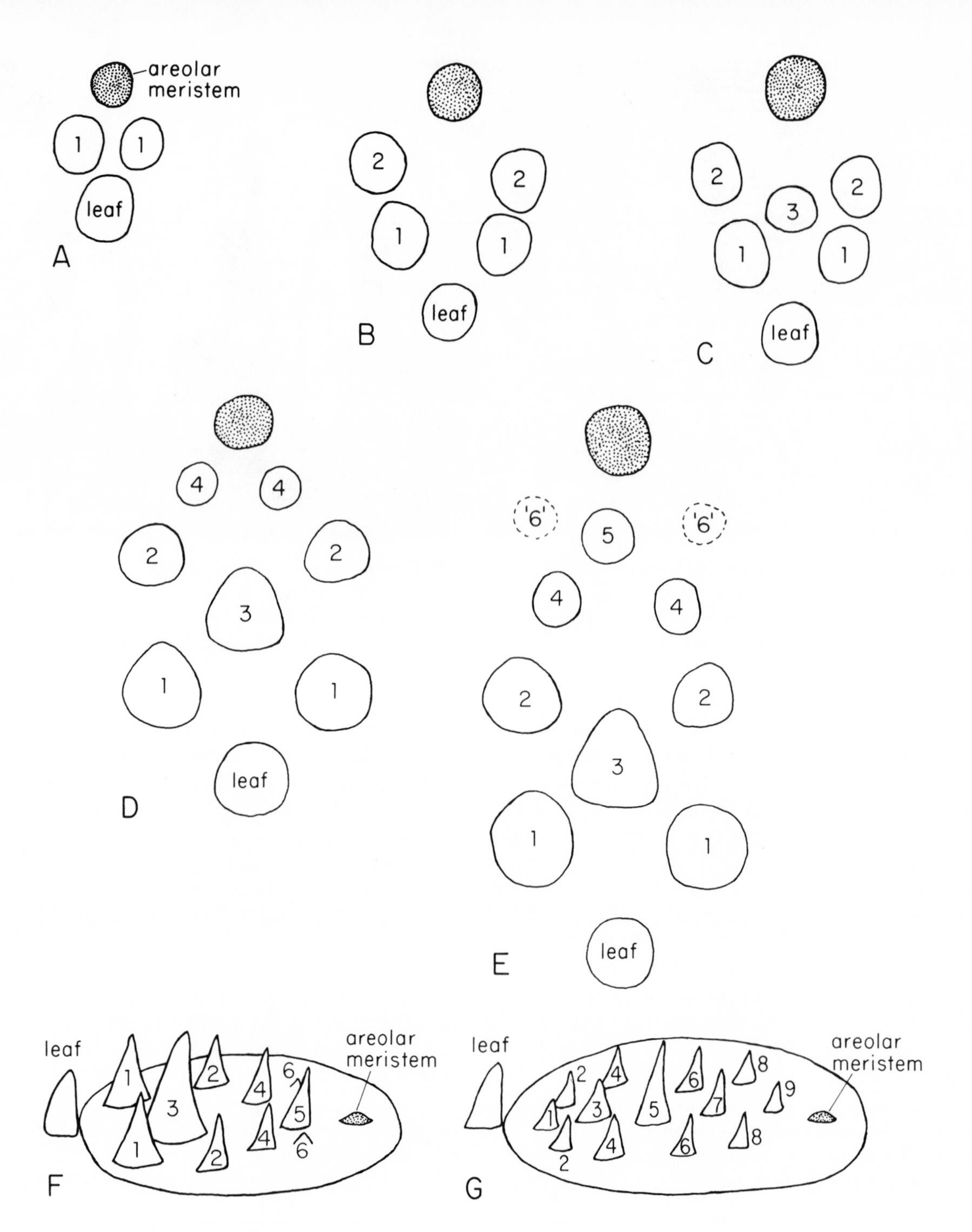

Figure 5.9. Sequence of spine initiation for the areole of *Echinocactus horizonthalonius (A–F)* and *E. grusonii (G).* Spines arise on the margin of the areolar meristem *(stippled)* on the side opposite the leaf. Numbers for the spines indicate the sequence of first to last (1 to 6 or 9); where two spines have the same number, they arose at the same time (see text). In successively older areoles (*A* through *E*), the leaf and areolar meristem are pushed apart by the addition of more spines between them. Spines that formed in a line between the leaf and the areolar meristem become central spines; the remainder develop as radial spines. In *E* and *F*, the sixth set of spines may not be present on some areoles. The final arrangement of young spines on areoles is shown in *F* and *G*.

from the side. For comparison we have also illustrated an areole of *E. grusonii* (Fig. 5.9*F*), which has only one lead spine primordium produced next to the leaf. In all general properties, the developmental sequences of both areoles are quite similar.

Several interesting variations in the sequence of spine initiation have been observed. For example, in *Pelecyphora aselliformis,* which has pectinate spine clusters (Fig. 1.7), the lateral spines appear to arise simultaneously rather than sequentially from bottom to top (acropetal, that is, toward the tip).

The sequence of spine initiation in *Pereskia* is particularly interesting because it may represent a primitive pattern. In *Pereskia grandifolia,* which is closely related to *P. sacharosa* (Chapter 2), the areolar meristem first produces the pair of spines next to the rudimentary leaf; but subsequently the spine primordia arise in a helical arrangement around the areolar meristem, an order that is reminiscent of the helical arrangement of leaves (see Chapter 6). Nonetheless, even on this areole the most robust spines, and, indeed, most of the spines, are produced on the abaxial side, a pattern showing that the primitive-looking cacti have a strong tendency toward forming spines only on the abaxial side of the areolar meristem. *Pereskia aculeata* is unusual—and some workers think it is primitive—because it has relatively few spines and because the first two spines are broad and curved backward (Fig. 2.1). These spines act as hooks for the climbing stems. Because these first two spines often form directly on the edge of the leaf, some workers have called them stipules, which are structures present at the base of leaves in some other dicotyledons.

Boke also learned that spines can mature in at least four different sequences once the primordia form on the areole. Species in all three subfamilies of cacti show that spine elongation can proceed in the same order as spine initiation—an acropetal sequence—thus, the first-formed primordia are the first ones to elongate. But three other maturation sequences have also been observed in the subfamily Cactoideae: (1) species may have spine elongation that proceeds in a direction the reverse of that for initiation (that is, from the tip to the base, or basipetal; youngest to oldest); (2) they may have simultaneous elongation of all spines—thus, elongation does not begin until all the spine primordia have been formed, as in species with pectinate spines; or (3) they may have central spines that elongate out of sequence, either before or after the lateral spines (the so-called mixed pattern). This information helps us to understand how the differences in the mature spine arrangements arise. Specifically, some primordia form as central spines and some do not; some spine primordia emerge and some do not; some primordia grow simultaneously and evenly and some grow in an unexpected order.

Variations in Other Spine Features

We have already discussed some of the factors that affect spine form, for example, length, arrangement, and persistence. Color is another variable. When spines are young, they may range in color from white or nearly translucent to yellow, golden, brown, pink, orange, red, gray, or nearly black. As spines age they characteristically fade or weather to duller shades. Individual spines also can be multicolored. Unfortunately, no one has investigated the factors determining color. The red colors are likely to be produced by pigments, such as betalains (Chapter 9); the red pigment in young spines is water soluble. Neither has anyone determined a function of spine coloration. We must therefore turn our attention to other structural features such as spine shape, texture, and cellular design for which there is more information.

In the same cactus garden where we have studied growth habit and spine arrangement, we can also walk along the path one more time and observe the different shapes of spines (Figs. 5.10 and 5.11). Certainly there will be specimens on which the central spine is strongly hooked (Figs. 5.10 and 5.12) or on which the spines are slender, ascending "fishhooks" (certain species of *Mammillaria* or *Parodia*). The central spines of most species of *Ferocactus* are very broad, are flattened on the upper surface, and are corrugated with transverse ridges. These spines are similar to the extremely rigid, daggerlike central spines of species like creeping devil, *Stenocereus eruca* (Fig. 1.25). In contrast to these rigid spines are the long, broad, flat and twisted, paperlike central spines, such as those in *Tephrocactus glomeratus* and *T. turpenii* (Fig. 5.11). Flat spines or those with a flat side are produced from a basal meristematic zone having the same shape. Unfortunately, no one has investigated the factors that stimulate a spine primordium to take on a different shape.

Another common spine type in cacti is referred to as a bristle. This term describes a spine that is long, pointed, fairly thin, and wiry, but not rigid (Figs. 5.5*C* and 5.13–5.15). Excellent examples of golden-colored bristles are found on the shoot tips or flowers and fruits of certain Mexican columnars, such as *Backebergia militaris, Pachycereus pecten-aboriginum* (Fig. 5.13), and *Mitrocereus fulviceps* (Fig. 5.14). Not too distantly related to these is a Mexican senita, *Lophocereus schottii* (Fig. 5.15), which has gray bristles on the flowering stems. Closer examination reveals that there is a transition from spines on the

5.10
5.11
5.12
5.13
5.14
5.15

Figures 5.10–5.15. Types of spines.
5.10. *Ferocactus* sp.; the typical cluster of spines in this genus includes a broad, strong, central spine with a prominent hook, several strong, straight, central spines, and numerous thinner radial spines.
5.11. *Tephrocactus turpenii;* the spines of the cultivated form are broad but thin; thus, they are flexible and become twisted into various shapes.
5.12. *Coryphantha clava;* a common type of areole with a strong central spine that is somewhat curved and numerous thin radial spines that tend to lie flat against the stem.
5.13. *Pachycereus pecten-aboriginum,* or hecho; the fruits of this species and several of its relatives are covered with long, flexible, golden bristles.
5.14. *Mitrocereus fulviceps;* the bat-pollinated flowers of this species are covered on the outside by thick, tawny pubescence through which emerge flexible, golden bristles.
5.15. *Lophocereus schottii,* or senita; long gray bristles form from the areoles on the flowering portion of the stem. The flowers of senita open at night and are pollinated by nocturnal moths; these flowers remain open the next morning. In this species several flowers are produced on an areole.

lower portions of the plant to bristles on the upper portions of the plant. Hair-spines, which occur on the South American senita, *Espostoa lanata,* are soft, very thin, twisted structures. Unlike regular spines, the hair-spines seem to lack lignin. The plumose spine, so called because it has very fine, featherlike lateral projections like the pinnae of a feather (Fig. 5.16), has trichomes that develop from the epidermal cells. Plumose spines, found on small, specialized cacti of tribe Cacteae, are merely highly derived forms of the regular spines found on the members of that tribe that has very short trichomes, for example, *Echinocactus* (Fig. 5.17) and *Ferocactus.*

One of the most curious spine features is the

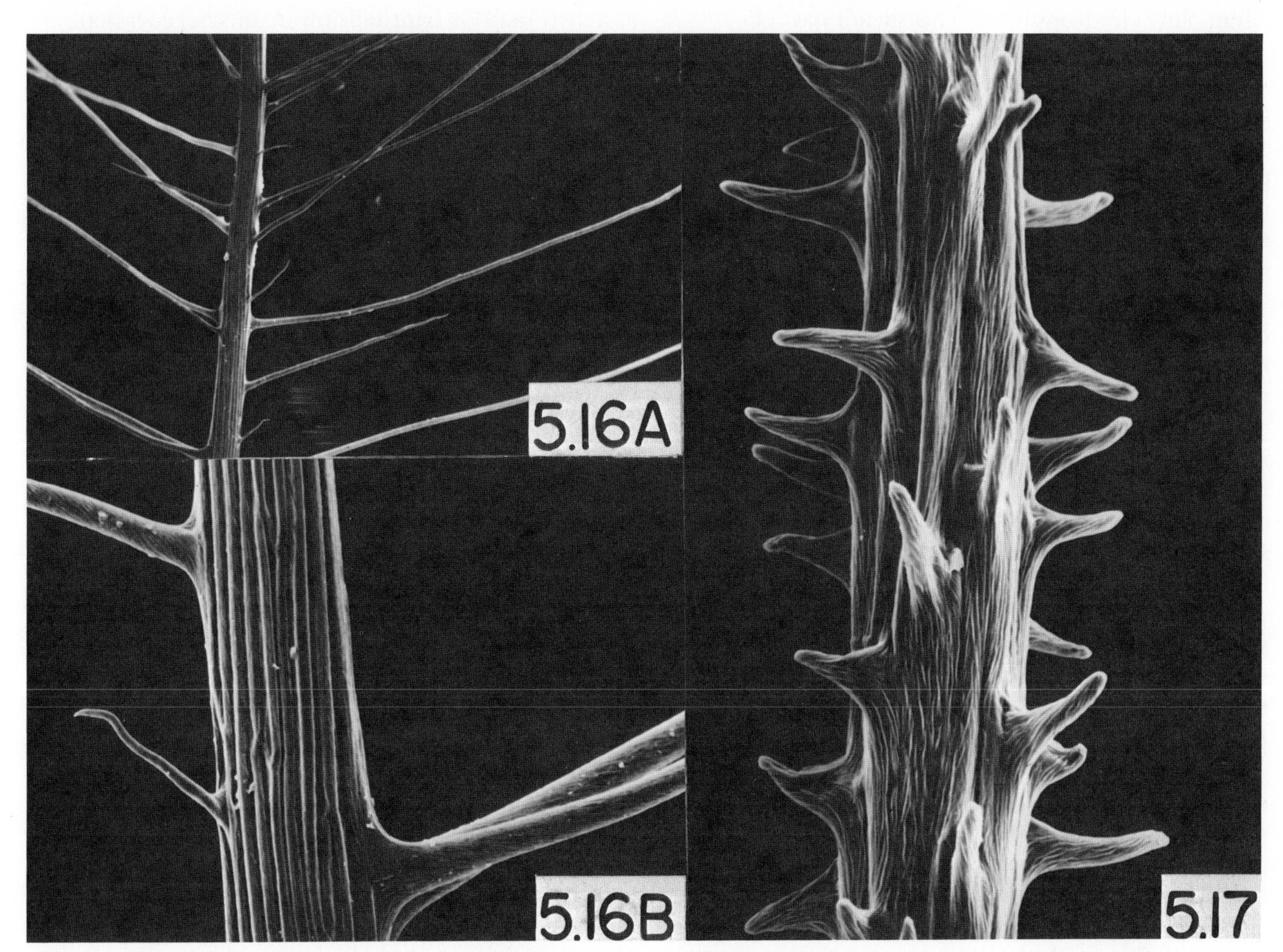

Figures 5.16 and 5.17. Scanning electron photomicrographs of spines in the tribe Cacteae.
5.16. *Mammillaria plumosa,* a species with very long hairs (epidermal projections) along the spine axis, as seen at very low *(A)* and higher *(B)* magnifications.
5.17. *Echinocactus grusonii,* or golden barrel cactus, showing the commonly observed type of epidermal projection on a species in this tribe (visible with a low-powered hand lens).

Figure 5.18. *Opuntia bigelovii,* or teddy bear cholla, from southern Arizona.
(A) A typical stand that has shed many of the terminal joints, from which new plants can grow.
(B) New plants have begun to develop by producing erect shoots from joints that have already formed adventitious roots.

yellowish sheath that covers the spines of the North American cylindropuntias. This sheath has the consistency of very thin paper. It fits loosely over the spine, either covering the spine to the base or, in a few species, covering only the upper portion of the spine. John Poindexter of Carlsbad, California published a partial description of sheath development in which he concluded that this sheath forms from the epidermis of the primordium, which separates from the central core during development.

Investigation of the cellular details of spines has been facilitated by the use of the scanning electron microscope, an instrument that enables us to see minute surface detail. With it scientists have uncovered a wealth of new information that may be important for determining the interrelationships of the cactus genera (Chapter 10) and also for providing clues on the functions of spines. Although spine surfaces of Pereskioideae and many species in the subfamily Cactoideae are smooth and unspecialized, some of the remaining species have unusual spines with interesting cellular characteristics. For example, some spines in the subfamily Cactoideae have epidermal hooks, such as the projecting upper cell ends in *Parodia.*

Both the regular spines and the glochids of the subfamily Opuntioideae have epidermal cells with pointed bases that project downward (Fig. 5.3). The projections act like hooks which make the spines barbed, or retrorse. The barbs enable the whole glochid to be pulled from the areole when it catches in skin or clothing. Glochids can also break along their length, thereby leaving their tips embedded in the skin. The mature spines are also barbed; thus, they too can catch and hold. These persistent spines can be useful for disseminating joints and fruits; the spines snag a vertebrate passerby and later the attached joint or fruit falls off in another location. Any visitor to the desert who has touched and then accidentally transported joints of cylindropuntias (for example, those of the teddy bear cholla, *Opuntia bigelovii;* Fig. 5.18) will attest to the excellent design of these spines for dissemination. Cattle that mistakenly eat chollas and platyopuntias can experience severe discomfort from the spines.

Scanning electron microscopy also shows that the surface cells of spines are not always sclerified in the subfamily Cactoideae; for example, species in *Pelecyphora, Turbinicarpus, Epithelantha,* and *Mammillaria* can have nonsclerified spine cells. Experiments were conducted by Ranier Schill and Wilhelm Barthlott of the University of Heidelberg to determine whether these species could absorb water through their spines. Schill and Barthlott showed that radioactively labeled water could be taken up by the spines and carried into the stem. Similarly, *Discocactus horstii* has a peculiar epidermal surface that may also be capable of permitting water to enter the stem. Actually, the uptake of water through spines was also studied by Harold Mooney and co-workers at Stanford University for *Copiapoa cinerea* var. *haseltonii* from an arid part of Chile. This species has finely grooved spines, and the water condenses on the spines and collects at their bases. Their short-term dye experiments showed that water did not enter the succulent tissues. Indeed, water absorption is not a function of most spines, because they are waterproofed by virtue of their cuticle and sclerified cells, because they lack xylem and thus cannot effectively transport water to the stem, and because the base of the spine is blocked by cork. The uptake of water through spines and areoles would seem to be an unwise strategy for most cacti, because if water can diffuse from outside to inside, then it also can diffuse back the other way. Thus, "leaky" areoles might permit more water loss than water gain and would be a useful "adaptation" only in special circumstances, for example, where dew condensation is the major or the sole source of water for the plant.

Experiments on the Axillary Meristem

To explore the regulation of plant development, investigators must devise experiments that enable them to determine which substances or physical factors affect plant form and how. One of the most useful techniques of the developmental biologist is tissue culture. An organ, a tissue, a single cell, or even a cell organelle such as a nucleus can be isolated very carefully from a plant and placed on a solid (agar) or liquid growth medium in a test tube, petri dish, or flask. By providing the correct nutrients and environmental conditions, these explants from an organism can be encouraged to grow. Then by varying the concentrations of chemicals in the medium or the physical conditions, an investigator can cause such explants to develop in a number of different ways and can identify those substances and physical factors that exert the greatest influence on mature form.

In a typical plant the axillary bud has the capability of becoming many different structures; and depending on environmental conditions and position, an axillary bud may become a vegetative shoot, a flower, an underground shoot, or a root. We want to know why the axillary bud of a cactus sometimes produces a leafy shoot, sometimes an areole with spines, and sometimes a root primordium. To get at this problem, experiments using tissue culture were conducted by James D. Mauseth and Walter Halperin at the University of Washington. These workers isolated undifferentiated axillary buds from young pads of *Opuntia polyacantha* and placed the buds into a sterile agar medium that supported cell division. In some cultures they also added sugar and representative compounds from the three major classes of plant hormones: gibberellic acid (gibberellin), auxin (naphthylacetic acid), and cytokinin (benzylaminopurine). These explants of axillary buds were examined 2 and 3 weeks after the experiment began for structures formed from the axillary meristems under various conditions.

Table 5.1 contains data from the tissue culture experiments on *O. polyacantha.* When a medium was

Table 5.1. Growth response of cultured buds of *Opuntia polyacantha* to plant hormones.[a]

A. Media with one hormone added

Concentration of hormone in culture medium (ppm)	Growth response		
	GA	BAP	NAA
0.0	0% spines	0% shoots	0% roots
0.1	80% spines	9% shoots	11% roots
1.0	81% spines	40% shoots	40% roots
10.0	81% spines	100% shoots	100% roots
50.0	98% spines	90% shoots	100% roots

B. Media with two or three hormones added

Hormones in culture medium (conc. in ppm)	Growth response	Comments
GA (50) + NAA (10)*	100% roots	Auxin inhibits the action of GA
GA (50)* + NAA (0.1)	100% spines, 10% roots	—
GA (50) + NAA (0.1) + BAP (10)*	100% shoots	—
BAP (10) + NAA (10)*	66% roots	Auxin inhibits the action of BAP
BAP (10) + NAA (10) + GA (10)*	50% spines	—
BAP (10)* + NAA (0.1)	100% shoots	—
GA (50) + BAP (10)*	100% shoots	—

Source: Mauseth and Halperin (1975).

a. GA, Gibberellic acid; BAP, cytokinin; NAA, auxin (naphthylacetic acid). Percentage values refer to percentage of cultures having that plant part. The asterisk indicates the hormone producing the major effect for each medium.

used on which cell division can occur but to which no hormones were added, no primordia were produced. But when hormones were added to a similar medium, the explants produce conspicuous primordia, a result showing that some plant hormone was necessary for active development to proceed. Cytokinin (BAP) used by itself, especially at a concentration of 10 ppm by weight, caused the axillary bud meristem to increase in size and then to produce a leafy shoot, just like a shoot that you would find at the start of the growing season of a pad. Gibberellic acid (GA) used by itself over a wide range of concentrations (0.01 to 100 ppm) induced the formation of areoles with spines, but the axillary meristem did not increase noticeably in size. In contrast, auxin (NAA) used by itself (5 to 100 ppm) induced the formation of adventitious root primordia, and the axillary bud meristem became inactive. Including 5 to 10% sucrose as an energy source in the agar caused the shoots and leaves in the cytokinin treatment to grow very rapidly and the spines to grow rapidly in the gibberellin treatment.

Intact plants normally have all three of these hormones along with sugars and many other organic substances. Therefore, tissue culture experiments can also help isolate the effects of a single factor and can indicate how the presence of one hormone is affected by the presence of the others. As we can see from Table 5.1, whenever any two or all three hormones were placed in the growth medium, an axillary meristem was induced to form roots, leafy shoots, or spiny areoles, but the product depended on which hormones were present and in what concentrations. When a growth medium contained a "normal" level (10 ppm) of auxin along with a normal concentration of either gibberellin (10 ppm) or cytokinin (50 ppm), roots were initiated and the axillary bud meristem became inactive. But when the auxin concentration was lowered to 0.1 ppm, the gibberellin treatment produced spiny areoles and the cytokinin treatment produced leafy shoots. From these observations Mauseth and Halperin concluded that auxin inhibits the action of gibberellins and cytokinins, which are expressed only when the auxin concentration is very low. When explants were grown on media with cytokinin and gibberellin, leafy shoots were produced, a result suggesting that cytokinin interferes with gibberellins and spine production. But when all three hormones were placed on agar at their most active concentrations, explants produce spiny areoles and adventitious roots. This result suggests that auxin and cytokinin neutralize each other's effects on gibberellin and thus allow gibberellin-mediated spine production.

On the basis of these results, four generalizations can be made: (1) the axillary bud meristem is initially uncommitted and develops primordia based on the messages that it receives; (2) this meristem will produce a root primordium when auxin is present in sufficiently high concentration; (3) a leafy shoot will form under the influence of cytokinin; and (4) spiny areoles will form in the presence of gibberellic acid. Therefore, we can expect to find an auxin-dominated system wherever adventitious roots form, as on the lower cut surface of a stem cutting or on the downward-facing side of shoots when they are propagated vegetatively. Likewise, new shoots will be produced where auxin concentration is low and where cytokinin level is high, as when a shoot is decapitated and the auxin source at the shoot tip has been removed. These are only speculations, of course, because no one has yet assayed hormone concentrations and hormone activity in cactus tissues. Nonetheless, these experiments help us to appreciate that the cactus areole — which is used to distinguish this family of plants — may have evolved in a fairly simple way, by adjustments in the hormonal systems, from a leafy shoot to a spine-bearing, condensed shoot. The major step in spine formation must certainly be related to the action of the gibberellins, which are known to stimulate cell divisions in intercalary meristems that are similar to the basal meristematic zone of cactus spines.

The Nature of Trichomes

Nearly all vascular plants have trichomes somewhere on the shoot. This may surprise someone who is not accustomed to looking for hairs on leaves, stems, flowers, or fruits; but they are common. In any case, trichomes are certainly not strangers to the cactologist. Indeed, every specific and generic description of cacti, if done properly, must describe trichomes. Trichomes are quite conspicuous on most vegetative areoles. At the shoot apex, where the areoles are very close together, the trichomes can form a solid mat so thick that they could be used as a cushion — if the spines were removed. When a cactus plant is cut in half lengthwise, the thickness of the hair becomes apparent; at least one scientific paper has reported apical pubescence greater than 4 cm thick in barrel cacti. The areoles of the pericarpel and floral tube may also bear hairs, so trichomes may also show up on the mature fruit. As mentioned earlier, the spine epidermis may have trichomes, and numerous species form surface trichomes on their skin (Chapter 6).

Typical trichomes (Fig. 5.19) are multicellular structures in which the cells are attached end to end to form a chain (uniseriate, or one-rowed). Each trichome develops from a single epidermal cell (Fig.

Figure 5.19. Scanning electron photomicrograph of trichomes on an areole of *Stenocereus alamosensis.* The dark bands on each trichome are cell walls. The ends of the trichomes are worn and broken because the cells are dead and abraded at the tip. ×150.

5.20). First, the outer wall of the epidermal cell bulges outward, and the nucleus and the other cell contents divide (Fig. 5.20*A*). Following this, a new cell wall is laid down between the two nuclei, and the projection becomes a cell (Fig. 5.20*B*). This initial cell is called the basal cell; it divides repeatedly to produce cells that are pushed outward (Fig. 5.20*C*–*E*) and eventually form a chain of very elongate cells. Growth of a hair may stop at any time; thus, hair length is developmentally variable. The terminal cells may die and collapse, but the basal cell can remain alive for weeks or months. Eventually all cells of the trichome die.

The best place to study trichome development is on the shoot tip—specifically, very close to the shoot apex. Trichomes begin to form on the areole while the first-formed spine primordia are being produced; and by the time all the spine primordia are present, the trichomes are also in place. They generally remain intact until the areole produces a flower, at which time the meristem dies; the surface tissues then dry out. Finally, trichomes become dry and brittle and are worn away by the physical elements.

A trichome originates in a different way than does a spine. As we have described, a trichome forms as a projection of a single, epidermal cell, whereas a spine arises from several subepidermal cells, which divide to form a dome of cells. Therefore, developmental information is required to settle arguments about the nature of a particular surface projection.

There are a few reported cases of so-called glands that are formed on the areole. Boke studied these structures in *Toumeya papyracantha* and concluded that they are primordia and are not trichomes. These broad-headed structures secrete a substance when the epidermis on the top of the gland disintegrates. Similar areolar glands occur in *Ancistrocactus, Thelocactus,* and *Coryphantha;* they may function as extrafloral nectaries, which attract ants away from the flowers.

The area with trichomes (pubescence) in Cactaceae varies in trichome frequency (number of trichomes per square millimeter), depth, and color. Many species have young trichomes that initially are white, tan, tawny, or brown and weather to gray. One group of columnar cacti in the genus *Stenocereus* (which includes organ pipe cactus, *S. thurberi,* and

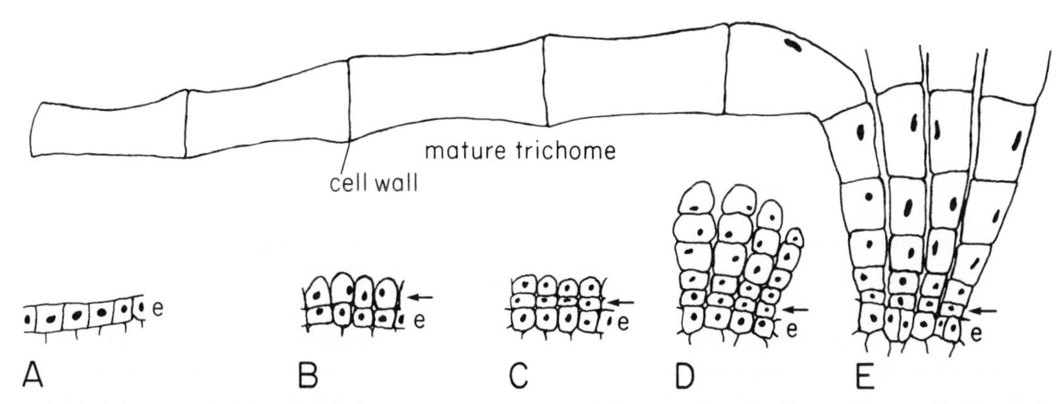

Figure 5.20. Development of trichomes (hairs) in cacti.
(A) Cells of the epidermis (e) before trichomes form.
(B) Each epidermal cell has divided to give rise to an outer cell, which will form the trichome *(arrow).*
(C) The trichome-forming cell *(arrow)* has divided and thereby produced a cell to the outside.
(D) The same basal cells continue to divide, each forming a chain of cells. The cells in the chain enlarge as they become older.
(E) The cells very distant from the basal dividing cell have become very large and elongated. The terminal cells have died and therefore lack nuclei.

the widely cultivated species *S. beneckei*) have red areolar trichomes that are described in the literature as "glandular" and attractive to ants. No one has isolated the chemicals for the pigment or for the glandular substance. When these red glandular trichomes die, they turn black.

Possible Functions of Spines and Trichomes

Ask any person on the street, "Why do cacti have spines?" and the response will be, "To protect them from being stepped on or eaten." One cannot deny that this is correct because it intuitively makes sense. The fact is that most vertebrates do not eat cacti. It is easy to document the serious injury inflicted by spines to domestic animals and to humans, animals that usually learn from their mistakes. Certainly, there are some interesting exceptions: the giant land tortoises of the Galápagos Archipelago feed on the tree platyopuntias, and some birds feed on cactus fruits and can excavate a home in a large succulent stem. Nonetheless, with some exceptions—mostly rodents—mammals do not eat cacti. However, if spines are removed from *Coryphantha vivipara* or *Mammillaria dioica,* then damage by rabbits and rodents can be seen in a matter of days. Spines are not effective defenses against small invertebrates, which can walk between the spines to reach the skin; defenses against small invertebrates must therefore be present within the skin (Chapters 9 and 10).

Biologists believe that spines provide a defense against herbivores and that herbivores are responsible for the evolution and perfection of spines. But there is no solid evidence to support the second assertion. First of all, nobody knows which herbivores to credit. The cactus family arose at least 50 million years ago—maybe more—apparently in the dry tropical areas of northern South America, and the first spines apparently arose on the shoots of *Pereskia*-like cacti that had no spines on the flowers or on the fruits (Chapter 2). The dominant and large mammalian herbivores that lived in South America at that time no longer live on earth: notoungulates, xenungulates, pyrotherians, litopterns, and astrapotheres. These groups were replaced over millions of years by different herbivores, some of which were very large. Looking at North America (from Panama northward) over the last 50 million years, we find an equally intriguing but different fauna of mammalian herbivores, including members of the families of camels, horses, rhinoceroses, deer, tapirs, peccaries, and pronghorn antelope. Rodents were present on both continents throughout much of the history of the cactus family. No one can really show which of this wide variety of potential herbivores, if any, were responsible for promoting spine development in the cacti. Certainly no direct evidence exists to demonstrate that animals helped to shape, through natural selection, the various spine arrangements and types mentioned earlier in this chapter.

Surely we can demonstrate that spines function some of the time for dissemination of shoots and fruits. Many of the common types of spines can catch hold of mammalian skin. When the fruit or joint is easily detached from the plant or is lying on the ground, the spines enable that part to be transported somewhere else; thus, they assist in the dispersal of cacti. Today we can show that this type of dissemination of joints and fruits is particularly common in the opuntias. In fact, a very widespread species of North America—the teddy bear cholla, *Opuntia bigelovii*—reproduces in nature largely by vegetative propagation (apomixis) by detached shoots and not by sexually produced seeds (Fig. 5.18). Nevertheless, most genera of cacti do not reproduce vegetatively in the wild, and they often do not have obvious adaptations for dissemination of shoots. Although spiny fruits characterize some genera, usually it is only the seeds and inner pulp that are scattered by birds, mammals, and ants, and not the entire fruit.

Recently investigators have begun to analyze how spines and trichomes affect the plant's microclimate (local climate) and thereby influence plant temperature, PAR interception, and water relations. These are topics that can be submitted to scientific testing and give us some alternative ideas for the possible functions of areolar features. To discuss such topics, we first need to develop a conceptual framework indicating how plant morphology interacts with environmental factors to influence plant temperature, especially temperatures of the chlorenchyma and the apical meristem. This activity leads us to a consideration of the energy budget of a cactus stem, a topic that we shall discuss now and again in Chapter 7 when we consider plant distributions.

Energy Budget Concepts

During the 1970s Donald A. Lewis and Park S. Nobel at UCLA put together an energy budget computer model to predict temperatures at different locations for the stem of the barrel cactus *Ferocactus acanthodes.* This model allowed a quantitative evaluation of the effects of ribbing (considered in Chapter 6), spine frequency, and particular daily patterns of water vapor conductance on surface temperatures and water loss. Later, Nobel extended the model and focused on the influences of spines and apical pubescence on the temperatures near the apical meristem; these features were found to be related to the latitudinal and elevational limits of cacti.

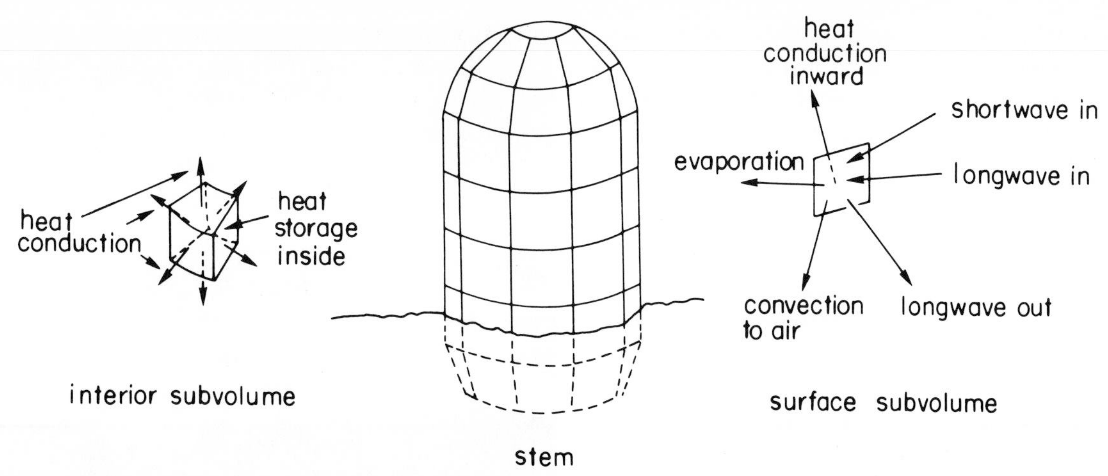

Figure 5.21. Schematic illustration of the division of a cactus stem into a number of subvolumes, each of which has a balanced energy budget (energy in − energy out = energy stored as heat). The various subvolumes interact to lead to the plant temperature at various locations. The specific energy budget terms for a thin surface subvolume and a massive interior subvolume are indicated.

The basic idea of the model was to divide the stem volume into 100 to 200 subvolumes (Fig. 5.21). The temperature was predicted for each of these simpler subvolumes, and then heat exchanges between the various subvolumes were allowed to proceed in order to predict the temperature at various locations of the entire stem. Certain of the subvolumes occurred on the plant surface and so were exposed to radiation and wind. Two types of radiation were considered: (1) shortwave radiation of sunlight, which includes the photosynthetically active radiation (PAR) discussed in Chapter 4 and other wavelengths such as ultraviolet and infrared; and (2) longwave or thermal radiation, which as its name implies occurs at wavelengths much longer than infrared. Although invisible to the human eye, longwave radiation is emitted by all bodies, including cactus stems, other vegetation, clouds, and even a clear sky. The amount of longwave radiation emitted varies tremendously with temperature, doubling as the temperature of an object is raised from −35°C to 10°C and doubling again by 64°C. Thus, because a clear sky can have an effective temperature of −35°C, much less longwave radiation is emitted by it than by an object at, say, 20°C.

We also need to consider the effect of the evaporation of water accompanying transpiration. Water evaporates mainly from the outer part of the chlorenchyma and the inner part of the guard cells and then diffuses out the stomatal pores, as discussed in Chapters 1 and 4. Such evaporation of water is a cooling process, which affects the temperature of the subvolumes near the plant surface. Although subvolumes toward the inside of the stem are not involved with radiation, wind, or water evaporation, two other energy terms must be considered for them (Fig. 5.21). Specifically, heat can be conducted from one subvolume to another subvolume, and heat also can be stored or released by changes in temperature, for example, an input of energy is required to raise the temperature of a part of the cactus stem.

Let us consider the effect of wind in a little more detail. Heat can be exchanged between the plant surface and the surrounding air—a process generally termed convection. Such convection occurs across the boundary layer of air adhering to the plant surface (Chapter 4). We have all experienced the effects attributed to the wind-chill factor—on a cold, windy day we feel colder than on a cold, calm day. This effect occurs because the increase in wind speed decreases the thickness of the air boundary layer surrounding our bodies (the boundary layer thickness is inversely proportional to the square root of the wind speed). Heat can then be more readily conducted from our warm bodies to the cold air beyond the boundary layer.

Larger objects tends to have thicker boundary layers (boundary layer thickness is approximately directly proportional to the square root of diameter for a smooth cylinder or sphere). In addition, a rough surface (as might be caused by ribs or spines) creates air flow patterns that also affect heat exchange.

These engineering relationships can be used to analyze the effect of spines and pubescence on stem temperature. But first, we shall use the model to do a particularly easy experiment that is very difficult to do without it. Specifically, we shall ask how transpiration affects the surface temperature of the stem. This is hard to do in the field because it might mean smearing petrolatum all over the surface of the stem to prevent any loss of water. Aside from the mess, petrolatum has reflectance properties different from those of the stem; thus, a cactus covered with petrolatum would have a response to radiation different from its response when in its natural state. The beauty of the energy budget model is that we can change one parameter at a time and let the computer tell us the consequences this change has for plant temperature.

When the water vapor conductance in the model is reduced from the values observed in the field to zero, then transpiration stops (recall Eqs. 1.1 and 4.2). Cessation of transpiration causes the average stem surface temperature to increase only about 0.7 °C during the winter and 0.3 °C during the summer; this result implies that transpiration does not lead to much cooling for the stem of *Ferocactus acanthodes.* The largest effect caused by eliminating transpiration occurs at night when stomates are open; at this time cessation of transpiration causes local surface temperatures on the stem to rise at most 1 °C or 2 °C. This rather small effect that transpirational cooling has on the surface temperatures of cacti reflects the relatively low transpiration rates of *F. acanthodes* (Fig. 4.10) and other cacti. Such low transpiration rates result primarily from the nocturnal opening of stomates when tissue temperatures are much lower than during the daytime. Thus, water loss is less, as we discussed in Chapters 1 and 4.

Stem Temperature

Armed with the computer thermal exchange model developed by Lewis and Nobel, we are now able to turn our attention to the effects that spines and pubescence have on cactus stem temperature. We can also conduct some simple manipulations on real plants to evaluate the results predicted by the computer model.

One of the species that has been used to study a cactus energy budget was the fishhook cactus of California—*Mammillaria dioica* (Fig. 5.22). This is a small cylindrical plant less than 20 cm in height, and it has a dense covering of thin, whitish, lateral spines (about 30 to 50 per areole). There are trichomes on the areoles, but there is no apical mat of pubescence. So dense are the spines that they shade 62% of the incoming solar irradiation from the stem surface on a side tubercle and about 85% at the top of the plant.

The dense covering of spines obviously reduces the interception of shortwave radiation by the stem surface. The longwave radiation received by the stem is also affected by the spines. In particular, the thin spines of *M. dioica* are very close to air temperature, because objects with such a small diameter have a very thin air boundary layer, which allows rapid heat convection to or from the air. Thus, most of the longwave radiation intercepted by the stem surface

Figure 5.22. *Mammillaria dioica,* a plant with female flowers, from a population in northern Baja California, Mexico; this particular plant has relatively few radial spines.

comes from the spines, which are at air temperature, instead of the sky, which can have an effective temperature far below that of the air. The main effect of removing the spines, therefore, is to increase the shortwave radiation reaching the stem surface during the daytime, thereby increasing the stem surface temperature, and to reduce the incoming longwave radiation at night, thereby decreasing the stem surface temperature (secondary effects are caused by reduced longwave interception during the day, changes in the air boundary layer next to the stem surface, and temperature influences on transpiration). Indeed, both the model and measurements on plants without spines (spines were removed by clippers) showed that greater daily extremes of stem surface temperature occurred for the spineless condition. For a spineless tubercle of *M. dioica* located at midheight on a stem 9.4 cm tall, the maximum surface temperature was 5 °C to 6 °C higher and the minimum was about 1 °C lower than the temperature of the untreated stem. The effects on the apex are even more dramatic; the maximum daytime summer temperatures were raised by over 7 °C and minimum nighttime temperatures were lowered by over 2 °C. Therefore, spines lower the risks of overheating the apex during a summer day and of freezing the apex during a winter night.

Two other species have also been analyzed in some detail: the barrel cactus, *Ferocactus acanthodes,* and saguaro, *Carnegiea gigantea.* Both species have a very thick mat of tannish pubescence covering the apex—in contrast to *M. dioica,* which has none. The mat of pubescence traps air and hence acts as an insulating layer of low heat conductance—rather like a knit cap on a skier. This layer of pubescence is absolutely crucial for predicting the tissue temperatures in the region of the apical meristem and, of course, is incorporated into the computer model. *Carnegiea gigantea* (Fig. 5.23) has less spine shading of the stem than does *F. acanthodes,* although the apical pubescence is fairly similar on these two plants. In simulations that removed the apical pubescence of *C. gigantea,* the daytime maximum temperature in springtime was raised 6°C and the nighttime minimum temperature was lowered 2°C. When the spines were removed instead of the pubescence, maximum and minimum temperatures were changed by only about 1°C, a result showing that in saguaro pubescence has the greater effect on extreme temperatures. For *F. acanthodes,* which has a greater spine frequency, removal of spines had a slightly greater effect on stem surface extremes than did the removal of apical pubescence.

The presentation of the data has been simplified here, but the reader can certainly see some interesting consequences of the results. Cacti that occur in regions where they risk being damaged by high and low temperatures, particularly temperatures that could damage the apical meristem, would be expected to have some form of apical covering such as dense spination or thick pubescence (Fig. 5.24) or both. Nobel tested this hypothesis by studying how the apical characteristics of columnar cacti of the Sonoran Desert relate to their northernmost distributions. In brief, the northern limits of the three most northern species, *C. gigantea, Stenocereus thurberi,* and *Lophocereus schottii* (34°56′N, 32°38′N, and 31°55′N, respectively) could be predicted by the computer model on the basis of apical pubescence and spine coverage. However, in addition to morphological and environmental influences on temperature, we must also consider the tissue tolerances of these cacti to extreme temperatures. We shall return to such considerations in Chapter 7, where temperature effects on cactus distributions will be discussed.

Figures 5.23 and 5.24. Apical pubescence.
5.23. *Carnegiea gigantea,* or saguaro, from Arizona; top view of a 30-year-old stem, the apex of which is hidden by a prominent covering of tawny pubescence.

5.24. *Copiapoa cinerea* var. *haseltonii,* from Chile; very thick, grayish pubescence covers the entire apical region of this barrel cactus.

Effects on PAR and Net CO_2 Uptake

Even though spines are useful for defense and, with pubescence, help to moderate temperature extremes of the stem, spines also carry some costs. Of course, one of the obvious costs is the energy or carbon used in making spines, which could have been used to make more stem or reproductive tissue. This structural cost is not trivial; for example, spines constitute 38% of the dry weight of the stem for a small *F. acanthodes* 4.6 cm in diameter and fully 8% for a stem 26 cm in diameter, both of which are substantial investments. Under the rules of natural selection, such expensive structures would be lost if they did not have significant benefits. But another cost is the effect that spines have on plant productivity. Because spines and pubescence absorb some ambient short-wave radiation and reflect some away, they decrease the PAR received by the stem surface. In Chapter 4 we showed that CO_2 uptake is best predicted by measuring the total amount of PAR received during the daytime for CAM plants; thus, you can see that a decrease in PAR could mean a decrease in net CO_2 uptake. This would occur only if PAR were limiting.

Is PAR ever limiting in a desert? Our intuition says no, but in fact the answer is yes. As indicated in Chapters 4 and 8, total daily PAR measured in a desert location such as the University of California Philip L. Boyd Deep Canyon Desert Research Center near Palm Desert, California on cloudless days averages 21 mol m^{-2} day^{-1} on the vertical sides of the stems over the course of a year. Examining two species that live in that habitat, *F. acanthodes* that can have 78% shading of the stem surface by spines and *Opuntia bigelovii* with 32% spine shading, we find that the stems would need 45 and 23 mol m^{-2} day^{-1}, respectively, to achieve 90% saturation of nocturnal acid accumulation and net CO_2 uptake. Hence, there is just about enough natural PAR on clear days for 90% saturation of the opuntia but less than half the amount needed for the barrel cactus. When the spines are removed from these two species, the 90% saturation values become 22 and 16 mol m^{-2} day^{-1}, respectively (Fig. 4.15), close to the average occurring naturally. The plant is therefore caught in a dilemma between productivity, which would be highest without spines and pubescence, versus the problems associated with temperature stress and herbivory. For instance, Nobel has shown that periodically removing the spines from *O. bigelovii* growing in an environmental chamber with a total daily PAR of 9 mol m^{-2} day^{-1} incident on the joints caused an increase in stem volume over 2.5 years that was 60% greater than that occurring for plants having the normal complement of spines. The presence of spines must therefore be regarded as an evolutionary compromise between the benefits of protection and the losses in net carbon uptake.

Lack of spines and trichomes would, of course, help to maximize PAR reception by the stem. Many of the taxa in which spines have been reduced or lost are the epiphytes, which grow in low light environments in woodlands and forests. There are also some spineless small cacti, such as peyote (Fig. 1.14) and the species of living rocks (Chapter 8) that live half buried in the soil with only the tips of the tubercles exposed. These small cacti tend to have many alkaloids (Chapter 9), which discourage mammals from eating them.

Other Possible Benefits of Spines and Trichomes

Over the years various authors have suggested other benefits that could be derived from spines and trichomes. Even though we shall mention them here, it is also necessary to indicate that the importance of these suggested benefits has not been critically examined using sophisticated models and experiments in the field.

Some authors have pointed out that the presence of spines (and trichomes) on a cactus can create a thicker boundary layer, thereby reducing transpiration. However, the effect is actually quite small, because the gas phase conductance is mainly determined by the stomates (Chapter 4). Also, a thicker boundary layer would slightly decrease uptake of carbon dioxide at night and thereby slightly decrease productivity. Movement of air and the behavior of wind on a spine-covered surface has not been carefully analyzed, and the orientation of spines may increase and not decrease air movement next to the stem, especially on the upwind side.

Spines and pubescence may also play a role in protecting the stem from damaging wavelengths of solar irradiation, such as ultraviolet radiation. For example, species growing above 4000 m in the Andes, such as *Tephrocactus floccosus* (Fig. 5.25) and *T. rauhii*, can have long, silvery white hairs that cover the entire plant. Whereas it can be shown that ultraviolet radiation is about 40% higher at these elevations than at sea level, there are no data indicating that these hairs in cacti protectively absorb or reflect ultraviolet radiation from the stem.

Spine function has to be judged on the basis of what spines do for the plant; but in addition to this, spines can play a major role in the structure of natural communities. Spiny cacti serve as a home for a variety of animals. For example, nesting birds and pack rats *(Neotoma)* use spiny cacti to construct a home that discourages predators from entering it. Spine clusters that have fallen beneath a plant

Figure 5.25. *Tephrocactus floccosus,* a caespitose cactus that looks like sheep. This specimen is growing at an elevation of 4600 m in the Andes near La Oroya, Peru. These plants grow in puna vegetation and are the tallest plants in this location even though the mound is less than 40 cm high. Plants are covered by long, silvery white hairs.

forming a spiny mat may also assist the seedling establishment of that species or others, either because the spines protect the seedlings from being eaten by herbivores or because they provide a favorable microhabitat for seedling growth, for example, higher soil moisture or more moderate air temperatures.

Areoles and Flowers

So far we have concentrated our attention on the vegetative aspects of areoles, but, of course, the areolar meristem is also capable of producing a flower. After the areolar meristem has produced spines, it can be stimulated to become broader and then develop directly into a sessile cactus flower. Once the areole becomes a flower, however, the meristem is used up, and no more primordia can be formed from that particular dome of cells; the areole is therefore retired. In fact, in many cacti cork forms beneath the fruits, either before or after the fruit is completely ripe; the cork seals the areole from the stem and prevents future water loss and infection. Consequently, if the meristem has produced a flower, then no new shoots can ever arise from that particular areole, unless there is another meristem (for example, an axillary bud of a spine) that is hidden in the areole.

After planting a cactus seed, great patience is required to wait for the first flower to appear. Hundreds or even thousands of vegetative areoles will be formed before one of them is stimulated to produce a flower. In the case of saguaro, *Carnegiea gigantea,* the first flower does not appear until the plant is over 40 years old and about 2 m tall. All the lower vegetative areoles can retain a dormant areolar meristem, which can be stimulated to develop into a new shoot if the top of the plant is injured. However, the "arms" (branches) of saguaros typically arise in the region where flowering has already occurred but from an areolar meristem that did not produce a flower. No experiments have apparently been conducted on any cactus to determine what factors cause a particular areole to produce either a flower or a shoot.

Position of the Flowers

Saguaros also can be used to illustrate a common mode of flowering that is called apical or subapical flowering. The flowers, one per areole, are produced like a ring near the stem tip. First, the flowers arise from areoles probably formed a year earlier and previously hidden beneath the thick mat of apical pubescence. Second, most areoles in the ring produce a flower, but there are a few unused areoles each year that can later in life be stimulated to form branches. Moreover, each year there are several solitary flowers that arise on lower portions of the stem from unused areoles of previous seasons. More flowers tend to form on the warmest side of the stem, which is the south-facing side in northern latitudes and the north-facing side in southern latitudes.

Apical or subapical flowering occurs in a large majority of cacti. For the subfamily Opuntioideae, in which the flowers are regularly produced near the tips of the joints and along the thin upper edge of the pad (platyopuntias), the most common exception to apical or subapical flowering is *Opuntia leptocaulis,* which is the widespread pencil cholla of North America and which bears most flowers from old areoles along the old stems. When the fruits are ripe, the stem is draped with many small, red fruits. In some species of the subfamily Cactoideae, such as some small cacti in *Rebutia* and *Lobivia* of western South America, flowers are most frequently produced from areoles on the lower half of the plant. Especially in the epiphytes one may observe that flowers are produced a considerable distance from the growing tip; and in some ribbed species, for example, *Stenocereus alamosensis* of the Sonoran Desert (Fig. 1.34) and species of *Borzicactus* (Fig. 1.35) and *Cleistocactus* from South America, which are pollinated by hummingbirds, flowers commonly occur both at the shoot tip and for a meter or more along the stem.

Specialized Types of Flower-Bearing Areoles

Table 5.2 lists the different types of flower production found in tribe Pachycereeae of North America. Notice from this table how some species have apical flowering, some have apical and lateral flowering,

Table 5.2. Positions of flowers in the columnar species of the North American tribe Pachycereeae.

Flower position	Examples[a]
Flowers apical to subapical, 90% of the flowers within 30 cm of the shoot tip; reproductive and vegetative areoles similar; one flower per areole, and areole used only once	*Carnegiea gigantea* (P) *Escontria chiotilla* (S) *Mitrocereus fulviceps* (P) *Polaskia chende* (S) *Stenocereus stellatus* (S) *Stenocereus treleasei* (S)
Flowers produced in a cephalium[b]; reproductive and vegetative areoles very dissimilar; one flower per areole, and areole used only once	*Backebergia militaris* (P)
Flowers apical, always clustered at the shoot tip; reproductive and vegetative areoles dissimilar, but not a true cephalium; one flower per areole, and areole used only once	*Cephalocereus totolapensis* (P)
Flowers apical in a pseudocephalium[c]; reproductive areoles with long, silvery white hairs, and areole used only once	*Cephalocereus chrysacanthus* (P) *Cephalocereus hoppenstedtii* (P) *Cephalocereus palmeri* (P) *Cephalocereus senilis* (P)
Flowers on upper stems, mostly within the upper 2 m of each shoot; reproductive and vegetative areoles similar; one flower per areole, and areole used only once	*Neobuxbaumia euphorbioides* (P) *Pachycereus grandis* (P) *Pachycereus pringlei* (P) *Stenocereus quevedonis* (S) *Stenocereus queretaroensis* (S) *Stenocereus thurberi* (S)
Flowers apical to lateral, mostly within the upper 2 m of each shoot; reproductive and vegetative areoles very dissimilar; two or more flowers per areole, and areoles produce flowers for many years	*Lophocereus* spp. (P)
Flowers on the upper half of the plant; reproductive and vegetative areoles similar; two or more flowers per areole, and areoles used many years	*Myrtillocactus* spp. (S) *Pachycereus marginatus* (P)
Flowers on most areoles of the plant; reproductive and vegetative areoles similar; one flower per areole, and areoles flower many years	*Neobuxbaumia mezcalaensis* (P) *Neobuxbaumia polylopha* (P) *Neobuxbaumia tetetzo* (P) *Stenocereus dumortieri* (S)
Flowers mostly subapical to lateral, but sometimes also apical; reproductive and vegetative areoles similar; one flower per areole, and areole used only once	*Stenocereus alamosensis* (S) *Stenocereus gummosus* (S)
Flowers subapical to lateral and never apical; reproductive and vegetative areoles similar; one flower per areole, and areole used only once	*Stenocereus eruca* (S)

a. Wherever appropriate, species from both subtribes, Pachycereinae (P) and Stenocereinae (S), have been used.

b. A cephalium is a dense, terminal reproductive portion of the shoot.

c. Flowers restricted to one side of the shoot.

Figures 5.26–5.31. Special types of flower clusters in cacti.
5.26. *Backebergia militaris,* or grenadier's cap, showing a cephalium consisting of numerous areoles with long, golden bristles.
5.27. *Cephalocereus* sp., a columnar cactus from Mexico that forms a lateral pseudocephalium by producing long, silvery white hairs on the areoles, where flowers are formed in vertical rows.
5.28. *Espostoa lanata,* a columnar cactus from Peru that also forms a lateral pseudocephalium. In this specimen, the fruits are still present.
5.29. *Discocactus horstii,* a small globular cactus from southern Brazil that forms a cephaliumlike mound covered by long, white hairs. The areolar trichomes are unlignified and may assist in water uptake.
5.30. *Melocactus* cf. *peruvianus,* growing on a hillside in western Peru and showing the typical spine-bearing areoles of this small barrel cactus.
5.31. *Melocactus matanzanus,* a cultivated specimen showing a dense cephalium covered by tightly packed areoles that bear dense pubescence and short, straight spines. Ripened fruit are present on the cephalium.

some have only lateral flowering, and some have flower clusters. One of the special features is a terminal cap (cephalium; Latin for head), found in *Backebergia militaris* (Fig. 5.26). For many years each upright stem has about eight ribs but does not produce flowers. When the stem is about 6 m tall, each shoot tip is converted into a dense head of helically arranged tubercles, some of which produce flowers, and the crowded areoles and flowers are protected by numerous long, golden bristles. This bristly cephalium grows from the tip, where new areoles and flowers are produced, and the oldest bristles blacken with age. An analogous condition appears in senita, *Lophocereus* (Fig. 5.15); but in senita ribs are present on the vegetative and reproductive portions of each stem and the bristles are gray.

A pseudocephalium (Latin for false head) occurs in several of the massive species of *Cephalocereus* in Mexico; in these species the flowers form on only one side of the shoot apex (Fig. 5.27), and each flowering areole is covered by long, silvery hairs. The pseudocephalium of North American species of *Cephalocereus* also occurs for several South American species in the genus *Espostoa* of the tribe Trichocereeae (Fig. 5.28). Consequently, on a plant that has flowered for many years, there is a long, vertical panel of hairs where the flowers were located. In contrast, *Pachycereus marginatus* and the species of *Lophocereus* and *Myrtillocactus* have flowers on young and old green stems, and these species produce more than one flower per areole. This is made possible because new buds are formed on the areole; hence, flowering areoles can produce several flowers per year for a number of years. Most remarkable is *Neobuxbaumia mezcalaensis,* which produces flowers at all levels on the stem. Only one flower is produced each year from an areole, but no one has determined whether this areole flowers year after year.

Cactophiles have the highest regard for species with unusual types of flower-bearing (floriferous) areoles. For example, *Discocactus* (Fig. 5.29) is a marvelous cactus with very thick apical pubescence that hides the bases of the flowers and fruits. Especially revered are true cephalia, which occur in species of *Melocactus* (Figs. 5.30 and 5.31); also, the cephalia of *Backebergia* are grafted on short, green, columnar stems so that they can be kept in a small greenhouse. *Melocactus,* often called Turk's cap, is very unusual; the bottom portion of the plant is a ribbed, hemispherical barrel cactus with heavy spines (Fig. 5.30), whereas the cephalium, which eventually can be longer than the barrel cactus, is a dense, narrow column of helically arranged areoles, often with reddish spines (Fig. 5.31).

In typical cacti the solitary flower forms from an areolar meristem that is positioned just above the spine cluster on the tubercle (Figs. 1.11 and 5.32–5.34). There are, however, some noteworthy exceptions, especially in the small North American cacti. One specialization is found in the hedgehog cactus *(Echinocereus),* on which the flower erupts through the skin above the areole instead of arising next to the spine cluster. A second specialization occurs in *Mammillaria* (Figs. 1.12 and 5.22), on which the flower is not produced next to the spine cluster but instead is produced at the base of the tubercle from a meristem that is completely separate from the areole. Developmental studies by Boke showed how this separation of the spine and floral meristems occurred: a portion of the meristematic region near the base of the leaf is elevated to form the spine-bearing areole and the remainder forms an axillary bud at the base of the tubercle and later grows into the flower.

Another case of dimorphism (in general, two forms of a structure in one species; in this case, two forms of meristem) occurs in *Epithelantha,* which has spine-bearing and flower-bearing areoles that are both located on the elevated tubercle. For this genus, Boke demonstrated that the division that produced the two meristems resulted when the spine primordia were just forming; and the floral meristem therefore has time to develop a double series of spines before it produces a solitary, sessile flower. Most difficult to understand are the "grooved" or "furrowed" tubercles of *Coryphantha* (including *Escobaria*). Coryphanthas have a set of spines positioned at the top of each tubercle; but a furrow develops from the areole toward the stem, and thus the flower is actually produced in the middle or at the lower end of the tubercle but not in the axil, as in *Mammillaria.* Boke noted that in tubercles of seedlings the furrow may be absent or very short and that the furrow is formed after the spine primordia have already been produced. At this time there is some growth of cells between the areolar meristem and the closest spine primordium, growth that causes the two to become separated. The more the cellular division in the tubercle, the greater the separation between the areolar meristem and the tip of the tubercle.

A very unusual and specialized flowering structure occurs in *Neoraimondia* of Peru. This genus has perennial short shoots (Fig. 5.35) that lack internodes and on which the flowers are produced year after year. No scientific study has been conducted on the growth and development of these special structures.

Although the variations in flowering areoles are well known, their function in the evolution of cacti remains unknown. Is a cephalium an adaptation for protecting flowers and fruits from foraging animals, for producing many flowers close together to ensure pollination, or for providing a favorable thermal budget for floral development; and is the cephalium "parasitic" on the lower green stem? Are there any

Figures 5.32–5.34. Position of flowers on the areole.
5.32. *Stenocereus eruca,* or creeping devil. The flower bud has arisen from the upper side of the areole for stems that lay on the ground.

5.33. *Escontria chiotilla,* a columnar cactus from southern Mexico. The flowers and buds are protected by some translucent, stiff bracts on the floral tube.
5.34. *Opuntia phaeacantha* var. *discata,* or Engelmann prickly pear. Flower buds have arisen on the thin edge of the cladode.

Figure 5.35. *Neoraimondia arequipensis* var. *roseiflora,* a columnar cactus from western Peru. Fruits borne on long spur shoots produce flowers every year. (From Gibson, 1978.)

advantages in producing flowers near the base of the tubercles rather than at the tips?

SELECTED BIBLIOGRAPHY

Boke, N. H. 1944. Histogenesis of the leaf and areole in *Opuntia cylindrica. American Journal of Botany* 31:299–316.

——— 1951. Histogenesis of the vegetative shoot in *Echinocereus. American Journal of Botany* 38:23–38.

——— 1952. Leaf and areole development in *Coryphantha. American Journal of Botany* 39:134–145.

——— 1954. Organogenesis of the vegetative shoot in *Pereskia. American Journal of Botany* 41:619–637.

——— 1955a. Development of the vegetative shoot in *Rhipsalis cassytha. American Journal of Botany* 42:1–10.

——— 1955b. Dimorphic areoles of *Epithelantha. American Journal of Botany* 42:725–733.

——— 1956. Developmental anatomy and the validity of the genus *Bartschella. American Journal of Botany* 43:819–827.

——— 1957a. Comparative histogenesis of the areoles in *Homalocephala* and *Echinocactus. American Journal of Botany* 44:368–380.

——— 1957b. Structure and development of the shoot in *Toumeya. American Journal of Botany* 44:888–896.

——— 1958. Areole histogenesis in *Mammillaria lasiacantha. American Journal of Botany* 45:473–479.

——— 1960. Anatomy and development in *Solisia. American Journal of Botany* 47:59–65.

——— 1961. Areole dimorphism in *Coryphantha. American Journal of Botany* 48:593–603.

——— 1967. The spiniferous areole of *Mammillaria herrerae. Phytomorphology* 17:141–147.

Buxbaum, F. 1950. *Morphology of cacti.* Section I. *Roots and stems.* Abbey Garden Press, Pasadena.

——— 1953. *Morphology of cacti.* Section II. *Flowers.* Abbey Garden Press, Pasadena.

Freeman, T. P. 1970. The developmental anatomy of *Opuntia basilaris.* II. Apical meristem, leaves, areoles, glochids. *American Journal of Botany* 57:616–622.

Gibson, A. C. 1978. Dimorphism of secondary xylem in two species of cacti. *Flora* 167:403–408.

Hemenway, A. F., and M. J. Allen. 1936. A study of the pubescence of cacti. *American Journal of Botany* 23:139–144.

Lewis, D. A., and P. S. Nobel. 1977. Thermal energy exchange model and water loss of a barrel cactus, *Ferocactus acanthodes. Plant Physiology* 60:609–616.

Mauseth, J. D. 1977. Cytokinin- and gibberellic acid-induced effects on the determination and morphogenesis of leaf primordia in *Opuntia polyacantha* (Cactaceae). *American Journal of Botany* 64:337–346.

——— 1983. Introduction to cactus anatomy. Part 6. Areoles and spines. *Cactus and Succulent Journal* (Los Angeles) 55:272–276.

——— 1984. Effect of growth rate, morphogenic activity, and phylogeny on shoot apical ultrastructure in *Opuntia polyacantha. American Journal of Botany* 71:1283–1292.

Mauseth, J. D., and W. Halperin. 1975. Hormonal control of organogenesis in *Opuntia polyacantha. American Journal of Botany* 62:869–877.

Mooney, H. A., P. J. Weissler, and S. L. Gulmon. 1977. Environmental adaptations of the Atacaman Desert cactus *Copiapoa haseltoniana. Flora* 166:117–124.

Nobel, P. S. 1978. Surface temperatures of cacti—influences of environmental and morphological factors. *Ecology* 59:986–996.

——— 1980a. Morphology, surface temperatures, and northern limits of columnar cacti in the Sonoran Desert. *Ecology* 61:1–7.

——— 1980b. Influences of minimum stem temperatures on ranges of cacti in southwestern United States and central Chile. *Oecologia* 47:10–15.

——— 1983a. Spine influences on PAR interception, stem temperature, and nocturnal acid accumulation by cacti. *Plant, Cell and Environment* 6:153–159.

——— 1983b. *Biophysical plant physiology and ecology.* W. H. Freeman, San Francisco, New York.

Poindexter, J. 1951. The cactus spine and related structures. *Desert Plant Life* 24:7–14.

Robinson, H. 1974. Scanning electron microscope studies of the spines and glochids of the Opuntioideae (Cactaceae). *American Journal of Botany* 61:278–283.

Ross, R. 1981. Leaf and spine initiation in *Echinocereus reichenbachii* and *pectinatus* complexes. *Cactus and Succulent Journal* (Los Angeles) 53:255–258.

Schill, R., and W. Barthlott. 1973. Kakteendornen als wasserabsorbierende Organe. *Naturwissenschaften* 60:202–203.

Schill, R., W. Barthlott, and N. Ehler. 1973. Cactus spines under the electron scanning microscope. *Cactus and Succulent Journal* (Los Angeles) 45:175–185.

6

Tubercles and Ribs

An entire book could be devoted to describing and illustrating the tubercles and ribs of cacti. One thing that really fascinates us is the geometry of cactus designs: the orderly positioning of separate tubercles into graceful helices, the arrangement of tubercles into vertically oriented ribs or ribs that wind slowly around the plant, the shapes and outlines of tubercles and ribs, and the formation of flattened, leaflike cladodes. We are similarly fascinated by the geometric forms of mollusk shells along the seashore—turrets, cones, augers, miters, tops, helmets, turbans, and whelks. Interestingly, the helical gyres of gastropod shells and the arrangement of cactus tubercles can be described by a mathematical relation first described by an Italian named Fibonacci in the early thirteenth century. Fibonacci never saw cacti, which are essentially restricted to the New World, so he could never ponder why they have tubercles and ribs and how they are produced.

Phyllotaxy and the Fibonacci Sequence

Tubercle Arrangement

Who was Fibonacci? Leonardo Fibonacci (c. 1170 to c. 1240), also known as Leonardo of Pisa, was the most distinguished mathematician of the Middle Ages. He was the person who made the Arabic numeral systems that Europe accepted. But biologists remember Fibonacci because he discovered a sequence of numbers, the Fibonacci sequence or the Fibonacci summation series, that has broad applicability to the description of helical structures. The sequence is as follows: 1, 1, 2, 3, 5, 8, 13, 21, 34, 55, ... Close examination of this set of numbers reveals several patterns. First, when you add two adjacent numbers, you obtain the next larger number in the sequence. Another thing you can do with this sequence is to make fractions from the numbers by using one of them as a numerator, say, 3, then skipping one and using the next, 8, as the denominator, that is, $\frac{3}{8}$. In all but the first two cases ($\frac{1}{2}$ and $\frac{1}{3}$), these fractions are approximately equal to each other: $\frac{2}{5} \cong \frac{3}{8} \cong \frac{5}{13} \cong \frac{8}{21} \cong \frac{13}{34} \cong 0.382$. Finally, when you multiply 0.382 times 360 degrees (the number of degrees in a circle), you obtain the Fibonacci angle, approximately 137.5°.

To understand the geometry of cactus tubercles, the reader must understand how the Fibonacci sequence relates to leaf production. As described in Chapters 1 and 2, leaf arrangement or leaf position (phyllotaxy) in cacti and in most land plants is helically alternate.* This means that the shoot apex produces only one leaf primordium at a time and produces them along a helically ascending curve. Let us follow the events of leaf initiation. First, a growing apex, which already has several primordia, produces a leaf primordium at time zero (Fig. 6.1*A*), and this primordium begins to elongate. The next day or so a new leaf primordium arises; but because the shoot tip has meanwhile experienced some elongation, the first leaf primordium is no longer as close as it was to the apex, and therefore the second primordium is higher than the first. This primordium is located 137.5°—the Fibonacci angle—around the stem (in our example, to the right) from the first one (Fig. 6.1*B*). The next day another primordium is formed above the second, again 137.5° to the right (Fig. 6.1*C*), and this pattern continues until the shoot apex stops producing leaf primordia (Fig. 6.1*D*). Now, if we draw a curve

* Some people describe leaf arrangement as spiral, although spiral is sometimes restricted to two-dimensional structures whereas a helix is clearly three dimensional. Thus, the term *double helix* is used to describe the arrangement of the genetic information of a cell, DNA.

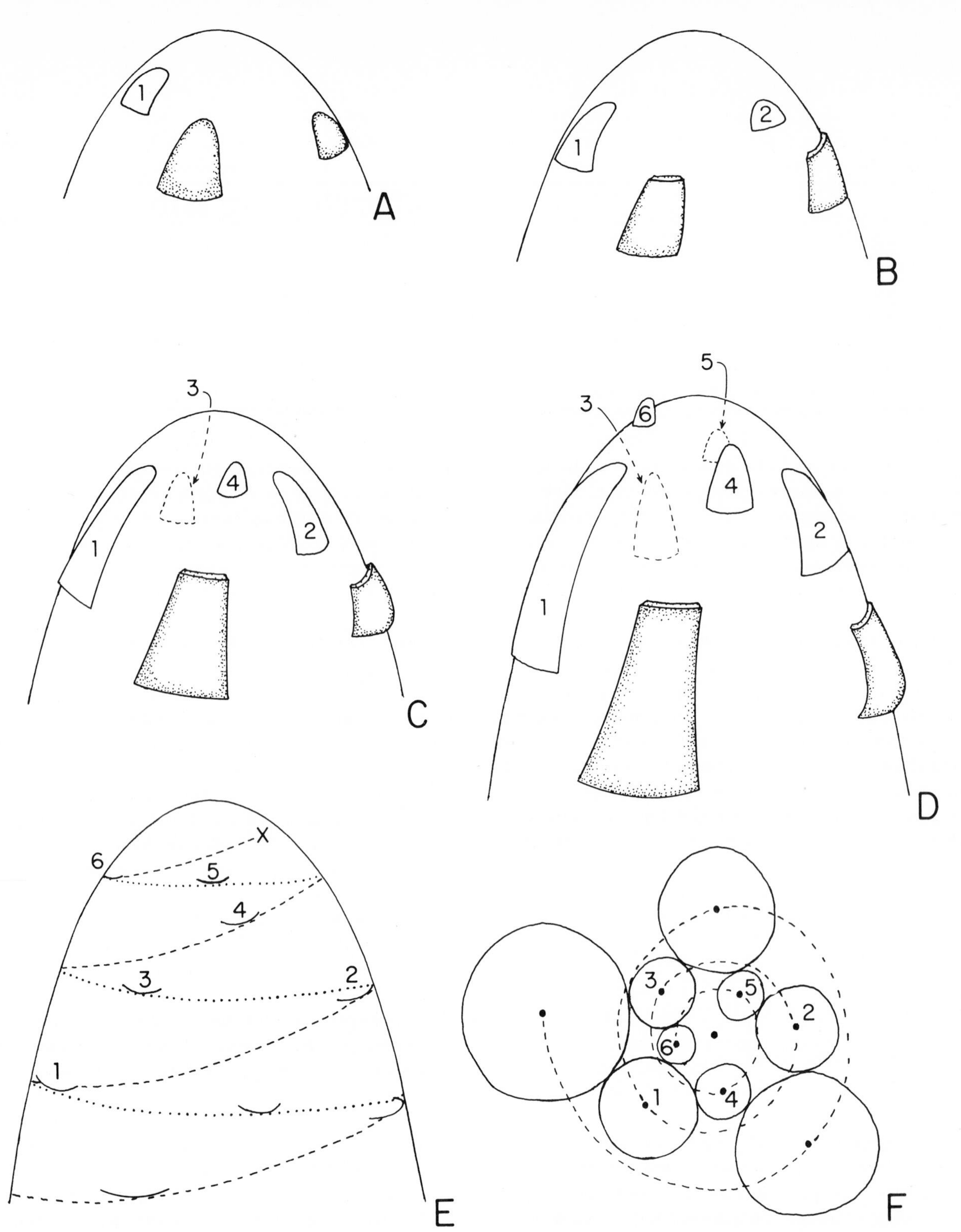

Figure 6.1. Model of the formation of leaf primordia on a shoot tip.
(A) A shoot tip with two recently formed leaf primordia (stippled) and a new one (1) that just arose on the margin of the apical dome.
(B) Leaf primordium 1 has elongated and moved away from the apical dome while the next primordium (2) has arisen 137.5° to the right. Stippled leaves have been cut so they do not obscure our view of the new primordia.
(C) Leaf primordia 3 (backside) and 4 have formed while the others have continued to elongate.
(D) Leaf primordia 5 (backside) and 6 have formed. Primordium 6 has formed above and slightly to the left of primordium 1 because the total number of degrees traversed in five leaf primordia was less than 720°.
(E) Same shoot tip as Figure 6.2*D* with a spiral drawn through the leaf bases in the order of their formation (parastichy).
(F) Top view of Figure 6.2 with leaf primordia drawn as circles and a spiral (parastichy) drawn to show the succession of leaf formation.

through the bases of the primordia in the order that they were produced, we will obtain a helix (Fig. 6.1*E*). Workers call this helix a parastichy or a genetic spiral. From a top view, this same pattern looks like a spiral or flattened helix (Fig. 6.1*F*). We have drawn this shoot with a helix that turns to the right toward the apex; but in any population of a plant species, about half of the individuals will possess a left-handed helix. Plants can be either left-handed or right-handed.

Looking down on the apex of a small cactus, preferably one that lacks spines and pubescence so that we can see the tubercles (Fig. 6.2*A* and *B*), we can observe that the tubercles in view are positioned every 137.5° around the stem. Some researchers have proposed that this angle is either the most efficient way to pack structures in a helix with the minimum amount of interference or it is the optimal way to position leaves around a stem to receive the most solar radiation from directly above the plant. Although these conjectures are interesting, we shall focus on the consequences, because this helps to describe the designs of tubercle arrangement in groups such as cylindropuntias, *Mammillaria, Coryphantha, Parodia,* and *Rebutia.*

The number of young tubercles that can be observed at the top of a cactus depends on the size and shape of the shoot tip. When the tip is long and narrow, the older primordia quickly leave the shoot tip and become lateral organs on the stem. Moreover, new primordia are produced in such a way that they hide the bases of the older tubercles. For example, when the narrow apex forms primordia, the sixth primordium (5 × 137.5° or 687.5° from time zero) will be situated above and a little to one side of the first primordium (720° equals two full turns around the stem); and 687.5° or nearly two full turns later, the eleventh primordium will occur above and slightly to one side of the sixth primordium. Likewise the twelfth primordium is approximately above seven and two, thirteenth above eight and three, fourteenth above nine and four, and fifteenth above ten and five. Thus, there are five, almost vertical rows of primordia, which are called orthostichies. The phyllotaxy is said to be 2/5 because for every two full turns around the stem there are five primordia formed, and there are five orthostichies.

On apices that are broader and flatter, more young primordia can fit in a ring around the apex. The Fibonacci angle is, of course, the same, but the circumference is greater. These apices (Fig. 6.2*C*) can have 3/8 phyllotaxy with 8 orthostichies (8 primordia for every 3 turns), 5/13 phyllotaxy with 13 orthostichies, 8/21 phyllotaxy with 21 orthostichies, and rarely 13/34 or 21/55 phyllotaxy with 21 or 34 orthostichies, respectively. In cacti, as well as many other plants, the number of orthostichies can change over time as they pass through a series of phyllotactic changes from seedling to adult as the primordium increases in size. A seedling and a young, slender plant may have a 2/5 phyllotaxy that changes to 3/8 phyllotaxy and later to the higher numbers of orthostichies.

Looking down on an apex with an 8/21 clockwise phyllotaxy and 21 steep, counterclockwise orthostichies (Fig. 6.3*A*), we can also see a set of less steep helices that wind clockwise. These are termed contact parastichies (Fig. 6.3*B*). In this plant, a contact parastichy is made by connecting tubercles 1, 14, 27, 40, . . . or 2, 15, 28, 41, . . . and so forth—every thirteenth primordia. In a plant with 3/8, it would be every fifth; and for 2/5, every third. This figure reemphasizes the geometric design of these shoots and the relation of phyllotaxy to the Fibonacci sequence. One set of contact parastichies has been shown for the shoot of *Borzicactus madisoniorum,* which has a 3/8 phyllotaxy (Fig. 6.2*D*).

These geometric patterns are easy to see in cacti because in most cactus species the internodes are very short or entirely lacking. Other dicotyledons experience growth (elongation) in the uninterrupted region between successive leaf primordia (internode), as do the pereskias, which have a 2/5 phyllotaxy. Therefore, on the mature shoot, there is a substantial amount of cylindrical stem separating the two leaves located at the nodes. Some cylindrical cacti with 2/5 phyllotaxy also form long internodes; but generally the number of orthostichies is high, the tubercle number high, and internodal growth suppressed. Thus, the tubercles are closely superimposed, an arrangement making the orthostichies and contact parastichies much more apparent.

Rib Geometry

The cactus plant already has a built-in way to make ribs—the orthostichy. By causing the tubercles in an orthostichy to coalesce or to become elevated as a unit, a cactus could produce 5, 8, 13, 21, 34, or 55 ribs. But life as a cactus is not that simple, because, as most cactus books show, these plants can have 2 to over 40 ribs and every number in between. For example, in *Lophocereus schottii* of the Sonoran Desert rib number ranges from 4 to 14; and a single plant can have a variety of rib numbers on different stems.

To determine how rib number is related to the Fibonacci sequence, Nobel counted rib number on 100 randomly selected plants of the barrel cactus *Ferocactus acanthodes* at each of two sites in the northwestern Sonoran Desert. This project was soon joined by Ronald Robberecht and expanded to include other species. Figure 6.4 shows that 51% of

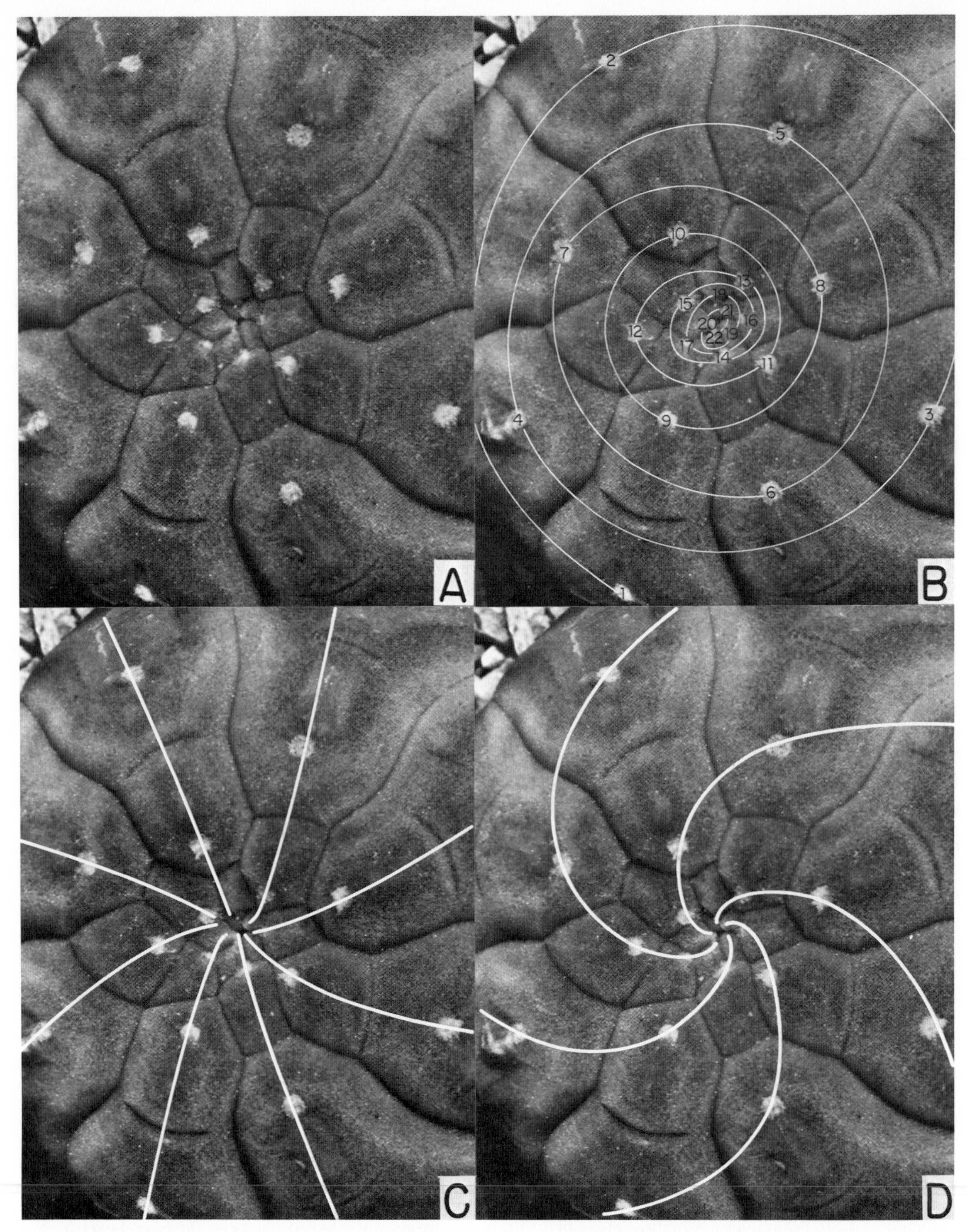

Figure 6.2. *Borzicactus (Matucana) madisoniorum,* a Peruvian cactus; top view of the shoot tip on which one can easily see the arrangement of the tubercles unhidden by spines or hairs.
(A) An unmarked photograph, showing the flat tubercles separated by horizontal and vertical depressions.
(B) Tubercles are numbered chronologically from youngest to oldest and connected by a spiral.
(C) The eight orthostichies.
(D) One set of contact parastichies are shown, connecting primordia separated by five leaf primordia.

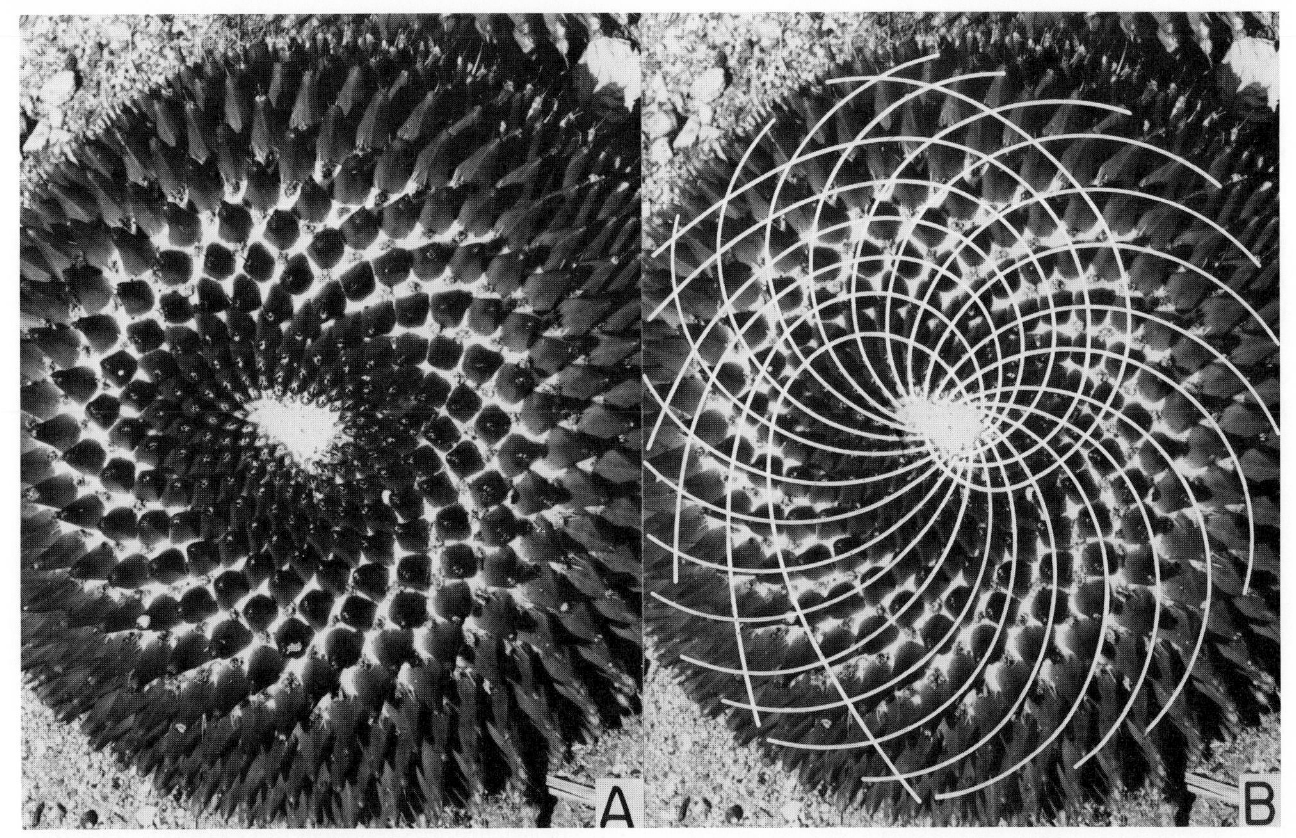

Figure 6.3. Phyllotaxy in *Mammillaria zuccariniana* from Hidalgo, Mexico.
(A) Unmarked top view, showing helical arrangement of tubercles in a plant with 8/21 phyllotaxy.
(B) Top view on which curves have been drawn to show the two sets of contact parastichies. The orthostichies (not shown) are nearly vertical.

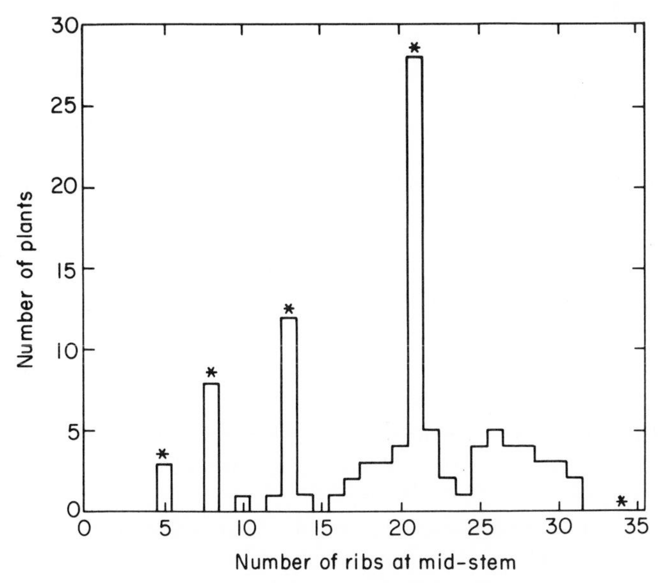

Figure 6.4. Number of ribs at midheight on 100 stems of *Ferocactus acanthodes.* Data, previously unpublished, were obtained by Nobel at the University of California Philip L. Boyd Deep Canyon Research Center (site at 33°39′N, 116°22′W, 300 m elevation). Asterisks indicate Fibonacci numbers.

the *F. acanthodes* surveyed had rib numbers at midstem in the Fibonacci sequence. Indeed, plants with Fibonacci rib numbers were far more numerous than were those with the adjacent rib numbers (Fig. 6.4); indeed, there were eight times more plants of *F. acanthodes* with a Fibonacci number of ribs at midstem than would be expected by chance alone. Rib number generally increased suddenly and predictably from one Fibonacci number at midstem to the next higher number at the top of the stem, for example, from 8 to 13 or from 13 to 21. For instance, about half of the plants with 8 ribs at midstem had 13 ribs near the apex; this situation occurred for a mean plant height of 8 cm. About half of the plants that had 13 ribs at midstem had 21 at the apex; this situation occurred for a mean height of 18 cm. The highest rib number in the study for *F. acanthodes* was 31 and was observed at the top of a few plants that were about 1 m tall.

Data were also collected on rib number for 15 species of *Ferocactus* from 208 wild and cultivated specimens and 139 specimens of barrel cacti belonging to other genera cultivated at the Huntington Botanical Garden in San Marino, California (Fig.

6.5). *Ferocactus* tended to have rib numbers in the Fibonacci sequence, whereas plants of other barrel genera, such as *Echinocactus* of Mexico and several genera from southern South America, do not. Also, there can be much variation within a species. Rib number in all barrels tends to increase from midstem to the top of the stem. Also the rib number at the base is generally lower than that of midstem; for *Ferocactus* the lower Fibonacci numbers toward the base represent the various rib stages of the small plants.

When rib numbers of shrubby and arborescent columnar cacti were investigated, these also were found to deviate from the Fibonacci numbers (Fig. 6.6). Robberecht and Nobel observed in a sample of 443 columnar plants that only 32% (144) had a rib number in the Fibonacci sequence, although half of the stems measured (219) had a rib number within one of a Fibonacci number. Moreover, rib number in these columnar forms does not increase monotonically from base to top as it does in barrel cacti. Instead, a set of ribs is formed before the stem is 10 cm tall, and then the number of ribs remains fairly constant. Of course, exceptions can be found, as in saguaro, *Carnegiea gigantea,* which is a club-shaped plant and must add ribs whenever the stem increases sufficiently in diameter (Fig. 1.17*A*).

These data imply that factors other than simple mathematical patterns have influenced the evolution of rib number in columnar cacti that have straight, vertical ribs. Perhaps the reason is fairly simple. An

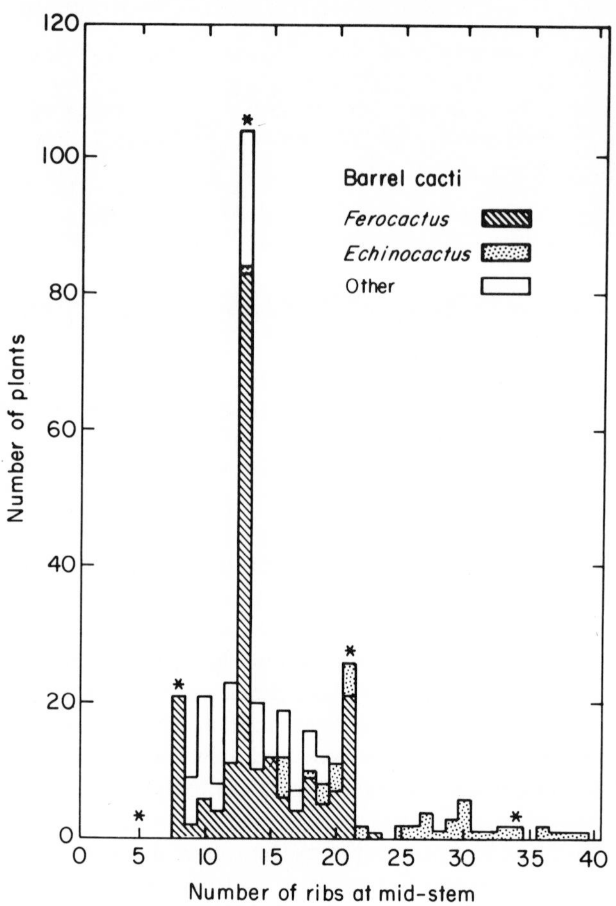

Figure 6.5. Rib number at midstem for various barrel cacti. The hatched bars represent 15 species of the genus *Ferocactus (F. acanthodes, F. alamosanus, F. covillei, F. echidne, F. fordii, F. glaucescens, F. gracilis, F. hamatacanthus, F. herrerae, F. pilosus, F. rafaelensis, F. robustus, F. townsendianus, F. viridescens,* and *F. wislizenii).* The stippled bars indicate species of *Echinocactus (E. platycanthus* and *E. grusonii).* The open bars indicate the species *Lobivia grandis, L. huascha, Neoporteria microsperma, Notocactus ottonis,* and *N. tephracantha.* The total number of plants measured was 357, with 15 ± 5 plants per species. Asterisks indicate Fibonacci numbers. (Data were collected by Nobel in the field for *F. acanthodes, F. covillei, F. viridescens,* and *F. wislizenii* and by Robberecht at the Huntington Botanical Garden, San Marino, California for the other species.)

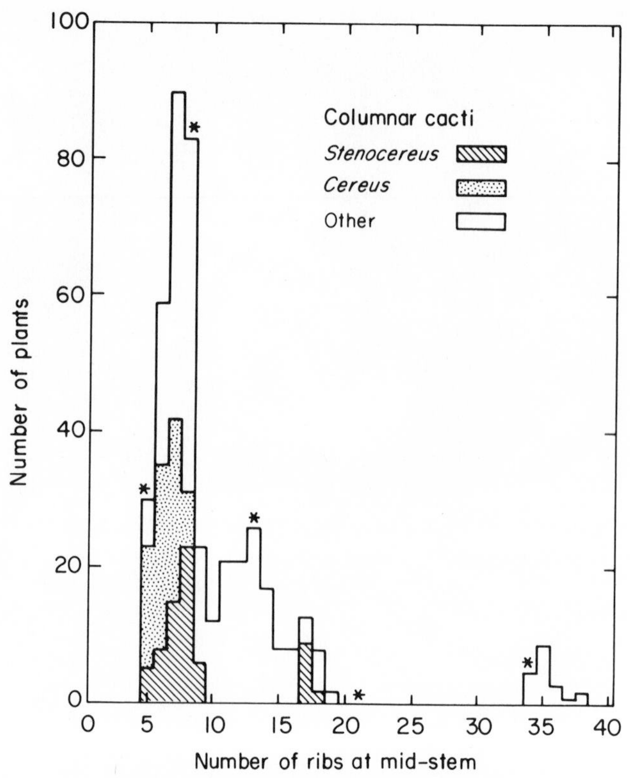

Figure 6.6. Rib number of midstem for various columnar cacti. The hatched bars represent five species of *Stenocereus (S. dumortieri, S. pruinosis, S. queretaroensis, S. stellatus,* and *S. thurberi).* Four species of *Cereus* are indicated by the stippled area *(C. forbesii, C. glaucus, C. peruvianus,* and *C. xanthocarpus).* The open bars indicate the species *Astrophytum ornatum, Cephalocereus palmeri, Echinopsis obrepanda, Harrisia tephracanthus, Lophocereus schottii, Myrtillocactus geometricans, M. schenckii, Notocactus leninghausii, N. tephracanthus, Pachycereus marginatus, P. pringlei, Stetsonia coryne, Trichocereus candicans, T. spachianus, T. thelegonus,* and *Weberbauercereus albus.* The total number of plants measured was 443, with 15 ± 5 plants per species. Asterisks indicate Fibonacci numbers. (Data were obtained by Robberecht for cultivated plants as for Figure 6.5.)

Figures 6.7 and 6.8. Features of ribs on barrel cacti. **6.7.** *Ferocactus* species, from southern Sonora, Mexico. This specimen has 13 ribs that gradually wind around the stem.

6.8. *Echinocactus platyacanthus (=E. ingens),* from Hidalgo, Mexico. The ribs are vertically oriented, and the photograph shows where certain ribs have been converted into two to accommodate the increase in plant diameter.

orthostichy is supposed to be a vertical row of leaf primordia. But as we mentioned before, the tubercles of cacti are not aligned truly vertically but instead are about 30° off; so the rib winds slowly around the stem, either to the right or to the left. The ribs of *Ferocactus* (Figs. 1.5 and 6.7), the numbers of which correspond to the Fibonacci sequence, tend to wind around the stem. In contrast—for reasons that will become apparent later in this chapter—most tall species have truly vertical ribs; therefore, it may be important that to guarantee vertical organization ribs form from tubercles that are not members of the same orthostichy. Note that the ribs of *Echinocactus* (Figs. 1.6 and 6.8), which do not conform closely to the Fibonacci sequence, are vertically oriented.

Development of Tubercles and Ribs

Earlier in this chapter we mentioned that leaf primordia are produced on a shoot tip, and in Chapter 5 we followed the growth of the areole. Let us return to the young primordium to discover how the base of the primordium develops into a tubercle or a rib. A close examination of the cellular structure of tubercles and ribs not only shows us how succulence has been produced but also gives us an opportunity to identify potential internal structural adaptations of cacti.

Tubercle Formation

First let us consider the leaf base (Fig. 5.1*A*). This structure is a dome of cells less than 1 mm in height (Fig. 6.9*A*). Atop the leaf base is a very tiny leaf blade about 0.1 mm in length that may soon shrivel and fall off, and in the axil between the leaf base and the stem there is an areole, that is, a modified axillary bud. After this the leaf primordium does not grow very much; but much cell division occurs in the leaf base and produces a tubercle (Fig. 6.9*B*). When the tubercle is several millimeters long, a major change occurs in its form of growth (see next section).

At this stage, four important sets of cells that later compose the mature tubercle are present (Fig. 6.9*C*). The outermost layer is called the protoderm—a term meaning early skin—because this layer of cells later becomes the epidermis. Beneath the protoderm lie one to several layers of cells that become the hypodermis if collenchymatous cell walls are laid down. Directly inside the future hypodermis is a single layer of actively dividing cells known as a peripheral meristem (also called the peripheral primary meristem and the subprotodermal meristem). This layer of cells produces the tissues of the outer cortex, which appear in straight files. The core of the tubercle is composed of large parenchyma cells, which were part of the original leaf base and became stretched by the elongation of the tubercle.

Protoderm and Epidermis

Cell divisions in the protoderm are essentially over when the tubercle is as little as 1 mm tall. Protodermal cells are very small and cuboidal. When the tubercle enlarges, these cells become enlarged and stretched. Therefore, in a cross section of mature cactus skin (Figs. 3.3–3.5), the epidermal cells may appear flat, like a tabletop. The outer wall of an epidermal cell may be convex when it bulges outward or it may even produce a small, nipplelike projection called a papilla (Chapter 3). Papillae have been

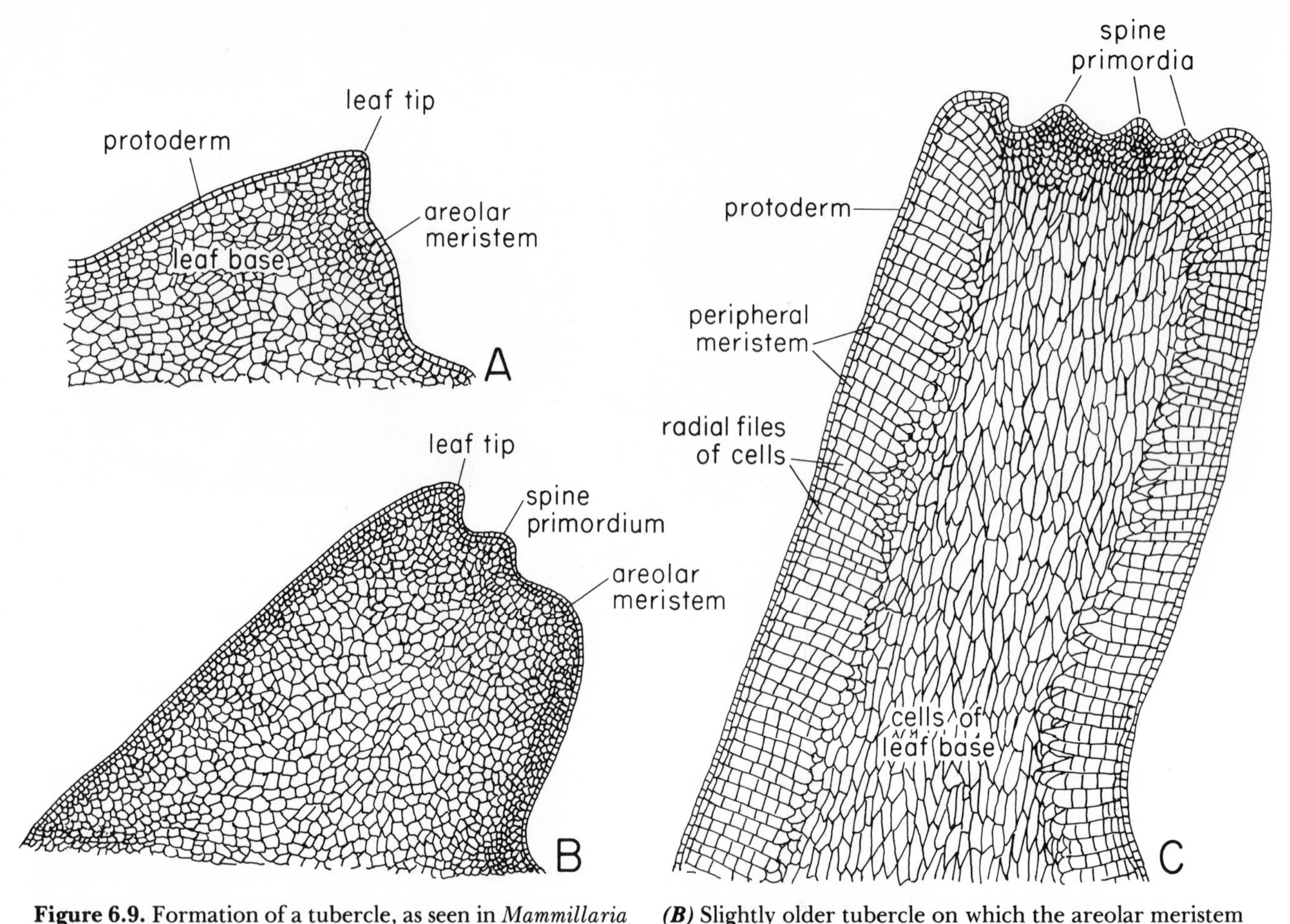

Figure 6.9. Formation of a tubercle, as seen in *Mammillaria heyderi.* (Adapted from Boke, 1953.)
(A) Longitudinal section through a very young leaf and areolar meristem, which is within 1 mm of the shoot apex. The leaf tip, although small, has nearly achieved its final size, and the leaf base has just begun to increase in volume. The surfce is covered by the undifferentiated cells of the protoderm.
(B) Slightly older tubercle on which the areolar meristem has formed a spine primordium and the tubercle has greatly increased in size by cell division.
(C) A tubercle that has started to form its final structure from cell divisions of the peripheral meristem (just under the protoderm), which produce radial files of cells. Cells of the radial files become chlorenchyma, whereas the original core cells of the tubercle are not green and become enlarged and stretched as the tubercle increases in length.

observed in *Opuntia, Pterocactus, Peniocereus, Lophocereus,* and numerous species of the tribe Cacteae, for example, *Lophophora, Turbinicarpus,* and *Ferocactus.* Many papillae can be seen with a hand lens, and they give the plant a scaly appearance. Some epidermal projections are long enough to be called trichomes, and they are used as a diagnostic feature to classify different cacti, for example, recognizing the series Tomentosae, a group of Mexican platyopuntias. In a few cases, the epidermis of a cactus is composed not of tabular cells but rather of tall cells, as in *Ferocactus robustus* (Fig. 6.10) and related species in *Ferocactus* and *Thelocactus.*

Regardless of the shapes of typical epidermal cells, there are some cells that must divide again before the tissue matures. These are the "mother cells" (meristemoids) that produce the guard cells and the subsidiary cells that surround the stomate (Figs. 1.29–1.31 and 3.6). Usually, one subsidiary cell is present on the outside of each guard cell (paracytic arrangement), but sometimes two parallel cells are present on one or both sides (parallelocytic arrangement). No one has carefully studied the sequence of cell divisions by meristemoids in cacti to determine exactly how the subsidiary cells are produced.

Also in a number of species the groundmass cells of the protoderm continue to divide even as the tubercle or rib is enlarging. This can be observed by looking for special clusters of epidermal cells on the mature plant. Figure 1.30 shows that one cell has divided repeatedly to produce a small cluster of cells that is distinct from the next cluster of cells. This type of late cell proliferation of the epidermis, which expands the surface, is found in numerous ribbed cacti, for example, in some species of columnar cacti in the tribe Pachycereeae and in many epiphytes.

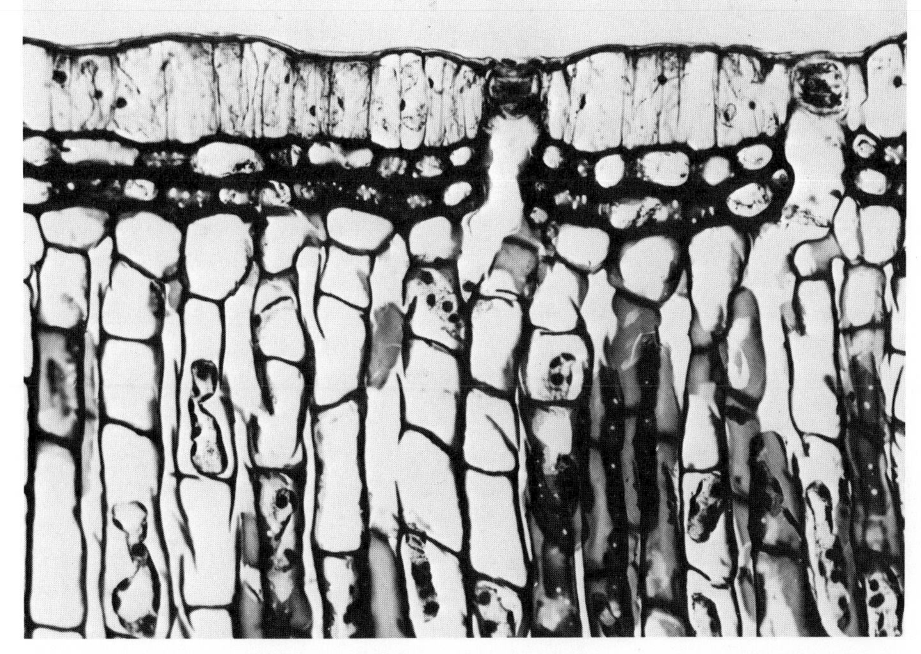

Figure 6.10. *Ferocactus robustus,* a large, caespitose plant from southern Mexico. Epidermal cells, which cover a two-layered hypodermis, are vertically elongate, and the guard cells are not deeply sunken below the stem surface. Cells of the chlorenchyma (beneath the hypodermis) are arranged in straight flies, because they were produced in these files from meristematic cells; note the conspicuous intercellular air spaces between the files of chlorenchyma cells. ×500.

Some species of Pachycereeae (Fig. 3.5) even have two or three layers of epidermal cells — a multiple epidermis — that are produced by divisions of the original protodermal layer.

Hypodermis

The hypodermis provides the integrity of the tubercle or rib, and so this tissue must form at an early stage of development. Cells of the hypodermis, like those of the epidermis, do not divide much once the layers have initially been formed. These cells are stretched as the tubercle enlarges. Typically, this pulling causes flattening of these cells in the same direction as the epidermal cells. But if they are also pulled in the direction perpendicular to the surface, the cells may become large and cuboidal or even very tall, as in *Pachycereus weberi* and *Mammillaria matudae.* The collenchymatous wall starts forming while the cell is still enlarging, so the thickness of the wall can change throughout the life of the cell. In fact, old skin may have thinner hypodermal walls than does young material.

The division of the meristemoid to produce the guard and subsidiary cells naturally increases the surface area of the epidermis. This expansion creates a pressure on the layers of cells beneath the stomate. Thus, deposition of material to form thick walls in the hypodermis cannot really start until the stomates begin to form. The cell-to-cell connections that lie directly beneath the stomate break; this breakage pulls the cells apart and results in the formation of a substomatal canal through which gas will diffuse between the atmosphere and the cortex. Subsequently, thick walls form in the hypodermis, thus leaving a very sturdy canal (Figs. 3.3 – 3.5, 3.7*A*). If the hypodermis matured before the stomates formed, then the substomatal canals could not be formed because the cells would be tightly cemented together and could not be pulled apart.

The hypodermis is a difficult tissue to cut properly for examination because its walls are so tough and thick. Nonetheless, within the last two decades, the skin of over 600 species of cacti have been examined by several researchers, who have searched for features to use in reconstructing evolutionary relationships (Chapter 10). In some cacti the collenchymatous walls are so thick that they virtually "choke" the living cell contents; but, in fact, every hypodermal cell is alive at maturity, having a fully functional nucleus. The thickness of the hypodermis along the tubercle — if it has hypodermis — is very uniform, but the hypodermis may be twice as thick between adjacent tubercles or in the crease between two ribs. It is common to find more collenchyma in the regions that undergo more bending. Some small,

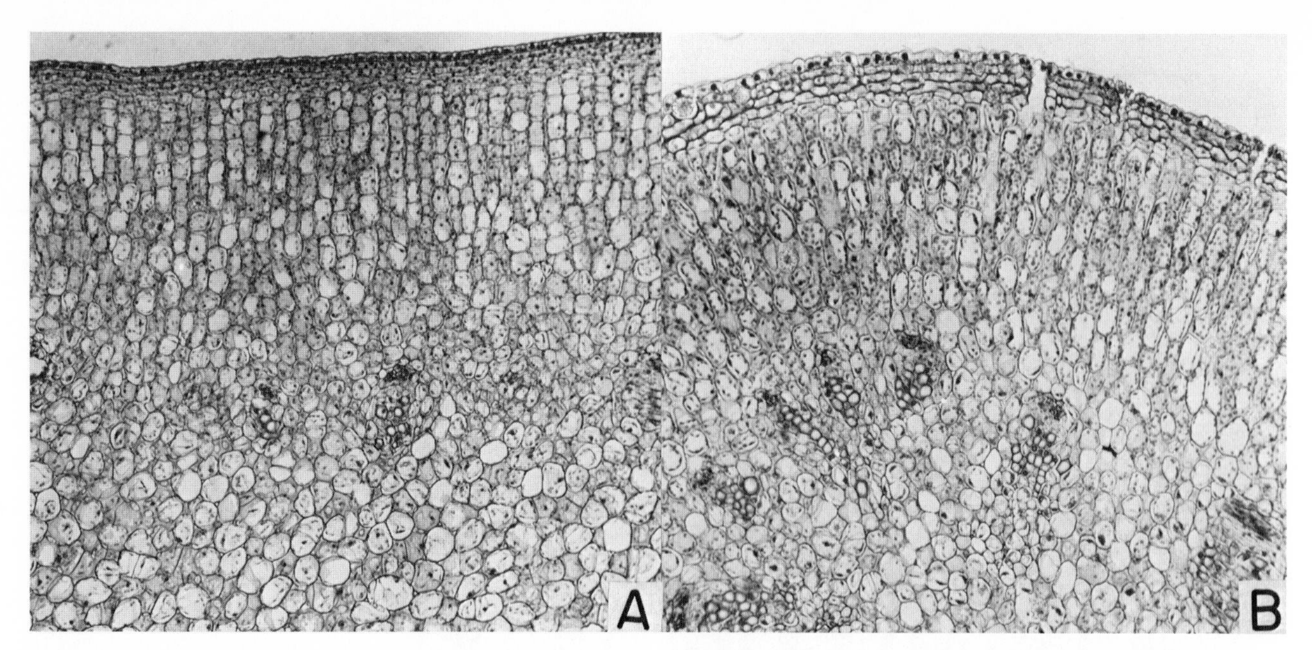

Fugure 6.11. Formation of the outer cortex in a tubercle of *Opuntia spinosior.*
(A) Radial files of cells produced for the outer cortex by cell divisions of the peripheral meristem, just beneath the outermost layer of cells (protoderm). ×170.
(B) Cells produced from the peripheral meristem have matured as thick-walled cells of the hypodermis (three to four layers) and chlorophyll-bearing cells of the chlorenchyma. ×170.

soft-skinned cacti lack collenchymatous hypodermis entirely, for example, species of *Disocactus* and *Nopalxochia.*

Perhaps one-third or more of all species of cacti have crystals in the skin, in either the epidermis or the hypodermis or in both. Most of these crystals are composed of calcium oxalate, but those of the genus *Stenocereus* (tribe Pachycereeae) are opaline, that is, they are made of silica (see Chapter 9). Crystals develop fairly late in the maturation of the skin and thus may be missing in very young samples. Many authors would be tempted to say that these crystals are present to discourage invertebrates with chewing mouth parts from eating the skin, although there has been no study showing the crystals in cacti function in this way. Crystals certainly must affect the passage of PAR through the skin to the chlorenchyma, and there is a need to study how light passes through or is reflected by skin that has crystals and by skin that lacks them.

Perhaps the most curious development of hypodermis occurs in the genus *Uebelmannia,* which was first described in 1967. Vertically elongate hypodermal cells that cause the surface to become very bumpy, like a cobblestone street, have been described for *U. pectinifer.* Other cases where the hypodermis causes the skin to bulge outward are found in the epiphytes *Lymanbensonia micrantha* of Peru and *Aporocactus conzattii* of Mexico.

Peripheral Meristem of a Tubercle and Cell Maturation

The most important aspect of tubercle succulence is related to the peripheral meristem (Figs. 6.9*C* and 6.11). Each primary cell (initial) of this cell layer divides by forming a new wall parallel to the surface of the tubercle (a periclinal wall). The outer "daughter cell" remains as the initial, and the inner "daughter cell," the derivative, becomes a cell of the outer cortex—specifically, a chlorenchyma cell. Figure 6.11 shows how each initial produces a neat file or chain of cells to the inside. Whenever you make a thin section of a young tubercle or a young rib and examine it under a microscope, these files are readily observed. Each cell in the file has a large central vacuole; and as the vacuole increases in size, the cell must also increase in size. Eventually the hydrostatic pressure within the cell (Chapter 3) is high enough to cause a cell to round up and thus lose the file arrangement seen in younger tissues. If a tubercle or rib is very old, you may not see files of cells even in the outer cortex because the cells commonly round up right to the hypodermis once cell division is completed. No one has actually documented when the peripheral meristem stops producing cortical cells, but it can be surmised that cortical cell division must stop about the time that the hypodermis develops its thick walls and the stomates become functional. It makes good sense to have fully functional mature chlorenchyma cells by the time CO_2 uptake becomes possible with the formation of stomates. In typical dicotyledons, the onset of photosynthesis and stomate development are very closely coupled.*

Cell enlargement is much more important than cell division in determining the ultimate size of most plant organs. Botany students learn that most of the cells in a leaf or a fruit are already preformed when the organ is very small. For example, a grapefruit experiences comparatively few cell divisions after it is only 2 cm in diameter. The same is true for a tubercle, because the cells are essentially all present when the thick walls of the hypodermis form. Water diffuses into the cells of the tubercle because they have high concentrations of dissolved substances, and most of this water ends up in the vacuoles of the tubercle cells (Chapters 3 and 4). Consequently, the cell walls expand to accommodate the increased hydrostatic pressure accompanying the increased protoplast volume, and the whole structure enlarges. A cactus is well designed to permit great cell enlargement without experiencing damage to the plant body because the collenchymatous hypodermis is a flexible, stretchable tissue that can easily be pushed outward.

The neat files of derivatives from the peripheral meristem are also disturbed when individual cells, called idioblasts, develop to perform special functions. The two common examples are mucilage cells and crystal-bearing cells (Chapter 9). Mucilage cells can be spotted easily in thin sections, even with a hand lens, because they lack chloroplasts and are colorless, whereas the surrounding cells are green because of the presence of numerous chloroplasts (Fig. 3.7*B*). Crystal-bearing cells and mucilage cells are generally larger than normal parenchyma cells; in fact, the diameters of some of these idioblasts may be more than four times that of typical cortical cells. Thus, there is a tremendous difference in their volumes, because volume is proportional to the diameter cubed.

Crystals are formed within the cell vacuole, but mucilage accumulates in quite a different way. As described by Mauseth, a cactus mucilage cell begins as do all the surrounding cells that develop into chlorenchyma, except that chloroplasts never develop in the mucilage cell. Instead, this cell makes many large cell organelles that synthesize mucilage (dictyosomes) and are also important in all plant cells for producing new cell walls. Small units (called vesicles) that are loaded with chemicals for mucilage bud off from the dictyosomes and migrate to the edge of the cell, where they fuse with the plasma membrane (Chapter 3). When this occurs, the mucilage is released outside the plasma membrane but inside the cell wall. Therefore, mucilage builds up in this space as the living cell pulls away from the walls and shrinks to the center. Water diffuses into the cell because mucilage is very hydrophilic ("water-loving"); thus, the cell wall is obliged to expand to yield a larger space for the mucilage. Eventually, the living cell dies, but the stretched cell wall packed with mucilage remains intact.

The simple, single mucilage cell is found throughout the cactus family. Some species also have large mucilage reservoirs, where all or nearly all the cells in a particular region of the stem or root produce mucilage. Mauseth has discovered long canals in the nopaleas *(Opuntia)* that have masses of single, floating mucilage cells within the canal. Mucilage is produced in canals in the famous species of *Ariocarpus,* which are well known to natives who use the mucilage as a type of glue. In the tribe Cacteae—specifically in the "milky" and "semi-milky" mammillarias (for example, *Mammillaria gummifera*), the canals produce thick, sticky, white or off-white "latex." In these species the canals are produced from long rows of cells, several cells across, that break down to form a wide tube, which contains the mucilage and all of the cellular debris.

Rib Formation

Whereas tubercles are present in most cacti, fewer than half of the species have ribs. Earlier we discussed how phyllotaxy could determine the number of ribs; now we shall discuss the anatomical formation of ribs. Unfortunately, very little scientific research has been conducted on the developmental stages of rib formation. Most of Boke's developmental studies were done on plants with helically arranged tubercles that lacked ribs, and almost everything published on rib development occurs in his brief description of *Echinocactus.*

A simple way to explain rib formation might be to say that the tubercles are fused into a vertical series to become a rib. This is an image that is created by studying external patterns (Figs. 6.12 and 6.13). Notice how the tubercle of each species of *Notocactus* in the figure has around the apex helically arranged tubercles that fit the Fibonacci sequence described earlier. On the edges of the shoot tip, the tubercles are organized into vertical ribs. You could say that they are fused together, but this is technically not correct. Actually, a vertical strip of tissue underneath

* An excellent description of the tissue development of a dicotyledonous leaf is found in an article on cottonwood: J. G. Isebrands and P. R. Larson. "Anatomical Changes during Leaf Ontogeny in *Populus deltoides,*" *American Journal of Botany* 60 (1973):199–208.

Figures 6.12 and 6.13. Arrangement of tubercles on the shoot apex of *Notocactus.*
6.12. *Notocactus uebelmannianus;* a plant with 8/21 phyllotaxy, in which the contact parastichies wind in both directions from the pubescent shoot apex. The tubercles are formed in a helical arrangement at the shoot apex, but as they emerge from the apical depression, they are raised with other tubercles in the same orthostichy into a rib.
6.13. *Notocactus crassigibbus,* a plant with 5/13 phyllotaxy. As in Figure 6.12, the tubercles are produced in a helical order, but eventually tubercles in each orthostichy form a rib.

these tubercles is produced, so that the tubercles above this strip are being uplifted. Boke noted that the meristem responsible for rib elevation is like a primary thickening meristem similar to those found in palms with very thick shoot tips.

Studies are needed to determine how ribs begin to form on a young plant, how tubercles are added to a rib near the shoot apex, and how ribs branch to form additional ribs. In addition, many interesting developmental questions remain unanswered; when the answers are found, they will help to explain the origins of rib structural diversity, such as the "chins" of chin cacti (*Gymnocalycium,* Fig. 1.11), of different types of rib margins such as straight or sinuous, of the differences in rib cross sections (triangular to rounded), and of various types of transverse and vertical grooves.

Although we find it convenient to identify a tubercle as a leaf base, a tubercle includes tissues that were produced by the peripheral meristem. Thus, some of the new tissue is really part of the stem, and therefore, the tubercle is actually a combination of leaf and stem tissue. Ribs, on the other hand, if they are produced by a primary thickening meristem internal to the leaves, are mostly stem structures. Nonetheless, despite this combination of tissues, the cactologist continues to refer to a cactus as a succulent stem even though it has tubercles, which arise as leaf bases.

Aging of a Stem

Tubercles and ribs can persist for many years in their original green state, but they may also undergo aging. As a cactus stem ages, the tubercles or ribs at the very base of the plant may develop bark. Because light does not penetrate bark very well, the chlorenchyma is put out of service, and the plant then seals up these old regions where water could be needlessly lost. Typically, old regions become so stretched that the ribs and tubercles become flattened, the spines break off, and the areolar meristem dies.

Bark is formed in plants from a layer of meristematic cells called the phellogen, or cork cambium. In most plants other than cacti, phellogen commonly forms from old cortical cells, especially in a layer just beneath the epidermis. A stimulus causes a cell just under the surface to divide to produce two cells, and one of these becomes the initial. Cells in the same layer and adjacent to the first cell also divide sequentially, and eventually a ring of cells is formed around or on one side of the stem or root. Each cell divides

Figure 6.14. *Ferocactus robustus,* a view and magnification of the outer stem similar to that shown in Figure 6.10. Periderm has formed on the cells at the right. Divisions of epidermal cells parallel to the surface of the stem have been responsible for the formation of a stack of flat cells, which will eventually become cork cells. The actively dividing cell on the inside of each stack is the initial, which continues to divide (cork cambium). It is easy to see that the epidermis gives rise to the cork cambium because the original two-layered hypodermis is still intact; the gas exchange canal has been blocked by cell divisions in the epidermis, limiting the entry of CO_2 into the chlorenchyma. ×500.

repeatedly, producing many flat cells to the outside and an occasional parenchyma cell to the inside. The outer cells (cork cells) develop thick walls impregnated with suberin, which is a special wax that is waterproof and unappetizing to insects.

Cacti form a phellogen from the epidermis itself, but cactus phellogen almost never forms as a continuous structure around a plant. Figure 6.14 shows the early stages of bark formation of *Ferocactus robustus.* In this figure, one can still see the intact hypodermis, the presence of which demonstrates that the phellogen arose from the epidermis proper.

Bark formation in cacti also tends to be patchy. The phellogen in Figure 6.15 was initiated wherever injury had occurred—wherever the skin had split or been damaged by freezing or sunscald. In general, green tissue is gradually replaced by brown patches (young periderm) and then by thick gray bark; but islands of green stem may persist, especially on the sides of ribs or tubercles. On the trunks of arborescent cacti, for example, the tree platyopuntias of the Galápagos archipelago, *Opuntia echios* (Fig. 6.16), the entire stem is surrounded by thick, flaky bark that resembles the bark of a pine tree.

The famous cornucopialike structures that form within the stems of saguaro in response to wounding are also bark. When a bird excavates a hole in a saguaro for a nest, the plant responds by sealing its tissues by forming a phellogen in the subsurface cells, an action that produces cork cells to the outside.

Fasciation or Crests

Cactus growers will pay a high price for forms that are "crested," or fasciated. Fasciated cacti are atypical specimens that have broad, malformed tops (Figs. 6.17–6.20) produced on typical cylindrical bases. The cylindrical stem develops into a stem that has two flat sides and is commonly fan-shaped or large and undulating. Abnormal stems that are fasciated may occur in any plant family following injury to the shoot apex; but in the most prized specimens, the abnormal form of the shoot is presumably heritable and not environmentally induced. A fasciated plant often has some normal as well as some fasciated shoots (Figs. 6.17 and 6.19).

The most thorough structural study of fasciation of a cactus was done by Boke and Robert G. Ross at the University of Oklahoma. These workers found that fasciated specimens of *Echinocereus reichenbachii* have apical meristems that are linear (very long and narrow), whereas the normal shoot apex is circular in outline and forms a typical, fairly flat dome. In the fasciated specimens, not only is the meristem long

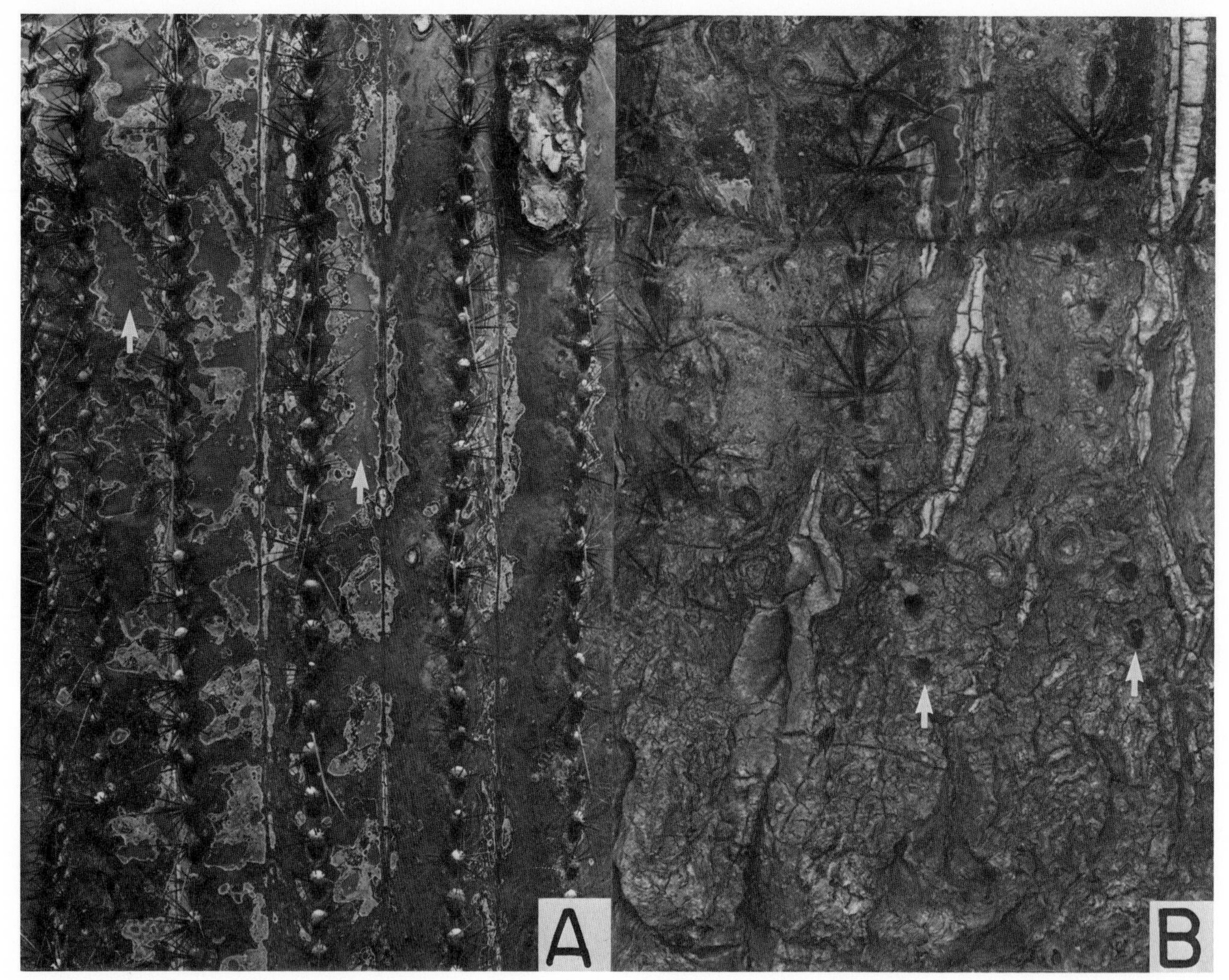

Figure 6.15. Old stems of *Carnegiea gigantea,* or saguaro, with periderm.
(A) Periderm has formed beneath the skin on most areas of the ribs, but patches of photosynthetic tissue are still present *(arrows),* for example, to the right of the areoles.
(B) Bark formation on the trunk on an old specimen; photosynthetic tissues have been replaced by thick bark, and many of the areoles have consequently been covered over by the corky tissues *(arrows).*

and narrow, but also the mother cells of the apex are arranged in linear files that are oriented in the same axis as the meristem. The growth of these cells appears to produce the elongation of the shoot tip and the resulting fasciated form.

Adaptive Significance of Ribs

Over the years, numerous writers have speculated on the adaptive significance of ribs in cacti and their varied surface features. Of course, cacti are not the only plants with ribs — witness, for example, the stem succulent species of *Euphorbia* mentioned in Chapter 1 — so discoveries made on cacti can possibly be extrapolated to explain convergent evolution. Most stem succulents are not simple, smooth spheres or cylinders, so some benefits must accrue to a plant that has projections such as tubercles and ribs, otherwise one would not expect the great diversification of external designs that has occurred. Indeed, biologists have mused over ribs, wondering about their role or roles. Are they a better way to support the plant; are they like fins for cooling the plant; are they adaptations for seasonal expansion and contraction; or are they for increasing the photosynthetic surface? Until very recently, the study of ribs and tubercles has not included experimental tests; but as for spines, ribs and tubercles are now being studied in careful ways to learn their actual consequences.

Effects of Ribbing

Ribbing may affect PAR interception, CO_2 uptake, and surface temperature. One of the earliest experimental observations on cactus ribs was made at the

Figure 6.16. A stand of *Opuntia echios* var. *barringtonensis,* a tree platyopuntia growing on Santa Fé (Barrington), Galápagos Islands. The trunk and main branches of the older trees have bark (periderm) somewhat like that of a pine tree.

Desert Laboratory of the Carnegie Institution in Tucson, Arizona. Effie Southworth Spalding and Daniel T. MacDougal, who pioneered in the study of water relations of desert plants, examined the daily and seasonal patterns of water uptake and water loss in succulents. Their favorite subject was saguaro, *Carnegiea gigantea* (which was named in honor of the support by the Carnegie Institution for cactus research). Spalding and MacDougal were able to demonstrate that the ribbed stem of a saguaro behaves like an accordion. As water is lost from the plant during dry periods, the tissues contract, and the ribs get thinner, more acute in outline, and closer together. After a heavy rainfall, the tissues rehydrate, plant volume increases, and the ribs tend to bow outward. Spalding and MacDougal also observed that the ribs on the north side of a saguaro plant characteristically are wider and shallower than those on the south and southwest sides. No completely satisfactory answer has ever been found for this pattern.

Ribs and tubercles and the flexible collenchymatous skin that covers them are ideal structures to permit accordionlike responses of plant tissues to changes in hydration without incurring physical damage to the cells. A cylindrical or spherical plant develops wrinkles as the stem loses water during dehydration; but a plant with ribs or with separate tubercles has many places to accommodate, like hinges, the decrease in volume without affecting the integrity of the skin. As indicated in Chapter 3, a cactus stem can tolerate up to 82% loss of its water without fatal consequences; and after rehydration, the same plant may show few outward signs that it was ever wizened and close to death. Because the skin has porous walls and no lignin, its cells can, through intercellular connections, continue to receive vital supplies from the cortical cells; these supplies maintain the skin as a living tissue.

Of course, the skin can only be stretched so far, and you can hurt a cactus with too much kindness. Many fine specimens of saguaro and barrel cactus have been watered so frequently in cultivation that they have developed high turgor pressure in the stem, which has caused the skin to split. The strength of cactus skin, specifically its tensile strength, has never been measured, but it would be interesting to calculate the amount of force exerted on the skin of a cactus that is fully hydrated. The rigidity of cactus skin appears to increase with age, because very old

Figures 6.17–6.20. Fasciated (crested) stems of cacti.
6.17. *Mammillaria* sp.; a crested form with one normal shoot (lower right).
6.18. *Lophophora williamsii,* peyote; a brainlike crest.
6.19. *Gymnocalycium bodenbenderianum;* a plant with typical cylindrical stems and also some that are strongly flattened.
6.20 *Cephalocereus chrysacanthus,* a columnar cactus on which the number of ribs increased rapidly by the formation of a broad, fasciated stem.

ribs and tubercles near the base of a plant no longer have their original shapes and become hard and flattened and eventually covered by bark.

PAR Interception and Nocturnal CO_2 Uptake

We can see at a glance that a ribbed or a tuberculate cylinder has significantly greater surface area than a smooth cylinder of the same diameter. Assuming that a rib is triangular in cross section makes it fairly simple to calculate the surface area. Or we can measure surface area directly by removing the spines, wrapping the ribs tightly with a sheet of paper, and then measuring the area of the paper. Calculating or measuring the surface area of a tuberculate stem is much more difficult—no one has ever done this—but if it were done, the value certainly would be much greater than that of a smooth surface. The question arises, therefore, "Why have cacti increased their surface area?" The general dogma of desert biology states that desert plants are trying to save as much water as possible, a hypothesis that is consistent with the large volume-to-surface ratios of cacti. But increasing surface area per volume by ribbing or by forming tubercles actually increases potential transpiration and therefore could send a plant backward to a strategy of less water conservation.

It is arguments like the previous one that show the reader why it is so important to understand all aspects of cactus biology in order to understand cactus design and strategy. A plant's productivity is dependent on the amount of surface area available for PAR interception and CO_2 uptake (Chapter 4); so if a plant keeps its stomates closed or limits surface area severely, this strategy also restricts the amount of carbohydrate available for growth and reproduction. Cacti already have excellent water reserves per unit of transpiring surface (Chapter 3) and Crassulacean acid metabolism, which includes nocturnal opening of stomates and very high water-use efficiency (Chapter 4). Meanwhile, PAR interception can be decreased by the presence of spines and pubescence (Chapter 5), both of which can buffer the plant from extreme temperatures and defend it against herbivores (see Chapter 5). The development of greater surface area by the formation of numerous ribs or tubercles leads to more area for PAR interception and for CO_2 uptake, both of which can affect overall plant productivity.

To understand how ribs affect plant productivity, in 1980 Nobel used a computer model to investigate the influence of ribs on the interception of PAR and on nocturnal CO_2 uptake by *Ferocactus acanthodes*. Ribs increase the stem surface area but decrease the PAR per unit surface area; thus the presence of ribs results in less CO_2 uptake per unit stem area. In fact,

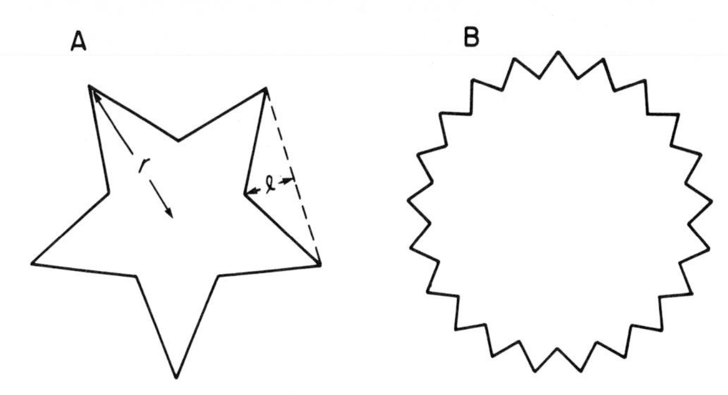

Figure 6.21. Cross sections of idealized stems with various rib numbers and depths.
(A) Stem with 5 ribs of fractional rib depth 0.35.
(B) Stem with 21 ribs of fractional rib depth 0.15. Fractional rib depth equals l/r.

for the particular combinations of rib number and rib depth used, the increase in stem area produced by the ribbing more-or-less compensates for the lower CO_2 uptake per unit area. Thus, the overall effect of ribbing on productivity per unit height was deemed to be minor. This study was followed up by a much more detailed computer and field study by Gary N. Geller working with Nobel at UCLA; he analyzed the effects of a broader range of rib characteristics on PAR interception and CO_2 uptake. Geller considered the number and angular spacing of ribs, the depth of the ribs, and the amount of shading from spines, surrounding vegetation, topography, or other sources. From this he tried to ascertain which factors were most important in determining PAR interception by the cactus stem.

To do his analyses, Geller determined the fractional rib depth (defined as the distance from the base of a rib trough to the midpoint of a line between two adjacent rib crests, l, divided by the maximum radius, r, of the stem; Fig. 6.21*A*). A stem with deep ribs would have a relatively large fractional rib depth (for example, 0.5). A stem like the one shown in Fig. 6.21*A* would have a fractional rib depth of 0.35, a value appropriate for *Lophocereus schottii* (Fig. 5.15). Many species, such as *F. acanthodes* (Fig. 4.8) and *Carnegiea gigantea* (Fig. 1.17), have shallower ribs; for *F. acanthodes* and *C. gigantea*, l/r is often about 0.15 (Fig. 6.21*B*). Finally, a plant with no ribs (a plant that is a regular polygon in cross section) would have an l/r of 0.00. Nearly all ribbed cacti fit within these two extremes—except some of the flat-stemmed epiphytes—and in most species the fractional rib depth is between 0.1 and 0.3.

Simulations were designed primarily to test the effects of rib number, rib depth, and shading on PAR interception and CO_2 uptake, both over the course of a day and at different times of the year (at one equinox and both solstices). Geller first pre-

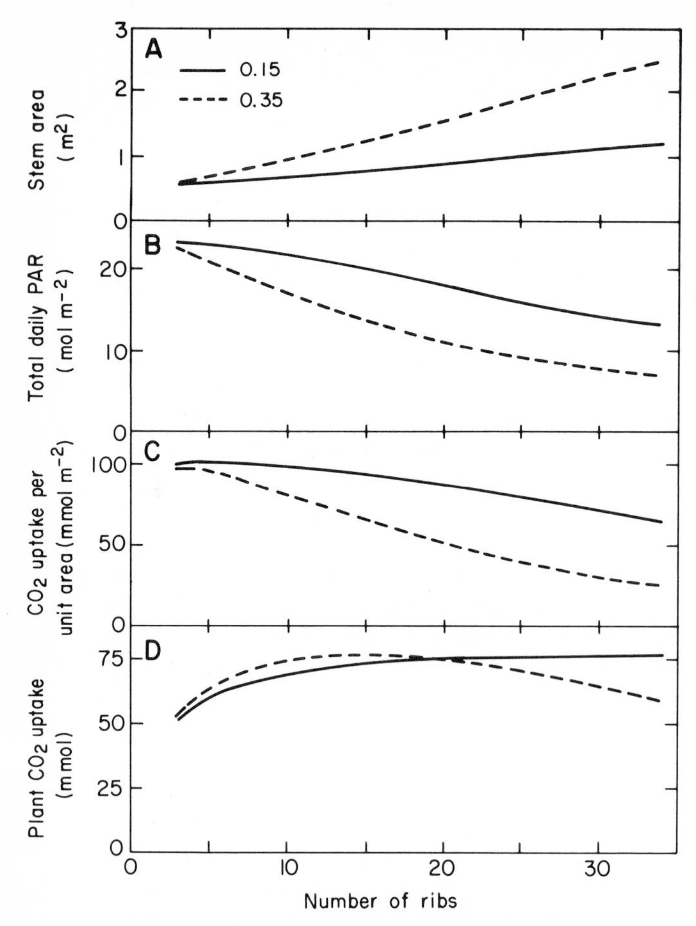

Figure 6.22. Influence of rib number on stem surface area *(A)*, average total daily PAR per unit area *(B)*, nocturnal CO_2 uptake per unit area *(C)*, and nocturnal CO_2 uptake for the whole plant *(D)*. Data are for a plant 1.0 m tall and 0.10 m in radius on an equinox at 34°N. The fractional rib depth was 0.15 or 0.35. (Data are adapted from Geller and Nobel, 1984, and from unpublished simulations.)

dicted the total daily PAR for each rib surface, then estimated the CO_2 uptake by using the relationship between CO_2 uptake and total daily PAR for *Ferocactus acanthodes* and *Opuntia ficus-indica* (Fig. 4.15).

First let us consider what happens when the number of ribs changes without changing any other rib characteristic, for example, depth. This is essentially the situation that occurs when an individual plant changes rib number, for example, from 13 to 21 to 34 (Fibonacci sequence). Figure 6.22*A* shows that the total stem area for a plant like *Ferocactus acanthodes* or *Carnegiea gigantea* (fractional rib depth of 0.15) increases substantially as rib number increases. However, the total daily PAR per unit surface area decreases because there is a greater surface over which the intercepted PAR is distributed. PAR interception is higher in the summer and lower in the winter than at the time of an equinox (Fig. 6.22), but the same general pattern prevails.

Nocturnal CO_2 uptake per unit area also decreases as rib number increases (Fig. 6.22*C*). Although at first glance this may seem disadvantageous, because a plant with many ribs has substantially *more* surface area, the nocturnal CO_2 uptake per meter of plant height actually increases substantially (Fig. 6.22*D*). Therefore, the simulations showed that more ribs lead to higher stem productivity under these conditions.

The importance of ribs is even easier to see for a plant with a fractional rib depth of 0.35, a value appropriate for *Lophocereus schottii* (Fig. 5.15). Stem area (Fig. 6.22*A*) increases but interception of total daily PAR (Fig. 6.22*B*) and CO_2 uptake (Fig. 6.22*C*) per unit area both rapidly decrease with an increase in rib number. Figure 6.22*D* shows that CO_2 uptake per unit height for a plant with deep ribs increases to a maximum around 13 ribs and then decreases. Interestingly, in southern populations of *L. schottii*, rib numbers of 13 and 14 have been observed.

What are the consequences of variations in fractional rib depth? A plant with five shallow ribs has less stem surface area but greater daily PAR interception per unit area than a five-ribbed cactus with deep ribs (Fig. 6.23*A* and *B*). CO_2 uptake per unit area (Fig. 6.23*C*) decreases with fractional rib depth, but total CO_2 uptake per unit stem height increases (Fig. 6.23*D*). When many ribs are present, the effects of varying rib depth on area, PAR per unit area, and CO_2 uptake per unit area are even greater. For a plant with 34 ribs, maximum CO_2 uptake per unit height occurs at an intermediate fractional rib depth of 0.15 (Fig. 6.23*D*). As the ribs become deeper than this, the PAR per unit area decreases faster (Fig. 6.23*B*) than stem area increases (Fig. 6.23*A*).

We should consider another factor: stem volume. Although stem volume does not change much as the number of ribs is varied, it does when fractional rib depth is varied, because stem volume is removed as the ribs get deeper. It is thus worthwhile to look at the CO_2 uptake expressed per unit of stem volume. For plants with five ribs, CO_2 uptake per unit stem volume increases 4.5-fold as the fractional rib depth goes from 0.0 to 0.6. Although it does not continuously increase for plants with 34 ribs (it increases 50% to reach a maximum at a fractional rib depth of 0.2), CO_2 uptake per unit stem volume at a fractional rib depth of 0.6 is only 20% less than it is at a fractional rib depth of 0.0, which is a much smaller decrease than for CO_2 uptake per unit stem height (Fig. 6.23*D*). Thus, deeper ribs can lead to a more advantageous distribution of plant biomass, at least with respect to net CO_2 uptake.

To help organize the interacting effects of rib depth and rib number, Geller and Nobel defined a new term, the perimeter ratio, as the ratio of total rib

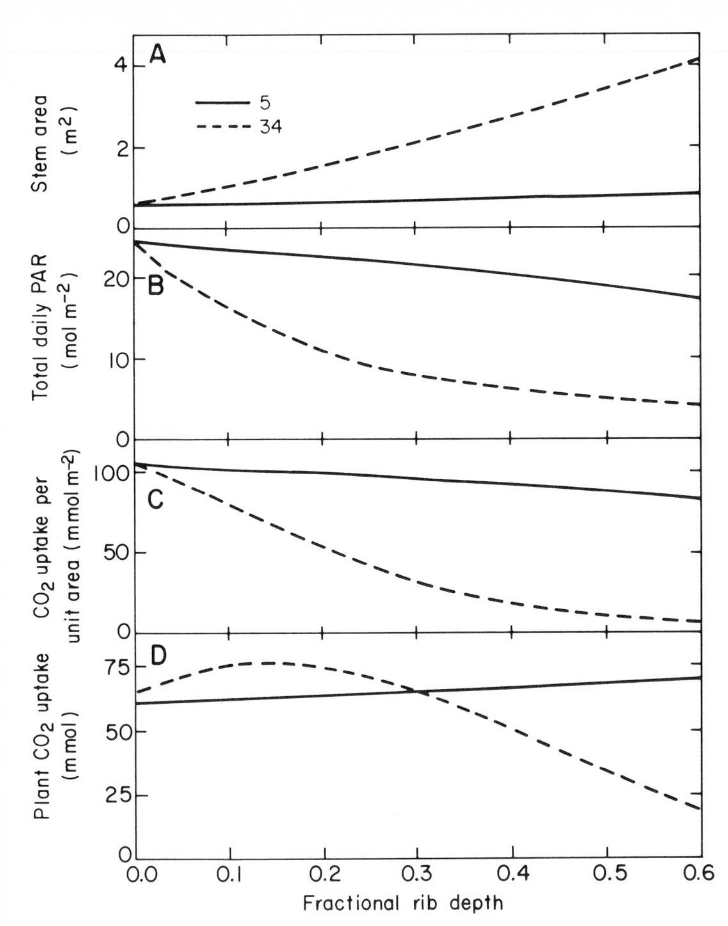

Figure 6.23. Influence of fractional rib depth on stem surface area *(A)*, average total daily PAR per unit area *(B)*, nocturnal CO_2 uptake per unit area *(C)*, and nocturnal CO_2 uptake for the whole plant *(D)*. Data, adapted from Geller and Nobel (1984), are for rib numbers of 5 or 34; simulations for plants with 16 ribs were intermediate between those for 5 and 34 ribs. Data are for a plant 1.0 m tall and 0.1 m in radius on an equinox at 34°N.

perimeter to stem diameter. For a smooth cylinder, this is simply the circumference (πd) divided by the diameter (d), which equals π, or 3.14. Plants that have the same perimeter ratio, for example, one with 13 moderately deep ribs and another with 34 shallow ribs (a perimeter ratio of about 6 in each case), had nearly the same CO_2 uptake per unit height (Fig. 6.24). Figure 6.24 indicates that CO_2 uptake is maximal at a perimeter ratio of about 6. Such simulations were for clear days at an equinox. The perimeter ratio maximizing CO_2 uptake is still near 6 on clear days at the winter solstice and at the summer solstice, but shading of the stem surface causes the optimal perimeter ratio to decrease.

As discussed in Chapter 5, shading a cactus stem can significantly decrease interception of PAR and productivity. This shading can be produced by spines, pubescence, other plants, mountains, clouds, and fog. Simulations showed that whenever shading is appreciable, the best design for a cactus is to have very shallow ribs or a low rib number. For moderate shading that reduces the PAR at the stem surface by 20%, the optimal perimeter ratio is just over 4 (Fig. 6.24). Based on rib morphology measured in the field, the perimeter ratio for *C. gigantea, F, acanthodes,* and *L. schottii* is 4.1 to 4.4, which would be optimal for CO_2 uptake under the moderate shading normally encountered in the field. Epiphytic cacti commonly live in shade and also have low values of perimeter ratio.

Ribs of cacti are not necessarily evenly spaced all around the stem, as indicated earlier. For instance, ribs on *C. gigantea* and *F. acanthodes* tend to be about 10% closer on the south (or southwest) side and about 10% farther apart on the north side than the average spacing. Also ribs on the south tend to be deeper than those on the north side, an arrangement that can affect the PAR per unit surface area. Nevertheless, although more rib area tends to occur in stem locations with higher available PAR, simulations indicate that the consequences of variable rib spacing and depth for net CO_2 uptake by the whole plant are rather minor.

The main effect of ribs relates to the distribution of PAR over the stem surface. The curvilinear relation between nocturnal CO_2 uptake and total daily PAR (Fig. 4.15) makes it advantageous to distribute the incident PAR over the stem surface in such a way as to avoid PAR saturation of CO_2 uptake by any of the surfaces. On the other hand, stem areas that are below PAR compensation and that

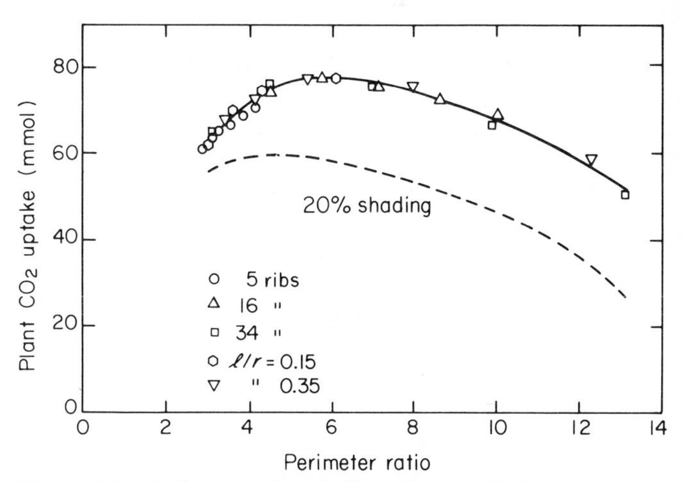

Figure 6.24. Influence of total rib perimeter divided by stem diameter (perimeter ratio) on nocturnal CO_2 uptake by the whole plant. Data are for various combinations of rib number and fractional rib depth. Also shown (– – –) is the pattern when the total daily PAR is reduced 20% from the unshaded condition. Data, adapted from Geller and Nobel (1984), are for a plant 1.0 m tall and 0.1 m in radius on an equinox at 34°N.

therefore have a net loss of CO_2 must also be minimized. These two factors underlie the utility of the perimeter ratio as a parameter predicting the configuration optimal for net CO_2 uptake. A secondary, complicating factor is that PAR is reflected from one rib surface to another. Thus, the total PAR absorbed by a ribbed surface is greater than that absorbed by a smooth circumscribing surface of the same diameter and reflective properties, that is, the ribbed surface would appear slightly darker because more of the incident PAR is trapped and hence not reflected back to the eye of the beholder.

It is hard to predict how these and other findings in the computer simulation study will be used to explain cactus ribbing in natural habitats. Desert biologists have observed that many cacti with different numbers and depths of ribs or tubercles coexist in a particular habitat; yet in another region the set of species and rib properties can differ. Perhaps the species are trying to maximize PAR interception and nocturnal CO_2 uptake in each environmental situation? The jury is still out, but the possible verdict is certainly tantalizing.

Reflectivity

A natural extension of our discussion of PAR absorption by stems of cacti is to examine the reflection of radiation—indeed we have already indicated that PAR can be reflected from one rib to another. Reflection is not only affected by the pigments in the stem but also by the surface microdetails such as roughness and thickness of the waxy covering. For instance, the main color reflected by the stems of most cacti and hence seen by us is green because wavelengths in the red and blue are strongly absorbed by chlorophyll. However, cacti exhibit many shades of green. Moreover, the corky stems of *Opuntia echios* (Fig. 6.16) are brown because essentially all of the chlorophyll has been lost. The red colors appearing in *Gymnocalycium* and *Neoporteria* are obviously due to red pigments, whereas in the related *Copiapoa cinerea* for reasons that are not clear in terms of pigments, the stems can be so dark they appear almost black. Purple or magenta stems colored by betalains (Chapter 9) are easily seen in some of the platyopuntias, most notably *O. violacea* (Fig. 5.5). Nonetheless, the percentage of PAR reflected from stems of cacti has rarely been measured or related to pigments.

Over 100 species of cacti have a surface that is gray or grayish green because their reflective properties are dominated by the nature of the surface rather than by pigment absorption. Some of these, such as *Ferocactus glaucescens,* appear to have very thick cuticles, whereas others, such as peyote, *Lophophora williamsii,* have a papillose epidermis. *Stenocereus beneckei,* a native of Mexico, is silvery white because of the presence of epicuticular wax, and the loose epicuticular wax of *Myrtillocactus eichlamii* makes its stems blue. Indeed, rough handling of the stem knocks the epicuticular wax off and the stem then appears green because of the chlorophyll in the underlying chlorenchyma. The gray color results when all wavelengths are reflected approximately equally, a common phenomenon that leads to the silvery color for the leaves of many desert plants. These reflective characteristics affect net CO_2 uptake because PAR cannot be used for photosynthesis if it is reflected away from the stem or absorbed by the cuticle, hypodermis, or pubescence, a situation analogous to our discussion of the influences of spines in the last chapter.

Reflectivity of the stem surface also directly affects the temperature of cacti. Specifically, only a certain fraction of the incident shortwave radiation (Chapter 5) is absorbed, generally about 0.6. If less energy were to be absorbed because of a highly reflective pubescence or cuticle, the stem would not heat up as much during the daytime but photosynthesis would also generally be less because of a lower PAR absorption. Stem properties also affect the absorption of longwave (thermal) radiation. As is the case with other plants, however, the fraction of longwave radiation absorbed by the stems of essentially all the cacti that have been measured ranges from 0.96 to 0.99; thus, variations in longwave properties apparently have only a minor influence on the thermal properties of cacti.

Thermal Effects of Ribs

With a little imagination we can see that ribs resemble heat-dissipating fins. Long ago a German worker named Franz Herzog concluded that ribs would increase convective heat loss and thereby decrease excessive plant temperatures. Lewis and Nobel investigated the thermal consequences of ribs in some detail for *Ferocactus acanthodes* using the model discussed in the previous chapter (Fig. 5.21). Although their model looked primarily at the effect of ribs on transpiratory surface area, they found that ribs appeared to have only a small secondary effect on stem surface temperatures. We must admit at the outset that the effect of ribs is quite complicated, and thus the conclusions must be accepted with caution.

Let us consider a cylindrical stem with 21 ribs of fractional rib depth 0.15 (Fig. 6.21*B*). What are the possible effects of the ribs on the exchange of longwave radiation? Because the longwave radiation received or emitted depends essentially only on the circumscribing surface area, rib number and depth

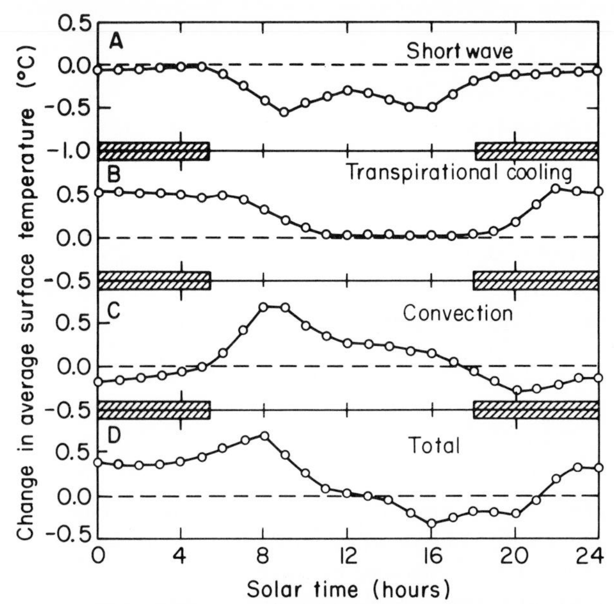

Figure 6.25. Effects of eliminating ribs on the average hourly surface temperature of a cactus stem with respect to shortwave radiation *(A)*, transpirational cooling *(B)*, convective heat exchange *(C)*, and all three factors together *(D)*. Previously unpublished data are for a 21-ribbed plant with a fractional rib depth of 0.15 (Fig. 6.21*B*) and are obtained by simulation for a plant 1.0 m tall and 0.10 m in radius on an equinox at 34°N. Environmental conditions are described in Nobel (1978).

have no detectable effect here. What is the effect of removing the ribs on the absorption of shortwave radiation? Removal of ribs eliminates the reflection of radiation from one rib to another and thus lowers the overall shortwave absorption of the plant by about 10%. This slight lowering of shortwave absorption lowers the average surface temperature of the stem somewhat, but only during the daytime (Fig. 6.25*A*). The effect is least near noon because the sun is then approximately overhead; hence less shortwave radiation is intercepted by a vertical stem. The maximum effect is only about 0.6°C. A similar thermal effect could be induced by a slight loss of chlorophyll (bleaching), a change that would cause the stem color to go from, for example, green to light green.

Increasing stem surface area by ribbing (Fig. 6.22*A* and 6.23*A*) allows not only more area for interception of PAR and CO_2 uptake but also more area for transpiratory water loss. For instance, eliminating the ribbing on a 21-ribbed plant with a fractional rib depth of 0.15 decreases the surface area by 26%. If the water vapor conductance (g_{wv} in Eq. 1.1 and Fig. 4.10*C*) were to remain the same, a similar decrease in transpiration and hence transpirational cooling would occur (Fig. 6.25*B*). The effect is rather small, however, and the loss of cooling is observed mainly at night because that is when stomates of CAM plants tend to be open (Figs. 4.9 and 4.10). Indeed, the loss of ribbing in this situation leads to a maximum rise in average stem surface temperature of only about 0.6°C, using g_{wv} in Figure 4.10 (a higher g_{wv} would lead to more cooling).

The situation for convective exchange is somewhat more difficult to model because any change in surface geometry affects the wind flow pattern around a cactus stem. Heat convection is proportional to the difference in temperature between the stem surface and the moving air just across the air boundary layer (Chapter 5) next to the surface. The coefficient of proportionality is known as the heat convection coefficient, which varies with position on a rib. For instance, it is highest near the rib crest where air movement is greatest, 24% lower at midrib, and fully 70% lower in the trough at the base of the ribs because the average air motion is much less there. For purposes of calculation, we shall assume that the elimination of ribbing causes the average heat convection coefficient to change from that suitable for a *Ferocactus acanthodes* with 21 ribs to a value appropriate for a smooth circumscribing surface of the same height and diameter as the cactus stem. For these calculations, values were for spineless surfaces—inclusion of spines lowers the heat convection coefficient about 15% for an *F. acanthodes* whose stem was shaded about 40% by spines. In this case elimination of ribs lowers the heat convection coefficient about 5%. Thus, the stem tends to heat up more above air temperature during the day and to cool down slightly more at night when ribs are eliminated (Fig. 6.25*C*).

Let us next consider the consequences of the simultaneous action of various factors influencing average stem surface temperature. The pattern of changes upon rib elimination depends on stem morphology, degree of stomatal opening, and environmental conditions. Figure 6.27*D* indicates that elimination of ribs can have relatively minor effects on the average surface temperature.

The effects of ribbing on interception of PAR, CO_2 uptake, and surface temperature of cacti that we have been discussing also apply to tubercles. For instance, it is clear that the projections represented by tubercles increase the stem surface area just as ribs do. But because of their more complicated geometry, the actual effects of tubercles have not been quantified in a model.

Origin of Ribs and Tubercles

In each of the three subfamilies of cacti, the origins of ribs and tubercles have been different. The primitive condition for the family is assumed to be the smooth, cylindrical stem because the leaf-bearing cacti—*Pereskia, Pereskiopsis,* and *Quiabentia*—as well as *Maihuenia* (Pereskioideae) and *Austrocylindropuntia* lack tubercles. In the subfamily Opuntioideae tubercles appear in the chollas, which live in open, high-light habitats in deserts and grasslands; but the species with narrow stems, few spines, and low or no tubercles, for example, *O. cinerascens, O. leptocaulis,* and *O. ramosissima,* occur also in open desert habitats. In very dry areas of California and Arizona, a tuberculate species like *O. echinocarpa* and species with flat tubercles, for example, *O. ramosissima,* can be found growing with a low, ribbed species, for example, *O. stanlyi,* and several platyopuntias, for example, *O. basilaris.* Based on this and other information, it would seem that leaf-bearing species with cylindrical stems living in partial shade gave rise to forms that have tubercles and are adapted for full-sun habitats, forms that in turn gave rise to new forms that have flattened tubercles (narrow-stemmed chollas), ribs (corynopuntias), nearly smooth, spherical joints (tephrocacti), or flattened stems (platyopuntias). Much research is needed to elucidate the reasons for these evolutionary events as well as the exact sequence of morphological changes.

In the subfamily Cactoideae, the primitive condition is a stem with ribs, as is found in species of Leptocereeae and Hylocereeae and in primitive members of each of the tribes. By and large, columnar forms have ribs, but in *Harrisia* a few species have helically arranged tubercles on long, fairly narrow stems; these species live in partially shaded habitats, such as scrubland or tall grassland. Elsewhere, helically arranged tubercles have arisen only on low, thick-stemmed cacti, the globular and caespitose growth forms (Figs. 1.7–1.9). Interception of radiation as well as heat storage must be much different for a low plant, such as a solitary stem of *Mammillaria* that is wedged in a rock crevice or half-buried in soil or for a tightly packed cluster of tuberculate stems in a caespitose plant. Where ribs occur in small, very low cacti, they are characteristically very flat, as in *Gymnocalycium, Lobivia,* and *Lophophora.*

Solving the mysteries of the many small cactus designs awaits new research. To be analyzed carefully are the living rocks of *Ariocarpus,* which have tubercles that are frequently half buried by soil (Chapter 8). In fact, all analyses of small cacti will have to consider how the morphology of the plant relates to that of its substrate, because for these the temperature of the substrate and plant are probably closely linked. These plants offer botanists an excellent opportunity to study the evolution of structural diversity in a group within an ecological context.

Cladodes

Nothing upsets a botanist more than to hear people talk about the cladode or pad and call it a leaf, because what they are actually discussing is a flattened shoot or stem. *Cladode* is the current term for a flattened stem that functions in place of a leaf, but another term, *phylloclad,* is also useful because this means leaflike stem. The cactus cladode consists of a stem, future shoots (that is, the axillary buds), and leaves; and it has the typical internal structure of a stem.

Origin of a Cladode

Plants have two modes of growth: indeterminate and determinate. Indeterminate growth could also be called indefinite growth because the shoot apex produces leaf primordia as long as the enviromental conditions are favorable for leaf production. Growth of the shoot stops when stress is imposed, for example, when watering is discontinued. In contrast, a plant with determinate growth produces discrete shoot units regardless of the environmental conditions. Examples of determinate growth are the cylindrical joints and cladodes of opuntias (Figs. 1.19–1.22). A joint arises from an areole of a dormant cladode and grows very rapidly. Within a week or two all of the leaves and areoles of the new shoot have been formed, after which the joint enlarges to its mature size and shape. Changing environmental conditions has little effect on that new shoot, and another whole joint must be produced to add more leaves and areoles.

Very little scientific information is available on the origin of the cladode in cacti. Several investigators have observed that the shoot apex of a platyopuntia has the normal pattern of apical zonation (Chapter 2), and a platyopuntia can have a phyllotaxy of 3/8, 5/13, or 13/21. Unfortunately, no details have been reported on the expansion of the stem, which is really caused by the elongation in one plane of the pith. Apparently the pith becomes elongated when its cells produce short chains of new cells, an arrangement that causes the areoles on the broad sides of the pad to be spread apart more than they are along the edges of the pad. A very detailed study of the developmental sequence in many apices is required to clarify how cladodes grow in each dimension.

Flattened Stems of Epiphytes

Many books and articles classify the flattened stems of cactus epiphytes as cladodes, for example, most

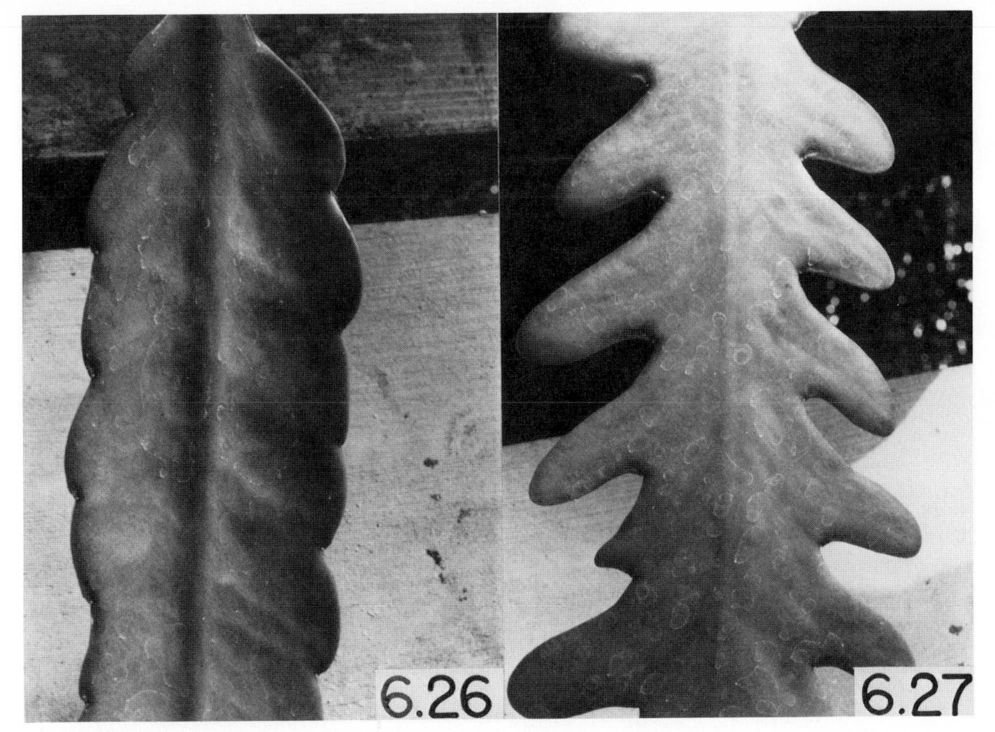

Figures 6.26 and 6.27. Two-ribbed stems of cactus epiphytes.
6.26. *Eccremocactus imitans.*
6.27. *E. rosei.*

stems of the species of *Epiphyllum, Disocactus, Eccremocactus* (Figs. 6.26 and 6.27), *Schlumbergera,* and *Nopalxochia,* and of some species of *Rhipsalis.* These are cladodes in the traditional sense, but they seem to have an origin that is very different from that of the flattened stems in platyopuntias. In flat-stemmed epiphytes of the subfamily Cactoideae, the pith is not expanded; instead, the pith is small and encircled by a vascular cylinder like that found in narrow-stemmed, ribbed, terrestrial cacti. The flattened portion is produced in a manner homologous to that of a rib. Moreover, the flat portion of many epiphytic stems is very thin—often less than 4 mm thick—whereas cladodes are typically 1 to 4 cm thick. With the exception of Christmas cactus, which is a species of *Schlumbergera* and has discrete joints (Fig. 1.27), most of these epiphytes have indeterminate growth. Finally, areoles on epiphytic stems are restricted to the two edges, whereas they cover the stems on a platyopuntia cladode. Therefore, it is easier to classify the epiphytes as two-ribbed cacti. Recognizing the differences between cladodes in platyopuntias and two-ribbed epiphytes makes sense when these cacti are studied carefully, because a single epiphytic stem can pass from a stage of having four ribs to three ribs to a totally flattened stem, all this without any pith enlargement (Fig. 6.28). In contrast, cladodes of platyopuntias never arise from stems with ribs because the cylindrical stems of the young plant have helically arranged tubercles. A few species classified as platyopuntias, for example, *Opuntia pumila,* actually have mostly cylindrical stems.

Functions of Flattened Stems

We have made a distinction between true cladodes and two-ribbed epiphytes because there seem to be two completely different strategies for producing flattened stems. Of course, both designs result in low volume-to-surface ratios, but this is where the similarity ends. Cladodes are not ideal structures for storing water, but they do have enough storage capacity to carry them through short periods of drought. Interestingly, platyopuntias are noticeably absent in extremely arid regions where drought lasts more than 6 months. On the other hand, flattened stems of epiphytes have very little water storage capability and must live in fairly moist environments (moist for a cactus, that is).

A thick, strong cladode is a necessity for building a tall plant. Specifically, all of the opuntioids with cylindrical stems are relatively small plants, rarely exceeding 4 m in the tallest forms, whereas many platyopuntias are large, massive trees. Given that

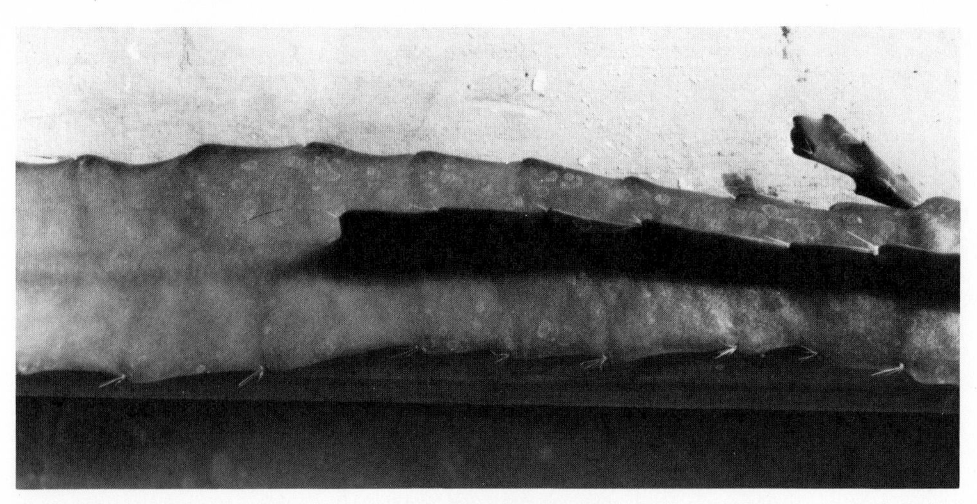

Figure 6.28. Variation in rib number in an epiphyte, *Nopalxochia horichii.* The old stem *(right)* has four ribs (one hidden), whereas the upper portion of that stem *(left)* has only two ribs. A young stem arising from the old stem *(right)* is tuberculate but not yet ribbed.

wood formation is not very great in opuntioids, a cylindropuntia with a narrow pith and a small woody cylinder lacks the mechanical strength to form a trunk and lacks lower branches capable of supporting a large canopy. The cladode, however, can be a much more massive structure; a series of vertical cladodes can be fused into a strong, central trunk often exceeding 20 cm in diameter. Therefore, a cladode can later be used to build a taller plant so that the canopy is higher off the ground and therefore shaded by less vegetation.

Ribbed stems of epiphytes are not designed for support, and so epiphytes must hang from host plants and rocks or cling to other structures by adventitious roots. In fact, when old parts of epiphytic plants die, the new roots are connected directly to the young stem bundles so that young ribs receive adequate water and minerals (Fig. 3.43). The adventitious roots of epiphytes erupt through the skin and directly feed into the vascular cylinder of the young stem.

Cladodes are so thick that PAR cannot pass through one side of the stem to be used on the other. Indeed, the presence of the woody cylinder and pith act as an opaque structure that blocks light passage through the cladode. In contrast, the very thin "rib" of an epiphyte mimics that of a typical leaf; therefore, light can enter from either side and supply at least some photons to chloroplasts on both sides of the stem. This epiphytic stem is a worthwhile type of photosynthetic organ for a plant that grows in shaded habitats, as do the forest epiphytes, and which may have stems turned in a variety of directions. A thin structure permits light entry, even from the tree side of the stem, whereas a cladode functions primarily as a stem with two completely independent photosynthetic faces; and as we shall see in Chapter 8, the faces are often oriented in a way that maximizes interception of PAR.

SELECTED BIBLIOGRAPHY

Boke, N. H. 1944. Histogenesis of the leaf and areole in *Opuntia cylindrica. American Journal of Botany* 31:299–316.

——— 1951. Histogenesis of the vegetative shoot in *Echinocereus. American Journal of Botany* 38:23–38.

——— 1952. Leaf and areole development in *Coryphantha. American Journal of Botany* 39:134–145.

——— 1953. Tubercle development in *Mammillaria heyderi. American Journal of Botany* 40:239–247.

——— 1954. Organogenesis of the vegetative shoot in *Pereskia. American Journal of Botany* 41:619–637.

——— 1955a. Development of the vegetative shoot in *Rhipsalis cassytha. American Journal of Botany* 42:1–10.

——— 1955b. Dimorphic areoles of *Epithelantha. American Journal of Botany* 42:725–733.

——— 1956. Developmental anatomy and the validity of the genus *Bartschella. American Journal of Botany* 43:819–827.

——— 1957a. Comparative histogenesis of the areoles in *Homalocephala* and *Echinocactus. American Journal of Botany* 44:368–380.

——— 1957b. Structure and development of the shoot in *Toumeya. American Journal of Botany* 44:888–896.

——— 1958. Areole histogenesis in *Mammillaria lasiacantha. American Journal of Botany* 45:473–479.

——— 1959. Endomorphic and ectomorphic characters in *Pelecyphora* and *Encephalocarpus. American Journal of Botany* 46:197–209.

——— 1960. Anatomy and development in *Solisia*. *American Journal of Botany* 47:59–65.
——— 1961a. Structure and development of the shoot in *Dolicothele*. *American Journal of Botany* 48:316–321.
——— 1961b. Areole dimorphism in *Coryphantha*. *American Journal of Botany* 48:593–603.
——— 1967. The spiniferous areole of *Mammillaria herrerae*. *Phytomorphology* 17:141–147.
——— 1976. Dichotomous branching in *Mammillaria* (Cactaceae). *American Journal of Botany* 63:1380–1384.
Boke, N. H., and R. G. Ross. 1978. Fasciation and dichotomous branching in *Echinocereus* (Cactaceae). *American Journal of Botany* 65:522–530.
Buxbaum, F. 1950. *Morphology of cacti.* Section I. *Roots and stems.* Abbey Garden Press, Pasadena, California.
Freeman, T. P. 1973. Developmental anatomy of epidermal and mesophyll chloroplasts in *Opuntia basilaris*. *American Journal of Botany* 60:86–91.
Geller, G. N., and P. S. Nobel. 1984. Cactus ribs: influence on PAR interception and CO_2 uptake. *Photosynthetica* 18:482–494.
Gibson, A. C. 1976. Vascular organization in shoots of Cactaceae. I: Development and morphology of primary vasculature in Pereskioideae and Opuntioideae. *American Journal of Botany* 63:414–426.
Gibson, A. C., and K. E. Horak. 1978. Systematic anatomy and phylogeny of Mexican columnar cacti. *Annals of the Missouri Botanical Garden* 65:999–1057.
Lewis, D. A., and P. S. Nobel. 1977. Thermal energy exchange model and water loss of a barrel cactus, *Ferocactus acanthodes*. *Plant Physiology* 60:609–616.
MacDougal, D. T., and E. S. Spalding. 1910. *The water balance of succulent plants.* Carnegie Institution of Washington, Publication no. 141, Washington, D.C.
Mauseth, J. D. 1984a. Introduction to cactus anatomy. Part 7. Epidermis. *Cactus and Succulent Journal* (Los Angeles) 56:33–37.
——— 1984b. Introduction to cactus anatomy. Part 8. Inner body. *Cactus and Succulent Journal* (Los Angeles) 56:131–135.
Nobel, P. S. 1978. Surface temperatures of cacti—influences of environmental and morphological factors. *Ecology* 59:986–996.
Robberecht, R., and P. S. Nobel. 1983. A Fibonacci sequence in rib number for a barrel cactus. *Annals of Botany* 51:153–155.
Spalding, E. S. 1905. Mechanical adjustment of the sahuaro *(Cereus giganteus)* to varying quantities of stored water. *Bulletin of the Torrey Botanical Club* 32:57–68.
Walter, H., and E. Stadelmann. 1974. A new approach to the water relations of desert plants. Pp. 214–302 in *Desert biology,* vol. 2, ed. G. W. Brown. Academic Press, New York.

7

Factors Affecting Distribution

Many environmental factors such as rainfall, temperature, soil type, and light dictate where plants can successfully grow. Cacti prefer some water, but not too much; they generally do not tolerate freezing; they generally do not grow well in deep shade; and they generally do not grow in saline soils (although some species are calciphiles, that is, lovers of limestone, or gypsophiles, that is, lovers of gypsum). In addition, biotic factors such as pollination, fungal and viral disease, competition, and herbivory can drastically affect the success of a plant species and thereby limit its abundance and distribution. Cacti, for instance, do not compete well in habitats where other plants can overtop them and thereby create a deep-shade environment. Other factors limiting plant distributions are related to human activities. Native habitats of cacti have been preempted for cultivation or for livestock grazing, and avid cactus collectors, for pleasure or for profit, have sometimes decimated accessible populations of interesting species.

The deserts of the New World are not the only places where cacti grow. Cacti are also common elements in many semiarid habitats, such as coastal bluffs, dry grasslands, scrubland or short-tree tropical forests. They occur at high elevations in mountains or cool habitats in lowland Patagonia, and the epiphytes of the family grow in moist, montane tropical forests. Nonetheless, the really well known places to see the great development of the family are the dry regions—places like western Peru and northern Chile; the Chaco and Monte of Argentina; Puebla, and especially the Valley of Tehuacán, in Mexico; the Chihuahuan Desert; and the Sonoran Desert.

Not surprisingly, the Sonoran Desert has been the traditional center for ecological studies on cacti. This arid zone is easily accessible to an assortment of academic institutions, and Saguaro National Monument near Tucson and Organ Pipe Cactus National Monument near the Arizona-Sonora border have attracted millions of visitors and numerous researchers. The Sonoran Desert has a great collection of cacti—over 125 species—largely because this desert has a variety of rainfall regimes in frost-free habitats and because many sites have a bimodal precipitation pattern, with significant rainfall in the winter and in the summer, which is well suited to the cactus life style. The Sonoran Desert also has regions where long summer dry seasons and freezing winter temperatures can limit the geographic distribution of species, giving researchers an opportunity to study how climatic factors influence seedling establishment, growth, productivity, and distributional range.

Certain environmental factors—water, low temperature, and high temperature—influence cactus physiology and distribution in the Sonoran Desert. We hope that the data presented in this chapter will give cactus collectors an insight into how cacti respond to environmental factors, which can be extrapolated to plants on windowsills, in greenhouses, and in outside gardens.

Water

We have already discussed many aspects of the water relations of cacti. The water-conserving strategy of transpiring at night was introduced in Chapter 1 and amplified in Chapter 4. We have seen that the succulence of the chlorenchyma cells (Chapter 3) is necessary for the storage of a large amount of water and hence malate (Chapter 4). The ribbing of the stem facilitates the changes in stem volume associated with storage of water (Fig. 1.28 and Chapter 6). In this chapter we shall examine water relations from a more ecological point of view, beginning with the establishment of the seedling, then examining root responses to soil water, then considering drought

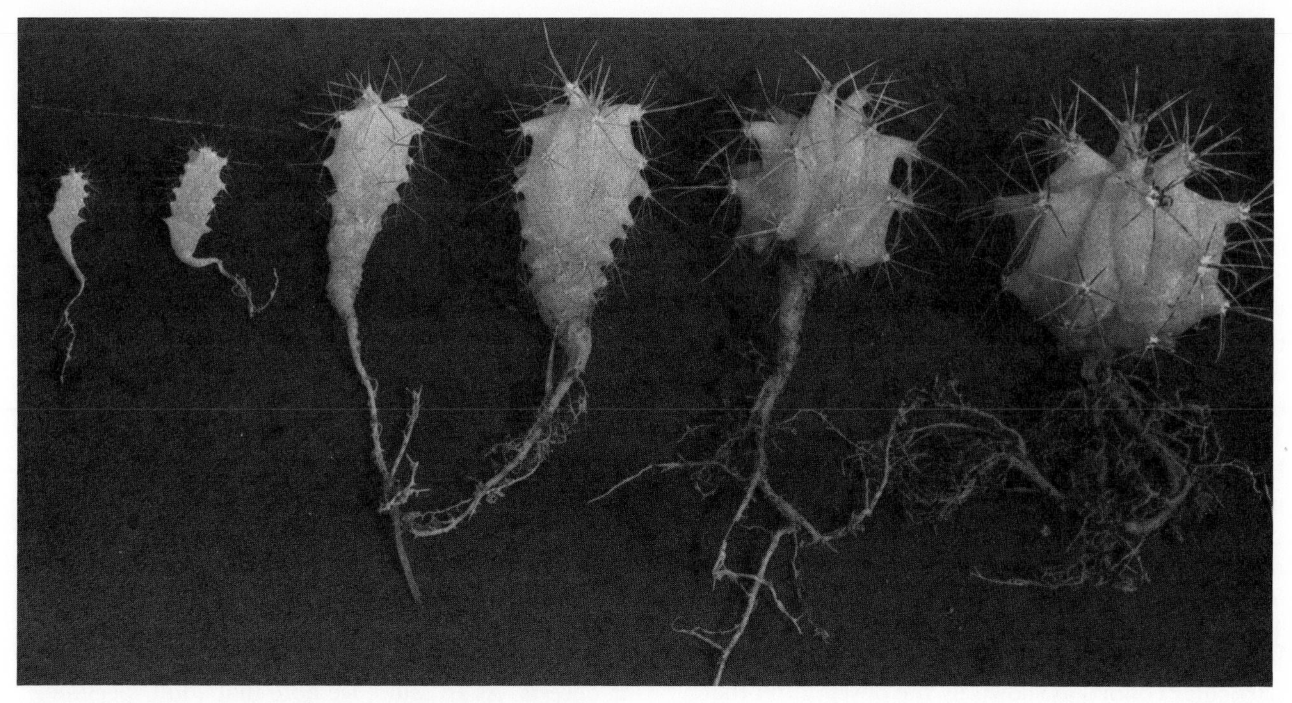

Figure 7.1. Seedlings of *Ferocactus wislizenii* grown under well-watered, glasshouse conditions. The ages of these young cacti *(left to right)* are approximately 3, 6, 12, 16, 20, and 24 months (the last seedling is 3.4 cm tall).

responses for the stem, and finally analyzing the water cost of reproduction. Indeed, the key environmental factor affecting the ecophysiology of cacti is the availability of water, so this is a logical starting point in our analysis of how cacti are affected by the environment.

Seedling Establishment

Water relations play a critical role in the survival of cacti during the seedling stage. To survive a particular drought, the seedling must develop sufficient water storage tissue during the preceding wet season. We have already indicated that a spherical shape is the best form for water economy, because a sphere maximizes the volume of water stored per unit area of transpiring surface; however, the reduced surface area of a sphere also limits CO_2 uptake and hence growth. For *Ferocactus,* the longer the growing period, the greater the volume-to-surface ratio becomes (Fig. 7.1 and Chapter 3). Thus, larger seedlings can successfully endure a longer drought. Under field conditions a seedling of *F. acanthodes* that has grown for 50 days in moist soil can tolerate 9 days of drought, but one that has grown for 150 days can tolerate 70 days of drought (Fig. 7.2). In years where short wet periods are followed by extensive droughts, successful seedling establishment for this species cannot occur. Because the volume-to-surface ratio for adults is very high, lethal desiccation of adults is very rare.

Peter W. Jordan and Nobel conducted a study of cactus seedling establishment in natural habitats using the "historical records" contained in the age structure of plant populations. The years for successful establishment can be determined by observing a new cohort of seedlings at a particular field location —a cohort is a group of organisms of the same age. But finding seedlings in the field can be tedious and difficult; it requires a trained eye with a very good

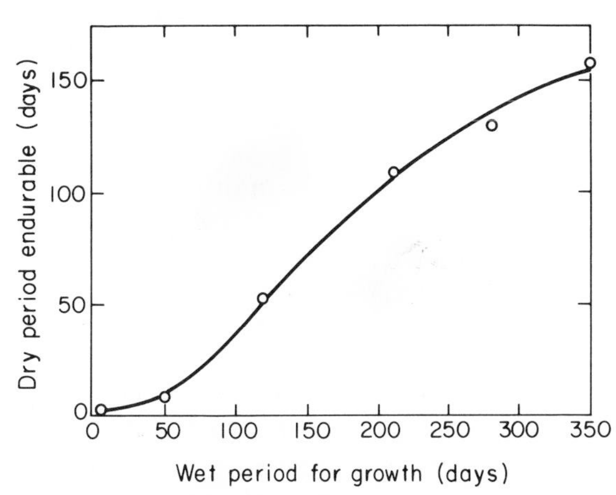

Figure 7.2. Tolerance to drought of seedlings of *Ferocactus acanthodes.* (Data are adapted from Jordan and Nobel, 1981.)

search image for small seedlings (Fig. 7.1). Alternatively, the heights of all small plants in a population at a particular site can be determined and discrete height classes, that is, peaks at which a large number of the plants occur, can be identified. Assuming that the annual change in height can also be calculated—taking into consideration the idiosyncratic nature of desert rainfall for the particular years involved—we can convert a height class into an age class. In this way, Jordan was able to determine in which years there was successful seedling establishment of *F. acanthodes* and *Carnegiea gigantea* in the western part of the Sonoran Desert.

The changes in all the height classes for a particular population can also be monitored with time. For instance, Jordan found that the most prominent peaks in the height distribution for seedlings of *F. acanthodes* at a particular site in the northwestern Sonoran Desert occurred at heights of 27 mm in May 1979 and 39 mm in November 1981. The shift from 27 to 39 mm represents the increase in height of a particular cohort, as shown by the observation that six individuals of average height 30 mm in May 1979 increased to 40 mm in November 1981. This is not a very impressive growth rate, but the annual increment in height can vary considerably. For instance, 1-year-old seedlings of *C. gigantea* can be 25 mm tall when grown in a commercial nursery, 13 mm tall when grown in a lath house, and only 3 mm tall in the field in Arizona.

The best time for seed germination for *F. acanthodes* and *C. gigantea,* as far as temperature is concerned, is in the late summer. Indeed, many experiments on seed germination, done in laboratories and in cactus nurseries, have shown that seed germination is highest at temperatures near 25°C. For germination to occur, adequate rainfall is necessary; but rainfall varies greatly from year to year. Jordan was able to correlate the summer wet period in various locations and in particular years with the observed cohorts of cactus seedlings. Indeed, most such seedlings germinating in response to summer rains are killed by droughts during the autumn or during the spring of the following year. Jordon examined weather records and calculated for a series of locations in southern California the percentage of years with wet periods long enough to enable seedlings of *F. acanthodes* to survive the ensuing droughts (Fig. 7.3). This species was found in regions where 10% or more of the years were suitable for seedling establishment. On the other hand, natural populations of *F. acanthodes* were absent from regions where less than 10% of the years were deemed suitable for establishment of seedlings (Fig. 7.3) based on their observed drought tolerance (Fig. 7.2).

Although common in Arizona, *Carnegiea gigantea* is almost completely absent in California. In California there is a tiny population of saguaro near Parker Dam on the western border of the Colorado River. But cross the Colorado River and drive several kilometers into Arizona, and you will find saguaro populations on rocky hillsides. The sharp western boundary of this species has interested scientists for years. Jordan studied the rainfall data for sites on both sides of the Colorado River, identifying suitable years for seedling establishment in each locality. Saguaro occurred near Quartzite, Arizona, where 23% of the years were deemed suitable for seedling establishment. However, saguaro populations were not present near Blythe, California, and Yuma, Arizona, where 6 and 3%, respectively, of the years were suitable for seedling establishment. Again, the importance of length of wet season versus length of drought is suggested here as a prime factor affecting the limits of this species; this factor may explain why saguaro does not occur throughout southern California deserts. As an aside, we note that adult specimens of saguaro are purchased and placed in private collections and botanical gardens in many areas of coastal southern California, where they do very well even without supplemental watering, although the dryness of the intervening regions does not allow them to range that far west naturally because of the failure of seedling establishment.

Of course, biotic factors also play a role in seedling establishment, especially herbivory of the seeds and seedlings by mammals, birds, and insects. The best studies on losses due to herbivory were again done on saguaro, for example, the studies by Warren "Scottie" Steenbergh and Charles H. Lowe conducted at Saguaro National Monument in Tucson. During its life span a healthy saguaro produces about 40 million viable seeds. There are thus 40 million chances for the production of more saguaros. But as any student of ecology knows, to maintain the population in a steady state only one seed leading to a reproductively successful plant is necessary. Only a small fraction, probably less than one fourth, of the viable seeds land in the proper microhabitat for germination. Of these, only about 0.4% actually germinate, mainly because of destruction by harvester ants, by birds such as white-winged doves and western mourning doves, and by rodents such as ground squirrels. Those seeds that survive foraging long enough to germinate still have a rough life ahead of them. Many rodents eat the tender young plants, as do a number of species of birds and the newly hatched larvae of many insects. Indeed, over 99% of the seedlings succumb to animals during the first year, and most of the seedlings that do survive are usually concealed beneath protecting plants or in rocky

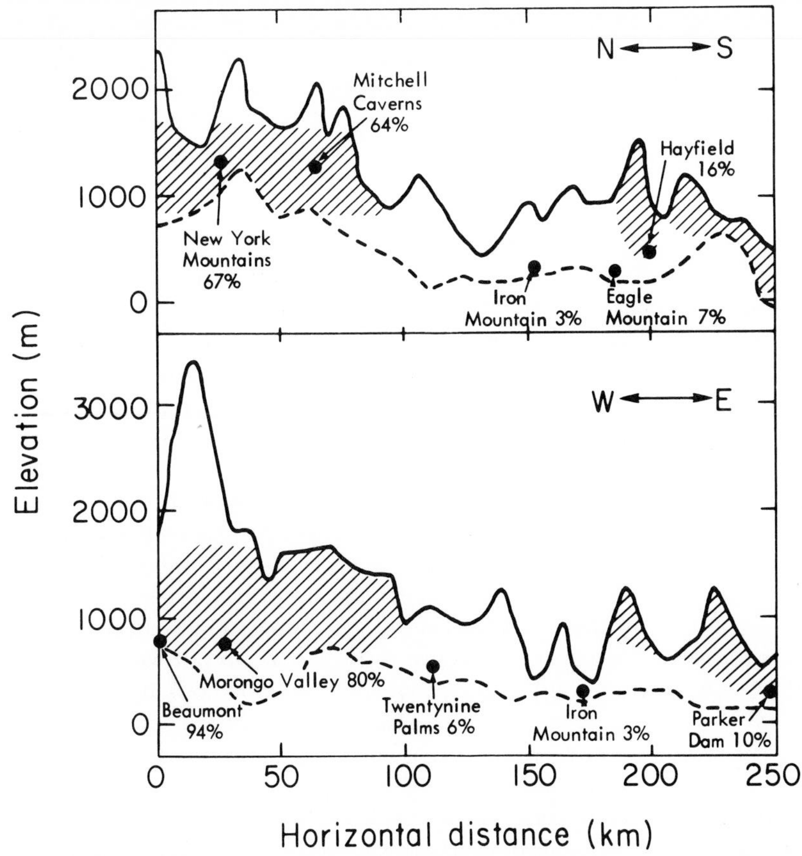

Figure 7.3. Elevational distribution of *Ferocactus acanthodes* at various locations in southern California. Elevational profiles indicate highest (—) and lowest (---) elevations within bands 20 km wide along transects running north–south *(upper panel)* or east–west *(lower panel).* The hatched areas represent known populations of *F. acanthodes.* Percentages accompanying the site designations represent the percentage of years suitable for seedling establishment for 1950 to 1980. (Data are from Jordan and Nobel, 1982.)

crevices. But the cacti soon begin fighting back against their predators. Oxalates poisonous to most animals begin to accumulate in the tissues, thus discouraging predation. Also, secondary chemical compounds (Chapter 9) that can deter herbivory occur, and spines develop, further deterring most animals.

Roots

Water that enters a cactus is taken up from the soil via the roots. At the same time, the minerals required for plant growth (to be considered in Chapter 9) also come from the soil. But root functions are much more difficult to understand than are those of leaves and stems because there is no known way to measure water and mineral uptake by a root that is still planted in its original soil and still attached to the plant. Removing the soil from roots causes breakage of young roots and root hairs, which are the fine structures that are responsible for much of the water uptake. Hence, we have little information on intact root physiology, and especially for cacti root observations are very meager.

A number of studies have indicated that cacti can take up appreciable amounts of water soon after a rainfall. For instance, *Opuntia puberula* can develop new roots a few hours after watering, and nocturnal stomatal opening for *O. basilaris* can occur about 1 day after a drought period is interrupted by a sizable rainfall. Just the appearance of new roots, however, does not mean that they contribute very much to the total water uptake by the plant, although the reestablishment of stomatal opening does indicate that appreciable water uptake can occur in a day or so.

One of the key aspects of the roots of cacti is that they are shallow. Even in large shrubby and arborescent cacti, most roots are found in the upper 30 cm of the soil, and the average depth for many species is only 10 cm. This statement is true for both established and rain-induced roots of *Ferocactus acanthodes* (Fig. 7.4). Fifteen days after a major rainfall, the root system of a barrel cactus 32 cm tall increased 27% in length, to a total of 230 m, when the lengths of all

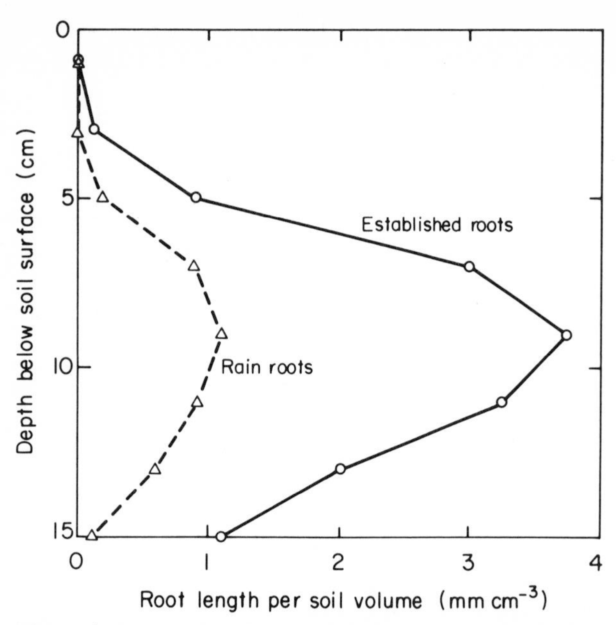

Figure 7.4. Lengths of roots of *Ferocactus acanthodes* per unit volume of soil at various depths below the soil surface. Data were collected for a plant 32 cm tall 15 days after a major rainfall (58 mm). (From Jordan and Nobel, 1984.)

the established and rain-induced roots were added. The shallowness of the roots means that they can take advantage of even light desert rains. In fact, rainfalls as low as 6 or 7 mm (about $\frac{1}{4}$ inch) on totally dry soil are sufficient to lead to water uptake by cacti and other desert plants, as indicated in Chapter 3 (see Fig. 3.51).

John Sanderson of the Agricultural Research Council Letcombe Laboratory in Wantage, England, has developed a technique for measuring water uptake by roots of agricultural plants such as barley (Fig. 7.5). Basically, the technique involves carefully attaching a small capillary to a severed root. Application of a partial vacuum to the upper end of the capillary, with the root at the other end dipped in water, causes water to be sucked from the bathing solution through the root to the capillary. By varying the negative pressure applied to the proximal end of a root, we can measure how much water flows in for a given drop in water potential. Because we know the diameter of the inside of the capillary, we can determine the volume of water coming out of the root per unit time by measuring how fast the air–water meniscus moves along the capillary (Fig. 7.5).

The water movement occurring in the Sanderson set-up is fully analogous to the process of water loss by transpiration, which causes a pulling or sucking on the water in the xylem. The sucking on the water was referred to in the older literature as *suction pressure* or *suction force* (*Saugkraft* in the original German papers), terms that are now replaced by the terms *negative hydrostatic pressure* (see Chapter 3) or simply *tension.* The tension in the water in the leaf or stem xylem caused by transpiration is transmitted all the way to the roots, where the negative hydrostatic pressure that is developed there facilitates the entry of water from the soil—recall from Chapter 3 that water will spontaneously move to regions of lower water potential (Eq. 3.3).

What does all this have to do with water uptake by cacti? Nobel took some *F. acanthodes* to England and made some measurements with Sanderson on water uptake by both established roots and rain-induced roots. When plants that had been deprived of water for 1 month were placed in an aqueous solution, new roots were visible in 8 hours. Although their rate per unit area was higher than the rate per unit area for old established roots, their small surface area led to very little total water uptake. As such rain roots grew day by day, they began taking over a larger fraction of the water uptake by the root system as a whole. However, it was not until 5 or 6 days of wet conditions had elapsed that the rain roots took up as much total water as the previously established roots.

Such studies on excised roots also provided information on one of the most perplexing questions in the water relations of desert succulents. Specifically, the stems of cacti are relatively wet, generally having water potentials of -0.5 to -1.0 MPa (range -0.2 to -1.7 MPa), and yet such plants can survive extensive droughts where the soil water potential can be below -9 MPa for many months. The water potential discussion in Chapter 3 indicates that water

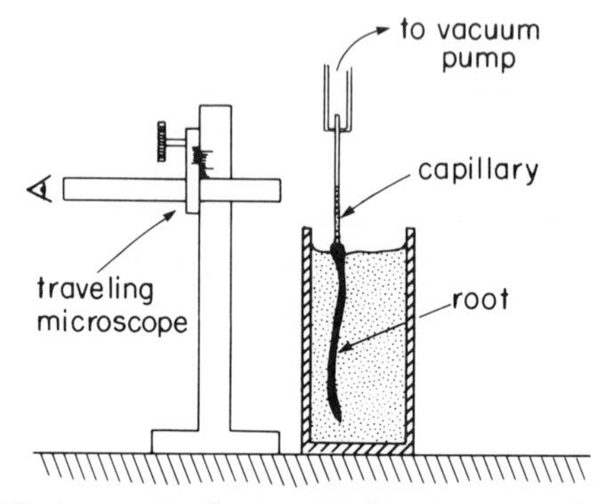

Figure 7.5. Apparatus for measuring water uptake by excised roots. The epidermis, cortex, and periderm were carefully removed under a dissecting microscope so that a small capillary could be sealed to the exposed stele (central tissue in root) using a low-melting-point wax/resin mixture. A vacuum could then be applied to the xylem contained within the stele, thus avoiding any artifactual flow of water along the root cortex. The excised root was placed in distilled water or another solution, and a partial vacuum was applied. A traveling microscope equipped with a fine vernier scale was used to monitor the resulting movement of the meniscus.

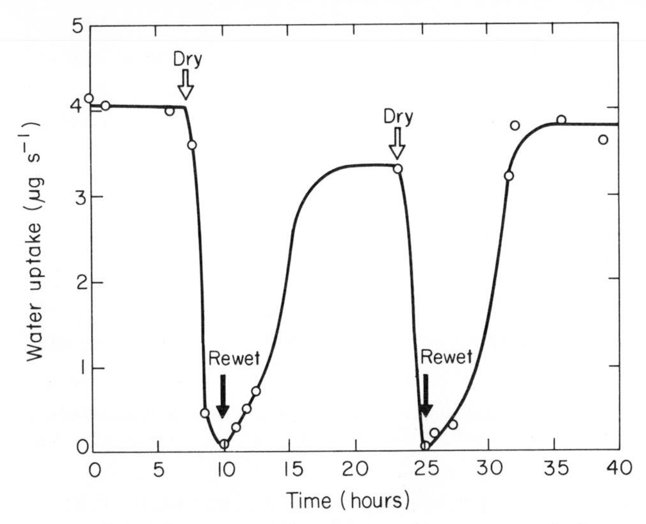

Figure 7.6. Time course for changes in water uptake by roots of *Ferocactus acanthodes* upon drying and wetting. A 26-mm terminal segment of a root was attached to a capillary as shown in Figure 7.5 and placed in an aqueous solution; a vacuum of 0.04 MPa was applied. The root was sequentially removed from (Dry) and replaced in (Rewet) an aqueous solution. (Adapted from Nobel and Sanderson, 1984.)

should then be moving from the relatively wet stems to the dry soil. Because water uptake from a wet soil to the stems can be quite rapid, water loss back to a dry soil would also be expected to be rapid if no change occurred in the roots. However, as the soil begins to dry, water is preferentially lost from the epidermis, cortex, and periderm of the roots, but the region that occurs radially inward from these tissues, known as the stele, remains hydrated. The loss of water from the external, or extrastelar, region greatly decreases the conductance of water out of the root and thus prevents loss of the precious water stored in the stem. The extrastelar dehydration somehow interrupts the water pathway from the soil to the stele, which also happens for roots of other plants during drought, although the exact anatomical explanation is not known. The roots also tend to shrink away from the soil during drying, thereby making it more difficult for water to move from the root to the soil. This shrinking is only a secondary factor, however, because the dehydrated roots have a very low water conductance, even when placed directly in an aqueous solution (Fig. 7.6).

We are now ready to appreciate one of the most fascinating aspects of the roots of cacti. When the dehydrated established roots are exposed to water, such as can occur from a light desert rainfall, the extrastelar region of the root becomes rehydrated and water uptake can begin again in a matter of hours (Fig. 7.6). For established roots of *F. acanthodes* that have been dried for a few hours and placed in an aqueous solution, the water uptake rate returns to the maximal value in about 10 hours (longer times are required if the drought is prolonged). This is how the stem water content can increase rapidly after a rainfall (Fig. 1.28). After a few days the contribution of new roots becomes important, as mentioned earlier, but the initial water uptake is by the rehydration of existing roots. What happens when a drought sets in? In actuality, this may be a more crucial aspect for the survival of cacti in desert regions than the uptake of water after a rain. The root conductance falls rapidly when roots are placed in air (Fig. 7.6). After 3 days of drought, the water conductance for the roots of *F. acanthodes* can decrease 700-fold, thereby minimizing water loss to a drying soil. Thus, the water potential of the stem can be safely maintained at a relatively high value even when the roots are in an extremely dry soil.

Stem Drought Characteristics

The storage of water in the stems of cacti is an obvious drought-enduring feature that we have already considered a number of times. The relatively high amount of water in the stem is sustained by nocturnal stomatal opening, which may continue to occur for 6 to 7 weeks after the soil is drier than the stem (Fig. 1.28), by the high volume-to-surface ratio of the adult plants (Chapter 3), and by the accordion-like swelling facilitated by ribbing (Chapter 6). Here we will return to these themes and shift our attention to certain metabolic aspects.

Some of the most comprehensive studies on the relation between drought and metabolic activity of cacti have been made by Stan R. Szarek and Irwin P. Ting at the University of California, Riverside. They found that *Opuntia basilaris* (Fig. 7.7*A*) could have a daily fluctuation of tissue acidity indicative of CAM activity even during long periods of drought, when the stomates are fully closed and the highly impervious cuticle essentially prevents any gas exchange with the environment. The technique involves using a cork borer to cut out stem portions from these platyopuntias at dawn and at dusk (Fig. 7.7*B*), after which the chemical contents of the removed samples are analyzed. (Indeed, if you were to visit the sites of their measurements at the University of California Philip L. Boyd Deep Canyon Desert Research Center near Palm Desert, California, you might occasionally observe curious little circular holes in certain cladodes of this species.) As soon as rainfall occurs, stomatal opening is induced, large amounts of CO_2 are taken up, and the relatively small daily oscillations in malate become large daily oscillations, as already mentioned in Chapter 4.

Figure 7.7. *Opuntia basilaris,* beavertail cactus. *(A)* Plant growing in its native habitat in southern California. *(B)* Holes were punched in the cladode by a cork borer to remove samples for analysis of tissue acidity.

The small malate oscillations during drought represent internally recycled CO_2. Specifically, CO_2 released by respiration is incorporated into malate at night via the CAM pathway (Chapter 4); during the daytime the malate is decarboxylated and the released CO_2 is fixed by the C_3 photosynthetic pathway, again behind closed stomates. Although no gas exchange with the environment is occurring, all the pathways are being used and all the enzymes are present. Thus, the cactus is metabolically ready to go as soon as the drought is lifted by a rainfall, an event that can happen during the late summer and the winter in the Sonoran Desert, which has a bimodal annual rainfall pattern (Fig. 4.16). During wet periods the stem water potential rises above −0.5 MPa, and it drops below −1.0 MPa toward the end of the ensuing droughts. When the stem water potential declines to −1.0 MPa or lower, stomates of cacti tend to remain closed.

Apparently the longest reported record for the persistence of nocturnal acid accumulation belongs to the cylindropuntia *O. bigelovii* growing at the same site as *O. basilaris.* Indeed, for plants that had been cut off at ground level and maintained in the desert for 3 years without the possibility for water uptake (Fig. 7.8), the acidity level of the tissue still increased throughout the night and decreased throughout the daytime. The magnitude of the fluctuation was reduced only 20% over the 3 years, although the water potential of the tissue had decreased from −0.8 to −1.3 MPa. Most of the joints had dried out, beginning at the bottom of the plant, so that only 5 out of an initial number of 40 remained green and living after 3 years. Nevertheless, by shifting resources toward the youngest joint the plant remained alive—if a bit reduced in size—over this 3-year drought period.

Reproduction

The number and timing of flowers can be affected by the water relations of the plants. For instance, absence of flowering by *Carnegiea gigantea* has been noted near Parker Dam, California, where only 10% of the years are suitable for seedling establishment. Thus, seed production may also be affected by low rainfall.

The flower petals and the other reproductive structures can also lose water by transpiration—another way that reproduction interacts with the water relations of the plant. For *Ferocactus acanthodes,* Nobel found that 8 g of water were lost per bud and that buds had an average duration on the plant of 26 days before they opened into flowers. The flowers lasted about 7 days and led to a water loss of 25 g each. The large surface area of the petals and sepals, which are leaflike and have stomates (in this case, nonfunctional stomates), accounts for the large daily water loss from the flowers. The fruit remained on the plant about 52 days (the longest period of the three stages being considered) and had a total water loss over this period of 12 g per fruit. After 52 days, the fruits were ready to fall off and thus sever their dependency on the stem. At this time they had an average water content of 2.2 g. Thus, the total water cost per floral structure was 47 g. For the approxi-

Figure 7.8. *Opuntia bigelovii,* or teddy bear cholla, severed at ground level *(A)* and then maintained on a ring stand for 3 years without water *(B).* In the dehydrated specimen the lower branches died but the terminal shoots were still able to grow when the plant was watered. (Photographs courtesy of Stan R. Szarek, Arizona State University.)

mately 12 floral structures on each of the barrel cacti considered (stems averaging 34 cm in height), the overall water cost for flowering was nearly 600 g. This represents about 4% of the total annual transpiration for the plants considered. Perhaps more important, flowering of *F. acanthodes* occurs during a period that is usually dry, and in the case considered, the water diverted to the reproductive structures represented 6% of the water that could be stored in a fully hydrated stem. Cacti may have to achieve a certain size before they have enough water to cope with the water expenditure of flowering.

The flowers on *Carnegiea gigantea* are larger than those on *F. acanthodes,* and each flower may transpire as much as 11 g of water during its single day of opening. The flowers contain an additional 33 g of water, and a considerable amount of water can also be lost during the bud and fruit stages. Nevertheless, the fractional diversion of water to reproduction is again relatively low compared with overall plant transpiration. In contrast, water demands for reproduction may be substantial for certain small cacti with numerous large showy flowers such as species of *Echinopsis,* or for plants with huge flowers, for example, plants in the genera *Hylocereus* and *Epiphyllum,* epiphytes with relatively low stem volumes.

Low Temperature

One very important factor influencing the distribution of cacti and other plants often is the lowest wintertime temperature. For instance, in 1911 the pioneering desert biologist Forrest Shreve, working at the Desert Laboratory of the Carnegie Institution of Washington in Tucson, stated that "the line which marks the extreme southern limit of frost is the most important climatic boundary in restricting the northward extension of perennial tropical species." Indeed, of the 65 species of tropical and subtropical arborescent cereoid cacti in the Sonoran Desert, only three—*Carnegiea gigantea, Stenocereus thurberi,* and *Lophocereus schottii*—occur further north than the frost line, and frost damage is common on all of these columnar cacti at the northernmost part of their ranges in Arizona. Actually, nearly all the damage is caused during particularly severe cold spells, such as those in January of 1913, 1937, 1962, and 1971 and in December of 1979 for Arizona. Even today, damage attributed to these specific periods can be seen on specimens of *C. gigantea;* frost damage causes a reduction in stem growth and often leads to a substantial constriction; because a saguaro may live for more than 200 years, a record of particularly catastrophic events over the past two centuries can be indelibly etched on the stem.

Although some cacti occur in southern Canada and the northern United States and at high elevations in Chile and Argentina where snow is common, cacti are generally not native to regions with appreciable freezing. Cacti tend to have a high cellular water content. For example, *C. gigantea* has relatively low osmotic pressures (Chapter 3) ranging from 0.3 to 1.0 MPa, and therefore its stored water should freeze at about the same temperature as pure water (a solution with an osmotic pressure of 1.0 MPa freezes at −0.8°C). But *C. gigantea* and other cacti can actually cool to −3°C to −12°C without freezing. This common situation is referred to as supercooling and occurs for nearly all plants; however, even the ability to supercool does not prevent freezing damage to cacti at subzero temperatures. Another aspect related to tolerance of low temperatures is the ability to harden, that is, to become tolerant of low temperatures after the plants are exposed to moderately low temperatures.

Cactus Thermal Model

By using the thermal model that we introduced in Chapter 5, we can quantitatively evaluate the influence of various morphological features on the temperatures of cacti, an evaluation that will help show how temperature affects their distributions.

Richard W. Felger and Lowe at the University of Arizona made some interesting observations in the mid-1960s on the variations in stem diameter of senita, *Lophocereus schottii* (Fig. 5.15). Felger and Lowe found that the stems of senita, like those of *C. gigantea*, became progressively larger in diameter the farther north they went. In the late 1970s, Henry J. "Harry" Thompson and Nobel reexamined the change in stem diameter with latitude at the field sites used in Felger and Lowe's studies. Nobel then used the computer model to evaluate the influence of stem diameter on surface temperatures, especially in the sensitive region near the apical meristem, which is generally the coldest portion of the stem and also is involved in stem growth.

The field sites examined for the three cereoid cacti considered are presented in Figure 7.9. Freezing damage, evidenced by necrotic tissue near the apex, was found at the northernmost sites for all the species. For mature stems of *C. gigantea* the diameter at midheight changed from 33 cm at the southernmost site to 44 cm at the northernmost site. Using the model, Nobel showed that such a change in diameter would cause the minimum temperature at the apical meristem to increase by about 3°C. For

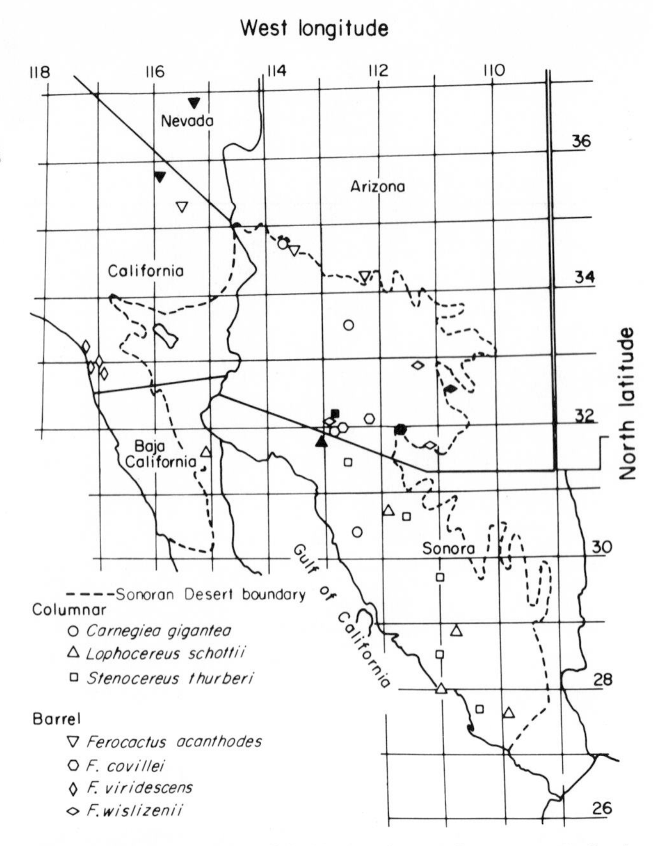

Figure 7.9. Location of field sites in southwestern United States and northwestern Mexico for columnar and barrel cacti. Filled symbols indicate sites where freezing damage was evident on at least 10% of the stems. (Data adapted from Nobel, 1980a, for the columnar cacti and from Nobel, 1980b, for the barrel cacti.)

the sites presented in Figure 7.9, the minimum annual air temperature decreases about 1.7°C for each degree latitude moved northward. Thus, the change in diameter by itself enables saguaro to extend its range northward about 2° latitude (just over 200 km; see Fig. 7.9).

Changes in diameter also occurred for the other two species: from 10 cm at midheight at the southernmost site to 15 cm at the northernmost site for *Stenocereus thurberi* and from 7 to 14 cm for the immature (nonfertile) stems of *Lophocereus schottii*. But in these species the diameter changes occurred within a size range that had relatively little effect on the simulated minimum surface temperatures of these cacti; indeed, all the ecological consequences of diameter changes are not yet fully understood.

In addition to latitudinal changes in diameter, major differences could be observed in spine shading of the apex. For instance, the amount of shading increased from 9% at the southernmost site to 41% at the northernmost site for *C. gigantea* and from 41 to 71% for *S. thurberi*. Spine shading of the apex remained near 57% for the mature stems of *L.*

schottii and 18% for the more frost-sensitive immature stems. Apical pubescence was essentially absent for *S. thurberi* and *L. schottii;* but it was quite thick for *C. gigantea* (9 to 10 mm), although the thickness did not vary with latitude. When all the morphological features are incorporated into the computer model, the surface temperature could be predicted for a particular set of environmental conditions. On one particular winter day, the predicted minimum apical temperatures were 7.7°C for *C. gigantea,* 5.9°C for *S. thurberi,* and 3.9°C for immature stems of *L. schottii.* Thus, larger stem diameter, greater apical spine shading, and very deep apical pubescence causes *C. gigantea* to be much warmer than the other two species under the same environmental conditions, a finding that can account for the much more northerly distributional limit of *C. gigantea* (34°56′N) than for the *S. thurberi* (32°38′N) or *L. schottii* (31°55′N).

The computer model was also used to evaluate another aspect of the northern limit of *C. gigantea,* namely, why seedlings of this species are often found under protective nurse plants (Fig. 7.10). At a site in northern Arizona near Wikieup, most of the seedlings that occur in exposed locations have evidence of freezing damage, whereas only a small proportion of those protected by nurse plants have such damage. The thin branches of nurse plants are close to air temperature, just as are the spines discussed in Chapter 5. Thus, rather than being fully exposed to the cold sky, the seedlings receive some longwave radiation from the warmer branches. At night this additional energy input raises the temperature of seedlings under nurse plants compared with the temperature of otherwise identical seedlings at exposed locations and thereby can prevent freezing damage for the protected seedlings. For a 25-cm tall seedling of *C. gigantea,* the minimum apical surface temperature can be raised just over 1°C by the nurse plant, which may be important at the northern or high elevation limits of the species. The effect is most apparent on the smallest plants because the increase in diameter accompanying growth causes the minimum apical temperatures to increase.

The distributional limits of the four species of *Ferocactus* occurring in the southwestern United States (Fig. 7.9) can also be related to spines and pubescence occurring in the apical region. In particu-

Figure 7.10. A small *Carnegiea gigantea* growing under the protective influence of a nurse plant in northern Arizona.

lar, *F. acanthodes* has thick apical pubescence (8 mm) and can have up to 98% shading of the stem apex by spines (Fig. 7.11), features consistent with the fact that its range extends to the coldest sites of these four *Ferocactus* species (Fig. 7.9; the 36°52′N site in Nevada is at 1590 m and the 35°47′N site is at 1680 m). *Ferocactus wislizenii* has a similar thickness of apical pubescence, but its apical spine shading averages only 49% (Fig. 7.12) and it does not range as far north (Fig. 7.9). *Ferocactus covillei* also has thick pubescence (8 mm), but its spine shading near the apex is not very great, only 15% (Fig. 7.13), and its range barely extends into the United States (Fig. 7.9). Finally, *F. viridescens* (Fig. 7.14) has apical pubescence that is only about 2 mm thick; it has 23% apical spine shading and is restricted to much warmer coastal habitats (Fig. 7.9). Indeed, the respective range boundaries are in agreement with predictions made by the computer model.

Another feature became apparent in studying the upper elevational limits of two cacti at a series of sites in the Andes of central Chile. In particular, the barrel cactus *Eriosyce ceratistes,* which goes by the common name of sandia, or watermelon, which it somewhat resembles (Fig. 7.15), was studied from 29°S to 33°S, and the common columnar cactus *Trichocereus chilensis* (Fig. 7.16) was studied from 30°S to 35°S. Using the computer model to adjust for the slight morphological differences from site to site, the change in upper elevational limit with latitude was shown to be completely dependent on the low temperature experienced by the stem apex — namely, the upper elevation limit of each species decreased 90 to 100 m per degree latitude moved southward away from the equator. This change is just what was expected on the basis of the slight differences in morphology, the elevational temperature lapse rate of a 0.54°C decrease in temperature for each 100 m increase in elevation, and the latitudinal change of a 0.61°C decrease at a

Figures 7.11 – 7.14. Spination and pubescence on the stem apices of barrel cacti in the genus *Ferocactus.* Photographs are for the northernmost sites indicated for each species in Figure 7.9.

7.11. *F. acanthodes* from southern Nevada.
7.12. *F. wislizenii* from central Arizona.
7.13. *F. covillei* from southern Arizona.
7.14. *F. viridescens* from coastal southern California.

Figures 7.15 and 7.16. Chilean cacti studied by Nobel for their responses to cold temperatures.
7.15. *Eriosyce ceratistes,* a very wide barrel cactus, and Nobel.
7.16. *Trichocereus chilensis,* near its upper elevational limit in central Chile.

given elevation per degree latitude moved southward. As before, low temperature was seen to be crucial for affecting the range boundaries of these cacti.

Tissue Tolerances and Hardening

The cactus energy budget model accurately predicts the temperature of cacti, but that is only part of the information required because we also need to know the tissue sensitivity to subzero temperatures. One of the ways to test for viability of plant cells is to see whether they take up a vital stain, such as neutral red, which is taken up only by living cells. Loss of the ability of the cells to accumulate this stain—a loss that can be induced by subzero temperatures—is followed by tissue necrosis; this result indicates that a substantial decrease in the ability to take up stain is a useful test for low-temperature damage in cacti. Nobel has therefore used vital staining to examine the low-temperature sensitivity of numerous species of cacti, including some that occur in habitats where they are snow-covered for much of the winter.

As the air temperature was lowered, the tissue temperature of *C. gigantea* decreased to −6°C and then rose to about −2°C (Fig. 7.17). Because the air temperature was continuously lowered, the rise in tissue temperature means that heat was somehow released within the stem tissue, a result that has been observed for many cacti and other plants. This interesting phenomenon can be interpreted in the following way. First, the plant is cooled by the lower air temperature outside, and the water in the tissue gets tricked into cooling below the temperature at which it would normally freeze, that is, it supercools. Then some of the water begins to freeze. We all know that it takes energy to melt ice—we commonly put frozen food into a microwave oven to defrost it. When the reverse process happens, namely, when water freezes, heat is released. Exactly the same amount of heat is released when a given weight of water freezes as is required to melt that weight of ice. The freezing of water in *C. gigantea* that causes the release of heat and hence the rise in chlorenchyma temperature (Fig. 7.17) apparently occurs outside the cell, that is, extracellularly, because the sites for ice formation are on the outer sides of the cell walls in the chlorenchyma.

The extracellular ice crystals grow in size as more and more water is frozen onto them, but initially the cells are not damaged. The healthy condition is indicated by stain uptake, which occurs when the cells are still alive. As cooling proceeds, more and more water diffuses out of the cells and joins the growing extracellular ice crystals. Eventually a temperature is reached at which the fraction of cells taking up stain precipitously drops, for example, near a chlorenchyma temperature of −8°C for *C.*

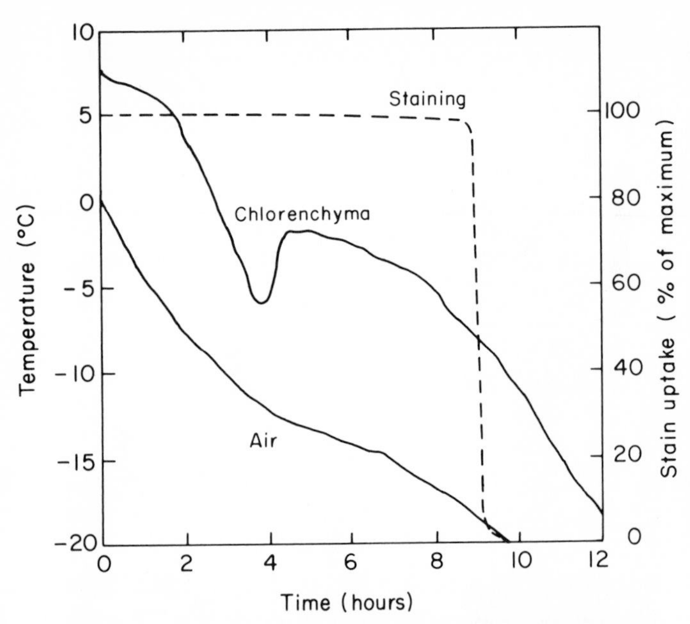

Figure 7.17. Cooling curve for *Carnegiea gigantea,* showing how the temperature of the chlorenchyma and its stainability changed as the air temperature was progressively decreased during the night. Temperature was measured using thermocouples placed in the chlorenchyma 1 mm below the stem surface, and stainability was assayed by removing sections of the chlorenchyma approximately every 30 minutes and determining the percentage of chlorenchyma cells that took up neutral red. Plants were grown at day/night air temperatures of 30°/20°C. (Data are previously unpublished observations of Park S. Nobel and Terry L. Hartsock. For further details see Nobel, 1982.)

gigantea (see cooling curve in Fig. 7.17). At this stage much water has been lost from the cells and the protoplasts have pulled away from the cell walls and appear very shrunken. Indeed, death is due to cellular dehydration and is generally known as frost plasmolysis. When such a tissue is warmed, cellular activity does not recover, membrane integrity is apparently destroyed, and the tissue becomes necrotic, usually in a matter of days. It is not clear whether any ice forms inside the cells during the lethal freezing episode, but such an event would undoubtedly prove fatal, just as it does for most other plants.

Increased cold hardiness, or cold hardening, which occurs for many plants in response to cooler temperatures in the fall and early winter, can increase the survival of cacti during the cold season. This was investigated by gradually reducing the day/night air temperatures for plants in the laboratory and seeing what happened to the low temperature response of stain uptake. For nearly 20 species of cacti, studied by Nobel, the low temperature at which stain uptake was reduced 50% decreased an average of 0.7°C for each 10°C lowering of the air temperature used for their growth. Such shifts can have major effects on range boundaries because 0.6°C is about the temperature change corresponding to 100 m of elevation or 1° of latitude, as indicated earlier. The ability to cold harden also tended to be greater for species from higher elevations. Moreover, it was fairly rapid, for example, the half-time (time to go one-half of the way between the initial and final conditions) for the shift when *Denmoza rhodacantha* and *Trichocereus candicans* from north-central Argentina were transferred from day/night air temperatures of 50°C/40°C to 10°C/0°C was only 3 days. Cold hardening may represent the development of increased tolerance to intracellular dehydration, protection of certain membrane proteins, or perhaps changes in membrane permeability reflecting changes in lipid composition.

Now that we have discussed the cellular events that eventually lead to low temperature damage for cacti and introduced the concept of cold hardening, let us consider the differences in low temperature tolerance that occurred for the species whose ranges were considered using the thermal model. When the plants were grown at low day/night air temperatures, such as 5°C/−5°C, 50% inhibition of stain uptake by *Carnegiea gigantea* and *Stenocereus thurberi* occurred at −9°C, and death occurred a few degrees lower. *Lophocereus schottii* was more sensitive and showed lethal signs at temperatures a few degrees higher. Thus, the computer model nicely accounted for the relative range boundaries for the former two cereoid cacti, but the restriction of *L. schottii* to warmer habitats also reflected a greater sensitivity to low temperatures.

When the low temperature tolerances of four species of *Ferocactus* (Figs. 7.11–7.14) were similarly compared, the tissue sensitivity of *F. acanthodes* and *F. wislizenii* were the same (stain uptake decreased 50% at −9°C when maintained at day/night air temperatures of 5°C/−5°C), a result indicating that the thermal model alone could predict the relative northern limits. But *F. covillei* was about 1°C more sensitive to low temperatures and *F. viridescens* about 2°C more sensitive; thus, their restriction to habitats with higher minimum wintertime temperatures reflected both morphological differences near the stem apex and inherent differences in tissue sensitivity to low temperatures.

Let us now turn to *Coryphantha vivipara* (Fig. 7.18), a species that occurs over a remarkably wide range from 30°N in northern Mexico all the way up to 50°N in southern Canada. It exhibited tolerance to very low tissue temperatures and also exhibited the greatest ability to cold harden among the nearly 20 species of cacti that have been investigated so far.

Figure 7.18. *Coryphantha vivipara.*
(A) Close-up of the spine-covered stem, showing two fruits.
(B) Field site at the Nevada Test Site near Mercury, Nevada (36°40′N, 116°1′W, 1110 m elevation) when the stem temperatures were being measured in January 1979 (plant is in the left foreground).
(C) The same field site a few weeks later. The low temperatures and snow cover caused no damage to the plant.

Specifically, as the day/night stem temperatures were reduced from 50°C/40°C to 0°C/−10°C in 10°C steps at weekly intervals, the temperature for 50% inhibition of stain uptake progressively decreased from −9°C to −19°C (Fig. 7.19). Hence, exposure to lower air temperatures enabled *C. vivipara* to tolerate lower tissue temperatures. Its ability to range into Canada is obviously related to its inherent tolerance of very low temperatures plus this ability to cold harden. Indeed, one high-elevation variety of *C. vivipara* did not show 50% inhibition of stain uptake until −23°C and did not die until near −30°C (−22°F). Similarly, three species of cacti from sites above 2200 m in southeastern Wyoming showed 50% inhibition of stain uptake near −20°C, whereas two opuntia species restricted to much warmer habitats tolerated only −6°C, even when cold hardening was taken into account. Thus, low temperature tolerances vary considerably among cacti and can be related to the minimum winter temperatures of their native habitats. Also, it is apparent that some cacti can occur in regions where considerable freezing occurs.

All these studies on cold tolerance of certain cacti should strike a familiar chord for cactus enthusiasts in places like Ohio and Germany, where attempts are made to grow "hardy" cacti outside year-round. Awaiting careful study of hardiness and cold acclimatization is another species, *Opuntia compressa,* that occurs throughout eastern temperate North America as far north as Ontario, Canada, and New England, but that also lives in southern locations where freezing is rare or absent.

High Temperature

One of the notorious features of deserts is high temperature. Indeed, cacti are often exposed to high air temperatures. But cacti are also massive compared with typical leaves, and they lack daytime stomatal opening, with its attendant transpirational cooling, because of their Crassulacean acid metabolism. Thus, stems of cacti often rise 15°C or more above air temperature. It should be no surprise, therefore, that cacti can tolerate high temperatures. It was quite a surprise, however, when Brigitte Didden-Zopfy, Stanley D. Smith, and Nobel found that cacti had the greatest high-temperature tolerance ever reported for vascular plants. The approach taken was similar to that used in the low-temperature studies. Cactus stems or parts thereof were subjected to high-temperature treatments, often for 1 hour, and the subsequent ability of the chlorenchyma cells to take up neutral red was then examined. Heat hardening was observed, and the increase in toler-

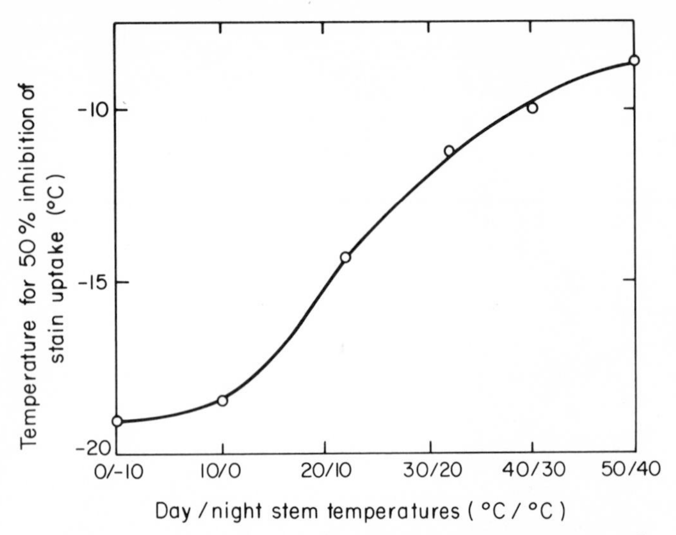

Figure 7.19. Relationship between stem temperatures and the subzero temperature at which the ability of the chlorenchyma cells of *Coryphantha vivipara* var. *deserti* to take up neutral red was halved. Plants were maintained at each temperature for 1 week. (Data are adapted from Nobel, 1981, and from previously unpublished observations.)

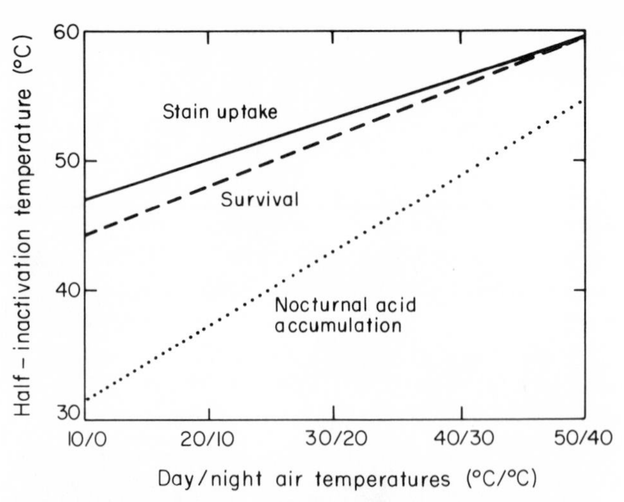

Figure 7.20. Influences of day/night air temperatures on heat tolerances of various processes for *Opuntia bigelovii.* Plants were maintained at 10°/0°C and then the temperature was increased in 10°C intervals every 2 weeks. Stems were subjected to various treatment temperatures for 1 hour and the 50% inactivation temperature determined graphically. (Data are adapted from Didden-Zopfy and Nobel, 1982.)

ance per 10°C change in growth temperature was much greater than for cold hardening. The high temperature tolerance of seedlings and of roots was also examined.

Tissue Tolerances and Hardening

The tolerance of *Opuntia bigelovii* (Fig. 5.18 and 7.8) to high temperatures was tested for plants at various growth temperatures (Fig. 7.20). The high temperature tolerated increased about 3.0°C per 10°C increase in growth temperature, that is, exposure to warmer growth temperatures allowed the plants to tolerate higher treatment temperatures, as measured by stain uptake. As the growth temperature was raised, the temperature that could be survived also increased, as did the temperature leading to 50% inhibition of nocturnal acid accumulation (Fig. 7.20). The temperatures that reduced the ability of the chlorenchyma cells to survive were similar to those that reduced their ability to take up neutral red (Fig. 7.20). At growth temperatures exceeding 50°C/40°C, the 1-hour treatment temperature leading to 50% survival of the plants more or less coincided with the treatment temperature reducing stain uptake to about 10% of maximum for *O. bigelovii* and other cacti, again a result indicating that stainability was a useful index of lethal temperatures.

Acclimation to high temperatures, or heat hardening, was found in all 14 species of cacti tested and amounted to an average increase of 4.9°C in the temperature at which stain uptake was reduced 50% as the day/night air temperatures were raised 10°C. The half-times for the shift in high-temperature tolerance upon transferring *O. bigelovii* and *Carnegiea gigantea* from day/night air temperatures of 30°C/20°C to 50°C/40°C averaged 2 days, a finding indicating that these species could react to changing weather fronts. The mean maximum temperature for 50% inhibition of stain uptake averaged 64°C (147°F!) for the 14 species at day/night air temperatures of 50°C/40°C; and *Ferocactus covillei* and *F. wislizenii* showed 50% inhibition at 69°C (a remarkable 156°F!) when grown at 55°C/45°C. These latter two species occur in some of the hottest habitats in the Sonoran Desert. The high temperatures tolerated by cacti are greater than those tolerated by other vascular plants, for which the high temperatures tolerated by the tissues rarely exceed 50°C. Indeed, high-temperature damage is hard to find on cacti in the field.

So far the mechanism for high-temperature tolerance and hardening is not understood for cacti. In other plants the cells appear to die when the proteins get "cooked"—the proteins agglutinate like egg white in a frying pan. Also, cellular membranes generally become leaky at higher temperatures and subcellular compartmentation breaks down. Cellular chemicals or the special structure of the proteins may prevent protein agglutination in cacti. Or the membranes may have a unique composition that changes as the growth temperature is varied, a change leading to the hardening effects. Further study of this phenomenon could lead to greater understanding of high-temperature tolerance for plants in general.

High-Temperature Tolerance of Seedlings

In addition to drought, which was considered at the beginning of this chapter, seedlings of cacti can be subjected to extreme temperatures, especially high temperatures. Indeed, surface temperatures for desert soils in excess of 80°C have been reported. The high temperature at which a 1-hour treatment reduced stain uptake by the chlorenchyma cells by 50% is 65°C for seedlings of *F. acanthodes* grown at 50°C/40°C and 1°C or 2°C higher for the adults. The slightly greater sensitivity to high temperature coupled with the greater influence of soil temperatures for the smaller plants suggests that high temperatures could be responsible for seedling mortality for *F. acanthodes* in exposed locations in hot habitats. Although slightly more damage, presumably resulting from high temperature, has been observed in exposed habitats, even the seedlings of *F. acanthodes* are remarkably tolerant of high temperatures. More studies of seedling microhabitats, including the effects of nurse plants, are critically needed.

Thermal Tolerances of Roots

Roots of *F. acanthodes* have about the same sensitivity to low temperatures as do the stems: 50% inhibition of stain uptake occurs at −8°C for plants grown at day/night air temperatures of 15°C/5°C. However, the roots are more sensitive to high temperatures than are the stems: 50% inhibition occurs at 57°C for roots and 63°C for stems when grown at 45°C/35°C. It is interesting, therefore, to observe that roots of desert cacti do not occur very near the soil surface. Specifically, except directly under the stem where soil temperatures are moderated, roots of *F. acanthodes* are absent at 1 cm below the soil surface (Fig. 7.4), where field temperatures have been measured at 65°C and can reach 70°C. Roots occur at 3 cm below the soil surface (Fig. 7.4), where they can take advantage of light desert rains but where the maximum temperatures expected would cause less than a 50% decrease in the root cortical cells taking up stain. Thus, the absence of roots of *F. acanthodes* from the upper part of the soil apparently reflects the lethal high temperatures that can occur in that region.

In summary, high temperatures may prevent roots of cacti from being very near the soil surface. Seedlings of cacti may be more vulnerable to high temperatures than adults; thus, seedlings may need to be protected in hot habitats by nurse plants or other microhabitat modifications. And although high temperatures can affect the water relations, and drought can limit the successful establishment of cacti, high temperatures may not directly limit the distribution of cacti because of their remarkable high-temperature tolerance and heat-hardening ability.

SELECTED BIBLIOGRAPHY

Brum, G. D. 1973. Ecology of the saguaro *(Carnegiea gigantea):* phenology and establishment in marginal populations. *Madroño* 22:195–204.

Despain, D. G. 1974. The survival of saguaro *(Carnegiea gigantea)* seedlings on soils of differing albedo and cover. *Journal of the Arizona Academy of Science* 9:102–107.

Didden-Zopfy, B., and P. S. Nobel. 1982. High temperature tolerance and heat acclimation of *Opuntia bigelovii. Oecologia* 52:176–180.

Ehleringer, J., H. A. Mooney, S. L. Gulmon, and P. Rundel. 1980. Orientation and its consequences for *Copiapoa* (Cactaceae) in the Atacama desert. *Oecologia* 46:63–67.

Felger, R. W., and C. H. Lowe. 1967. Clinal variation in the surface-volume relationships of the columnar cactus *Lophocereus schottii* in northwestern Mexico. *Ecology* 48:530–536.

Jordan, P. W., and P. S. Nobel. 1981. Seedling establishment of *Ferocactus acanthodes* in relation to drought. *Ecology* 62:901–906.

——— 1982. Height distributions of two species of cacti in relation to rainfall, seedling establishment, and growth. *Botanical Gazette* 143:511–517.

——— 1984. Thermal and water relations of roots of desert succulents. *Annals of Botany* 54:705–717.

Kausch, W. 1965. Beziehungen zwischen Wurzelwachstum, Transpiration und CO_2-Gaswechsel bei einigen Kakteen. *Planta* 66:229–238.

Levitt, J. 1980. Responses of plants to environmental stresses, 2nd ed., vol. 1: *Chilling, freezing, and high temperature stresses* and vol. 2: *Water, radiation, salt, and other stresses.* Academic Press, New York.

MacDougal, D. T., and E. S. Spalding. 1910. *The water-balance of succulent plants.* Carnegie Institution of Washington, Publication no. 141, Washington, D.C.

Niering, W. A., R. H. Whittaker, and C. H. Lowe. 1963. The saguaro: a population in relation to environment. *Science* 142:15–23.

Nobel, P. S. 1976. Water relations and photosynthesis of a desert CAM plant, *Agave deserti. Plant Physiology* 58:576–582.

——— 1977. Water relations and photosynthesis of a barrel cactus, *Ferocactus acanthodes,* in the Colorado Desert. *Oecologia* 27:117–133.

——— 1980a. Morphology, surface temperatures, and northern limits of columnar cacti in the Sonoran Desert. *Ecology* 61:1–7.

——— 1980b. Influences of minimum stem temperature on ranges of cacti in southwestern United States and central Chile. *Oecologia* 47:10–15.

——— 1980c. Morphology, nurse plants, and minimum apical temperatures for young *Carnegiea gigantea. Botanical Gazette* 141:188–191.

——— 1981. Influence of freezing temperatures on a cactus, *Coryphantha vivipara. Oecologia* 48:194–198.

——— 1982. Low-temperature tolerance and cold hardening of cacti. *Ecology* 63:1650–1656.

——— 1983. Low and high temperature influences on cacti. Pp. 165–174 in *Effects of stress on photosynthesis,* ed. R. Marcelle, H. Clijsters, and M. van Poucke. Nijhoff/ W. Junk, The Hague.

——— 1984. Extreme temperatures and thermal tolerances for seedlings of desert succulents. *Oecologia* 62:310–317.

——— 1985. Desert succulents. Pp. 181–197 in *Physiological ecology of North American plant communities,* ed. B. F. Chabot and H. A. Mooney. Chapman and Hall, London.

Nobel, P. S., and J. Sanderson. 1984. Rectifier-like activities of roots of two desert succulents. *Journal of Experimental Botany* 35:727–737.

Shreve, F. 1911. The influence of low temperatures on the distribution of the giant cactus. *Plant World* 14:136–146.

Smith, S. D., B. Didden-Zopfy, and P. S. Nobel. 1984. High-temperature responses of North American cacti. *Ecology* 65:643–651.

Steenbergh, W. F., and C. H. Lowe. 1969. Critical factors during the first years of life of the saguaro *(Cereus giganteus)* at the Saguaro National Monument, Arizona. *Ecology* 50:825–834.

——— 1976. Ecology of the saguaro. I: The role of freezing weather in a warm-desert plant population. Pp. 49–92 in *Research in the parks,* National Park Service Symposium Series, no. 1, U.S. Government Printing Office, Washington, D.C.

——— 1977. *Ecology of the saguaro.* II: *Reproduction, germination, establishment, growth, and survival of the young plant.* National Park Service Scientific Monograph Series, no. 8, U.S. Government Printing Office, Washington, D.C.

Szarek, S. R., H. B. Johnson, and I. P. Ting. 1973. Drought adaptation in *Opuntia basilaris. Plant Physiology* 52:539–541.

Szarek, S. R., and I. P. Ting. 1974. Seasonal patterns of acid metabolism and gas exchange in *Opuntia basilaris. Plant Physiology* 54:76–81.

——— 1975. Physiological responses to rainfall in *Opuntia basilaris* (Cactaceae). *American Journal of Botany* 62:602–609.

Turnage, W. V., and A. L. Hinckley. 1938. Freezing weather in relation to plant distribution in the Sonoran Desert. *Ecological Monographs* 8:529–550.

Turner, R. M., S. M. Alcorn, G. Olin, and J. A. Booth. 1966. The influence of shade, soil, and water on saguaro seedling establishment. *Botanical Gazette* 127:95–102.

Uphof, J. C. Th. 1916. Cold-resistance in spineless cacti. *University of Arizona Agricultural Experiment Station Bulletin* no. 79:119–144.

Walter, H., and E. Stadelmann. 1974. A new approach to the water relations of desert plants. Pp. 213–310 in *Desert biology,* vol. II, ed. G. W. Brown, Jr. Academic Press, New York.

8

Growth Habits

Our book began with a discussion of different growth habits without comment on why this diversity occurs. In this chapter we shall address variation in growth habits; but our discussion is limited by a paucity of research on the subject. When asking why a stem branches repeatedly on one plant and another remains unbranched as a solitary columnar, the investigator really is trying to determine what physiological or reproductive benefits accrue from many narrow stems, as a caespitose form has, or one wide, solitary stem, as a barrel cactus has. Why do many different growth forms coexist in a single location, and what factors determine which growth habits are able to survive in the same place? The answers to these questions are being prepared and will undoubtedly emerge before the next century begins. Meanwhile, a review of some interesting morphological patterns may stimulate imaginative studies by those who have the skills to investigate these exciting areas of research.

Diversification of Growth Habits

Growth Habits of Trichocerei

One way to appreciate the great diversity of growth habits found within the cactus family is to study the range of variation found within a single, closely related group of species. Thus far we have concentrated on the cacti of North America. Now we shall shift our attention to the South American species of *Trichocereus* and its close ally *Echinopsis,* which present a wide range of growth habits. So few features separate these taxa that several cactologists have very quietly hinted that all of these species could be classified in the same genus, which would have to be called *Echinopsis* because this is the oldest valid name.

Arborescent plants in the trichocerei include the massive plants of *T. terscheckii* of Argentina (Fig. 8.1) and *T. chilensis* of Chile (Fig. 7.16); these cacti have stems that are commonly over 20 cm thick because they have a very wide pith. Trichocerei also grow as large shrubs, for example, *T. pachanoi,* presumably of Ecuador, which has sturdy ascending stems that arise from near the base of the plant. In addition, there are several low, erect shrubs with fairly slender stems, for example, *T. spachianus* of Argentina (Fig. 8.2), which rarely exceeds 2 m in height. The shrubs grade into the narrow-stemmed, caespitose species, such as *T. huascha* of northern Argentina (Fig. 8.3). Also growing in northern Argentina is a large caespitose, thick-stemmed plant, *T. candicans* (Fig. 8.4), and a short barrel cactus, *Echinopsis leucantha* (Fig. 8.5). But most species of *Echinopsis* (Fig. 8.6) grow in thick clumps, superficially like a caespitose plant except that the small cylindrical plants are produced as offshoots; therefore each stem has its own roots. Some species of *Echinopsis* are solitary and cylindrical or nearly globose. Finally, a remarkable and special growth habit in this supergenus is displayed by *T. peruvianus;* it is rooted on ledges and its heavy stems grow to 5 m or more in length and simply hang down cliffs in western Peru (Fig. 8.7). The commonly cultivated *T. thelegonus* of Argentina (Fig. 8.8) also has a lax growth habit.

The evolution of growth habit in this supergenus has never been studied from a phylogenetic viewpoint. Nevertheless, because the tallest, the smallest, and the lax growth habits have highly derived reproductive features, we assume that either a small tree or a large shrub was the primitive growth habit. Larger plants became possible as the pith enlarged, an anatomical development that also produced an increase in rib number and rib width. Low caespitose plants evolved when basal branches formed early during plant growth, a development that was accom-

Figures 8.1–8.8. Growth habits of the trichocerei.
8.1. *Trichocereus terscheckii,* a tall arborescent cactus growing near Catamarca, Argentina.
8.2. *T. spachianus,* a shrub that is 1 to 2 m tall and has many stems arising from ground level.
8.3. *T. huascha,* a low caespitose, hedgehog-type species growing with the massive *T. terscheckii* (Fig. 8.1) in Catamarca, Argentina.
8.4. *T. candicans,* a massive, low caespitose plant growing in central Argentina.
8.5. *Echinopsis leucantha,* a barrel cactus common in desert regions of Argentina.
8.6. *E. eyriesii;* large clumps of small cylindrical plants appear as a caespitose growth habit.
8.7. *T. peruvianus,* a columnar form from Western Peru; the heavy stems of this plant hang down the steep rock faces.
8.8. *T. thelegonus;* cultivated specimens of a creeping columnar cacti from Argentina.

panied by a reduction in pith diameter and the formation of very low and narrow ribs. The barrel cactus growth habit evolved when basal branches did not form in a small cactus; thus, all of the energy could be directed into a single stem.

Convergent Evolution of Growth Habit

A person from Arizona standing in a population of *Trichocereus terscheckii* in central Argentina would feel right at home because the growth habit of this species is essentially identical to that of *Carnegiea gigantea* (Fig. 1.17*B*) of the Sonoran Desert. When the flowers appear the disguise is exposed, however, because the long, funnelform flowers of the trichocerei bear long, silky hairs on the floral tube (see the hairs present in *Borzicactus,* Fig. 1.35) and are considerably different from the flowers of saguaro in many structural aspects.

Reconstruction of the events in cactus evolution is complicated by the fact that cactus groups in different geographic regions have undergone similar types of radiation in growth habit. The striking resemblance of *Trichocereus terscheckii* and *Carnegiea gigantea* is just one of a hundred such similarities. Other examples of look-alikes are the low and sprawling shrubs *Cleistocactus baumannii* of Argentina (Trichocereeae) and *Bergerocactus emoryi* of Baja California (tribe Hylocereeae?) or the procumbent *Trichocereus thelegonus* (Trichocereeae, Fig. 8.8) and *Stenocereus eruca* (Pachycereeae; Fig. 1.25). The highly specialized solitary columnar growth habit (Fig. 1.4) occurs in six species of *Neobuxbaumia* and *Cephalocereus,* and *Carnegiea gigantea* can also be tall and unbranched. Likewise, barrel cacti appear in four tribes, and this is never the primitive condition for those tribes. Also, caespitose organisms appear independently in numerous tribes of Cactoideae, in *Maihuenia* of Pereskioideae, and in many different places in Opuntioideae, for example, in tephrocacti, chollas (*Opuntia tunicata* var. *davisii*), and platyopuntias.

The Nature of Branching

No one has been able to generalize much about how and when cacti form branches, although a real hallmark of cacti is that they branch much less often than typical dicotyledons. Even though a cactus has hundreds or even many thousands of axillary buds (areoles), lateral branching is very infrequent in most cacti and is totally eliminated in growth habits such as the solitary columnar trees, barrel cacti, and small cylindrical and globular forms. One likely explanation for the low frequency of branching is that axillary buds are immediately converted into areoles, that is, specialized short shoots; thus, they are already committed to a spine- and flower-producing function. Another explanation is that for most cacti the growing shoot tip exhibits tremendous developmental control on the entire shoot and is somehow able to suppress the outgrowth of lateral shoots, a property of plants called apical dominance.

Although rigorous experimental studies are still needed to understand branching in cacti, we can begin to appreciate apical dominance by analyzing how and when branches are initiated. The simplest case is the solitary shoot that lacks branches; however, even though these plants normally do not branch, they can form new shoots when the shoot tip of the cactus is destroyed, for example, by freezing or by decapitation. Several new shoots will then quickly develop at the top of the plant from healthy areoles in the vicinity of the wound or cut because the dominance of the shoot tip, presumably enforced by hormonal control, has been removed and the lateral shoots have resumed growth as leaf-bearing shoots. The most bizarre example of lateral branching is displayed by mutant (monstrose) cacti, which produce lateral branches from nearly every areole when the main shoot tip is removed (Figs. 8.9 and 8.10); in many monstrose plants apical dominance is never regained. In most situations no new branches arise once a new set of actively growing shoots is present because the plant is again under the control of dominant apices.

The second type of anecdotal evidence about apical dominance comes from observing cacti that normally produce relatively few branches, for example, 1 to 20 stems per plant. In these cases lateral branches characteristically arise from areoles at a considerable distance from the original and actively growing shoot tip. The branches may arise in an irregular or a regular pattern, but never much closer to the apex than 0.5 m. This pattern of branching suggests that apical dominance decreases as the distance from the shoot tip increases, until at some point the areoles are freed from apical control.

The third type of evidence comes from arborescent and shrubby species that have many lateral branches, particularly those that grow by producing discrete joints. The joints that are produced do not arise from areoles on the current shoot segment but instead develop from one or several of the upper areoles on the shoot of the previous growing season (Figs. 1.26 and 1.27). The areoles are probably stimulated to grow while the shoot tip is inactive, that is, during the dormant period or just before growth is initiated in the spring. Therefore, the growth of lateral shoots from areoles coincides with the period when the chemicals used to enforce apical dominance are not diffusing from the shoot tip to the lower stem.

Apical dominance is poorly developed in some

Figures 8.9 and 8.10. Monstrose forms that have early development of lateral shoots from many of the areoles.
8.9. *Austrocylindropuntia subulata.*
8.10. *Cereus peruvianus.*

cacti, especially in some of the leaf-bearing cacti (Chapter 2); for example, lateral branching is quite common in *Pereskia aculeata.* The absence of apical dominance enables *Maihuenia* (Figs. 2.22 and 2.23) to form a dense cushion plant; precocious development of leafy shoots on even the uppermost areoles produces the caespitose growth habit (Figs. 2.22*C* and 2.23*C*). Moreover, there are a number of columnar cacti that are considered to be fairly primitive Cactoideae, such as species of *Leptocereus,* and that have branching designs not unlike typical dicotyledons. Thus, apical dominance was not strongly expressed in the early cacti, but it became a controlling influence on cactus evolution after succulence appeared. We can also generalize by saying that caespitose plants have less apical dominance than do typical cacti; in these species, we generally find that branching can occur fairly close to the shoot tip. Therefore, in genera where some species are caespitose and others are barrel-shaped, the developmental explanation could be that apical dominance is altered to produce the different forms.

Intraspecific Variation in Branching Patterns

When growth habit of a cactus species is described, it is usually convenient to simplify the description and to portray the characteristics of the typical specimen of that species. Nonetheless, one of the first lessons for a beginning cactologist is that most species are polymorphic; that is, each species has several forms. One way to show polymorphism in a species is with photographs (Fig. 5.5*A*–*D*). But eventually the serious student of plant form must quantify the differences in structure and seek reasons for intraspecific variability, that is, variation found within a single species.

Galápagos tree opuntias. Intraspecific variation in growth habit can be studied on the Galápagos Archipelago, where a single species takes on different features on different islands. In 1972 Charles H. Racine and Jerry F. Downhower from Ohio State University studied the variations in vegetative structure and reproductive effort in the tree platyopuntias on these islands. This study quantified the morphological characteristics of three populations: *Opuntia echios* var. *gigantea* on Santa Cruz Island, *O. echios* var. *barringtonensis* on Santa Fé, and *O. galapageia* var. *macrocarpa* on Pinzon.

Mature specimens on Santa Cruz (*O. echios* var. *gigantea*) are 1 to 3 m taller than those on Santa Fé (*O. echios* var. *barringtonensis;* Fig. 6.16), but they are also different in other features. First, a tree on Santa Cruz has fewer pads (cladodes) than a tree of the

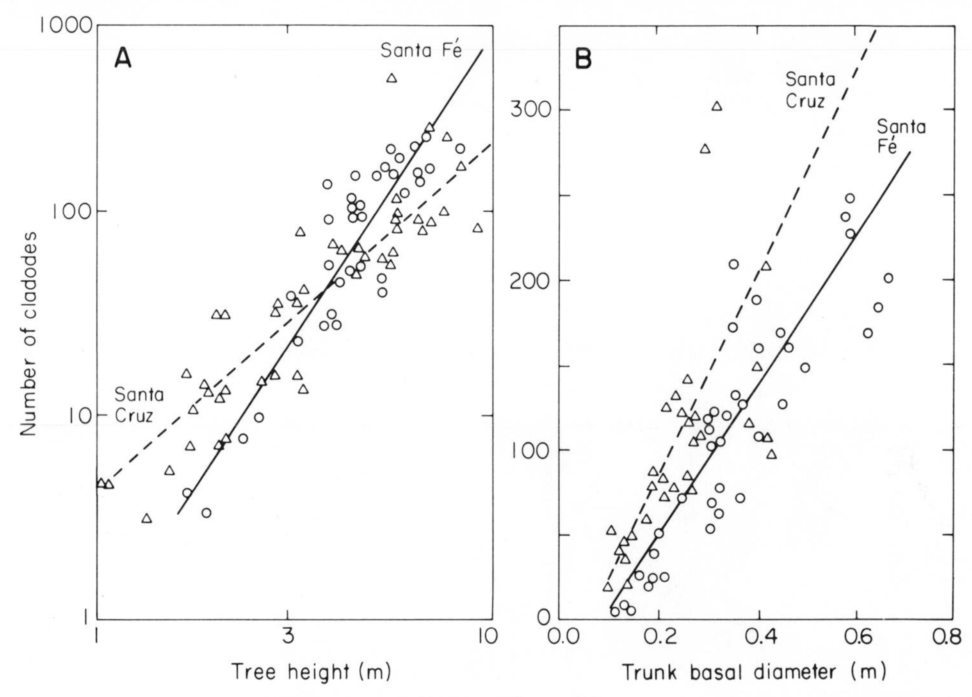

Figure 8.11. Relationship between cladode number and tree height (*A*; note logarithmic scales) or trunk basal diameter *(B)* in tree platyopuntias, *Opuntia echios* var. *gigantea* on Santa Cruz *(triangles)* and *O. echios* var. *barringtonensis* on Santa Fé *(circles)* in the Galápagos Archipelago. The lines indicate that tree height increases more rapidly with cladode number on Santa Fé, where trunk basal diameter is also greater.

same height on Santa Fé (Fig. 8.11*A*). For example, a tree on Santa Cruz with 100 cladodes is about 6 m tall, whereas one on Santa Fé is only about 5 m tall. The branching patterns for both varieties are similar, but the cladodes are longer on var. *gigantea* on Santa Cruz. The trunks of the platyopuntias on Santa Cruz are also more slender (Fig. 8.11*B*); for trees with 100 cladodes, a specimen on Santa Cruz has a mean trunk basal diameter of 23 cm, whereas one on Santa Fé has a diameter around 30 cm.

Racine and Downhower also showed that seedlings of the two taxa differ markedly: those from Santa Cruz have long and narrow cladodes, whereas those from Santa Fé have short, elliptical cladodes. One year after germination, greenhouse-grown seedlings of the platyopuntia from Santa Cruz were much taller than those from Santa Fé (Fig. 8.12). This finding supports the conclusion that plants from Santa Cruz grow upward more rapidly than those on Santa Fé.

The difference between the vegetative features of the two populations was interpreted in the following way. On Santa Cruz variety *gigantea* has evolved under strong selection for rapid vertical growth because this tree opuntia grows with many woody plants and therefore occurs in fairly dense vegeta-

Figure 8.12. Young plants of *Opuntia echios* from the Galápagos Archipelago; var. *gigantea (right)* has long, narrow cladodes and var. *barringtonensis (left)* has wider, broadly elliptical cladodes.

Figures 8.13 and 8.14. Branching patterns.
8.13. *Pachycereus pringlei,* or cardón; a relatively young, arborescent specimen with a trunk 3.5 m tall, growing in a thornscrub habitat near La Paz, southern Baja California, Mexico.

8.14. *Stenocereus thurberi,* or organ pipe cactus.
(A) This plant is growing in dense vegetation in southern Baja California and has relatively few branches near the ground.
(B) This plant is growing near Lukeville, Arizona, and has many branches that arise very close to the ground.

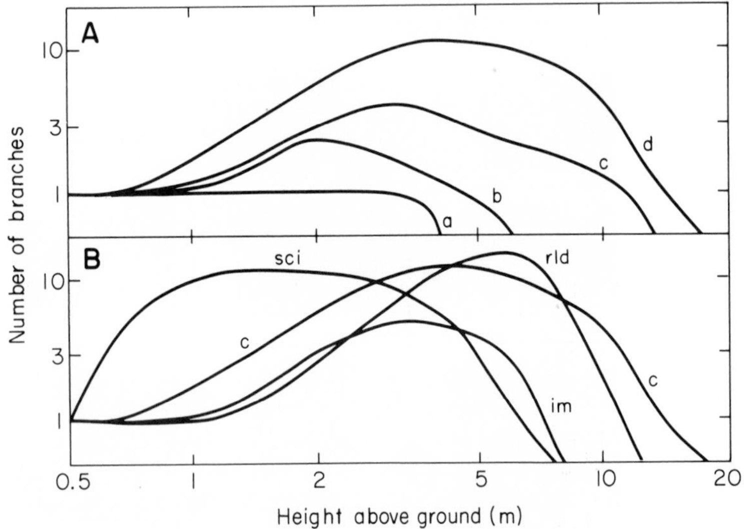

Figure 8.15. Branching patterns of cardón, *Pachycereus pringlei.* Note logarithmic scales. (Redrawn from Cody, 1984.)
(A) Number of branches in four size classes (a through d) at Cataviña, northern Baja California, Mexico. When the plant is over 4 m tall (b), one or several branches form about 1 m from the ground. Maximum branch number at Cataviña occurred around 4 m, and plants in this population can be almost 20 m tall.
(B) Branching patterns of old individuals at four sites in northwestern Mexico: Cataviña (c; same as d in *A*), Santa Cruz Island (sci), Rancho Los Divisaderos (rld), and Isla Monserrate (im). In each population, the mature growth habit is different.

tion. Moreover, the stands of tree opuntia themselves are also very dense, averaging 4100 stems per hectare. An opuntia plant therefore must compete for light with other opuntia plants (intraspecific competition) as well as with other leafy trees and shrubs (interspecific competition). Rapid vertical growth of the opuntia enables it to grow into full sun as soon as possible. Because the vegetation is not dense on Santa Fé, the tree platyopuntias do not have to grow as tall, a conclusion that we shall consider again toward the end of the chapter. Thicker trunks may be an advantage on Santa Fé, where the vegetation is much sparser and the winds can be strong.

Branching patterns of columnar cacti. Cactologists have traditionally described branching patterns by using very simple terminology: lateral branches many, few, or none; stems very long and indeterminant, long with lateral branches, and jointed; erect, ascending, decumbent (bending back toward the ground), pendent (hanging limp), and procumbent (lying on the ground); thick to very thin; and arising from the top, middle, or base of the plant. Martin L. Cody at UCLA has attempted to quantify branching of individuals in local populations of columnar cacti to compare the branching patterns and growth habits of various populations in the presence and absence of other plant species. Because columnar cacti have strong apical dominance and generally erect stems, Cody treated each plant as an ascending, vertical system. Therefore, to get a crude but useful profile of branching design, he measured the number of branches (the trunk being one) occurring at specific heights; he did this for plants of different size classes within each population.

Cody studied eight species of cacti in the Sonoran Desert from a variety of habitats, but we shall abstract only a few of the patterns observed. First, how does a plant add branches during its lifetime? To answer this question, we shall consider three populations: *Pachycereus pringlei* (cardón, Fig. 8.13) from Cataviña, Baja California, Mexico; *Myrtillocactus cochal* (cochal) from Colonet, Baja California; and *Stenocereus thurberi* (organ pipe cactus, Fig. 8.14) from Lukeville, Arizona.

Figure 8.15*A* shows branching patterns in four of Cody's eight size classes of cardón. Young plants up to 4 m tall (curve a) are solitary columnars (branch number = 1). A plant 6 m tall had an average of 2 or 3 erect stems (curve b); one 12 m tall had an average of 4 or 5 stems (curve c); and a giant 17 m tall had an average of 10 or 11 stems. The peak of curve d indicates that in giant specimens the greatest number of branches occurs at around 4 m, and the steepness of the first part of the curve tells us approximately how fast branches were added with increasing height. From these data, we can envision this cardón as very tall with a definite trunk and ascending branches that arise mostly between 1 and 5 m. The upper half of the plant consists of only a few stems.

Plants in the other two populations produce branches in very different ways. Cochal (Fig. 8.16) has a very short trunk, with several main branches near the ground; and there is a steady and very rapid increase in branching to the very top of the plant, so the canopy is very dense. Organ pipe cactus growing in Lukeville has no central trunk but instead produces lateral branches at ground level even before the plant is 0.5 m tall, and most future branches are added near the ground (Figs. 8.14*B* and 8.17*A*). This plant therefore has the greatest number of branches within the first meter and results in the organ pipe form; that is, a form with numerous, unbranched, erect stems that arise mostly from the base.

Another way to use these data is to examine how branching patterns change within a species from one population to another. For this, the curves of the tallest size class can be compared to highlight their morphological differences. Figure 8.15*B* shows the curves for *P. pringlei* from Cataviña and three other locations. At Cataviña the plants are exceedingly tall, branch fairly close to the ground, and have about a dozen branches at the maximum point. Plants in a thornscrub habitat at Rancho Los Divisaderos (south of La Paz), Baja California, produce branches much higher from the ground, and maximum branching occurs above the level of the surrounding plant

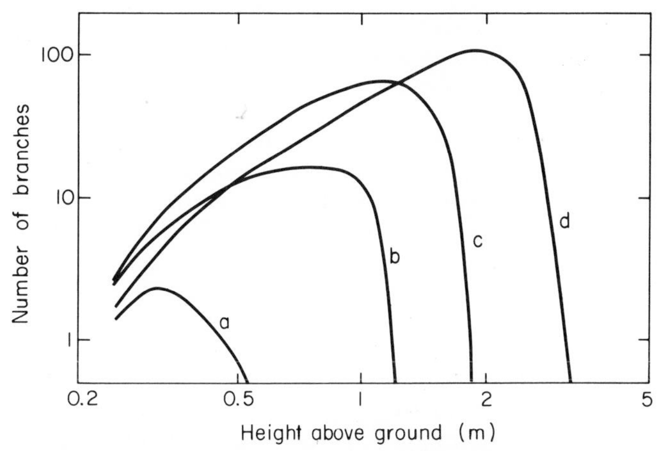

Figure 8.16. Branching pattern of *Myrtillocactus cochal* at Colonet, Baja California, expressed as the number of branches at various heights above the ground. Curves a through d show the branching patterns of plants that are short to tall, respectively. Notice that these plants have a main trunk with a few low branches that then repeatedly branch to yield a dense canopy less than 1.5 m from the ground. Each curve drops quickly to zero from the maximum point, a pattern signifying that the plant has a dense, flat-topped canopy. Note logarithmic scales. (Redrawn from Cody, 1984.)

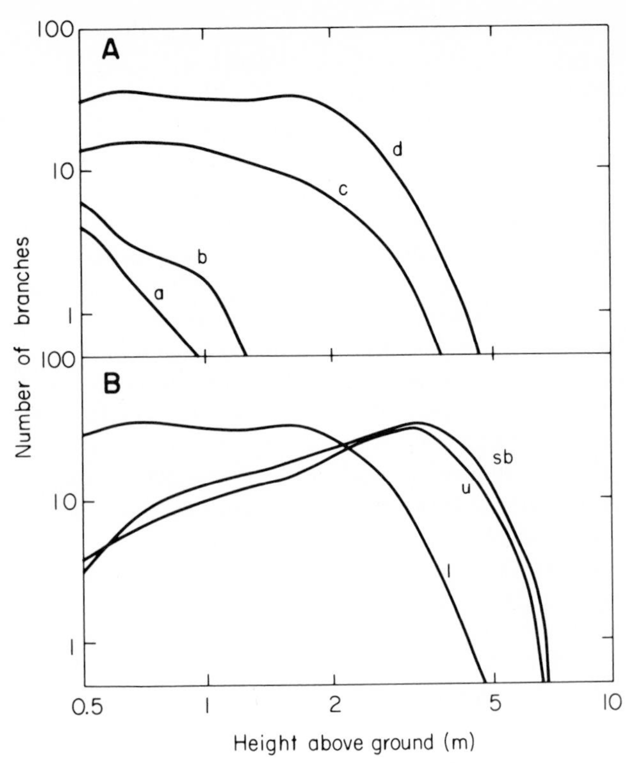

Figure 8.17. Branching pattern of organ pipe cactus, *Stenocereus thurberi,* expressed as the number of branches at various heights above the ground. Note logarithmic scales. (Redrawn from Cody, 1984.)
(A) Curves a through d show the branching patterns of plants that are short to tall, respectively, for a population at Lukeville, Arizona, in Organ Pipe Cactus National Monument. Branch number is greatest close to the ground.
(B) Comparison of the size class of mature individuals from populations at Lukeville (l; d of *A*); Ures, Sonora, Mexico (u); and San Bartolo, Baja California, Mexico (sb). In the two Mexican populations, which occur in tropical deciduous forest, branch number was low near the ground and highest around 4 m, which was the mean height of the surrounding vegetation. Although the populations at Ures and San Bartolo are quite separate, the growth habits of *S. thurberi* in the two locations were essentially identical.

canopy. These are tall plants but somewhat shorter than those at Cataviña. At the other extreme, plants on Santa Cruz Island in the Gulf of California (not the Santa Cruz in the Galápagos Islands) produce low branches and are relatively short individuals. Finally, on Isla Monserrate plants have a distinct trunk but produce relatively few branches throughout their lifetime, even though they grow to 9 m in height; and many specimens are solitary columnars.

A quick glance at organ pipe cactus from three locations also shows clear correlations with habitat (Fig. 8.17*B*). Whereas plants growing in the full-sun desert habitat near Lukeville branch mainly next to the ground (Fig. 8.14*B*), the plants of two populations growing in thornscrub, San Bartolo in Baja California (Fig. 8.14*A*) and Ures in Sonora, have relatively few branches near the ground and have maximum branch number at 4 m, which is just above the level of the canopy of the surrounding vegetation (Fig. 8.17*B*).

Cody's data demonstrate that branching patterns not only depend on the nature of the surrounding vegetation but also on the specific location, for example, on a hillside or a flat area or on different aspects of a hillside. On the basis of his data Cody has also hypothesized that the branching pattern for one species of columnar cacti can shift if the cactus occurs with a different species having the same branching pattern. To determine whether such biotic interactions actually occur, we need much more data and analysis than currently exists.

Epiphytes

About 120 species of cacti, roughly 7% of the family, are epiphytes. Every serious cactus collector grows a few epiphytic specimens in hanging baskets. In the New World before Columbus arrived, natives cultivated cactus epiphytes; and since their introduction into Europe during the 1700s, cactus epiphytes have attracted many admirers because of their splendid flowers. In fact, the largest and the smallest flowers in the Cactaceae belong to the epiphytic species (Table 1.1), and the red flowers of *Nopalxochia, Aporocactus,* and *Schlumbergera* have fancy structures that are adapted for hummingbird pollination.

In this chapter we shall reiterate some facts about the vegetative nature of epiphytes. We mentioned in Chapter 1 that epiphytes have juvenile and mature shoot morphology on the same plant (Fig. 1.24). In fact, the juvenile structure is actually more typical of cacti in general because it has spines and ribs. Most cacti begin shoot growth with 2/5 phyllotaxy and quickly form more than five vertical ribs or an elaborate design of prominent tubercles arranged in many helical series (Chapter 6). In contrast, epiphytes begin growth with 2/5 phyllotaxy and quickly become specialized with fewer than five ribs and poorly developed tubercles. It is common to find epiphytic stems that have three or four ribs at the base and only two ribs on the upper shoot (Fig. 6.28). Spination can also be reduced; most two-ribbed stems of epiphytes lack spines entirely, for example, stems in *Cryptocereus* (Fig. 8.18), *Disocactus, Eccremocactus* (Figs. 6.26 and 6.27), *Epiphyllum,* and *Nopalxochia.* A major exception is *Strophocactus wittii,* an Amazonian plant, the flattened stems of which grow closely appressed to and climb around branches and trunks; this species has many fine, short spines.

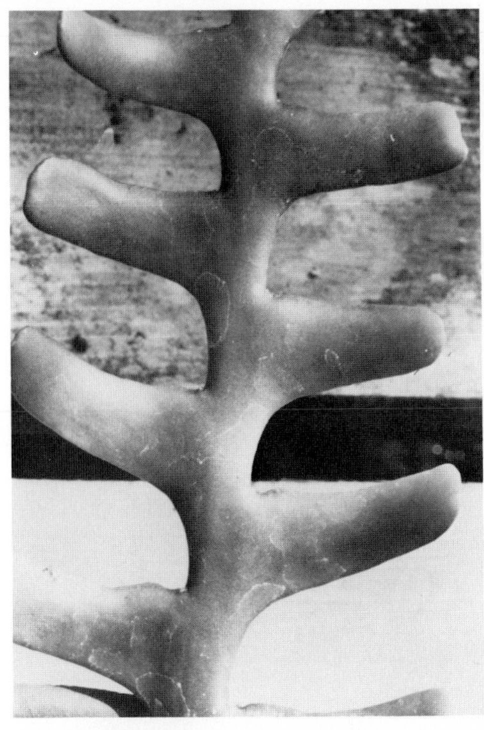

Figure 8.18. Deeply lobed, flattened, two-ribbed stem of *Cryptocereus anthonyanus,* an epiphyte from Chiapas, Mexico.

Radiation of vegetative forms in epiphytes has been studied, particularly in the genus *Rhipsalis.* Many naturalists have seen one species, *R. baccifera* (Fig. 1.24), growing in tropical areas of both the Old and the New World, where it hangs from crotches of tropical forest trees and may even emerge as a pendent plant from rock ledges. The stems of this organism and related species are cylindrical and spineless, although a small spine primordium is usually hidden in each areole. The juvenile stems of these species have weak ribs and thin, fairly soft spines. Species closely related to *R. baccifera,* such as *R. fasciculatus* and *R. aculeata* (Fig. 8.19), have mature stems that look like the juvenile stems of *R. baccifera.* Some other species, formerly classified as the genus *Hariota,* have cylindrical, spineless, jointed stems (Fig. 8.20). Still another section of the genus, including *R. rhombea* (Fig. 8.21), has flattened, two-ribbed stems; and another set of species, including *R. cruciformis* (Fig. 8.22) and *R. monacanthus,* has three or four well-defined ribs. Among the species with four ribs is *R. paradoxa* (Fig. 8.23); every time shoot growth starts, a new set of four ribs is initiated, but the ribs are shifted 45 degrees from the previous set. The genus *Rhipsalis,* which formerly included only epiphytes, now also includes several small, ribbed, terrestrial forms that were assigned to *Pfeiffera* (Fig. 8.24), and these are in turn closely related to the shrubby terrestrial genus *Corryocactus.*

The last point to be made is that epiphytism in Cactaceae has evolved several different times. For example, *Rhipsalis* belongs to the tribe Notocacteae, which is centered in eastern and southern South America. In this group, flattened, two-ribbed epiphytes have evolved in the genus *Rhipsalis* and in the related genus *Schlumbergera.* In North America, however, flattened stems have evolved in a number of genera of the tribe Hylocereeae, which is located primarily in Middle America and the West Indies. As far as scientists can tell, the two-ribbed species (six genera) of Hylocereeae have even evolved independently several times within that tribe from the nyctocerei and from the selenicerei. The evolutionary factors involved were discussed briefly in Chapter 6—a flattened, leaflike stem may be an adaptation to maximize PAR interception in shaded forest habitats.

Living Rocks

The word *mimicry* is widely used by biologists to describe an animal that resembles another organism or a background, such as tree bark or a leaf, to deceive either predator or prey. Botanists often find it hard to describe a plant as a mimic because there is frightfully little evidence that demonstrates how mimicry works in plant systems. Nonetheless, some of the "living stones" [*Lithops,* Fig. 8.25 (family Aizoaceae) and *Pseudolithops* (family Asclepiadaceae)] of South Africa, which are small leaf succulents, are cryptically colored and perfectly textured to blend in with the background soil and rocks. Likewise there are some "living rocks" in the cactus family, which live primarily in the Chihuahuan Desert and are exceedingly difficult to spot unless the collector gets on hands and knees to search carefully for them.

The living rocks of Cactaceae are small, somewhat flat-topped, spineless organisms belonging to the genera *Ariocarpus* (Fig. 8.26), *Aztekium, Lophophora* (Fig. 1.14), *Obregonia,* and *Strombocactus.* These are members of tribe Cacteae, but they are not necessarily all closely related within the tribe because, according to some researchers, the cryptic plant form has evolved independently several times. Another set of living rocks occurs in South America, particularly in the Atacama Desert (Fig. 8.27). All of these plants are cryptic because they project at most only a few centimeters above the soil surface or rock outcrop; thus, 90% of the plant is underground.

How the plant remains buried is a fascinating story. During the dry season much water is lost from the plant, and it shrinks in size. When the stem loses its turgidity, it is pulled downward into the soil because the root is firmly anchored in the soil. Consequently, the upper stem becomes flush with the soil surface

Figures 8.19–8.24. Stem designs in *Rhipsalis.*
8.19. *Rhipsalis aculeata;* this plant has long cylindrical stems and soft spines.
8.20. *Rhipsalis hernimiae;* this plant has short cylindrical stem segments.
8.21. *Rhipsalis rhombea;* this plant has strongly flattened (two-ribbed) stems.
8.22. *Rhipsalis cruciformis;* this stem has three ribs.
8.23. *Rhipsalis paradoxa;* this stem has four ribs, which are shifted 45° every time a new set is initiated.
8.24. *Rhipsalis ianthothele;* a terrestrial species with four, spine-bearing ribs.

and can never protrude very far because the growth rate is very slow.

The stems of these cryptic species are normally dull and gray and match the color of the substrate. There have been no direct observations to show that these plants have avoided predators by being hard to find. To add mystery to the crypsis hypothesis, we have shown that these species are loaded with toxic alkaloids (Chapter 9), compounds that are very effective against vertebrate herbivores because they are poisonous in even fairly small doses. The seedling has small spines (Fig. 3.49), but the mature structure has none. Two reasons have been suggested for the loss of spines: (1) the spineless plant has higher PAR

Figure 8.25. Four species of living stones in the genus *Lithops* of the ice plant family (Aizoaceae). The top surfaces are textured and colored to match the substrates in their native habitats in South Africa.

interception by the tubercles, which are mostly buried in the growth; and (2) the poisonous alkaloids protect them from potential herbivores and therefore the spines are not needed for defense. No work has been done to describe how these plants respond to winter minimum and summer maximum temperatures, although these may be the most important physiological stresses (Chapter 7).

Evolution of Wood Structure

As researchers investigate some of the ecological reasons for the evolution of growth habit, their attention is eventually drawn to the mechanical strength of the stem. Obviously, a tall tree must have enough mechanical strength in the trunk to support the weight of the canopy; and, conversely, a plant that creeps along the ground does not need much mechanical strength. Therefore, information on the amount, form, and quality of wood and the amount of succulent tissue are factors needed to explain the evolution of size and erectness in cacti. As it turns out, the cellular evolution of cactus woods was a very dynamic and exciting process that paralleled the evolution of growth habit. Thus, wood structure in a single genus may be very diverse, whereas unrelated plants with the same growth habit have the same wood structure.

In his studies on the leaf-bearing cacti (Chapter 2), Bailey documented the ancestral condition of wood for the Cactaceae. He determined that even in these least specialized members of the cactus family, the cellular structure of the wood is that of a highly specialized, advanced dicotyledon (Figs. 2.5–2.7), and the wood is fully lignified. This water-conducting tissue is composed of short vessel elements with a large hole on each end wall and many small pits on the lateral walls; the wood fibers are long and thin with pointed ends and few, very small pits (libriform wood fibers); and the wood parenchyma is highly specialized.

Bailey's pioneering papers on the leaf-bearing cacti inspired a graduate student named Art Gibson to make the evolution of cactus woods his area of study. Gibson was not a cactus enthusiast but wanted to study wood evolution. Why should an internal tissue like wood evolve at all? There was some evidence in

Figures 8.26 and 8.27. Living rocks of the Cactaceae. **8.26.** *Ariocarpus fissuratus;* a small plant about 10 cm in diameter with dull, flat tubercles lacking spines. The appearance of this plant helps to conceal it in the dry soil and rocks in the Chihuahuan Desert of Mexico.
8.27. *Neoporteria glabrescens* from the Atacama Desert of Chile; the pebblelike tubercles with very short spines barely protrude through the soil surface.

the literature, especially in publications of Sherwin Carlquist at Claremont Graduate School in California, that wood structure of a plant is adaptive for the habitat where the plant is found. Using cacti collected from many geographic areas and from many different genera and tribes, Gibson set out to determine what, if any, correlations exist between growth habit and cellular characteristics of each wood.

Wood Structure of Cactoideae

An initial study of 119 species in about 50 genera of the subfamily Cactoideae showed clear correlations between the growth habit of a plant and the type and dimensions of cells found in the stem wood (Table 8.1). Trees and erect shrubs have the longest and widest wood fibers as well as the longest and widest vessel elements. Low shrubs and epiphytes have short, narrow wood fibers and vessel elements; and barrel, caespitose, small cylindrical, and globular cacti have short, extremely narrow vessel elements and generally lack wood fibers, a condition referred to as nonfibrous wood. Hence, the length of the long cells in the wood appears to correlate with plant height.

Let us consider the members of the supergenus example that we used earlier in the chapter (Figs. 8.1–8.8), *Trichocereus* and *Echinopsis* (Fig. 8.28). The longest, widest, thickest-walled wood fibers occur in *T. terscheckii,* the tallest species; whereas the shorter trees and shrubs have narrower, shorter, thin-walled wood fibers. *Trichocereus thelegonus,* a procumbent species, has very few, extremely short, narrow, thin-walled fibers; and the barrel, caespitose, and small cylindrical species have very weak fiber development or lack fibers entirely. In place of wood fibers, the small, nonfibrous cacti have fusiform parenchyma cells with no secondary wall and some peculiar, empty, fusiform cells known as vascular tracheids, which have a helix of secondary wall on the inside (Fig. 8.29).

Figure 8.28 also shows the nature of the vessel elements in each species. As before, Gibson found that vessel element length decreases from the tall to the short species, and the pits on lateral walls get larger. Consequently, the nonfibrous woods have vessel elements with very weak walls that would not be able to support much weight. In fact, the woods of these small cacti mostly lack lignin and have no mechanical cells.

When we analyze the woods of Cactoideae in a different way, we can illustrate how unrelated species have woods that may be more similar to each other than they are to the woods of members in the same genus. In Table 8.2 we have presented groups of look-alike species, with the growth habits arranged in descending order of plant biomass. It is interesting to learn that the thickest wood fibers in the family occur in the solitary columnar cacti, which belong to two different genera, whereas wood fibers have only medium wall thickness in the more massive trees, which instead have a much thicker accumulation of wood in the trunk. Wood fibers in the small trees and large shrubs are shorter than in the arborescent species, and the weak and thin woods of the decumbent species and the epiphytes consist of very short and narrow wood fibers. The wood of the organ pipe cactus, *Stenocereus thurberi,* which can be up to 6 m tall (Fig. 8.14), is much different from that of the creeping devil, *S. eruca,* a procumbent member of the same genus (Fig. 1.25). The wood cells of organ pipe cactus are very similar to the cells of other species of its tribe that have the same growth habit, but all of them have shorter and narrower wood fibers than do their taller relatives in the same tribe.

Table 8.1. Growth habits and lengths of wood cells in mature, lower-stem woods of species in subfamily Cactoideae.

Growth habit[a]	Mean length (μm) of wood fibers (range of means)	Mean length (μm) of vessel elements
Large trees ($n = 35$)	949 (657–1207)	371
Small trees and large, erect shrubs ($n = 28$)	742 (510–961)	340
Arching or trailing shrubs ($n = 10$)	701 (606–828)	307
Decumbent shrubs ($n = 4$)	555 (527–579)	224
Epiphytes ($n = 13$)	416 (299–574)	228
Suffrutescent perennials, *Peniocereus* ($n = 2$)	454 (428–475)	216
Procumbent shrubs ($n = 1$)	408	208
Caespitose perennials with some wood fibers ($n = 4$)	387 (278–527)	198
Caespitose perennials without wood fibers ($n = 7$)	None	190
Small globular perennials ($n = 13$)	None	165
Herbaceous perennials, *Wilcoxia* ($n = 2$)	253 (252–254)	163

Source: Gibson (1973).

a. Growth habits are listed in the order of descending erectness and size of the plants. Notice how the lengths of wood fibers and of vessel elements decrease in roughly the same order as the changes in erectness of the growth habits. n, Number of plants examined.

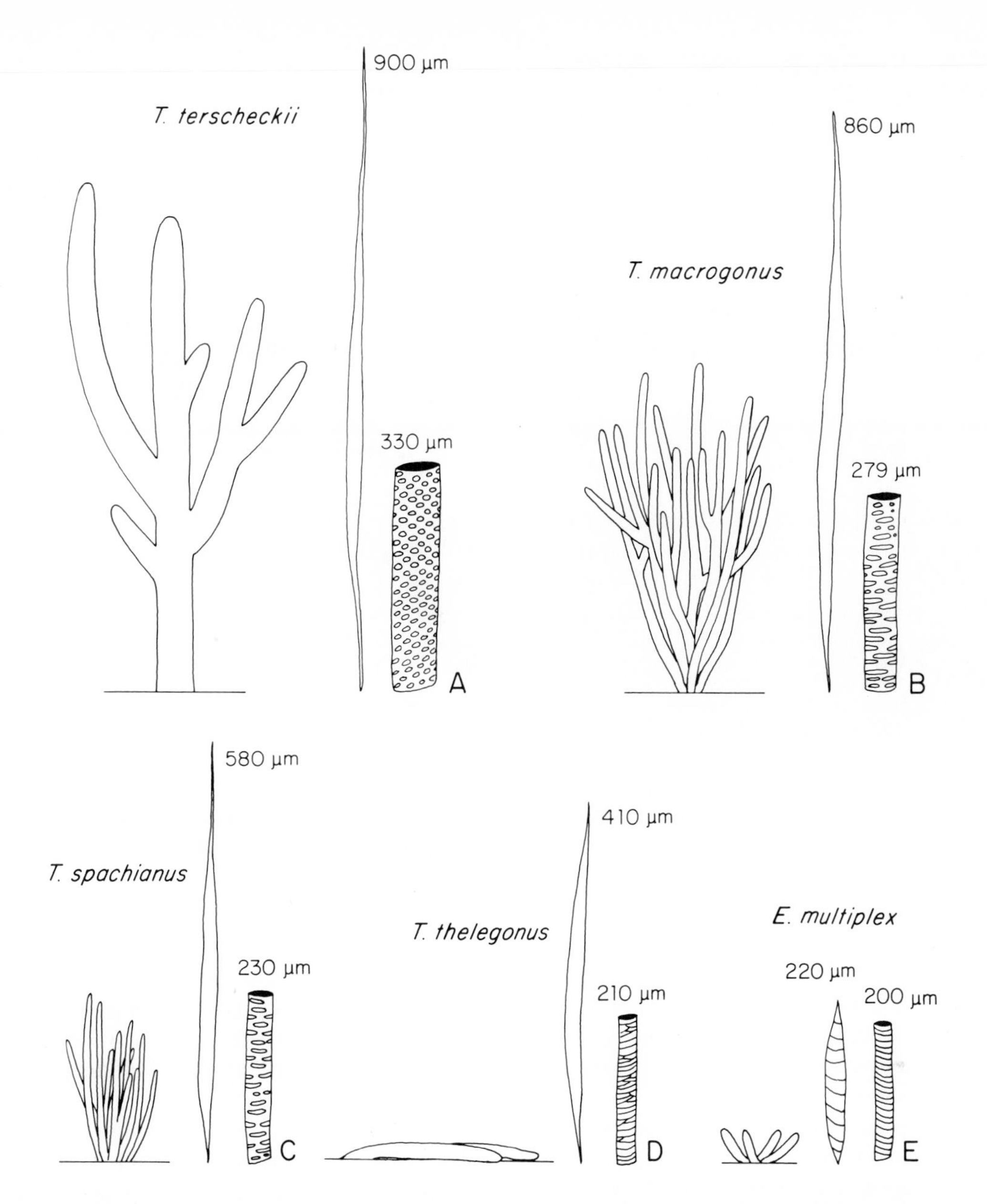

Figure 8.28. Relationship between growth habit (see Figs. 8.1 – 8.8) and the nature of the wood fibers and vessel elements in the wood in *Trichocereus* and *Echinopsis.* (Data from Gibson, 1973.)

(A) *Trichocereus terscheckii* is a large tree; it has the longest, widest, and thickest-walled wood fibers and the longest and widest vessel elements, which have many small pits on their lateral walls.

(B) *T. macrogonus* is a large erect shrub; it has narrower and slightly shorter wood fibers and narrower and shorter vessel elements, which have fewer, long pits on the lateral walls.

(C) *T. spachianus* is a low shrub with weak branches; it has short and narrow wood fibers with thin walls and short, narrow vessel elements with elongate pits.

(D) *T. thelegonus* is a procumbent plant; it has extremely short, narrow, and thin-walled wood fibers and vessel elements with thin, netlike (reticulate) thickenings of the lateral walls.

(E) *Echinopsis multiplex* is a small plant with small cylindrical stems that are loosely connected to form a caespitose plant; this species lacks wood fibers entirely but instead has vascular tracheids with helical thickenings and narrow, extremely short vessel elements with helical thickenings.

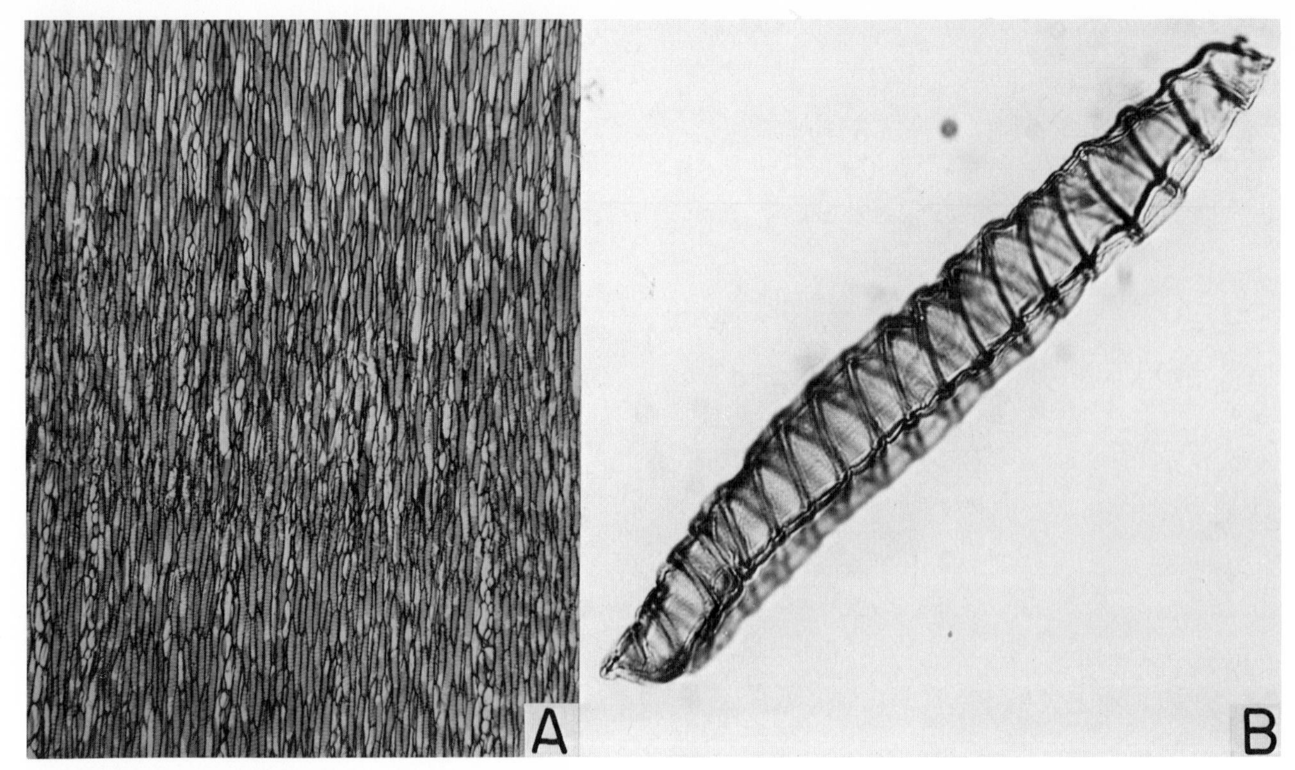

Figure 8.29. Vascular tracheids with helical thickenings in the wood of the barrel cactus, *Ferocactus townsendianus.* (From Gibson, 1973.)
(A) Longitudinal (tangential) view showing elongate cells with helical thickenings, which are the most abundant cells in the wood. ×35.
(B) A single, vascular tracheid with its internal, helical thickening; this cell has been isolated from other cells in the wood by dissolving the chemical (pectin) that cements the cells together. The vascular tracheid has no hole at the end of the cell, whereas the vessel elements in this wood do have holes (perforations). ×620.

The sample size in Gibson's study is too small to evaluate these data statistically, but the overall anatomical trends are easily observed—species with similar growth habits have similar woods.

Starting from an ancestral wood with very short vessel elements and wood fibers, as represented by the leaf-bearing pereskias, the woods of the Cactoideae have evolved in a very dynamic way. Plants that are tall and massive have longer wood cells than their ancestors, and those that are shorter with diminutive growth forms have shorter and narrower wood cells. Moreover, wood fibers and vessel elements have undergone many changes to match the growth habit and development of succulence.

Wood Structure of Platyopuntias

The woods of platyopuntias have experienced roughly the same evolutionary changes with respect to growth habit as have the plants just considered. As Bailey remarked, there are fairly comprehensive records of morphological divergence in cacti, and platyopuntias have species with a wide range of erectness.

In a study of 35 platyopuntia species, belonging to 17 different species groups, Gibson again found that the longest wood fibers occur in arborescent species and in erect shrubs that have well-defined trunks. These wood fibers are over 0.7 mm in length, and in several arborescent species the fibers are as long as those found in arborescent columnar cacti that are approximately as tall as the arborescent platyopuntias. Shrubs with less erect growth habits have shorter wood fibers; and low forms without a central stem, such as the beavertail cactus, *O. basilaris* (Fig. 7.7), as well as the Argentine *O. pampeana,* have stem woods that do not have fibers. Fiber diameter and fiber wall thickness decrease more or less linearly with a decrease in height and erectness of the plant. In plants with thinner wood fibers, fewer fibers are made, and the overall thickness of the wood decreases with decreases in height and erectness (Figs. 8.30–8.33). Concomitantly, the vascular rays that are lignified in arborescent species are converted into rays that lack lignin, have no thick walls, and store abundant starch. Hence, the woods of small platyopuntias, like the woods of small Cactoideae, completely lack mechanically strong cells.

The wood skeleton of the specialized species of platyopuntia discussed in Chapter 3 occupies a

Table 8.2. Cellular features in the lower-stem woods of species having different growth habits.[a]

Growth habit and species (tribe)	Mean length (μm) of wood fibers	Mean diameter (μm) of wood fibers	Average thickness (μm) of wood fiber wall	Mean length (μm) of vessel elements	Mean diameter (μm) of vessels
Large tree with many erect, heavy branches					
Cereus jamacaru	893	32.8	2.4	327	72.1
(Cereeae)	1155	31.4	2.4	445	92.5
Pachycereus pecten-aboriginum	956	32.8	3.6	391	70.4
(Pachycereeae)	1176	33.9	3.6	448	79.7
Large tree with relatively few, heavy branches					
Carnegiea gigantea	831	32.8	4.8	365	69.3
(Pachycereeae)	984	35.5	3.6	417	100.0
	791	35.2	4.2	374	81.6
	908	34.9	4.2	448	73.2
	885	25.7	4.2	414	84.6
Trichocereus terscheckei	790	33.9	4.2	320	92.2
(Trichocereeae)	906	28.9	4.2	332	85.7
Solitary columnar tree					
Cephalocereus totolapensis (Pachycereeae)	755	33.0	9.0	392	71.0
Neobuxbaumia mezcalaensis (Pachycereeae)	1021	34.7	8.4	420	76.2
Large shrub to small tree with many erect, vertical branches					
Lemaireocereus hollianus (Leptocereeae?)	741	26.2	3.6	346	81.9
Pachycereus marginatus (Pachycereinae, Pachycereeae)	747	30.6	3.6	314	76.2
Stenocereus thurberi	863	27.6	5.4	353	81.4
(Stenocereinae, Pachycereeae)	699	29.2	3.6	314	71.8
	726	25.7	3.6	325	57.3
Decumbent shrub					
Bergerocactus emoryi	566	31.1	3.0	238	43.1
(Hylocereeae)	579	37.1	3.6	194	57.1
Cleistocactus baumannii (Trichocereeae)	547	21.1	3.0	234	41.2
Prostrate columnar plant					
Stenocereus eruca (Pachycereeae)	385	21.0	2.4	190	41.5
Trichocereus thelegonus (Trichocereeae)	408	20.3	3.0	208	40.1
Hanging epiphyte					
Eccremocactus rosei (Hylocereeae)	398	16.8	2.4	219	27.6
Rhipsalis baccifera	337	17.2	3.6	194	23.8
Caespitose plant					
Echinocereus cinerascens	294	19.6	3.6	180	37.1
(Echinocereeae)	—	—	2.4	219	25.7
Mammillaria poselgeri (Cacteae)	—	—	—	158	38.5
Globular plant					
Lobivia torecillacaisis (Trichocereeae)	—	—	—	241	45.3
Lophophora williamsii (Cacteae)	—	—	—	169	42.9

Source: Gibson (1973) and unpublished data.

a. For each growth habit, two species were chosen from different genera and, if possible, from different tribes (parentheses). Wherever possible, measurements from several different collections have been used.

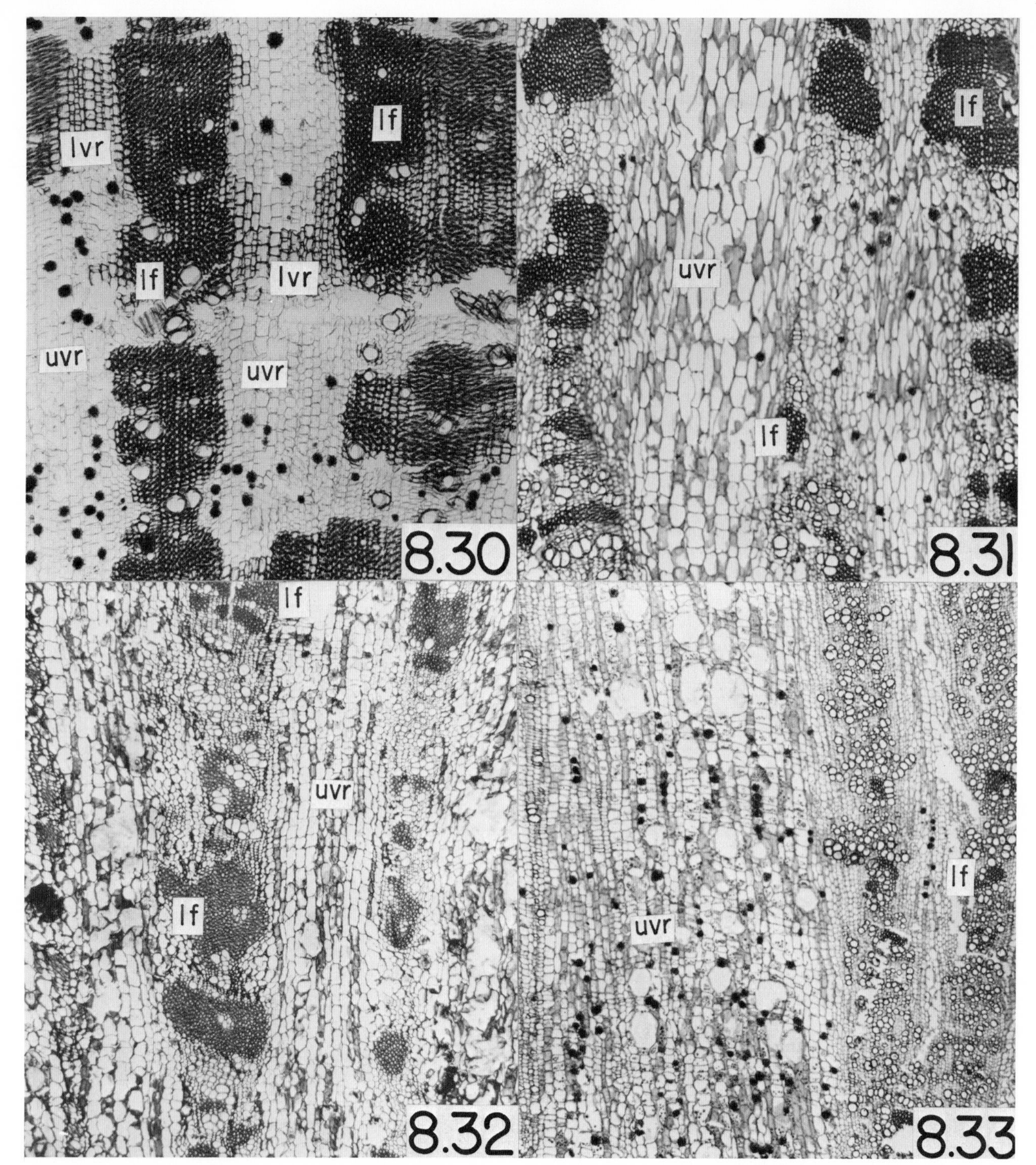

Figures 8.30–8.33. Woods of North American platyopuntias, showing the relative amount of lignified cells in cross sections of the same size (×33). (From Gibson, 1978b.)
8.30. *Opuntia atropes,* a short, arborescent platyopuntia with wide panels of lignified libriform wood fibers (lf) that are interrupted by unlignified patches of unlignified cells such as the unlignified vascular rays (uvr); however, portions of the vascular rays do have lignified cell walls (lvr).
8.31. *Opuntia rufida,* an erect shrub that has patches of libriform wood fibers (lf); but over half of the wood consists of unlignified vascular rays (uvr).
8.32. *Opuntia lindheimeri* var. *linguiformis,* a low shrub that has scattered patches of wood fibers.
8.33. *Opuntia fragilis,* a low plant that has short-lived shoots in which only very small patches of libriform wood fibers can be seen; thus, most of the wood is unlignified. The dark structures in the vascular rays of the figures are druses; the large, clear cells contain mucilage.

smaller volume than the unlignified part of the wood does; therefore, the woods of platyopuntias are quite succulent. In contrast, the woods of Cactoideae and of the cylindropuntias tend to exclude the succulent patches, an arrangement permitting succulent tissues to reside mostly in the primary rays. Especially in the chollas, the wood is generally solid with little succulent tissue. Cylindropuntias and tephrocacti are intriguing because their woods look more similar to the xylem of nonsucculent desert perennials than they do to cactus woods; they do not fit the trends in wood fiber sizes found in the other cacti except that the caespitose species (*Tephrocactus* and *Opuntia tunicata* var. *davisii*) have nonfibrous woods.

Convergent Evolution of Nonfibrous Woods

Even though the evolution of cell dimensions in cactus woods, as related to growth habit, should have commanded center stage, the nonfibrous woods have captured most of the interest. In 1845 Mathias Schleiden, coauthor with Theodor Schwann of the Cell Theory, observed in *Mammillaria* some strange, cigar-shaped (fusiform) cells with very prominent helical internal thickenings, the vascular tracheids (Fig. 8.29). At first glance these cells look like vessel elements that we would find in a young leaf or a flower petal, but there are three obvious differences: vascular tracheids are extremely short cells, whereas vessel elements in young leaves are fairly long; they have no perforations (holes in the ends), so they cannot be organized into pipes; and they are produced in the wood by the vascular cambium, not by the procambium (layer producing the first-formed xylem cells). Various authors have noted vascular tracheids in other small cacti, including some with ringlike (annular) rather than helical thickenings, and Bailey took great interest in vascular tracheids with annular thickenings in wood specimens of *Pereskiopsis* and *Quiabentia.* Bailey interpreted these as highly specialized end products of evolution—short vessel elements that have lost the ability to form perforations. As we shall see, this explanation probably was not correct.

Nonfibrous cactus wood is apparently unlike the secondary xylem of any other plant on earth. It has four components: (1) extremely short and narrow vessel elements that have a simple perforation at each end of the cell and very scanty helical or annular thickenings on the lateral walls; (2) empty, nonliving, fusiform vascular tracheids with scanty secondary thickenings that project far into the center of the cell (the cell lumen); (3) living (nucleate), fusiform, thin-walled parenchyma cells located among the vessel elements and vascular tracheids; and (4) living, large, squarish, thin-walled parenchyma cells organized into vascular rays. The wood lacks lignin, except in the secondary thickenings of the vessel elements and vascular tracheids.

The cactus wood studies of the 1970s revealed that nonfibrous woods have evolved independently in each of the three subfamilies (vascular tracheids occur in *Maihuenia* of the Pereskoideae, Fig. 8.34), independently in at least five different tribes of Cactoideae in both North and South America, and independently in North and South American lineages of opuntias with cladodes and with cylindrical to globular stems. In each case elimination of the wood fibers was matched by the appearance of vascular tracheids in the secondary xylem.

A great deal has been learned about the origin of vascular tracheids by comparing the wood structure of species with some wood fibers to their closest relatives that have fiberless wood. A group of dry-fruited species of platyopuntias have been studied in this way. The largest species is *O. rufida,* a shrub that often has a short trunk. This species has fibers in the wood from the beginning of wood formation to the very end, and the vessels have the same type of pitting of the lateral wall found in *Pereskia.* Very few vascular tracheids are present in the stem, and most of these are found near the pith. *Opuntia microdasys* is a smaller plant without a main trunk; it has a small percentage of wood fibers composing the wood, but wood fibers are typically absent in the early-formed wood, where some vascular tracheids occur. The vessels of this species have elongate pits on their lateral walls. Beavertail cactus, *O. basilaris,* has cladodes that arise at ground level and has no central axis. In this species wood fibers are absent in the stem, the vessel elements have scanty secondary thickenings, and the vascular tracheids occur in most parts of the xylem, except in the very last-formed increment. However, the relatively nonsucculent roots of all three species, including beavertail cactus, have wood fibers from beginning to end.

Much has also been learned about the origin of nonfibrous woods by studying the changes in cell types within a single stem. Gibson has found species in each of the subfamilies that produce nonfibrous wood for a number of years, but eventually the plant produces a fiber-bearing wood. In *Maihuenia* vascular tracheids and vessel elements with scanty secondary thickenings are no longer produced; instead the vascular cambium produces vessel elements with *Pereskia*-like pitting and wood fibers (Fig. 8.34*C*). In the transition region between nonfibrous and fiber-bearing xylem, some vascular tracheids are found in contact with wood fibers. In some barrel cacti, for example, *Ferocactus wislizenii,* a plant may produce nonfibrous wood for over 25 years (Fig. 8.35*A*). When the plant reaches a height of about 1 m

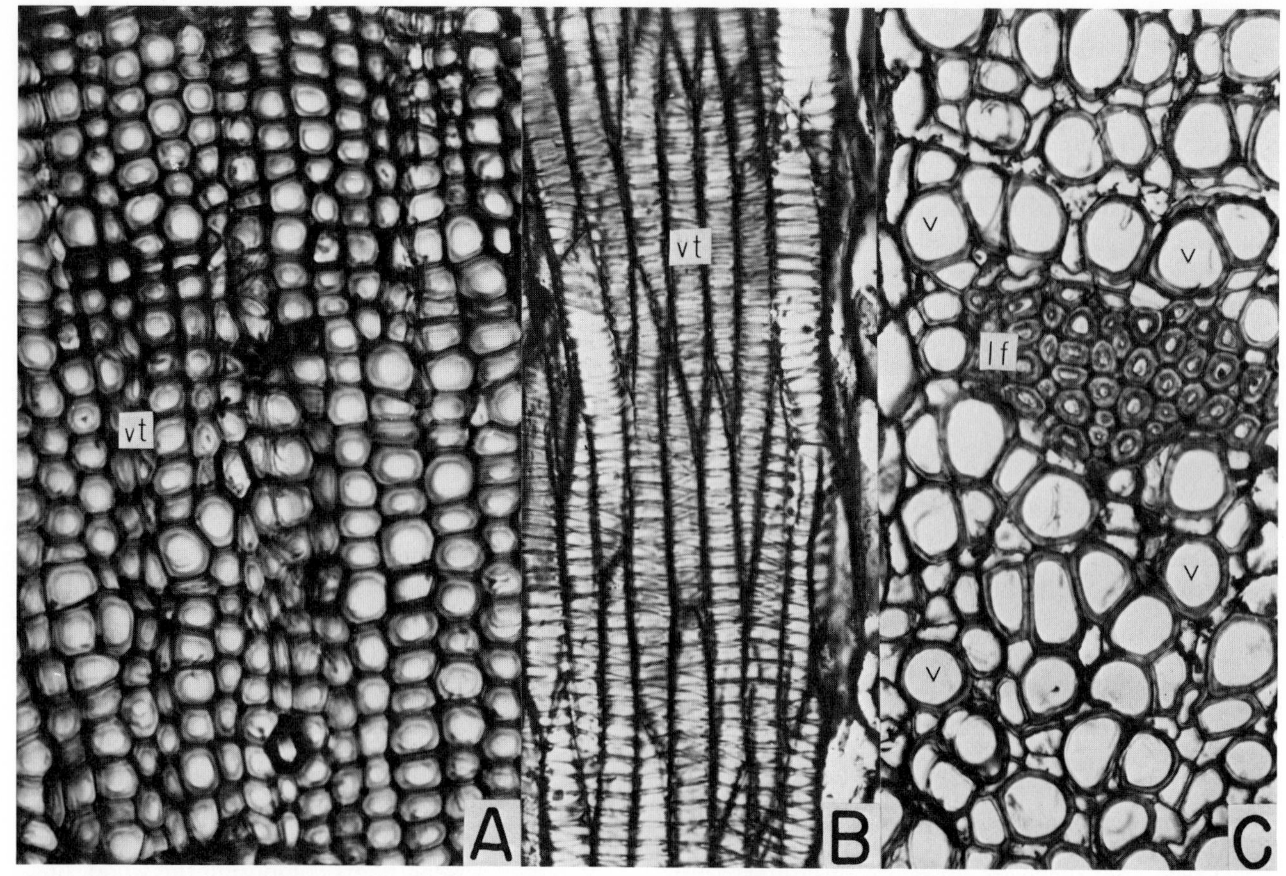

Figure 8.34. The composition of the lower stem wood (secondary xylem) of *Maihuenia patagonica* from Argentina. ×250. (From Gibson, 1977a.)
(A) Cross section through the inner portion of the wood, in which most of the cells are vascular tracheids (vt).
(B) Longitudinal (tangential) view of the vascular tracheids, which have helical thickenings.
(C) Cross section through the outer portion of the wood, where vascular tracheids are absent and the wood is composed largely of vessels (v) and occasional clusters of libriform wood fibers (lf).

(sometimes earlier), however, the wood structure changes so that the outermost xylem contains fully pitted vessel elements and thin-walled wood fibers but no vascular tracheids (Fig. 8.35*B*). Transformations of this kind have been observed in *Echinocereus* of North America (Echinocereeae), in *Echinopsis* of South America (Trichocereeae), and in a caespitose cholla *(Opuntia tunicata* var. *davisii).*

It is unfortunate that Bailey entitled his 1966 article, "The significance of the reduction of vessels in the Cactaceae." When he discussed the vascular tracheid, Bailey never made it clear to the casual reader that these cacti also have true vessels. Consequently, those who have read this article and have not carefully followed the recent research have accepted these as "vesselless" plants and have made incorrect statements about them. Cacti do not lack vessels—every cactus has plenty of vessels, in every growth ring of wood. More important, the evidence presented in the wood surveys of Cactaceae strongly suggests that vascular tracheids in cacti probably represent modified wood fibers or what would be wood fibers and not a vessel element that has lost the ability to form perforations. In general, cactus vascular tracheids occupy the positions of wood fibers in so-called normal cactus wood, and they disappear once wood fibers are produced.

The physiological significance of nonfibrous woods in cacti has not been studied, but it is interesting to speculate on the meaning of these strange woods, some of which are 3 cm thick. In the first place, nonfibrous wood is probably energetically less expensive to make because it contains less of the metabolically expensive cellulose and lignin. Second, mechanically weak woods can be tolerated in these small cacti because the plants are small and support can be provided by the large succulent tissues of the cortex and pith. Third, because there are almost no thick walls on the xylem cells, water can diffuse in any direction without much resistance. Finally, because the vessels and vascular tracheids have very scanty and flexible secondary thickenings, the wood, as well as the rest of the plant, can expand and contract with changes in water availability without

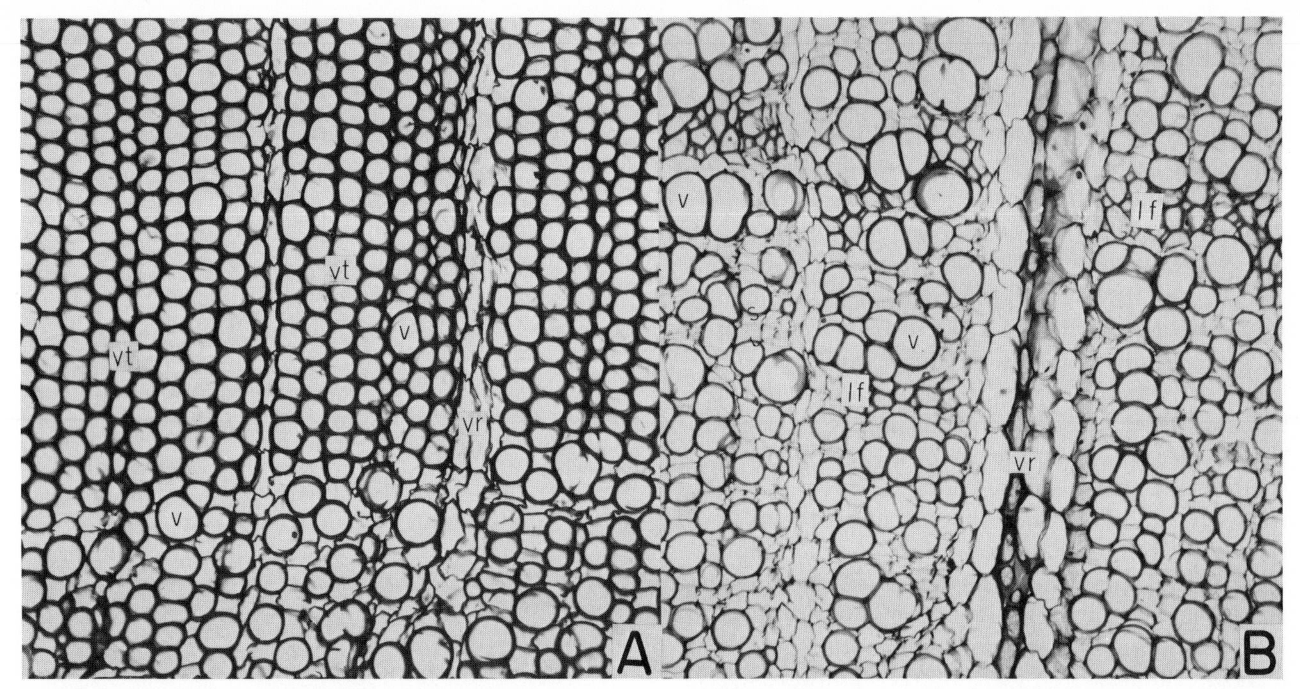

Figure 8.35. The wood (secondary xylem) of a barrel cactus, *Ferocactus wislizenii,* which was 90 cm tall. *(A)* Section from the inner wood at the base of the plant, showing that most cells are vascular tracheids (vt). At the start of each growth increment, the plant produces numerous vessels (v); but after that vessels are relatively infrequent. ×150. *(B)* View of the outermost region of wood from the base of the plant; in the outermost wood of a very old specimen such as this, vascular tracheids are no longer formed, and instead libriform wood fibers (lf) are produced. Both sections have thin-walled parenchyma cells in the vascular rays (vr) and in the vicinity of the vessels. ×150.

damage to the cells. Some innovative experiments are needed to study water movement in these tissues, perhaps using water-soluble dyes.

Gibson has also suggested a developmental explanation for the origin of these strange nonfibrous woods. Mature wood is actually xylem that bears wood fibers and no vascular tracheids, like the xylem formed in old specimens. It is reasonable to suggest, therefore, that nonfibrous wood is actually a juvenile form. For many years xylem is made by the vascular cambium under a developmental system that is insufficient to produce complete wall formation. When the stem becomes biochemically mature, that is, when the normal adult complement of stimuli (sugars and hormones) is present, a fibrous wood, typical of the ancestral cacti, is permitted to form. Thus, juvenilism, or the delay in the onset of biochemical maturity, may be the mechanism by which these incredible woods have evolved.

Plant Form and Interception of Photosynthetically Active Radiation

Each tribe (Table 1.2) and even most genera of cacti exhibit a range of forms, and the radiation of form and habit has been very dramatic in particular taxa, such as in the trichocerei, the North American columnar cacti, and the platyopuntias. We can explain this diversity by describing the changes in branching, changes in development of ribs versus tubercles, changes in plant height due to changes in strength of the wood, and so forth; and by this type of analysis we can eventually understand *how* this diversity has arisen. Nonetheless, we still need to explain in ecological and physiological terms *why* the diversity in whole-plant designs is so great in cacti. This field of research is just starting, and most of the published studies have considered how just one parameter, photosynthetically active radiation (PAR), has caused variations in plant form.

We have already indicated that the photosynthesis by cacti can be PAR limited, even in the high-radiation environments of deserts (Chapters 4 and 5). We have also considered the influence of ribs on PAR responses of cacti (Chapter 6). We shall now consider other ways that morphology of cacti influences PAR interception, and especially how stem orientation or height can affect the interception of PAR. Such studies have been performed primarily by Nobel, who has made observations on native cacti throughout the Western Hemisphere, from Canada to Argentina and Chile, and on cultivated cacti in Israel and even "down under" in Australia.

Seasonal and Latitudinal Effects on PAR Availability

We showed in Chapter 4 that nocturnal CO_2 uptake varies with the total daily PAR received during the daytime on the stems of *Ferocactus acanthodes* and *Opuntia ficus-indica* (Fig. 4.15). For instance, the PAR compensation point occurred just below 4 mol photons m^{-2} day^{-1} and 90% saturation occurred near 22 mol m^{-2} day^{-1}. Using computer simulations we can determine the PAR available on clear days at various latitudes and seasons. These simulations are done for vertical surfaces because, unlike the condition for the leaves of many plants (including pereskias, the leaves of which tend to be horizontal), the stems of most cacti primarily have a vertical orientation. These analyses show that vertical surfaces receive considerably less PAR than do horizontal surfaces and that some surfaces face in directions that get no direct sunlight at certain times of the year.

Simulations and measurements for studying the effects of orientation on PAR interception were used

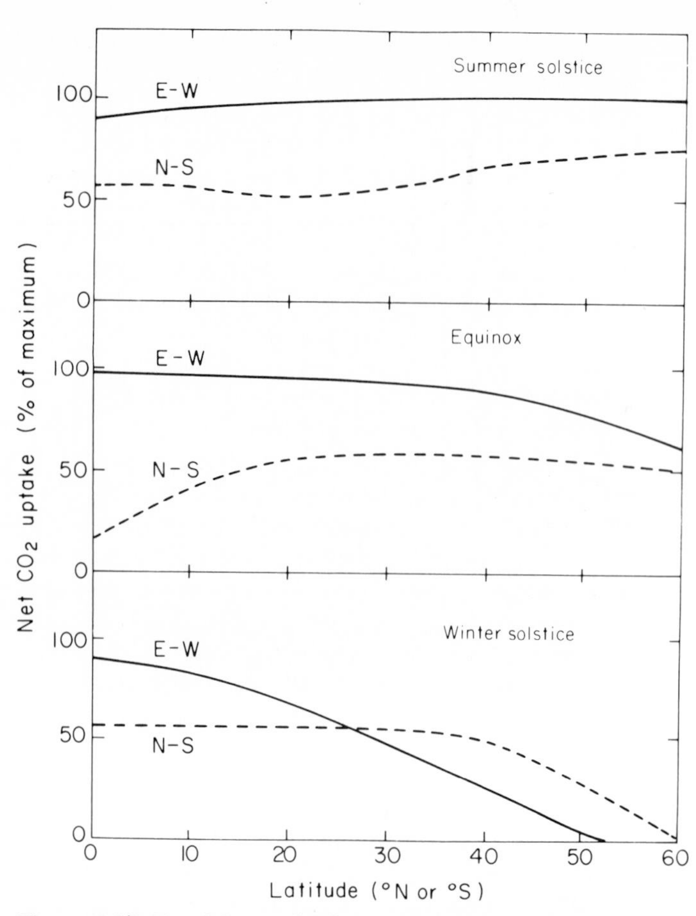

Figure 8.37. Net CO_2 uptake for vertical stem surfaces facing east–west or north–south, calculated from the PAR levels in Figure 8.36 and the response of nocturnal CO_2 uptake to total daily PAR in Figure 4.15. (From Nobel, 1985.)

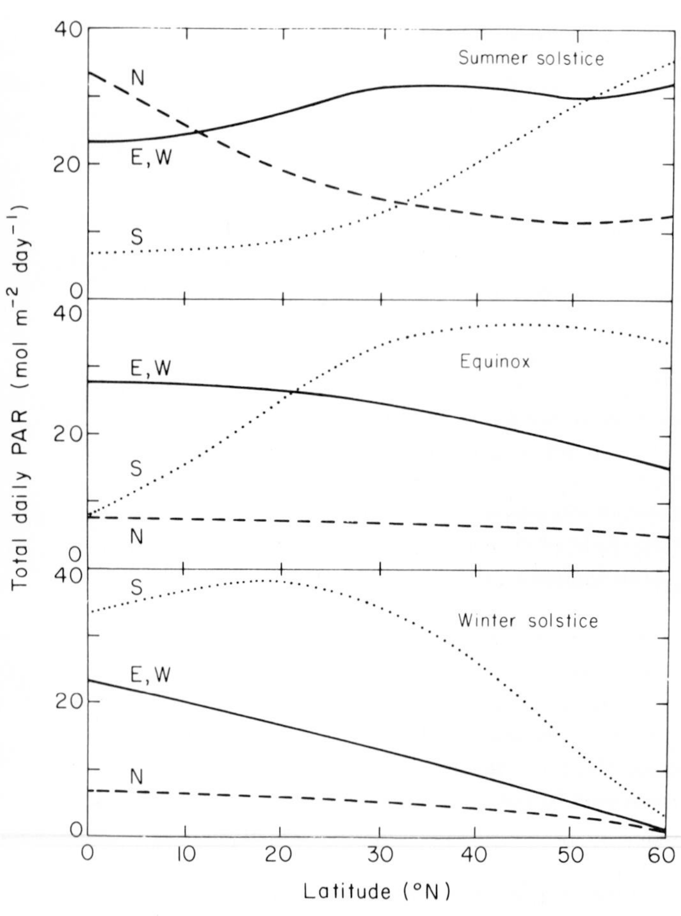

Figure 8.36. Total daily PAR, simulated by a computer model, incident on vertical surfaces facing in the indicated directions at various latitudes and times of the year on clear days. (From Nobel, 1985.)

to fit an intermediate latitude, 34°N, such as is appropriate for the University of California Philip L. Boyd Deep Canyon Desert Research Center, where many of the studies on the ecophysiology of desert succulents have been performed. On a clear day at the summer solstice there, the total daily PAR is 66 mol m^{-2} day^{-1} on a horizontal surface, 32 mol m^{-2} day^{-1} on an east- or west-facing surface, 15 mol m^{-2} day^{-1} for a south-facing surface, and 14 mol m^{-2} day^{-1} for a north-facing surface. Thus, the sides of a cactus stem facing in the four cardinal directions then receive an average of only 35% as much PAR as the horizontal surface. At the winter solstice, the sides receive 63% as much PAR as a horizontal surface. The low trajectory of the sun in the winter sky actually causes the south-facing surface to receive more PAR (31 mol m^{-2} day^{-1}) than does a horizontal surface (23 mol m^{-2} day^{-1}); whereas a north-facing surface receives no direct sunlight and therefore only intercepts diffuse radiation from the sky or the sunlight that has been reflected from the surroundings (5 mol m^{-2} day^{-1}). Averaged over the whole year and all orientations, vertical surfaces receive about half as much PAR as horizontal surfaces. This result has important implications for the growth and productivity of cacti.

The total daily PAR for the two solstices and the equinox at 34°N as well as all other latitudes up to 60° from the equator are presented in Figure 8.36 (to determine total daily PAR interception for latitudes in the Southern Hemisphere, simply switch N and S). For latitudes near the equator (0° to 20°N), south-facing surfaces received the lowest PAR near the summer solstice, north-facing surfaces near the winter solstice, and both north- and south-facing surfaces near an equinox. For midlatitudes (20° to 40°N), PAR interception by north- and south-facing surfaces was lowest near the summer solstice and by north-facing surfaces was lowest at the winter solstice and equinoxes. South-facing surfaces received the most PAR at the equinox and winter solstice at mid-latitudes, and this advantage was even greater at high latitude (40° to 60°N). It is no wonder, therefore, that a south-facing window is highly desirable during the winter at high latitudes in North America, Europe, and Japan, both for the comfort and cheerfulness of the human domicile and also for a cactus that might be placed near the window.

There are many ways to analyze the results in Figure 8.36. For example, if the values obtained for the solstices and equinoxes are averaged for an entire year, the total daily PAR on clear days for low to high latitudes ranges from 17 to 21 mol m^{-2} day^{-1}, or 20 mol m^{-2} day^{-1} when averaged over all latitudes considered (Fig. 8.36). Relating this to data presented earlier in Figure 4.15, we discover that such a total daily PAR would lead to just over 80% of maximum nocturnal CO_2 uptake. Thus, PAR is generally *limiting* for the growth of cacti. Moreover, this average is really an optimistic value that does not include days that are cloudy and that is only true if the value used is uniform at 20 mol m^{-2} day^{-1}, which it is not. Therefore, the estimates understate the extent to which cacti can be limited by PAR interception in natural environments.

Combining the available PAR information (Fig. 8.36) and the PAR responses of net CO_2 uptake by cacti (Fig. 4.15) enables us to predict CO_2 uptake for various orientations. We shall restrict our attention to surfaces that face east–west or north–south (if one side of a vertical cladode of a platyopuntia faces east, the other side must face west). Moreover, orientation effects that are not possible for cylindrically symmetric stems have been observed for such cladodes. Figure 8.37 shows that a cladode facing east–west will generally have a higher net CO_2 uptake than a cladode facing north–south. This observation relates to the generally low total daily PAR received by a north-facing surface over much

Figure 8.38. *Opuntia chlorotica,* or pancake pear. This shrubby platyopuntia occurs in similar habitats over wide regions of the Mojave and Sonoran deserts.

of the year in most latitudes north of the tropics, as pointed out earlier. A very interesting change occurs at middle latitudes during the winter (Fig. 8.37), however. The low trajectory of the sun in the winter causes a south-facing surface in the Northern Hemisphere and a north-facing surface in the Southern Hemisphere to receive plentiful PAR and thereby to make up for the poor showing of the other surface. Indeed, at latitude 27°N or S, the north–south combination actually exceeds the east–west combination for PAR interception and hence net CO_2 uptake (Fig. 8.37). This fact has important implications for the orientation of the vertical cladodes of platyopuntias.

Orientation of Platyopuntia Cladodes

Cladodes of a platyopuntia are arranged vertically on the plant—one on top of another—to form an erect plant. Sometimes the cladodes are randomly oriented, so that the surfaces face in all compass directions in a fairly even way, but in some cases the orientation of the pads shows a trend for certain directions. For many years botanists tried to determine why orientation of cladodes is observed.

Determining the orientation of the unshaded terminal cladode of platyopuntias is easy, but it requires a little discipline. If an eager group of students is taken into the field and instructed to form a ring around a plant, such as a specimen of *Opuntia chlorotica* (Fig. 8.38) or *O. ficus-indica* (Fig. 4.7) with many cladodes, and then asked to raise their hands if they think most of the cladodes are facing them, invariably all the students raise their hands. The eye sees the broad expanse of the surface of a cladode facing the observer and registers an impression of much area; if a cladode is edge-on to the observer, then that observation tends to be discounted. The same biased impression can be made by a solitary observer slowly circling around a large platyopuntia. Therefore, to get a more reliable measure of possible orientation, the angle that each cladode makes with some fixed direction, such as north, needs to be determined. Also, care must be taken to consider only cladodes that are in full sunlight, that is, those that are not shaded by other cladodes, other vegetation, or even buildings or topographical features, because only then will the implications of directional effects of PAR interception on CO_2 uptake be easily seen.

Some of the earliest measurements were made on *O. chlorotica* in the Mojave Desert, where rainfall occurs predominantly in the winter, and in the Sonoran Desert, where the annual rainfall pattern can be bimodal, that is, having two seasons of precipitation, winter and summer. For sites consider-

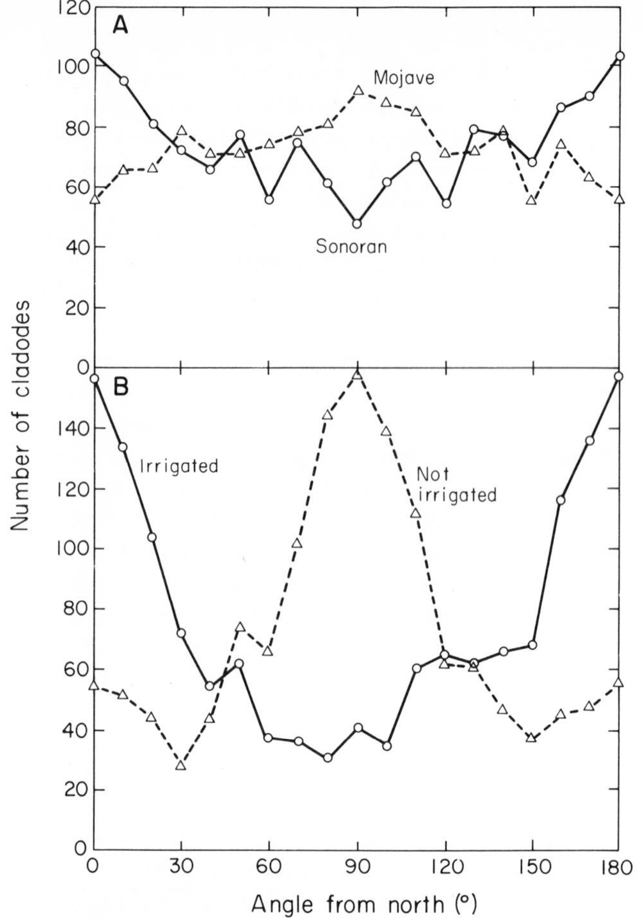

Figure 8.39. Orientation patterns for terminal, unshaded cladodes of two platyopuntias.
(A) *Opuntia chlorotica,* showing data averaged for two sites in the Mojave Desert in California (average latitude of 35°N) and two sites in the Sonoran Desert in Arizona (33°N).
(B) *O. ficus-indica,* showing data for two nonirrigated sites (35°N) and two irrigated sites (32°N) in Israel. Orientations of a total of 660 cladodes clockwise from true north were measured at each site. Data are presented in angle classes 10° wide centered on the angles indicated. (From Nobel, 1981, 1982b.)

ably north of 27°N latitude, there was a slight tendency for cladodes to face north–south in the Mojave Desert and east–west in the Sonoran Desert (Fig. 8.39*A*). In fact, in the Sonoran Desert about twice as many cladodes faced east–west as north–south. These orientations maximize PAR interception and net CO_2 uptake by the cladodes at times of the year most favorable for growth.

Nobel has made observations on cultivated stands of *O. ficus-indica* to determine what factors influence cladode orientation. For example, orientation was studied at sites in Israel, where about 80% of the annual rainfall occurs near the winter solstice. In nonirrigated sites at 35°N, there was a pronounced tendency for the terminal, unshaded cladodes to face north–south (Fig. 9.39*B*), an orientation that would enable the plant to take advantage of the lowness of the sun's trajectory in the winter. However, at irrigated sites, where plants grow year-round, the predominant orientation was east–west. Thus, seasonality of growth was again shown to be important.

In all locations where Nobel studied orientation, the orientation of terminal cladodes of platyopuntias were shown to be in accord with the predictions of maximizing net CO_2 uptake, as embodied in Figure 8.38. Even a survey of 23 species on four continents showed that cladodes tended to face east–west, unless growth occurred near the winter solstice and the site was more than 27° latitude from the equator, in which case a north–south orientation was preferred. There were, however, a few quirks in the story. In particular, the preferred orientation for terminal cladodes on 100-year-old cultivated plants of *O. ficus-indica* near Santiago, Chile, was not truly east–west but was rotated about 20°. Indeed, in the excitement of measuring orientations on century-old plants, shading from topographical features was not taken into account. Upon returning to that site the next day, Nobel found that a hitherto "unobserved" mountain was blocking considerable radiation from the west. Back in the laboratory, the computer model predicted that a rotation of 20° from east–west would lead to maximum CO_2 uptake when the western portion of the landscape was correctly represented. Such topographical effects on orientation patterns were subsequently observed for *O. littoralis* (Fig. 8.40) growing on steep hillsides facing the Pacific Ocean in southern California, and for *O. erinacea* growing in mountainous regions in inland, southern California. Moreover, seasonality of cladode development was observed, because cladodes that formed in winter months tended to face north–south whereas those formed in other months tended to face east–west.

The tendency of cactus cladodes to have preferred orientations has implications for the cultivation of platyopuntias. Observations in central Mexico indicate that east–west facing cladodes of *O. amyclaea* intercept more PAR and have greater dry matter production than north–south facing cladodes. Perhaps judicious pruning of these plants could be used to increase productivity. New cladodes tend to form in the same plane as the cladodes from which they develop; therefore, favorable orientations can be established at the time of planting the field by placing detached cladodes facing in the best direction for PAR interception. Even if this is not done, many of the new cladodes will be rotated slightly to expose the surface to higher PAR levels.

The tendency of new cladodes to have the same orientation as the underlying cladode can also be used to interpret the orientation found in natural populations. Specifically, cladodes facing in a direction leading to more PAR interception would be

Figure 8.40. *Opuntia littoralis* var. *vaseyi*, growing in its native habitat near the coast in southern California.

expected to have higher productivity, including the production of more new cladodes. In turn, these new cladodes would tend to perpetuate the favorable orientation of the underlying cladode. Indeed, orientation tendencies are more pronounced in species that have many cladodes occurring along a particular axis, for example, *O. ficus-indica* and *O. stricta,* than in species with just a few such cladodes, for example, *O. inamoena* of Brazil.

The evidence for accepting cladode orientation as a response to improving PAR interception is now very compelling; for cacti, PAR is on the verge of limiting productivity, even for unshaded cladodes on clear days, and platyopuntias orient their cladodes to receive as much PAR as possible. At the same time, orientation of cladodes affects thermal budgets because interception of more PAR also means intercepting more shortwave radiation (Chapter 5). Absorption of shortwave radiation will lead to higher tissue temperatures. Field measurements indicate that when a cladode faces the sun the average surface temperature is 1°C to 6°C higher than when a cladode is edge-on. But this temperature aspect has little effect because extremely high tissue temperatures can be tolerated by platyopuntias and because daytime temperature has relatively little effect on nighttime CO_2 uptake for cacti under natural conditions.

Stem Height and Tilting

The variations in stem height of certain cacti that are correlated with variations in the height of surrounding vegetation, as discussed earlier in this chapter, can also be interpreted as a response to PAR (Fig. 8.41). *Stenocereus gummosus* (Fig. 8.42) occurs in the open coastal scrub of northern Baja California, where its short stems often curve over to the ground and root, forming bramblelike thickets. In southern Baja California this species occurs in subtropical short-tree deciduous forest as plants with tall, upright trunks and vertical branches often exceeding 5 m in height. The variations in stem height are correlated with variations in the height of the neighboring vegetation (Fig. 8.41*A*). In particular, the surrounding vegetation reduces the available PAR, and *S. gummosus* apparently responds by evolving the appropriate plant form, that is, the stems tend to grow into the upper canopy positions with higher PAR.

Similar relations between available PAR, stem height, and height of the surrounding vegetation have been quantified by Nobel for *Opuntia echios* in the Galápagos Islands (Figs. 8.41*B* and 8.43) and *Trichocereus chilensis* in Chile (Figs. 8.41*C* and 8.44).

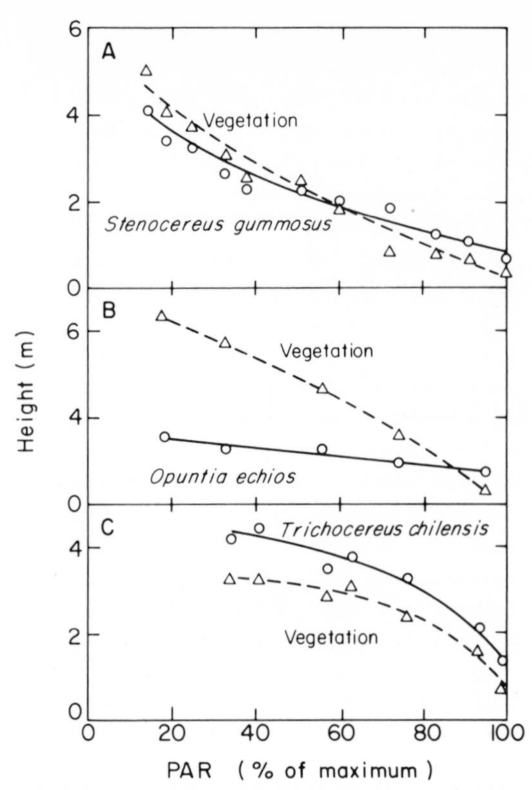

Figure 8.41. Correlation between the average PAR incident on the vertical stem surfaces and the height of cacti and their surrounding vegetation.
(A) *Stenocereus gummosus* observed over a 1300 km transect in Baja California, Mexico.
(B) *Opuntia echios* observed over a 16 km transect on Santa Cruz Island, Galápagos Archipelago, Ecuador.
(C) *Trichocereus chilensis* observed over a 12 km transect in central Chile. PAR data for *S. gummosus* represent the total PAR received from the four cardinal directions 1 m above the ground for 15 plants per site divided by the analogous PAR received in an exposed region where nothing blocked the sun's rays. PAR data were obtained 2 m above the ground for 30 plants of *O. echios* per site and 32 plants of *T. chilensis* per site. (From Nobel, 1980, 1981.)

In these cases the stem height of the cacti increased twofold to threefold as the height of the surrounding vegetation increased (Fig. 8.41*B* and *C*). As before, the cacti were taller where the PAR available to the stems was reduced more by the surrounding vegetation; that is, increased plant height apparently compensated for the PAR reduction. Moreover, *O. echios* occurred at a site where the PAR available on the surfaces of the cladodes was 18% of the maximum at an exposed (unshaded) site but not at a nearby site where available PAR was only 8% of the maximum. This latter site would have yielded a total daily PAR on a surface of only about 2 mol m^{-2} day^{-1}, a value that is below the PAR compensation point for cacti (Fig. 4.15). Therefore, no growth

Figures 8.42–8.44. Sites used for collection of data shown in Figure 8.41.
8.42. *Stenocereus gummosus,* or agria, near Bahia Concepción, Baja California (26°42′N, 111°53′W, 8 m).
8.43. *Opuntia echios* on Santa Cruz Island, Galápagos Archipelago (0°32′S, 90°17′W, 90 m).
8.44. *Trichocereus chilensis* in central Chile (33°2′S, 71°10′W, 600 m).

would be expected in that habitat. Such responses by cacti to shading underscore the point that PAR can limit CO_2 uptake; shading by other vegetation or even cloudiness can greatly reduce the productivity of these plants. Consequently, general increases in the height of vegetation from one locality to another can serve as an effective boundary limiting the range of certain species of cacti.

Another aspect that may be related to PAR interception by cacti is the equatorial tilting of the stem axis. Stems of species of the North American barrel cacti in the genus *Ferocactus* tend to point southward. For example, around 33°N latitude in Arizona, stem tilting is about 12° for *F. covillei* and 17° for *F. wislizenii.* The stem tilting for certain species of *Copiapoa* in Chile is even more striking (Fig. 8.45); near 27° S latitude, the declination of the stem axis from the vertical averages 40° for *C. cinerea* and *C. lembckei.* During the summer such a sharp equatorial tilt increases PAR interception by about 15%, but near the winter solstice PAR interception decreases by over 30%. Although the physiological basis and consequences for the compasslike behavior of these cacti is not fully understood, pointing toward the sun's trajectory can cause more warming of the apical region than occurs in a strictly vertical orientation. This temperature increase could favor stem development as well as the production of flowers during cool periods.

Theoretically, PAR can also affect the distribution of cacti by being too high. This effect is commonly observed when some favorite cactus plant kept in a dimly lit room is put outdoors into full sunlight. Such a change can lead to bleaching (loss of the green color) of the chlorophyll as well as other types of plant damage. The chlorophyll bleaching at high PAR levels is accompanied by a decrease in nocturnal CO_2 uptake (usually above PAR levels of 30 mol m^{-2} day^{-1}), although the mechanism for this inhibition by light has not been thoroughly investigated. The same signs of damage can be observed for cacti in the field upon removal of surrounding vegetation or even the spines. This observation suggests that cactus spines can serve a function that we did not discuss in

Figure 8.45. Equatorial tilting of *Copiapoa cinerea* near Pan de Azucar, Chile (26°S).

Chapter 5; they may also protect the stem tip from receiving too much total daily PAR. The next generation of studies on cactus growth habits will have to consider many different structural and physiological components.

SELECTED BIBLIOGRAPHY

Anderson, E. F. 1961. A study of the proposed genus *Roseocactus. Cactus and Succulent Journal* (Los Angeles) 33:122–127.

——— 1963. A study of the proposed genus *Neogomesia. Cactus and Succulent Journal* (Los Angeles) 35:138–145.

——— 1967. A study of the proposed genus *Obregonia. American Journal of Botany* 54:897–903.

Anderson, E. F., and S. M. Skillman. 1984. A comparison of *Aztekium* and *Strombocactus* (Cactaceae). *Systematic Botany* 9:42–49.

Bailey, I. W. 1962. Comparative anatomy of the leaf-bearing Cactaceae, VI. The xylem of *Pereskia sacharosa* and *Pereskia aculeata. Journal of the Arnold Arboretum* 43:376–388.

——— 1964. Comparative anatomy of the leaf-bearing Cactaceae, XI. The xylem of *Pereskiopsis* and *Quiabentia. Journal of the Arnold Arboretum* 45:140–157.

——— 1966. The significance of the reduction of vessels in the Cactaceae. *Journal of the Arnold Arboretum* 47:288–292.

Bailey, I. W., and L. M. Srivastava. 1962. Comparative anatomy of the leaf-bearing Cactaceae. IV. The fusiform initials of the cambium and the form and structure of their derivatives. *Journal of the Arnold Arboretum* 43:187–202.

Buxbaum, F. 1950. *Morphology of cacti.* Section I. *Roots and stems.* Abbey Garden Press, Pasadena.

Carlquist, S. 1975. *Ecological strategies of xylem evolution.* University of California Press, Berkeley.

Cody, M. L. 1984. Branching patterns in columnar cacti. Pp. 201–236 in *Being alive on land,* ed. N. S. Margaris, M. Arianoustou-Farragitakọ, and W. C. Oechel. *Tasks for vegetation studies,* vol. 13. W. Junk, The Hague.

Cody, M. L., R. Moran, and H. Thompson. 1983. The plants. Pp. 49–97 in *Island biogeography in the Sea of Cortez,* ed. T. J. Case and M. L. Cody. University of California Press, Berkeley.

Conde, L. F. 1975. Comparative anatomy of five species of *Opuntia* (Cactaceae). *Annals of the Missouri Botanical Garden* 62:425–473.

Darbishire, O. V. 1904. Observations of *Mamillaria elongata. Annals of Botany* 18:375–416.

Ehleringer, J., H. A. Mooney, S. L. Gulmon, and P. Rundel. 1980. Orientation and its consequences for *Copiapoa* (Cactaceae) in the Atacama Desert. *Oecologia* 46:63–67.

Gibson, A. C. 1973. Comparative anatomy of secondary xylem in Cactoideae (Cactaceae). *Biotropica* 5:29–65.

——— 1975. Another look at the cactus research of Irving Widmer Bailey. Pp. 76–85 in *1975 Yearbook, Supplement, Cactus and Succulent Journal* (Los Angeles), ed. C. Glass and R. Foster.

——— 1977a. Vegetative anatomy of *Maihuenia* (Cactaceae) with some theoretical discussions of ontogenetic changes in xylem cell types. *Bulletin of the Torrey Botanical Club* 104:35–48.

——— 1977b. Wood anatomy of opuntias with cylindrical to globular stems. *Botanical Gazette* 138:334–351.

——— 1978a. Dimorphism of secondary xylem in two species of cacti. *Flora* 167:403–408.

——— 1978b. Wood anatomy of platyopuntias. *Aliso* 9:279–303.

Mauseth, J. D. 1982. Introduction to cactus anatomy. Part I. Introduction. *Cactus and Succulent Journal* (Los Angeles) 54:263–266.

——— 1984. Introduction to cactus anatomy. Part 9. Primary and secondary growth. *Cactus and Succulent Journal* (Los Angeles) 56:181–184.

Müller-Stoll, W. R., and H. Süss. 1970. Über tüpfelfreie Tracheiden mit Spiral- und Ringverdickungen bei *Trichocereus schickendantzii* (Web.) Britt. et Rose (Cactaceae). *Berlin Deutsche Botanische Geschichte* 83:237–244.

Nobel, P. S. 1980. Interception of photosynthetically active radiation by cacti of different morphology. *Oecologia* 45:160–166.

——— 1981. Influences of photosynthetically active radiation on cladode orientation, stem tilting, and height of cacti. *Ecology* 62:982–990.

——— 1982a. Orientations of terminal cladodes of platyopuntias. *Botanical Gazette* 143:219–224.

——— 1982b. Orientation, PAR interception, and nocturnal acidity increases for terminal cladodes of a widely cultivated cactus, *Opuntia ficus-indica. American Journal of Botany* 69:1462–1469.

——— 1985. Form and orientation in relation to PAR interception by cacti and agaves. Pp. in *On the economy of plant form and function,* ed. T. J. Givnish. Cambridge University Press, Cambridge. Forthcoming.

Racine, C. H., and J. F. Downhower. 1974. Vegetative and reproductive strategies of *Opuntia* (Cactaceae) in the Galapagos Islands. *Biotropica* 6:175–186.
Rodríguez, S. B., F. B. Pérez, and D. D. Montenegro. 1976. Eficiencia fotosintética del nopal (*Opuntia* spp.) en relación con la orientación de sus cladodios. *Agrociencia* 24:67–77.
Schleiden, M. J. 1845. Beiträge zur Anatomie der Cacteen. Mémoires de l'Academie Imperiale des Sciences, St. Pétersbourg, sér. 6, 4:335–380.
Süss, H. 1974. Die Anatomie des Holzes von *Dendrocereus mudiflorus* (Engelm.) Britt. et Rose (Cactaceae). *Feddes Repertorium* 85:759–765.
Wiggins, I. L., and D. W. Focht. 1967. Seeds and seedlings of *Opuntia echios* J. T. Howell var. *gigantea* Dawson. *Cactus and Succulent Journal* (Los Angeles) 39:26–30.

9

Special Chemicals

Sooner or later anyone studying cacti becomes aware of the special chemicals that they produce. Few plants, for example, are as notorious as peyote, *Lophophora williamsii,* which contains the potent, hallucinogenic alkaloid mescaline. A collector who must cut open a cactus stem is certainly impressed by the slippery mucilage or the presence of latex or hard crystals. At other times the vibrant magenta, red, orange, pink, purple, and yellow pigments of the flowers and fruits attract attention. Many of these chemicals are referred to by botanists as secondary plant products. Secondary products are naturally occurring compounds that are often unique to a group of organisms. They are not the products or the intermediates of the biochemical pathways common to most living organisms (for example, those in Chapter 4). The term is not meant to suggest that these compounds are secondary in importance to the plant's survival, however, because these substances do have important functions. We know this because the plant expends considerable energy to make and store these chemicals, drawing upon carbohydrate and nitrogen sources, which are valuable commodities, to make these organic molecules. Cactologists can use information from these features to help determine the evolutionary histories of species, because closely related species generally have a common chemistry.

Betalain Pigments

When a red fruit or the yellow perianth of a cactus is ground in water, the solution becomes colored. Grinding up these tissues releases a water-soluble pigment that is present in the water-storing vacuoles of the cells. Many angiosperms have water-soluble pigments, so there has been a considerable amount of research on these chemical compounds.

One class of water-soluble, vacuolar pigments comprises the anthocyanins (Fig. 9.1), which produce the pink and red colors in carnations *(Dianthus),* the reds and blues of delphiniums *(Delphinium),* and the purplish colors of leaves of many cultivated plants, such as African violets *(Saintpaulia)* and zebrina *(Commelinaceae).* When plant chemists analyze the vacuolar pigments of cacti, however, they do not find anthocyanins.

The bright red and yellow colors found in cacti are produced by a class of pigments known as *betalains* (Fig. 9.1). Purple-red betalain pigments are abundant in the fleshy roots of the cultivated beet, which belongs to the genus *Beta*—hence the name betalains—and are responsible for the dark red color of the water used to boil beets. Beets belong to the goosefoot family, the Chenopodiaceae; and this and the other families that are closely related to cacti, for example, the portulacas (Portulacaceae) and bougainvilleas (Nyctaginaceae), have abundant betalains. In fact, the only other betalain-containing organisms besides the group of families mentioned above (see also Chapter 11) are some mushrooms, including the hallucinogenic and deadly fly agarics *(Amanita).* In all studies that have been done on plant pigments, no species has been found that possesses both betalains and anthocyanins, so investigators have concluded that they are mutually exclusive.

There are other plant pigments that produce red and yellow colors, for example, the carotenoids (Fig. 9.1). Carotenoids are relatively long molecules that, like chlorophyll, are found in all photosynthetic plants, from the smallest, single-celled green alga to the largest marine kelp and the tallest trees on land. Carotenoids can be stored in plastids called chromo-

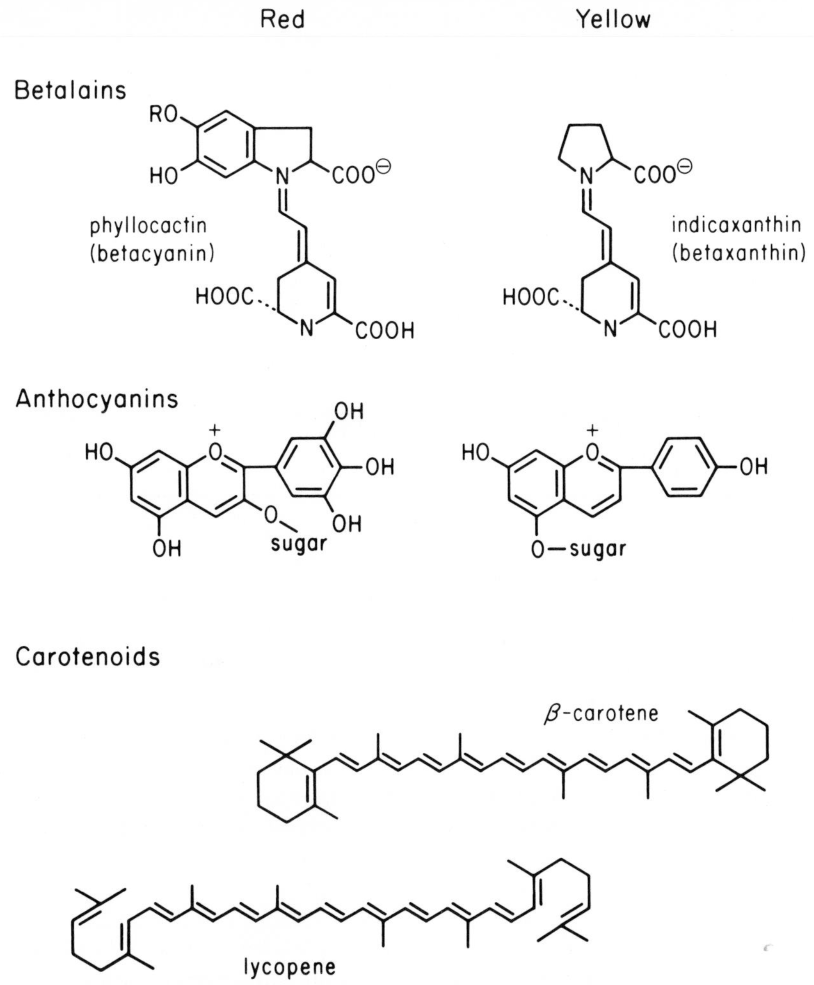

Red

Yellow

Betalains

phyllocactin
(betacyanin)

indicaxanthin
(betaxanthin)

Anthocyanins

Carotenoids

β-carotene

lycopene

Figure 9.1. Classes of red and yellow plant pigments: water-soluble betalains, water-soluble anthocyanins, and water-insoluble carotenoids. The betacyanin phyllocactin and the betaxanthin indicaxanthin are common in cacti, but anthocyanins are absent. The carotenoid *β*-carotene is the orangish pigment in carrots, and the red lycopene is in tomatoes. R, sugar moiety.

plasts but are not found in the central vacuole because they are not water soluble; thus, they are extractable only with chloroform or ether. (Carotenoids also occur as photosynthetic pigments in another type of plastid, the chloroplast.) Examples of carotenoids are the red pigments of tomato *(Lycopersicon)* and chili peppers *(Capsicum)*, the orange pigments of carrots *(Daucus)* and pumpkins *(Cucurbita)*, and the yellow pigments of squashes *(Cucurbita)* and bananas *(Musa)*. Carotenoids are the compounds from which humans make vitamin A. Of course, cacti also have carotenoids, but these compounds are not conspicuous and have not been studied very much in cacti because the betalain pigments predominate.

There are other plant chemicals that sometimes impart color to plant tissues. For example, certain alkaloids are yellow. Tannins, which are widespread in the plant kingdom, impart reddish brown and black colors to plant parts, but cacti do not seem to have many tannins.

Structure and Biosynthesis of Betalains

The properties of betalains were not well defined until the 1960s, when their structures were first elucidated and named. Betalains are complex organic molecules that have two or more atoms of nitrogen (Fig. 9.1); and the presence of nitrogen atoms in the betalain molecule places the betalains in another class of pigments, completely unrelated to the flavonoids and anthocyanins. The red compounds are called betacyanins, and the yellow ones are called betaxanthins.

To understand the uniqueness of betalains among pigments, the reader must appreciate that these

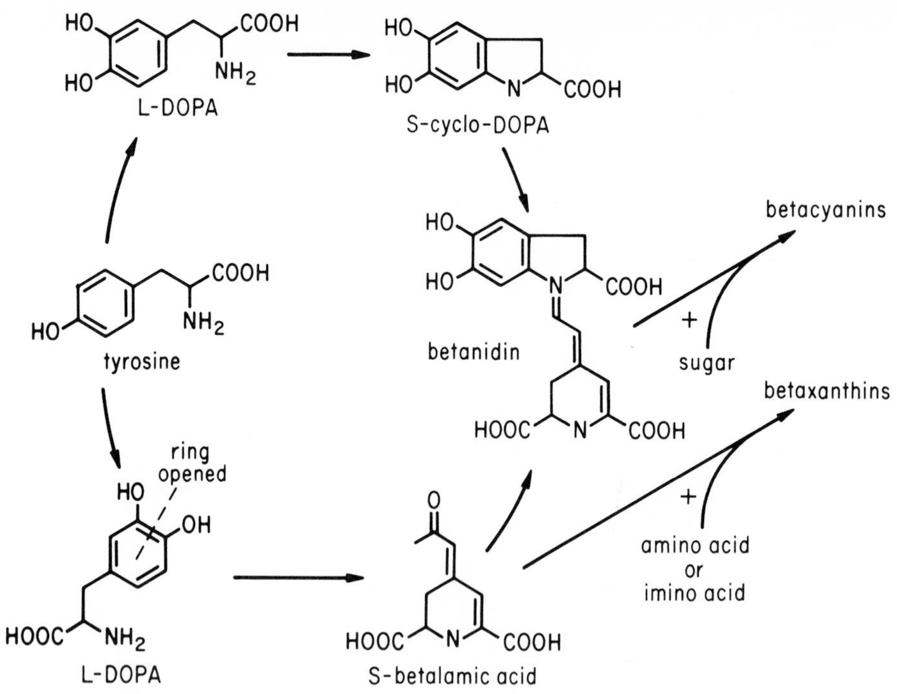

Figure 9.2. Biosynthetic pathway of betalains beginning with the amino acid tyrosine, which is made into L-DOPA and then either S-cyclo-DOPA or S-betalamic acid. When S-betalamic acid combines with an amino or imino acid, a betaxanthin is formed; and when S-cyclo-DOPA and S-betalamic acid are combined, a betacyanin is formed. The addition of sugar to the betacyanin molecule makes it very soluble in water. (After Mabry, 1976.)

nitrogen-containing pigments are synthesized via a special biosynthetic process (Fig. 9.2). Interestingly, the synthesis beings with a compound that is present in all living organisms, an amino acid called tyrosine. From tyrosine a compound called L-DOPA is made, and from L-DOPA two compounds are synthesized (L-DOPA is short for the L-form of β-(3,4-dihydroxyphenyl)-α-alanine). In one case, the nitrogen that has two hydrogen atoms gives up a hydrogen so that the nitrogen can be bonded to the carbon ring to form S-cyclo-DOPA. Simultaneously, in another pathway the six-carbon ring of L-DOPA is opened and through several chemical steps L-DOPA accepts two oxygen atoms and forms a new ring with the nitrogen in it to produce a compound called S-betalamic acid. After S-betalamic acid is modified slightly, this molecule is bonded to the nitrogen of S-cyclo-DOPA to form betanidin. When a sugar molecule is added to betanidin, a betacyanin is formed. Betacyanin is very soluble in water because its sugars are very soluble in water. Betaxanthins are formed by combining a molecule of S-betalamic acid with any amino or imino acid other than S-cyclo-DOPA. For example, the betaxanthin called indicaxanthin (Fig. 9.1), a pigment found in the fruit of *Opuntia ficus-indica,* is formed by combining S-betalamic acid with the amino acid proline, and the betaxanthin called vulgaxanthin is formed by combining S-betalamic acid with the amino acid glutamine.

Absence of Anthocyanins in Cacti

Unlike betalains, the anthocyanins contain no nitrogen. They are, instead, a type of flavonoid; and plants manufacture flavonoids via a completely different metabolic pathway (Fig. 9.3). Flavonoids are formed by combining four small molecules (esters of coumaric and malonic acids) with the help of a large molecule (coenzyme A) to form a 15-carbon skeleton. Therefore, both the components of the molecule and the steps in making flavonoids and betalains are vastly different.

Anthocyanins are ultimately formed from the simpler flavonoids when the skeleton is modified to yield an oxygen with a positive charge. This arrangement permits electrons to resonate around the molecule, and electron resonance of the proper type is what we see as color in the visual wavelengths of light. In anthocyanins the resonance produces reds, blues, yellows, pinks, and purples, whereas in betalains blues are absent. The pigments of *Disocactus amazonicus,* which are reported to be blue, have not been studied.

Many cacti and species in families related to cacti

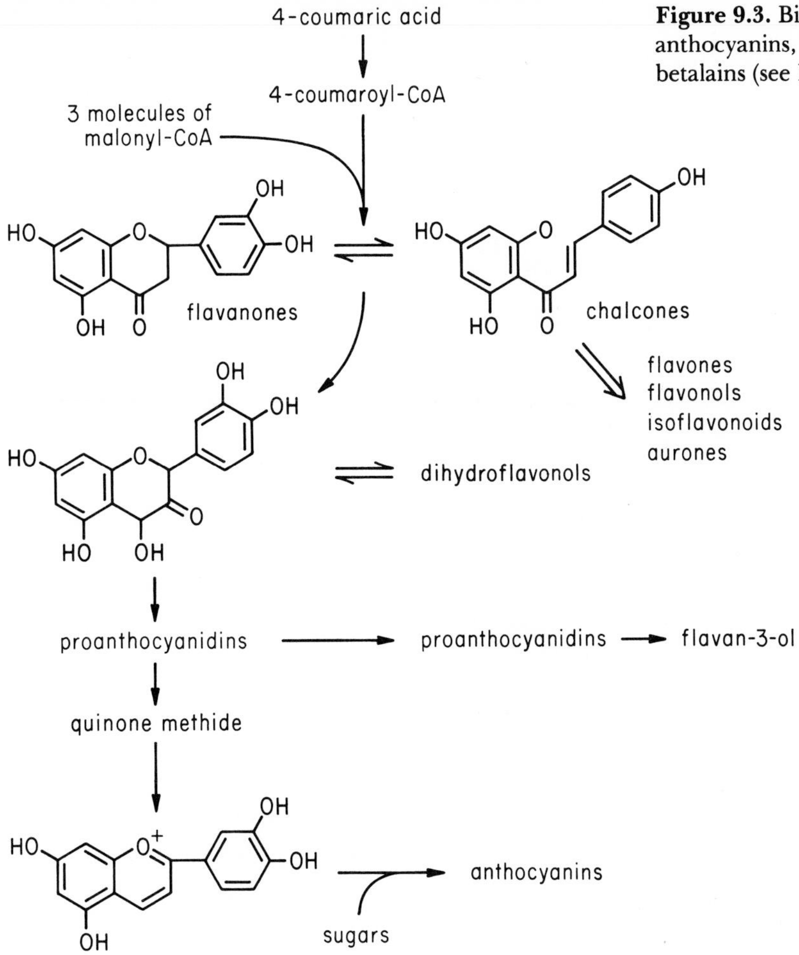

4-coumaric acid

4-coumaroyl-CoA

3 molecules of malonyl-CoA

flavanones

chalcones

flavones
flavonols
isoflavonoids
aurones

dihydroflavonols

proanthocyanidins

proanthocyanidins

flavan-3-ol

quinone methide

anthocyanins

sugars

Figure 9.3. Biosynthetic pathway of flavonoids and anthocyanins, which is unrelated to the biosynthesis of betalains (see Fig. 9.2). (After Mabry, 1976.)

(Chapter 11) have flavonoids, but researchers have found that betalains and anthocyanins do not occur in the same plant family. According to this research, cacti and the other betalain-producing plants have lost the ability to convert flavonoids into anthocyanins because they have lost the enzyme that catalyzes one of the crucial steps of anthocyanin synthesis.

Alkaloids

An alkaloid is a bitter-tasting, strongly or weakly basic (alkaline) compound and has at least one nitrogen that is usually incorporated into a carbon ring. Many alkaloids have been or have the potential to become drugs and medicines because they show significant physiological activity in humans and other animals, especially on the nervous system. Consequently, plant chemists have made a special effort to survey species of all major groups of plants in search of alkaloids. As we shall see, the cactus family has a generous assortment of alkaloids, so cacti have received considerable attention.

Tyramine Alkaloids and Phenethylamines

The most common alkaloids found in cacti are those that differ very little from the amino acid tyrosine, from which they are derived and with which they share a fundamental structure (Fig. 9.4). In all of these compounds the nitrogen atom is present on a side chain and not in the carbon ring.

The synthesis of the simplest cactus alkaloids (Fig. 9.4) begins in the same way as betalain biosynthesis does, that is, by the conversion of tyrosine to L-DOPA; or alternatively these alkaloids can also be manufactured from tyramine, which is produced directly from tyrosine. Tyramine is different from tyrosine because it lacks a carboxylic acid group (—COOH); thus, tyramine is a base, not an acid. The deletion of the carboxylic acid group from tyrosine results in the simple alkaloid 4-hydroxyphenethylamine, or tyramine.

There are already several dozen of these single-ring alkaloids known from cacti, and new ones are being identified every year. In Figure 9.4 we have illustrated some of these compounds to show how they differ in various ways but still share the same phen-

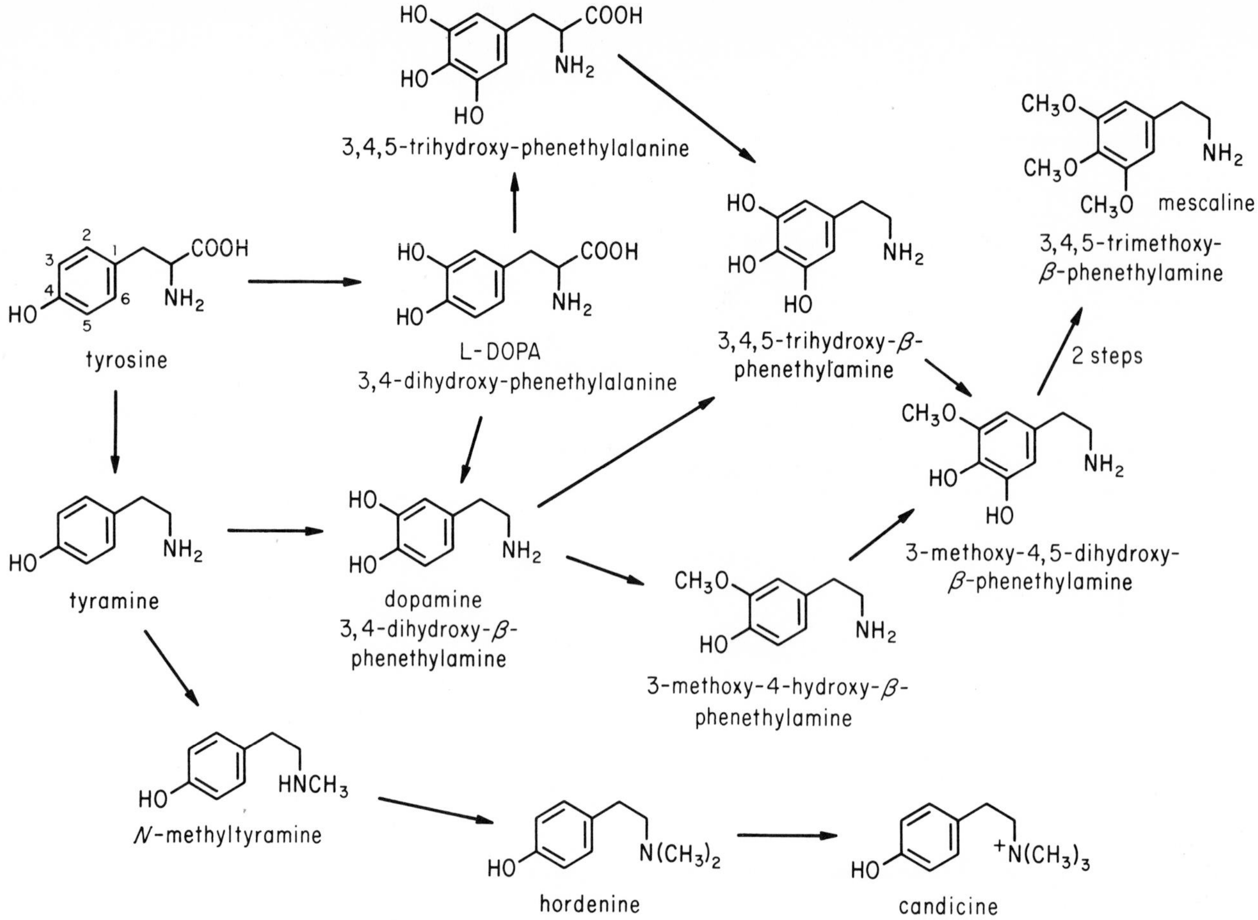

Figure 9.4. Probable biosynthetic pathway in cacti of the tyramine-based alkaloids, phenethylamines, which can be made from the amino acid tyrosine. Numbers on the carbon ring of tyrosine indicate the positions to which chemical radicals can be attached. At first a hydroxyl (—OH) radical is added to position 3; a methyl group (—CH_3) can then be substituted for its hydrogen. The most biochemically complex molecule is mescaline *(upper right).* (After Anderson, 1979.)

ethylamine skeleton. For example, let us examine the tyramine derivatives that lead to candicine. Each of these alkaloids has a hydroxyl group (—OH) attached to the carbon ring at the 4 position. By substituting a methyl group (—CH_3) for a hydrogen on the nitrogen, *N*-methyltyramine is formed. Then by substituting a second methyl group for a hydrogen, hordenine is formed; and by substituting —CH_3 for a third hydrogen on the nitrogen, candicine is made (bottom of Fig. 9.4). It is not uncommon to find two or more of these tyramine compounds in a single genus. For example, *Ariocarpus* has *N*-methyltyramine and hordenine, *Cereus* has tyramine, hordenine, and candicine, *Espostoa* has tyramine, *N*-methyltyramine, and hordenine, and *Trichocereus* has tyramine, *N*-methyltyramine, hordenine, and candicine. All four compounds have also been found in individual plants of *Lophophora williamsii* and *Trichocereus candicans.*

DOPA (an acid) and dopamine (a base) have an additional —OH located at the 3 position on the carbon ring. Commonly, one or both of hydrogens of the —OH groups at positions 3 and 4 are replaced by —CH_3, substitutions yielding methoxy groups (—OCH_3) on the ring. For example, 3-methoxy-4-hydroxy-phenethylamine, which occurs widely in cacti, is one of the simple derivatives of dopamine (Fig. 9.4). The 5 position on the carbon ring can also accept either —OH or —OCH_3, substitutions that make the compound more specialized in a biochemical sense and also provides the compound with a very long name, for example, 3,5-dimethoxy-4-hydroxy-phenethylamine.

Mescaline and Related Peyote Alkaloids

When the skeleton just described has —OCH_3 groups at positions 3, 4, and 5 (3,4,5-trimethoxyphenethylamine), the most "exciting" alkaloid—mescaline—is formed. Mescaline is the potent cactus hallucinogenic alkaloid, best known from peyote, *Lophophora williamsii;* but within the last two decades mescaline has also been confirmed to occur in numerous species belonging to genera as diverse as *Opuntia* and *Pereskiopsis* (Opuntioideae), *Stenocereus* and *Trichocereus* (Cactoideae), and *Pereskia* (Pereskioideae). In most species this compound is present only as a trace, whereas in peyote mescaline constitutes roughly 1% of the dry weight.

It is interesting to note that none of the cacti most closely related to *Lophophora* are hallucinogenic. Actually, the other cacti in tribe Cacteae that produce a "high" are the Dona Aña cactus, *Coryphantha macromeris* var. *runyonii,* and, to a lesser degree, *Mammillaria longimamma* and its closest relatives, the active ingredients of which are alkaloids that are similar to epinephrine and are significantly weaker than mescaline.

Peyote is also loaded with many other alkaloids. Fifty-six nitrogen-containing compounds that are derived from tyrosine have been isolated from *L. williamsii,* including 20 tyraminelike alkaloids. Some mescaline in the plant is also bound in unique acids, but most of the phenethylamines produced by peyote are converted into another group of alkaloids, the tetrahydroisoquinoline alkaloids, which have a different chemical skeleton.

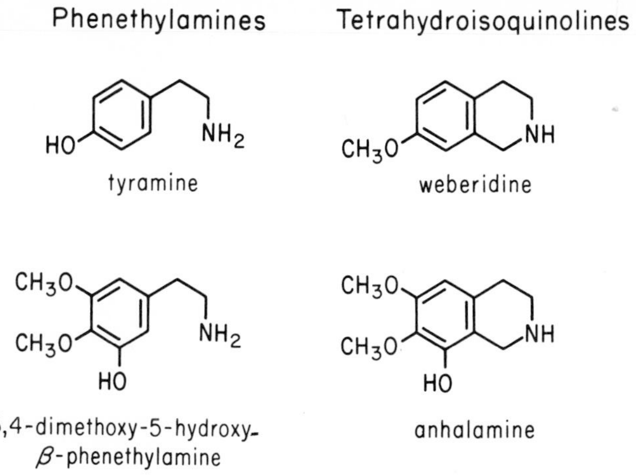

Phenethylamines | Tetrahydroisoquinolines

tyramine | weberidine

3,4-dimethoxy-5-hydroxy-β-phenethylamine | anhalamine

Figure 9.5. Classes of alkaloids found in cacti, phenethylamines (Fig. 9.4) and the tetrahydroisoquinolines. The latter have two carbon rings, one of which contains a nitrogen.

Tetrahydroisoquinolines

In peyote there are over 20 alkaloids that have two rings. These alkaloids, the tetrahydroisoquinolines, are produced when the dimethoxy- and trimethoxyphenethylamines accept one more carbon and thereby incorporate the nitrogen into a new ring (Fig. 9.5). Each compound has a common name, such as pellotine, which is an adaptation of a local name, pellote, and others such as anhalamine, anhalidine, anhalonine, anhalinine, and anhalonidine, which are derived from a former generic name of the plant, *Anhalonium.* By and large, most tetrahydroisoquinolines found in *Lophophora* are unique; that is, they are not found elsewhere in Cactaceae or in any plant group. But several have appeared in other species, especially in those species that also produce mescaline.

A second great diversification of tetrahydroisoquinoline alkaloids developed in the columnar cacti of Mexico, specifically in the species currently assigned to subtribe Pachycereinae (Pachycereeae). By 1985, 26 compounds had been identified in these columnar cacti, of which only four have been found in other species of cacti. Figure 9.6 shows the nature of the chemical radicals attached to the two-ring skeleton of the tetrahydroisoquinoline alkaloids. This figure also shows the interesting common names, which are based on taxonomic names, native names, or locations. Even when the common names are somewhat meaningless, such as tehuanine from a plant from near Tehuacán, Mexico, they are certainly easier to say than the formal name, 2-methyl-5,6,7-trimethoxy-1,2,3,4-tetrahydroisoquinoline. The most complex compounds are the large, polar alkaloids found in *Lophocereus* and *Pachycereus marginatus.* These large molecules are formed by bonding together three molecules of tetrahydroisoquinoline. Once again, the common names of these compounds are euphonious, but unfortunately the names pilocereine and pilocerediine are quite misleading because these compounds have not been found in any species of *Pilocereus* (=*Cephalocereus* in Table 1.2).

In 1982 Jerry L. McLaughlin and his co-workers at Purdue University discovered in cacti another class of alkaloids based on dihydroisoquinoline from saguaro. The discovery of these alkaloids, all of which occur in extremely low concentrations, was important, not only because the molecules are fluorescent, which is very unusual, but also because dihydroisoquinoline may be the biosynthetic precursor of the tetrahydroisoquinolines. This discovery may open the door for detailed studies of alkaloid production in columnar cacti.

Stem Darkening

Saguaro is one of the cacti in which the cut surface of a fresh stem turns dark over a 24-hour period.

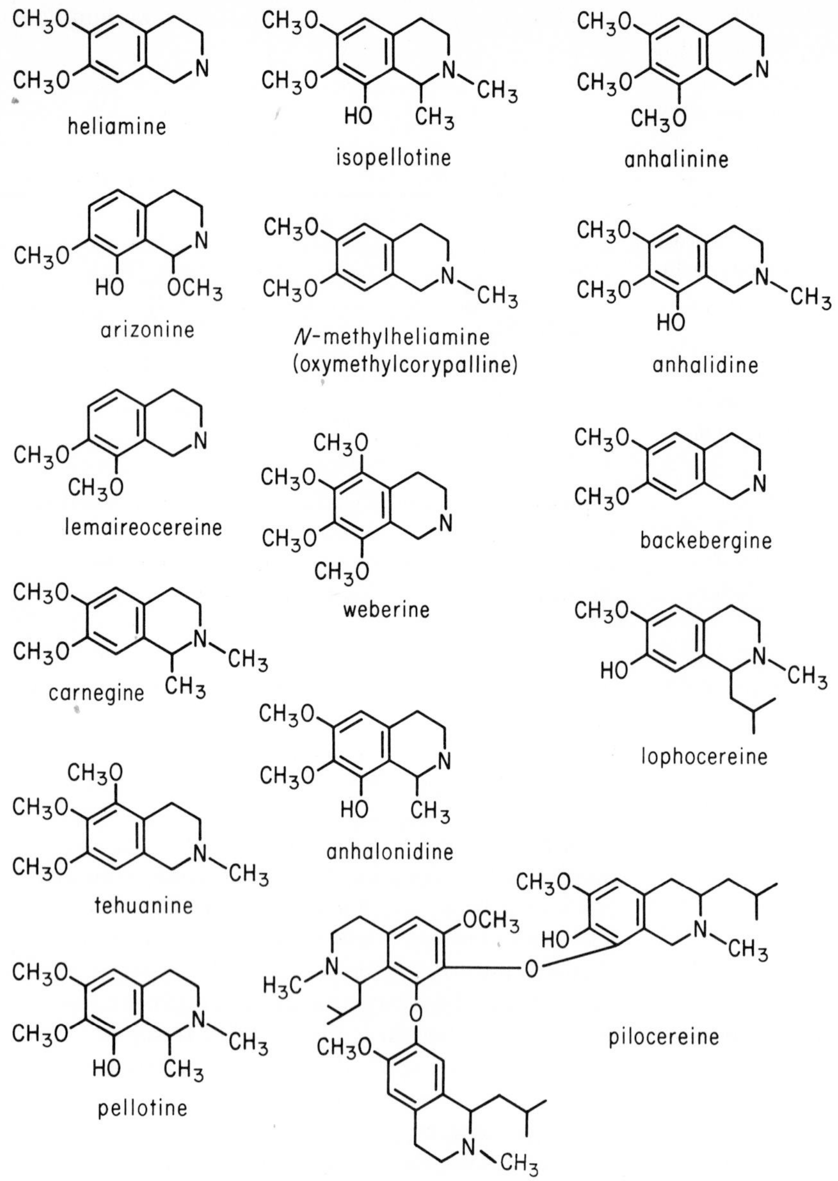

Figure 9.6. An assortment of tetrahydroisoquinoline alkaloids from the stems of columnar cacti in the subtribe Pachycereinae (Pachycereeae).

This darkening was believed to be caused by the oxidation of the alkaloid dopamine or of DOPA to the zoological pigment melanin.* But in 1980 Rachel Mata and McLaughlin studied the extremely rapid darkening of stems of *Pachycereus weberi* and discovered that a novel phenolic alcohol with a glucose attached is the responsible compound (Fig. 9.7). When the sugar is removed (hydrolyzed), a reaction that can be accomplished either by an enzyme or in the presence of a strong acid like malate, the sugarless form of the molecule (the aglycone) darkens rapidly in air. When cut open, all species of *Pachycereus* show a rapid change in cortical color from green to black, and lemairin is the likely agent for this wound response.

* C. Steelink, M. Yeung, and R. L. Caldwell, "Phenolic Constituents of Healthy and Wound Tissues in the Giant Cactus *(Carnegiea gigantea)*," *Phytochemistry* 6(1967):1435–1440.

Triterpenes and Sterols

Triterpenes and sterols are complex molecules that have a large carbon ring system. A small percentage of plants produce these compounds in large quantities, although all plants contain the pathways leading to their synthesis. Plants that have abundant triterpenes were sought by pharmaceutical companies after the 1940s, when Russell Marker, a biochemist, revolutionized the production of cortisone, steroids, and oral contraceptives by showing that one can economically synthesize these important chemicals from triterpenes in plants, especially yams *(Dioscorea)*.

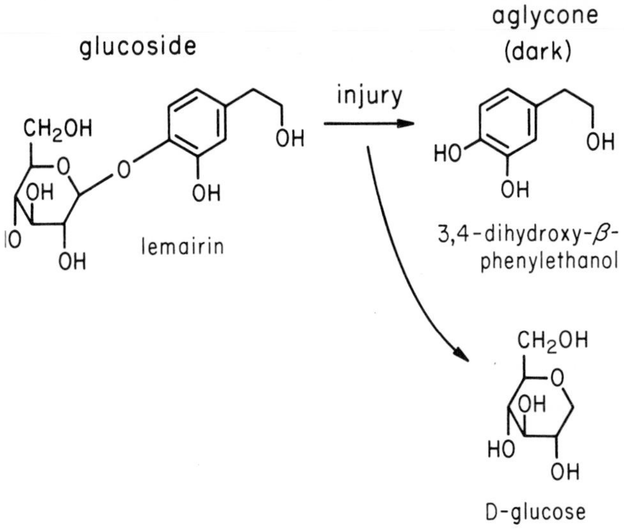

glucoside

aglycone (dark)

injury

lemairin

3,4-dihydroxy-β-phenylethanol

D-glucose

Figure 9.7. The glucosidic alcohol lemairin, found in *Pachycereus weberi,* is converted to a dark sugarless form (aglycone) when the stem is cut. (After Mata and McLaughlin, 1980b.)

The presence of triterpenes in succulents from numerous plant families has stimulated research on the chemistry of succulent plants.

Stem Triterpenes

One way to quickly determine whether a plant has abundant triterpenes is to grind up the plant material with water in a blender. Triterpenes are usually sapogenins — literally soap-formers — which readily produce suds when mixed with water. Early American Indians used plants containing triterpenes for soap and shampoo; and because triterpenes are biologically active and damage red blood cells, sapogenic materials were also used by them to stupefy fish for ease of capture.

It was known from folklore uses in Mexico that certain cacti have sapogenic compounds (saponins). In the 1950s Carl Djerrasi at Wayne State University in Michigan isolated and identified the triterpene skeletons of the columnar cacti of Mexico. The triterpenes described by Djerassi and co-workers belonged to two groups, the oleanane series and the lupane series. Figure 9.8 shows one plausible biosynthetic pathway for these two classes of triterpenes. Squalene, which initially has no carbon rings, is converted into a four-ringed steroid structure, from which plant sterols can be manufactured. (In humans, squalene is also the precursor of cholesterol.) To make a triterpene, a fifth carbon ring is formed, and from this compound the lupanes and oleananes are derived.

The way in which plants synthesize most complex molecules is difficult to study, so for triterpenes we can now only hypothesize a biosynthetic pathway for these compounds in cacti (Fig. 9.9). The triterpene molecules become biochemically more complex by adding hydroxyl groups (—OH). In the oleanane series there are compounds with one —OH (monohydroxy), two (diol), and three (triol); and oleanolic acid, a monohydroxy compound, is therefore biochemically less specialized than its numerous diols or the highly specialized triol treleasegenic acid. Likewise, the triol longispinogenin and its acid myrtillogenic acid are more specialized than the chemically similar diols. Oleanane triterpenes have sugars attached to the ring, a substitution that converts them into water-soluble glycosides, the suds-producing saponins. In the lupane series, fewer compounds have been discovered in cacti, and half of these (lupeol, betulin, and calenduladiol) are aglycones (without sugars) and therefore not saponins.

The total diversity of triterpenes in cacti is still unknown because so few species have been examined. Some recent surveys of triterpenes by Kevin Spencer of the University of Illinois at Chicago have revealed that the columnar cacti studied by Djerrasi may have over 50 distinct glycosidic compounds in a single stem. This is not unreasonable if there are several oleanane aglycones produced, each of which can accept several different sugars at many places on the ring system. Triterpenes are also present in other groups of cacti, for example, in something as different as *Mammillaria longimamma* (Fig. 1.13).

Sterols

Our knowledge of cactus chemistry is still rudimentary, and new sterols are still being isolated, especially from Mexican columnar cacti. For example, three of the five recently isolated dihydroxy-sterols bear the names of the cacti from which they were first isolated (*Peniocereus fosterianus* and organ pipe cactus, *Stenocereus thurberi*): peniocerol, macdougallin, stenocerol, thurberol, and cyclosterol.

Organ pipe cactus is quite a remarkable plant because 10 to 21% of the soft tissue by dry weight is lipid, whereas succulents in general do not store abundant fats or oils in their vegetative tissues. The organ pipe cactus lipids are mostly medium-length molecules made from the triterpene aglycones (lupeol, betulin, and calenduladiol) and dihydroxy-sterols. They are stored as droplets in the vacuoles of the chlorenchyma cells, a location that is quite unusual, because most plants and other cacti that have been examined store their lipids in the chloroplasts. Why organ pipe has special lipids in large quantities remains to be studied. It seems likely that

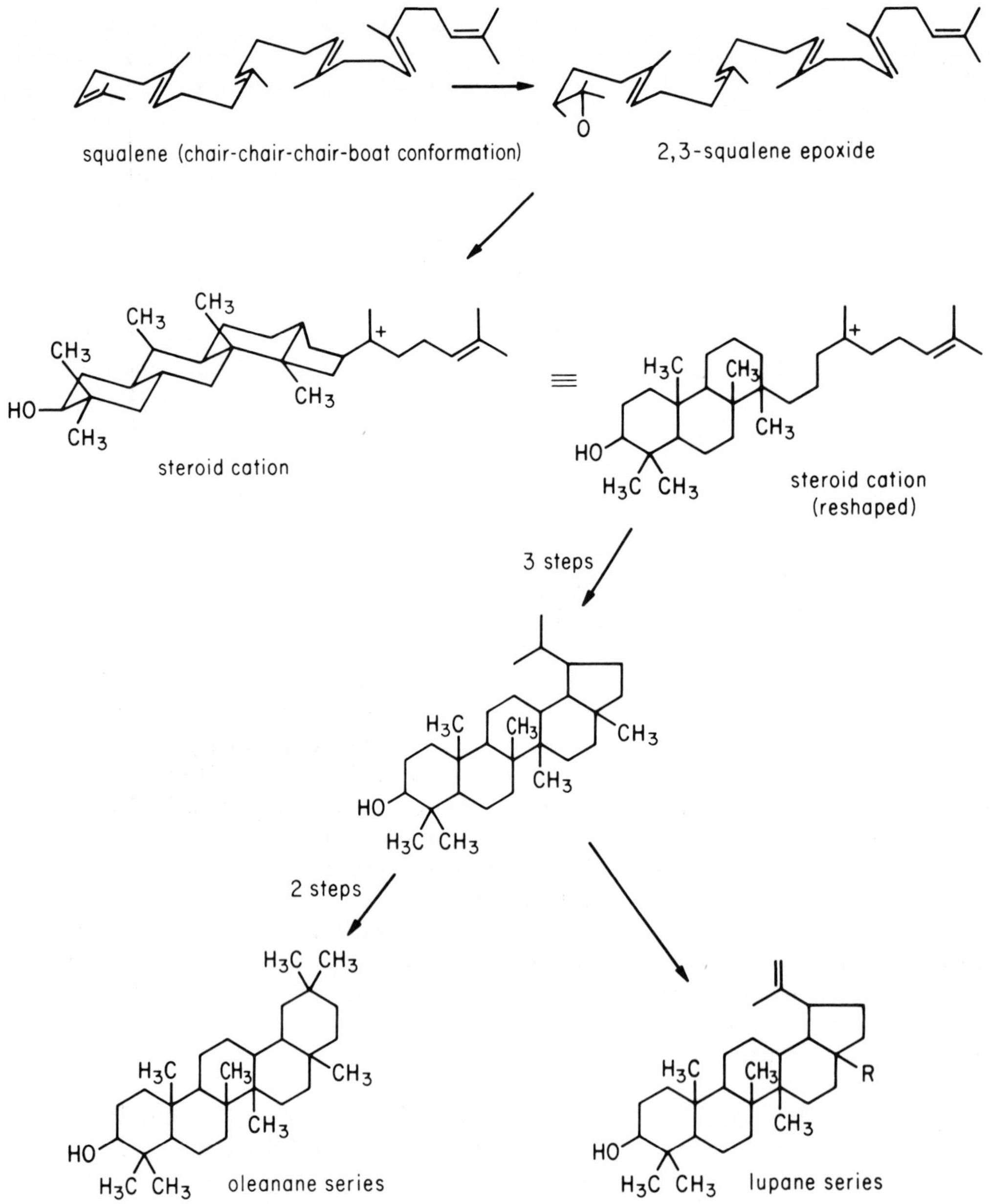

Figure 9.8. Probable pathway by which squalene is converted through several steps to triterpenes of the lupane series or the oleanane series. (After Luckner, 1972.)

other cacti will be found, particularly in *Stenocereus*, that have this unusual type of lipid.

Mucilage and Latex

Chemistry of Mucilage

Anyone who has cut a cactus stem or fruit has noticed its generally sticky, mucilaginous nature, which can coat anything that it touches. Although mucilage may appear to be produced throughout the cut structure, it actually is produced only in special cells or canals (Chapter 6). The molecule is assembled by the mucilage cell (Fig. 9.10) and then stored between the cell membrane and the cell wall, an arrangement that eventually causes the death of the cell when the plasmalemma pulls away from the cell wall.

Mucilage is a fibrous polysaccharide, meaning that it is a highly branched compound containing many sugars. Because this molecule is very large and complex, and also because it is sticky and therefore messy to handle, phytochemists have not shown great

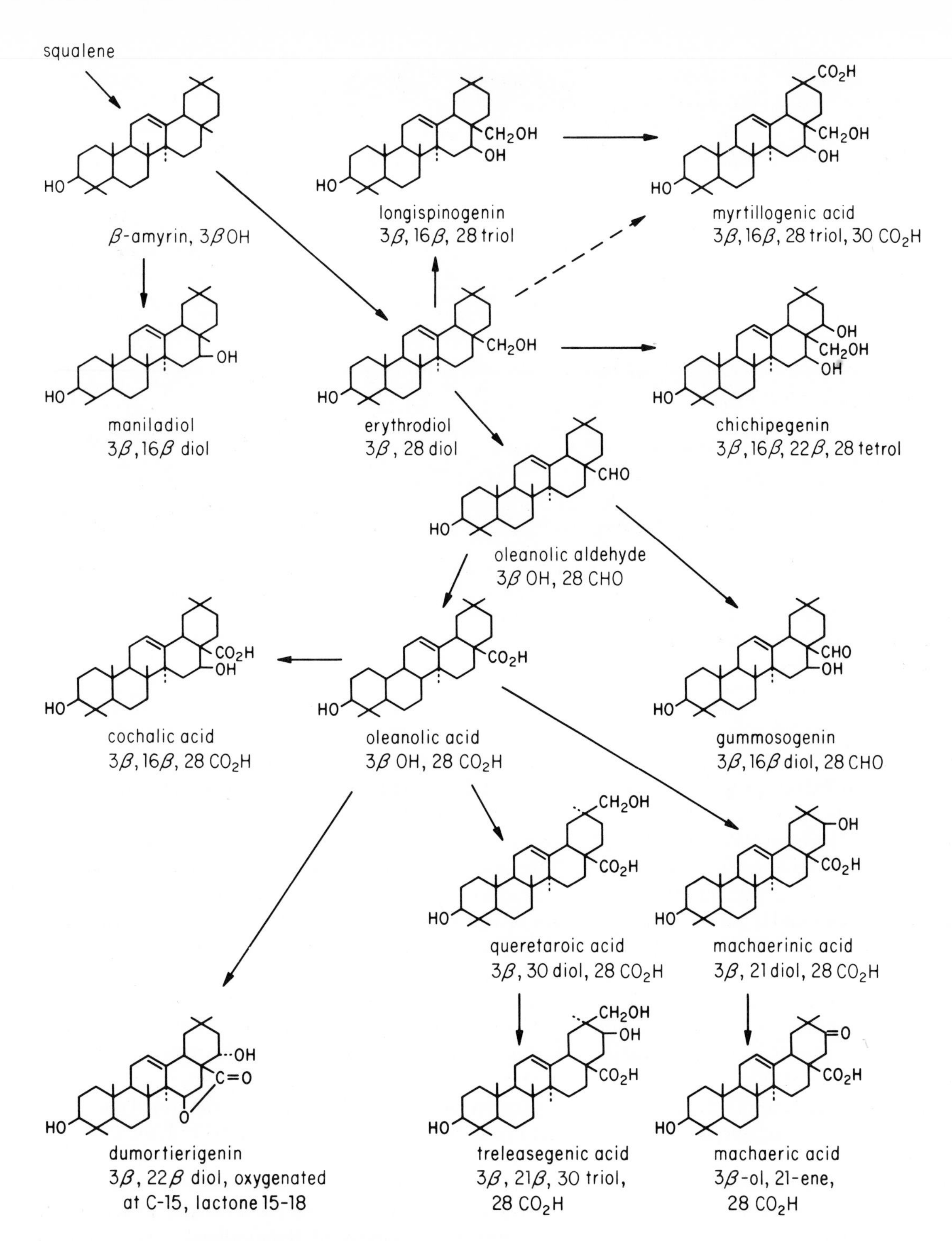

Figure 9.9. Possible biosynthetic pathways of oleanane triterpenes found in the columnar cacti of the subtribe Stenocereinae (Pachycereeae). (After Gibson, 1982.)

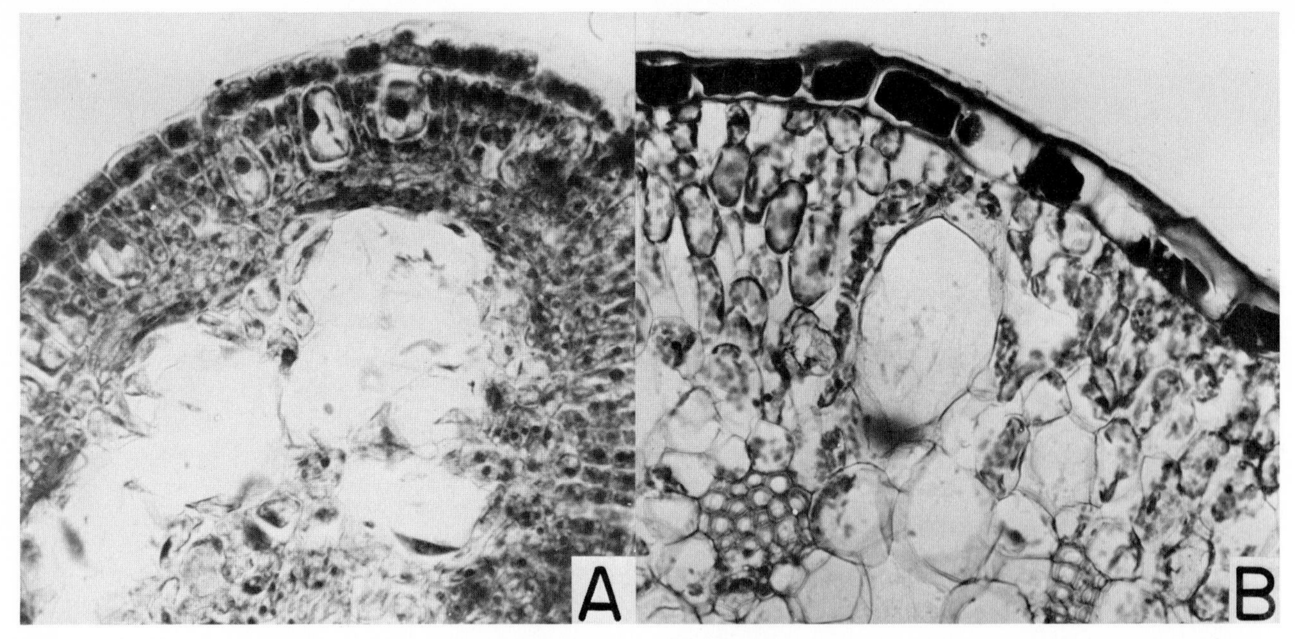

Figure 9.10. Development of mucilage cells in *Maihuenia poeppigii.*
(A) In the outermost layer of the leaf mesophyll, cells that have started to develop as mucilage cells have large nuclei and vacuoles. In the center of the leaf, mucilage cavities have already become very large. ×420.
(B) A mature mucilage cell is present in the outer chlorenchyma of the leaf; the epidermal cells contain a large quantity of tannin. ×210.

interest in cactus mucilage. However, Donald McGarvie and Haralambos Parolis of Rhodes University, Grahamstown, South Africa, and Shlomo Trachtenberg and Alfred Mayer of the Hebrew University, Jerusalem, Israel, have worked on this chemical mystery. Most published information on cactus mucilage has been obtained from *Opuntia ficus-indica,* which is widely cultivated for its edible fruits and cladodes.

For *O. ficus-indica,* each mucilage molecule may contain over 30,000 sugar subunits, or residues. The most abundant sugars have five or six carbon atoms (Fig. 9.11*A*) and possess the following rather strange-sounding names: arabinose, galactose, rhamnose, and xylose. Mucilage also contains a considerable amount of galacturonic acid, which is a carboxylated sugar. Because hydrogen ions tend to dissociate from the carboxy part (—COOH; see Eq. 4.8 for the analogous dissociation from carboxyl groups on malic acid), the galacturonic subunits, and hence the mucilage molecule as a whole, tend to be strongly negatively charged. This negativity leads to a binding of calcium ions (Ca^{2+}) to many places in the mucilage molecule, and as much as 20% of the insoluble calcium in a cactus stem can be associated with the mucilage.

In *Opuntia ficus-indica,* mucilage consists of about 35 to 40% arabinose, 20 to 25% each of galactose and xylose, and about 7 to 8% each of rhamnose and galacturonic acid. A hypothetical polymer that contains these sugar residues in approximately these ratios and was constructed on the basis of the available chemical evidence is depicted in Figure 9.11*B*. The backbone of the molecule consists of alternating rhamnose and galacturonic acid residues to which are attached side chains composed of three galactose residues. Arabinose and xylose residues are branches on the galactose side chains; xylose appears to be attached to arabinose, which is then bonded to the galactose. Some galactose side chains have arabinose but not xylose, and some have one xylose with two arabinose residues. The structure we have presented is actually only a minute portion of the mucilage molecule, because the unit portrayed in Figure 9.11*B* is repeated about 500 times. The mucilage from other species of *Opuntia* is composed of the same sugar residues, but the residues are present in different ratios than for *O. ficus-indica.*

Mucilage cells in the opuntias that have been studied can occupy about 3% of the stem volume and may affect metabolism in a number of ways. First, because so much Ca^{2+} is bound to the mucilage, mucilage cells act as a calcium reservoir for the stem. Second, water can be bound to the many sugar residues of the mucilage molecule, and thus the presence of mucilage affects the state of hydration of the stem. This water-storage role appears to be only a secondary aspect of the water relations of the stem,

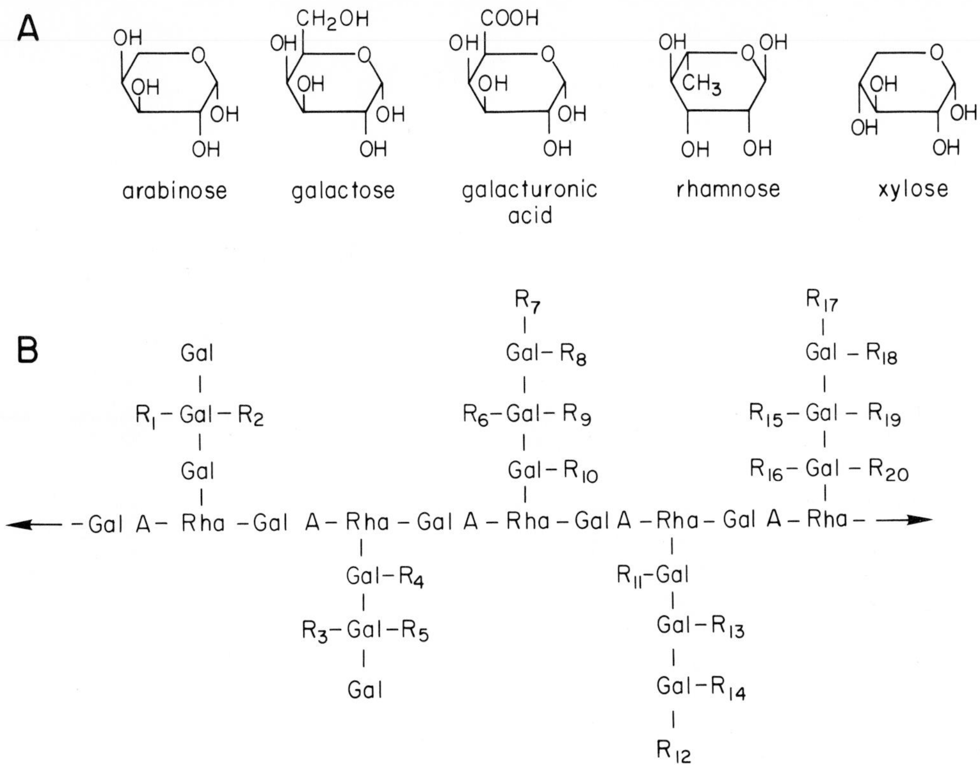

Figure 9.11. Chemical composition of mucilage. *(A)* Subunits in mucilage. *(B)* Tentative identification of repeating unit in mucilage from *Opuntia ficus-indica.* The 20 side chains R_1 through R_{20} contain a total of about 15 xylose residues and 25 arabinose residues. In the backbone Gal A is galacturonic acid and Rha is rhamnose. On the side chains Gal is galactose. (Adapted from McGarvie and Parolis, 1981a,b,c.)

however, because the stem can have nonlethal fluctuations in its water content of over 50%, fluctuations that reflect changes in water content of the parenchyma cells. Third, some workers have assumed that mucilage plays a role in protecting cacti from high temperature damage. But it is unclear how molecules as large as those of mucilage would be able to enter the chlorenchyma cells, where the protection from thermal damage is most likely needed. There is also no direct evidence that mucilage deters predation.

Although mucilage is generally considered to be a nuisance, one cactus, *Ariocarpus kotschoubeyanus,* has such a fine-quality mucilage that it was and still is used as a glue for repairing pottery by Indians in Mexico.

Latex

Latex is the common name for some slightly viscous fluids that exude from the cut surfaces of certain plants in copious amounts. Latex is generally white or chalky, but latex in some plants may be clear, yellow, orange, or even red. It is manufactured and contained in extremely long, microscopic tubes called laticifers (latex-bearers). The most famous latex-bearing plants are those from which rubber, chewing gum, and gutta percha are obtained and poppies, which have an alkaloid-bearing latex. However, latex is a familiar term to horticulturalists because many common plants, for example, euphorbs (Euphorbiaceae, Figs. 1.36 and 1.37), stapelias and milkweeds (Asclepiadaceae), and dogbanes (Apocynaceae), have copious latex. In addition, the cactus genus *Mammillaria* has species with "milky sap," "semi-milky sap," and "watery sap."

Because the word *latex* has been used as a name for all milky exudates of plants, there is considerable confusion about this material, which is chemically very different in the various groups of plants. In its strictest sense, latex refers to a class of long hydrocarbon chains known as polyterpenes that coagulate to form rubber or rubberlike products. This property distinguishes latex from white resins, such as those of sumacs (Anacardiaceae) and members of the frankincense family (Burseraceae). Usually latex is not pure because the laticifer also contains other cellular contents, such as nuclei and plastids, which remain when the laticifer is formed. Some polyterpene-bearing latex may have other special chemicals, such as alkaloids, diterpenes, and triterpenes. In the case of the cactus *Mammillaria*, the term *latex* is used to describe an exudate that appears to be a polysaccharide (complex carbohydrate) and is probably more similar chemicaly to mucilage than to either rubber or resin.

We must couch our discussion of cactus latex in very tentative terms because evidently no phytochemist has studied the structure of *Mammillaria* latex. No one knows the nature of the sugar residues or what causes the white color of the latex. Liquid polysaccharides are clear unless they have other substances, such as alkaloids, attached to them.

Excellent anatomical descriptions of the formation of laticifers in cacti have been done by Mauseth and George Wittler at the University of Texas. These studies showed that the formation of cactus laticifers is very different from that of plants in other latex-bearing families. In particular, two features distinguish cactus laticifers. First, the cactus laticifer is created by the breakdown of cell walls of at least two layers of cells. Second, the laticifer is encircled by parenchyma cells, which form a lining for the tube called an epithelium. The laticifers of *Mammillaria* contain many small, membrane-bounded sacs called vesicles, which may be the organelles that produce the special chemical making up the latex. The observations of Mauseth and Wittler suggest that cactus laticifers were evolutionarily derived from mucilage-producing structures.

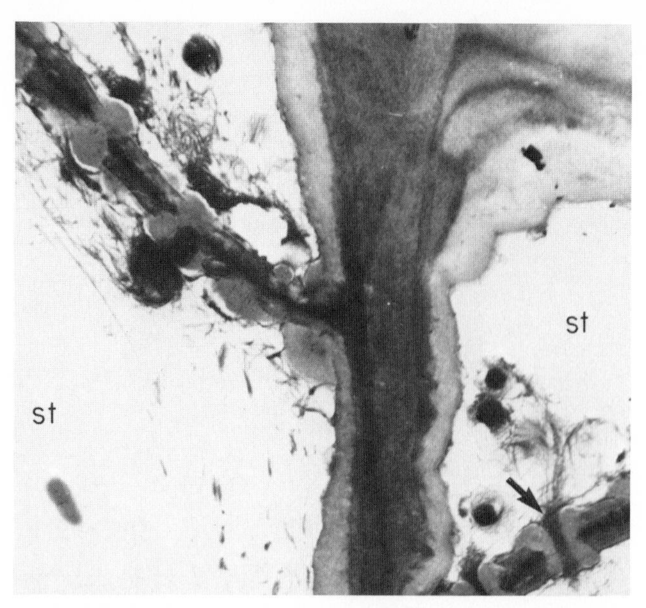

Figure 9.12. Sieve-tube elements (st) of the phloem in the barrel cactus, *Ferocactus wislizenii.* End walls of these cells have pores that are lined with the polysaccharide callose *(arrow).* Carbohydrates pass from one cell to the next through the pores. ×12,200.

Protein Inclusions in Sieve-Tube Elements

The elongate cells of the phloem that transport sugars from the site of production in the chlorenchyma cells (Chapters 3 and 4) to regions where sugars are used or stored are called sieve-tube elements (Fig. 9.12). These cells are so named because the narrow end wall is perforated and therefore looks like a sieve. In the late 1960s H.-Dietmar Behnke at the University of Heidelberg discovered that the colorless plastids occurring in sieve-tube elements have interesting inclusions. The plastids in the sieve-tube elements of some plants have starch grains, whereas some have deposits of protein or of both starch and protein. Behnke surveyed angiosperms for these inclusions and discovered that each family or each group of families has a particular type of plastid inclusion. The cactus sieve-tube-element plastid has a conspicuous, peripheral ring of protein —actually proteinaceous fibers—that is present next to the central proteinaceous globule (Fig. 9.13). According to his latest classification of plastid inclusions, the cactus plastid is labeled as Type PIII c′ f. This is the same plastid type that occurs in families that are closely related to the Cactaceae (Chapter 11).

The occurrence of proteinaceous inclusions in cacti is apparently not restricted to sieve-tube elements, because Bailey also observed filamentous protein bodies in the vessels of *Pereskia.* It is not known at this time whether these inclusions in the xylem are produced by the cactus or formed by a microorganism, such as a virus. If the xylary inclusions are formed by the cactus, the researcher then must determine whether they are homologous to those found in the phloem.

Crystals

When Schleiden described vascular tracheids in 1845 (Chapter 8), he also made the first observations on the occurrence of crystals in cacti. Other researchers working in that century followed with published accounts of cactus crystals. In form, size, and chemistry the crystals of Cactaceae appear to be as diverse as those found in any other plant family. In fact, the range of diversity seen in cacti is nearly the same as is observed in all the remaining dicotyledonous families put together.

Calcium Oxalate Crystals

Crystals composed of calcium oxalate (Figs. 9.14–9.20) can occur in nearly all parts of a cactus except

Figure 9.13. Leucoplast in a young sieve tube of the barrel cactus, *Ferocactus wislizenii.* This plastid has a very large proteinaceous globule (pg) and a small peripheral ring of protein *(arrow).* ×60,000.

spines, trichomes, and the vascular cambium. When a stem is examined with a special type of microscope that uses polarized light, the crystals preferentially transmit light polarized in certain directions and therefore can shine brightly (Figs. 3.11, 3.12, and 9.14), a condition called birefringence. Calcium oxalate is relatively insoluble in water and thus occurs in crystalline form, usually in the vacuoles. In a given cell, calcium oxalate may be precipitated as a solitary crystal (one per cell; Fig. 9.15), as several separate crystals, or as numerous crystals that are radially arranged or clustered. Crystals that are individual and not aggregated are called prismatic crystals, and they typically have a rhomboidal or polyhedral shape (Figs. 9.14–9.16). A solitary prismatic crystal may occupy over half the volume of a cell, as in the hypodermis of *Mitrocereus fulviceps.* A cell may have one large crystal with a number of smaller ones growing on its surface (Fig. 9.16), and it is common to find species with many small prismatic crystals in the epidermis, for example, *Cephalocereus* species and many of the epiphytic species of tribe Hylocereeae. Occasionally investigators have found cells in which there are numerous minute prismatics (crystal sand).

Figure 9.14. Prismatic crystals in the skin of *Cephalocereus hoppenstedtii* visualized with a light microscope using polarized light. The crystals are visible here because they rotate light as it passes through them. ×150. (From Gibson and Horak, 1978.)

The term *druse* refers to most types of spheroidal aggregates, in which numerous prismatic crystals in a vacuole form from a common center and project outward and radiate in all directions. Cactus druses may assume many forms, even within a single stem, because they can have narrow, needlelike crystals; long, blunt crystals; or polyhedral crystalline components (Figs. 9.19 and 9.20). There are some distinctive crystal aggregates that are round in one plane and strongly flattened in another (Fig. 9.18). Druses are the most common form of calcium oxalate in cacti and usually are found in idioblastic cells of the succulent tissues. Interestingly, druses are especially common in the outermost cell layer of the hypodermis in all species of subfamily Opuntioideae (Figs. 2.21 and 9.20) and in certain genera of tribe Cacteae.

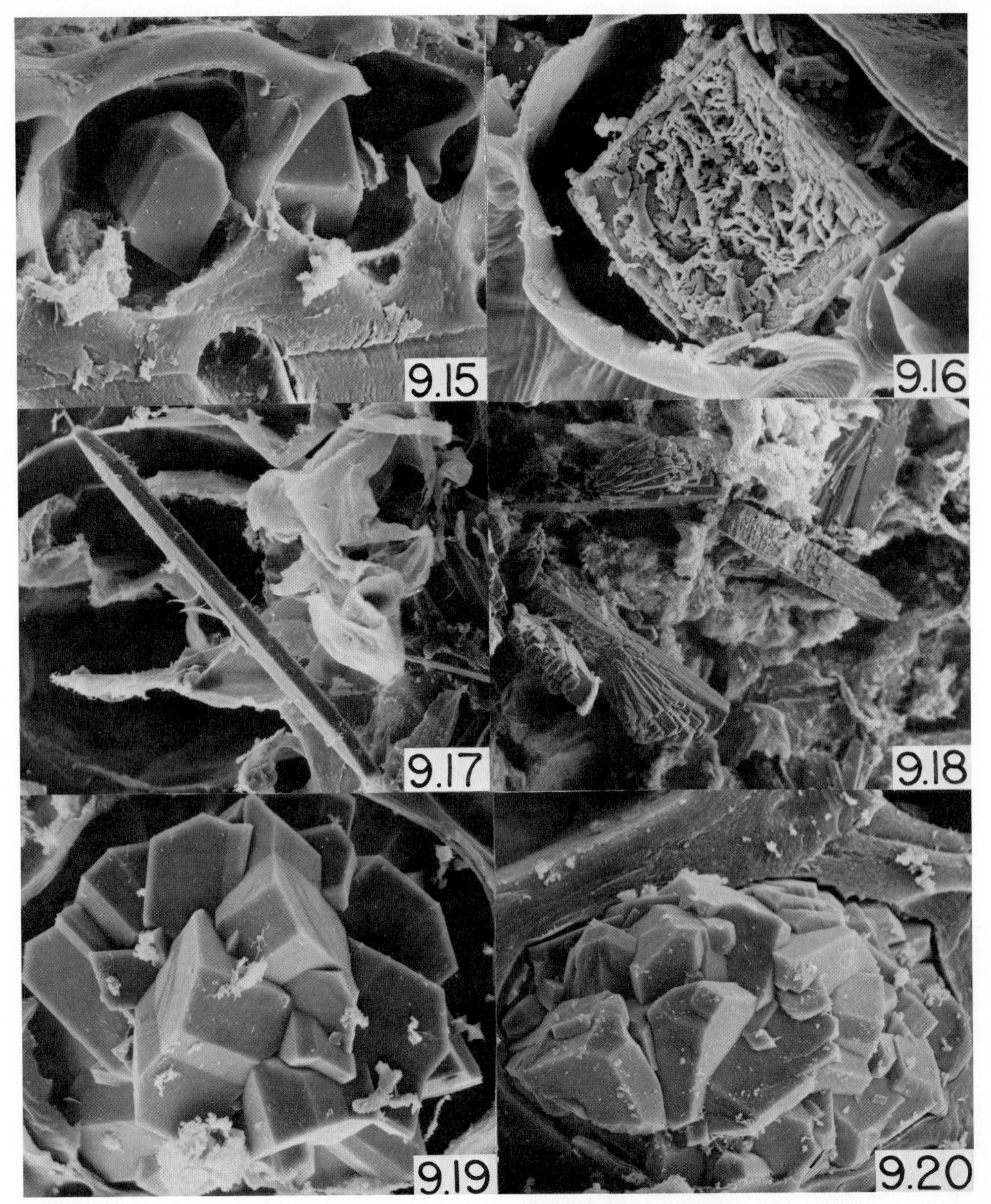

Figures 9.15–9.20. Calcium oxalate crystals in the stems of cacti.

9.15. *Leptocereus weingartianus,* a scandent shrub on the island of Hispaniola in the West Indies; a large, solitary prismatic crystal (about 0.02 mm in diameter) is present in each epidermal cell.

9.16. *Selenicereus (Deamia) testudo,* an epiphyte that climbs on tree trunks in Central America and Mexico; a large, cuboidal crystal (about 0.03 mm in diameter) with tiny prismatic crystals on its surface, present in a cortical cell.

9.17. *Disocactus himantocladus,* an epiphyte of Central America; a long, thin, needlelike crystal is visible; the crystals are typically present in large clusters in the cortical cells.

9.18. *Cephalocereus polycephalus,* a columnar cactus from Hispaniola; crystals are aggregated into flat, wheellike structures that are about 0.1 mm across.

9.19. *Pereskia sacharosa;* a large druse (crystal aggregate) almost fills a cortical cell.

9.20. *Opuntia schottii,* a low cylindropuntia from the sand dunes of western Texas; a large druse completely fills the lumen of a cell in the outermost layer of the hypodermis.

In some epiphytes of tribe Hylocereeae, we find numerous elongate, needlelike crystals called raphides (Fig. 9.18) that are produced in a single vacuole and then aggregated into bundles. At the other extreme, many columnar cacti have large, irregular crystal aggregates formed when numerous very large rhomboidal crystals fuse in the same vacuole.

In the past cactus systematists have entertained the idea that the distribution patterns of distinctive calcium oxalate crystals can be used to determine whether two species or two genera are closely related; and there have been a number of successes in this area of research (Chapter 10). On the other hand, the fact that we do not understand the significance of these differences in crystal morphology is an obstacle to the use of crystal morphology in cactus systematics. From studies by Bailey on the leaf-bearing cacti, we have learned that the form of a crystal is strongly dependent on the nature of the cell wall. For example, in a wood cell with a relatively thick, lignified cell wall, a solitary prismatic crystal can form; but when the same cell has a thin, unlignified cell wall, only a druse can form. Differences in the size and composition of idioblastic druses seem to be determined by position in and the health of the plant. Consequently, much research is needed to identify the factors that control crystal deposition in cacti before these observations can be used as a reliable tool in classification of the species.

Water-Soluble Calcium Crystals

In his investigations of the leaf-bearing cacti, Bailey studied a group of calcium crystals that are soluble in cold water. Like calcium oxalate crystals, these water-soluble structures can shine brightly (are birefringent or anisotropic) under polarized light; but after they are soaked in dilute sodium hydroxide, these crystals are converted into brown, dark (isotropic) bodies. Bailey was intrigued by these flat crystal aggregates, which consisted of much-elongated needlelike crystals that grew evenly around the edges. He attempted to show that in *Pereskiopsis* these unusual crystals were not polysaccharides (such as starch, inulin, and mucilage) but rather were made of calcium malate. Formation of such crystals might be a way in which these succulent cacti store malate, which is produced via the CAM pathway, for later use. Careful physiological investigation is needed to determine whether these calcium malate crystals are important in cactus physiology.

Silica Bodies

We do not expect to find crystals of silica, SiO_2, in plants, because silicon is not generally cited as a crucial element in plant metabolism and because SiO_2 is very insoluble in water. Nevertheless, some plant species are able to accumulate silica in their tissues. In the wet tropics certain woods contain crystalline deposits of silica, presumably where locally high levels of silica in the water are passively concentrated and then precipitated. Horsetails *(Equisetum),* also known as "scouring rushes," actively encrust the cellulosic cell walls of the outer stem with silica, an arrangement that makes them rigid. But probably the most exciting cases are where species actually build a silica body as an amorphous, opallike crystal within the cell. These structures are called silica bodies.

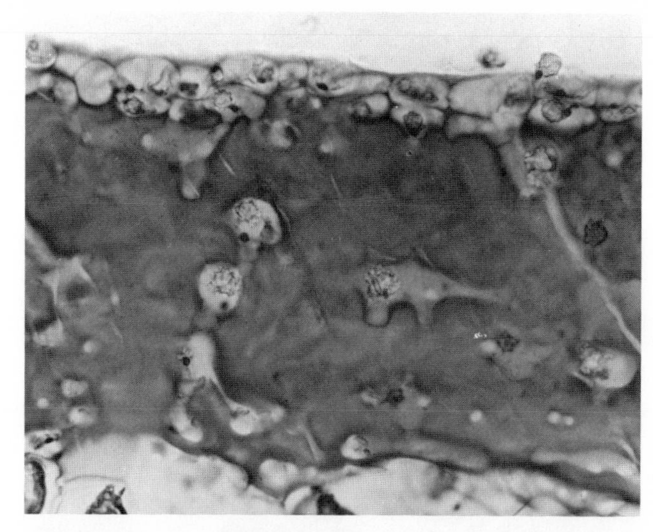

Figure 9.21. Close-up of the skin of organ pipe cactus, *Stenocereus thurberi,* as seen with a compound (light) microscope, showing large silica bodies in hypodermal and epidermal cells. ×180.

Silica bodies have been observed in over 20 plant families, including the palms (Arecaceae), grasses (Poaceae), sedges (Cyperaceae), gingers (Zingiberaceae), bananas (Musaceae), and rapateads (Rapateaceae). In 1975 the Cactaceae were added to this list when Gibson and Karl E. Horak discovered silica bodies in the skin of certain Mexican columnar cacti (Figs. 9.21 and 9.22). Eventually it was found that silica bodies occur in all species of *Stenocereus,* a genus centered in Mexico. To date no other cacti have been shown to have silica bodies.

Because silica bodies are hard and indigestible structures, they may protect the plant against foraging by larvae of insects with chewing mouth parts. This hypothesis, though reasonable, is untested.

Elemental Analysis

The preceding discussions of prominent calcium oxalate crystals and of the importance of calcium in mucilage lead us directly into an overall considera-

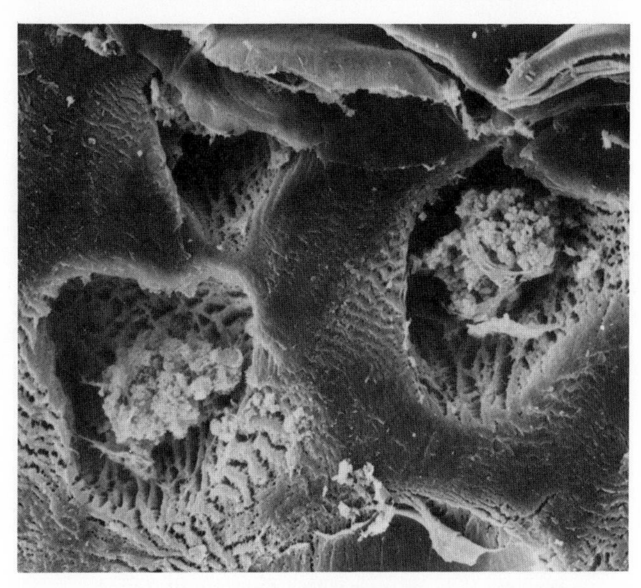

Figure 9.22. Silica bodies in the hypodermal cells of *Stenocereus gummosus,* or agria. ×3600.

tion of the elemental contents of cacti. Consideration of the levels of specific elements in a cactus plant permits us to think about how they got there and leads to the broad and much-neglected subject of mineral nutrition of cacti. We can then consider what growth conditions are most favorable for cacti and whether cacti are excluded from certain regions because levels of specific minerals are too low (deficient) or too high (toxic).

Nutrient Level in Cacti

Nutrients are traditionally classified into two categories: the macronutrients, which are present at fairly high levels in the tissues; and the micronutrients, which are present in low levels. The macronutrients include nitrogen, phosphorus, and potassium. Indeed, these three elements (N, P, and K, respectively) are the basis for the fertilizer industry. For example, one can buy a 5-10-5 fertilizer, which contains 5% N by weight, 10% PO_5, and 5% K_2O. Other macronutrients are calcium (Ca), magnesium (Mg), and sulfur (S). Macronutrients are required to build important molecules and to form the ionic milieu (environment) of cells. For instance, molecules of DNA contain nitrogen and phosphorus, proteins contain nitrogen and sulfur, and potassium is the most common ion in plant cells and is very important in the opening and closing of stomates. We have already indicated that calcium is present in the crystals and mucilage of cacti, but it is also a key ingredient in the "cement" that holds plant cells together. One magnesium atom is incorporated into each chlorophyll molecule, and magnesium is not only an important general ion in plant cells but also is needed for the activity of Rubisco, the enzyme that catalyzes the incorporation of CO_2 into photosynthetic products during the daytime (Eq. 4.5), and of PEP carboxylase, the enzyme that leads to the initial fixation of CO_2 at night in CAM plants (Eq. 4.7).

The major functions of micronutrients are in proteins, where their presence is crucial for the function of those enzymes. For example, copper (Cu), zinc (Zn), and manganese (Mn) are important micronutrients in many essential enzymes. Iron (Fe), which is sometimes classified as a macronutrient but usually as a micronutrient, is also an important constituent of many enzymes.

The elemental contents of cacti (Table 9.1) are similar to those of most other plants. One major difference is a much higher calcium level in cacti, because of the presence of numerous calcium oxalate crystals. Also, the calcium level seems to increase with age; for example, it is three times higher in 1-year-old cladodes of *Opuntia ficus-indica* than in 7-week-old cladodes. Manganese also tends to be higher in cacti, but the sodium level is much lower (Table 9.1) than the average for other plants. Although sodium is found in all plants, it may not actually be required. The micronutrient chloride, which is necessary for photosynthesis, often occurs at about the same concentration as sodium and potassium combined.

Physiological Implications

Cacti can be confronted with saline conditions every time drought occurs because the loss of soil water tends to concentrate the ions in the remaining soil water. Thus, understanding the salt tolerance of cacti can be important for evaluating the distribution patterns of the species.

When seedlings of *Carnegiea gigantea, Ferocactus acanthodes,* and *Trichocereus chilensis* were grown in aqueous solutions for 6 months, growth was greatly reduced when the sodium chloride (NaCl) concentration exceeded 0.1 *M*, and growth was completely inhibited at 0.2 *M* NaCl (by comparison, sea water is about 0.5 *M* NaCl). The above salt studies by Nobel were expanded when he worked in the labortory of Ulrich Lüttge in Darmstadt, Germany. *Cereus validus,* a cactus that can grow in periodically saline areas near Salinas Grandes in Argentina, tolerated fairly high levels of salinity (0.4 *M* NaCl) for a few weeks. But the sodium and chloride levels built up in the chlorenchyma and caused the nocturnal CO_2 uptake and malate accumulation to decrease. Moreover, the finer roots died at high salt levels—one way, if a bit expensive in terms of plant material, of preventing excess salt accumulation. For *Opuntia ficus-indica,* sodium levels are highest in the root and

Table 9.1. Elemental contents of chlorenchyma in cacti.[a]

Element	Symbol	Level in chlorenchyma	Function
Macronutrients			
Nitrogen	N	1.7%	Component of DNA, RNA, proteins
Phosphorus	P	0.18%	Component of DNA, RNA, proteins
Potassium	K	1.4%	Main cation in cells
Calcium	Ca	4.0%	In crystals, ion in cells
Magnesium	Mg	1.1%	In chlorophyll, important ion in cells
Micronutrients			
Boron	B	44 ppm	Growth factor (mechanism unknown)
Copper	Cu	6 ppm	In enzymes (proteins)
Iron	Fe	160 ppm	In enzymes
Manganese	Mn	190 ppm	In enzymes
Zinc	Zn	28 ppm	In enzymes
Other			
Sodium	Na	210 ppm	Ion in cells

Source: Nobel (1983).

a. Data are averages for the chlorenchyma of 10 species obtained in the field in North America and South America. Data are expressed on a dry weight basis, and 1% corresponds to 10,000 ppm (parts per million by weight).

become progressively lower in the cladodes that are farther from the ground. However, the sodium level was always low, a result indicating that this species does not accumulate sodium. Indeed, the cacti studied so far tend to be sodium excluders and do not tolerate high sodium levels in the tissues.

For many plants growing under natural conditions, nitrogen is the most limiting element, and the same generalization holds for cacti. For field specimens of 10 species the nitrogen level averaged 1.7% of the dry weight. The species that had the highest nitrogen levels in the chlorenchyma tended to have greater nocturnal acid accumulation. This circumstantial evidence that tissue nitrogen level affects cactus growth was directly tested by varying the nitrogen levels in aqueous solutions used to grow seedlings (Fig. 9.23*A*). When the nitrogen level was very low (0.15 m*M* or 1/100 that of a typical full-strength nutrient solution), almost no change in stem volume occurred over a 6-month period. When the nitrogen concentration of the solution was 4 m*M*, average stem volume doubled in 6 months; and stem diameter increased slightly as nitrogen level was further increased. The nitrogen level in the stem tissue also increased as the nitrogen in the nutrient solution increased (Fig. 9.23*B*). Such higher nitrogen levels reflect higher chlorophyll levels (each chlorophyll molecule contains four nitrogen atoms) and also higher protein levels. Thus, nitrogen application should enhance the growth of cacti. This is apparently the case for certain species of *Opuntia* that are

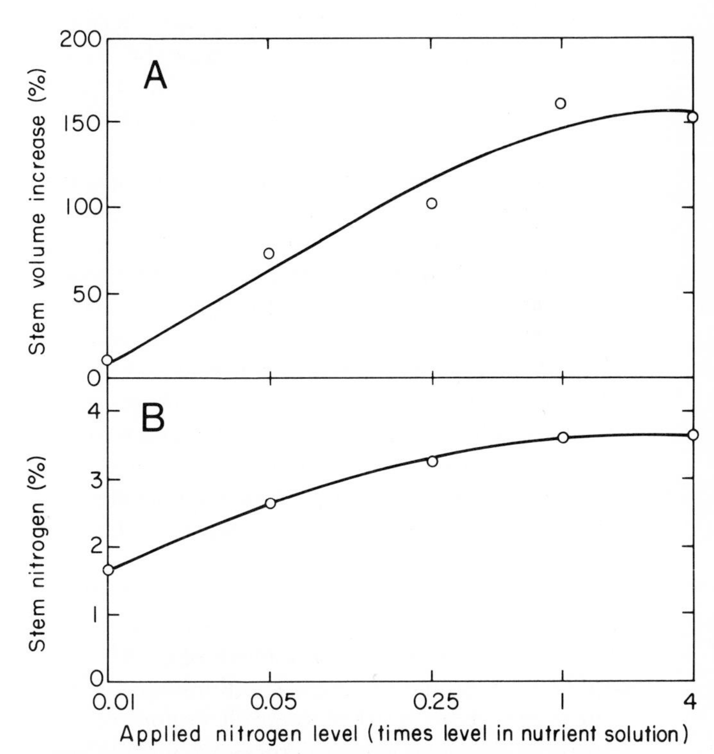

Figure 9.23. Influence of nitrogen in an aqueous nutrient solution on the growth rate of cactus seedlings *(A)* and on the nitrogen level in the stem *(B)*. Data represent averages for *Carnegiea gigantea, Ferocactus acanthodes,* and *Trichocereus chilensis* growing for 6 months in a standard nutrient solution developed for plants (Hoagland's solution, which contains 15 m*M* nitrogen as nitrate). (Adapted from Nobel, 1983.)

raised worldwide for their fruits and in some locations for the young cladodes, which can be used as green vegetables (Fig. 4.7).

Future experiments with cactus nutrition must determine both deficiency and toxicity levels of elements for cacti, about which little is presently known. Such information can help ecologists to relate the distributional patterns of cactus species in the field to the characteristics of the soil. These types of studies may explain why, for example, some species are only found on very strange soil types, for example, those species of the Chihuahuan Desert that only live on gypsum (calcium sulfate).

Plant Defenses — An Overview

Even though many cactus enthusiasts frequently talk about how cacti defend themselves from being eaten by animals, there has been very little scientific study on the subject. Cacti obviously have structures and chemicals that other plants living with them do not have; and because these are energetically expensive features to make, we can reason that they must have some purpose. Our intuition tells us that defenses are needed to protect a cactus from being eaten because in a dry habitat, where water and food can be seasonally very scarce, a fat, juicy, undefended cactus plant would be a welcome waterhole and restaurant for animals.

Let us consider the following observations on the putative defenses of a cactus against predators and diseases. Spines are the first line of defense deterring large predators from eating the stem; but certainly spines have no effect on invasions by small insects and microbes that can easily avoid the armament. A thick cuticle provides a line of defense against microbes, which cannot digest wax. The tough cell walls of the skin prevent most insects with piercing or sucking mouth parts from tapping in or gravid females from ovipositing eggs in the juicy tissues. And the crystals present within the skin presumably deter insects with chewing mouth parts.

Chemical defenses are operative within the soft cortex. First, a very healthy cactus stem has a low pH and a diurnal change in pH as a result of CAM metabolism (Chapter 4). Acidic conditions and diurnal fluctuations of pH are biologically important factors that may stop many microbes from growing in CAM tissues. Second, the soft cortex of a cactus has very low levels of nutrients and proteins because so much of the plant is occupied by vacuoles filled with water and malate. Consequently, many organisms could not live on this meager diet. The stem is also full of mucilage, which is apparently indigestible to most organisms and may actually interfere in physical ways, for example, by blocking the breathing pores of insects. Finally, many cacti have either abundant triterpenes or abundant alkaloids, compounds that can discourage and perhaps fatally poison a predator or microbe.

Most of the laboratory studies on the effects of cactus chemicals on other organisms are not relevant, for example, the tests of hallucinogens on laboratory rats or for medicinal uses. But there are a few critical studies showing that cactus chemicals are biologically active in natural situations. An early study of this type was done by Henry W. Kircher and William B. Heed and their co-workers at the University of Arizona; this team studied *Drosophila,* yeasts, and bacteria that live in the rotting tissues of cacti. In senita, *Lophocereus schottii,* of the Sonoran Desert, only one species of fruit fly, *D. pachea,* is found, and this organism cannot live in other cacti. In the 1960s, Kircher and Heed discovered that there is a good reason for this host specificity. To reproduce, the fly requires unique sterols that are made by the plant, and the reason that no other species can use the rotting cactus stem is because this is the only species tolerant of the tetrahydroisoquinoline alkaloids present in the tissues. They also determined that another species, *D. mettleri,* which is found feeding but not breeding on decaying stems of senita, is also tolerant of the alkaloids, a characteristic that gives it an ecological advantage over other cactophilic flies. By extrapolation we expect that alkaloids are reasonably effective poisons against certain groups of insects that try to inhabit desert plants.

There have been some experiments performed on the biological activity of cactus triterpenes, with few conclusive results. But William "Tom" Starmer at Syracuse University has shown that when organ pipe cactus lipids are added to fly food in high concentration, the survival of *D. mojavensis* larvae is depressed. Moreover, a yeast is able to detoxify the lipids of organ pipe cactus; consequently the fly can develop faster.

The reader can appreciate how primitive our knowledge is of the biological and evolutionary significance of cactus chemicals. Experiments are run to test the effects of natural chemicals on bacteria, yeasts, and insects that currently use the plants, but there is no way to determine whether the chemicals arose to defend the cactus against groups that have perhaps evolved to a new food source or, more likely, become extinct. In many ways scientists are groping for answers on the adaptive significance of chemical compounds without knowing either the villains or the phylogenetic relationships of the plants. Because we cannot rigorously test many hypotheses on the historical reasons for secondary plant products, future research on these cacti will have to concentrate on the effects of these chemicals on the potential present-day herbivores.

SELECTED BIBLIOGRAPHY

Anderson, E. F. 1979. *Peyote, the divine cactus.* University of Arizona Press, Tucson.

Bailey, I. W. 1961. Comparative anatomy of the leaf-bearing Cactaceae, III. Form and distribution of crystals in *Pereskia, Pereskiopsis* and *Quiabentia. Journal of the Arnold Arboretum* 42:334–340.

——— 1965a. Comparative anatomy of the leaf-bearing Cactaceae, XIII. The occurrence of water-soluble anisotropic bodies in air-dried and alcohol-dehydrated leaves of *Pereskia* and *Pereskiopsis. Journal of the Arnold Arboretum* 46:74–81.

——— 1965b. Comparative anatomy of the leaf-bearing Cactaceae, XV. Some preliminary observations on the occurrence of "protein bodies." *Journal of the Arnold Arboretum* 46:453–461.

——— 1966. Comparative anatomy of the leaf-bearing Cactaceae, XVI. The development of water-soluble crystals in dehydrated leaves of *Pereskiopsis. Journal of the Arnold Arboretum* 47:273–283.

Barker, J. S. F., and W. T. Starmer, eds. 1982. *Ecological genetics and evolution: the cactus-yeast-*Drosophila *model system.* Academic Press, Sydney.

Behnke, H.-D. 1981. Sieve-element characters. *Nordic Journal of Botany* 1:381–400.

Behnke, H.-D., and W. Barthlott. 1983. New evidence from the ultrastructural and micromorphological fields in angiosperm classification. *Nordic Journal of Botany* 3:43–66.

Clark, W.D., G. K. Brown, and R. L. Mays. 1980. Flower flavonoids of *Opuntia* subgenus Cylindropuntia. *Phytochemistry* 19:2042–2043.

Clark, W. D., and B. D. Parfitt. 1980. Flower flavonoids of *Opuntia* series Opuntiae. *Phytochemistry* 19:1856–1857.

Djerassi, C. 1957. *Cactus triterpenes.* XXVI. *Festschrift Arthur Stoll.* Birkhauser-Verlag, Basel.

Dodd, J. L., and W. K. Lauenroth. 1975. Responses of *Opuntia polyacantha* to water and nitrogen perturbations in the shortgrass prairie. Pp. 229–240 in *Prairie: a multiple view,* ed. M. K. Wali. University of North Dakota Press, Grand Forks.

Doetsch, P. W., J. M. Cassady, and J. L. McLaughlin. 1980. Cactus alkaloids XL. Identification of mescaline and other B-phenethylamines in *Pereskia, Pereskiopsis,* and *Islaya* by use of fluorescamine conjugates. *Journal of Chromatography* 189:79–85.

Everitt, J. H., and M. A. Alaniz. 1981. Nutrient content of cactus and woody plant fruits eaten by birds and mammals in south Texas, U.S.A. *Southwestern Naturalist* 26:301–306.

Fellows, D. P., and W. B. Heed. 1972. Factors affecting host plant selection in desert-adapted cactiphilic *Drosophila. Ecology* 53:850–858.

Fogleman, J. C., W. B. Heed, and H. W. Kircher. 1982. *Drosophila mettleri* and senita cactus (*Lophocereus schottii*) alkaloids: fitness measurements and their ecological significance. *Journal of Comparative Biochemistry and Physiology, Comparative Physiology, A,* 71:413–418.

Gershenzon, J., and T. J. Mabry. 1983. Secondary metabolites and the higher classification of angiosperms. *Nordic Journal of Botany* 5:5–34.

Gibson, A. C. 1982. Phylogenetic relationships of Pachycereeae. Pp. 3–16 in *Ecological genetics and evolution: the cactus-yeast-*Drosophila *model system,* ed. J. S. F. Barker and W. T. Starmer. Academic Press, Sydney.

Gibson, A. C., and K. E. Horak. 1978. Systematic anatomy and phylogeny of Mexican columnar cacti. *Annals of the Missouri Botanical Garden* 65:999–1057.

Gibson, A. C., K. C. Spencer, R. Bajaj, and J. L. McLaughlin. 1985. The ever-changing landscape of cactus systematics. *Annals of the Missouri Botanical Garden.* In Press.

Heed, W. B., W. T. Starmer, M. Miranda, M. W. Miller, and H. J. Pfaff. 1976. An analysis of the yeast flora associated with cactiphilic *Drosophila* and their host plants of the Sonoran Desert and its relation to temperate and tropical associations. *Ecology* 57:151–160.

Kircher, H. W., and H. L. Bird, Jr. 1982. Five 3-*β*, 6-*α*-dihydroxysterols in organ-pipe cactus. *Phytochemistry* 21:1705–1710.

Kircher, H. W., and W. B. Heed. 1970. Phytochemistry and host plant specificity in *Drosophila.* Pp. 191–209 in *Recent advances in phytochemistry,* vol. 3, ed. C. Steelink and V. C. Runeckles. Appleton-Century-Crofts, New York.

LaBarre, W. 1975. *The peyote cult,* 4th ed. Shoe String Press, Hamden, Connecticut.

Luckner, M. 1972. *Secondary metabolism in plants and animals.* Academic Press, New York.

Mabry, T. J. 1976. Pigment dichotomy and DNA-RNA hybridization data for centrospermous families. *Plant Systematics and Evolution* 126:79–94.

Mata, R., and J. L. McLaughlin. 1980a. Tetrahydroisoquinoline alkaloids of the Mexican columnar cactus, *Pachycereus weberi. Phytochemistry* 19:673–678.

——— 1980b. Lemairin, a new glucoside from the Mexican cactus, *Pachycereus weberi. Journal of Natural Products (Lloydia)* 43:411–413.

——— 1982. Cactus alkaloids. 50. A comprehensive tabular summary. *Revista Latinoamericana de Química* 12:95–117.

Mauseth, J. D. 1978a. The structure and development of an unusual type of articulated laticifer in *Mammillaria* (Cactaceae). *American Journal of Botany* 65:415–420.

——— 1978b. Further studies of the unusual type of laticiferous canals in *Mammillaria* (Cactaceae): structure and development of the semi-milky type. *American Journal of Botany* 65:1098–1102.

——— 1980a. Release of whole cells of *Nopalea* (Cactaceae) into secretory canals. *Botanical Gazette* 141:15–18.

——— 1980b. A stereological morphometric study of the ultrastructure of mucilage cells in *Opuntia polyacantha* (Cactaceae). *Botanical Gazette* 141:374–378.

——— 1983. Introduction to cactus anatomy. Part 5. Secretory cells. *Cactus and Succulent Journal* (Los Angeles) 55:171–175.

McGarvie, D., and H. Parolis. 1981a. The acid-labile,

peripheral chains of the mucilage of *Opuntia ficus-indica. Carbohydrate Research* 94:57–65.
——— 1981b. The mucilage of *Opuntia aurantiaca. Carbohydrate Research* 94:67–71.
——— 1981c. Methylation of the mucilage of *Opuntia ficus-indica. Carbohydrate Research* 88:305–314.
Mohamed, Y. A. H., C.-J. Chang, and J. L. McLaughlin. 1979. Cactus alkaloids XXXIX. A glucotetrahydroisoquinoline from the Mexican cactus *Pterocereus gaumeri. Journal of Natural Products (Lloydia)* 42:197–200.
Nobel, P. S. 1983. Nutrient levels in cacti—relation to nocturnal acid accumulaton and growth. *American Journal of Botany* 70:1244–1253.
Nobel, P. S., U. Lüttge, S. Heuer, and E. Ball. 1984. Influence of applied NaCl on Crassulacean acid metabolism and ionic levels in a cactus, *Cereus validus. Plant Physiology* 75:799–803.
Piattelli, M. 1981. The betalains: structure, biosynthesis, and chemical taxonomy. Pp. 567–575 in *The biochemistry of plants: a comprehensive treatise,* vol. 7: *Secondary plant products,* ed. E. E. Conn. Academic Press, New York.
Starmer, W. T. 1982. Associations and interactions among yeasts, *Drosophila* and their habitats. Pp. 159–174 in *Ecological genetics and evolution: The cactus-yeast-*Drosophila *model system.,* ed. J. S. F. Barker and W. T. Starmer. Academic Press, Sydney.
Starmer, W. T., W. B. Heed, M. Miranda, M. W. Miller, and H. J. Pfaff. 1976. The ecology of yeast flora associated with cactiphilic *Drosophila* and their host plants in the Sonoran Desert. *Microbial Ecology* 3:11–30.
Starmer, W. T., H. W. Kircher, and H. J. Pfaff. 1980. Evolution and speciation of host-plant specific yeasts. *Evolution* 34:137–146.
Stewart, O. C. 1980. Peyotism and mescalism. *Plains Anthropology* 25:297–309.
Trachtenberg, S., and A. Fahn. 1981. The mucilage cells of *Opuntia ficus-indica* (L.) Mill.—development, ultrastructure, and mucilage secretion. *Botanical Gazette* 142:206–213.
Trachtenberg, S., and A. M. Mayer. 1982a. Mucilage cells, calcium oxalate crystals and soluble calcium in *Opuntia ficus-indica. Annals of Botany* 50:549–557.
——— 1982b. Biophysical properties of *Opuntia ficus-indica* mucilage. *Phytochemistry* 21:2835–2843.
Unger, S. E., R. G. Cooks, R. Mata, and J. L. McLaughlin. 1980. Chemotaxonomy of columnar Mexican cacti by mass spectrometry/mass spectrometry. *Journal of Natural Products (Lloydia)* 43:288–293.
Wittler, G. H., and J. D. Mauseth. 1984a. The ultrastructure of developing latex ducts in *Mammillaria heyderi* (Cactaceae). *American Journal of Botany* 71:100–110.
——— 1984b. Schizogeny and ultrastructure of developing latex ducts in *Mammilaria guerreronis* (Cactaceae). *American Journal of Botany* 71:1128–1138.
Yatsu, L. Y., T. J. Jacks, and H. W. Kircher. 1979. Visualization of saturated lipid bodies in organ pipe cactus by a novel cytochemical technique. *The Microscope* 27:61–65.

10

Phylogeny and Speciation

Interested observers of cactology and biology probably wonder why scientists are so particular about names. Anyone can appreciate, of course, that for identification purposes each species must have a unique scientific name; and the scientific name should be the oldest one available so that the first author gets full credit for the original discovery. But why is a species moved back and forth between genera, and why should generic names also change? There are, after all, over 350 generic names available for cacti, and yet we used only 121 of them (Table 1.2).

A phylogeny is a model of the evolutionary relationships between an ancestor and all of its known descendants. Scientists adjust generic names because they want to have a phylogenetic classification system. Just as a person might want to know the precise relationships of individuals within his family, that is, the genealogy and all of the branches and marriages, biologists want to draw a family tree for the species, showing the origin of each. Various methods have been used to produce such family trees of plants and animals, and the results can look very different even when they are based on the same information. Ultimately, one reconstruction must be more accurate than the others because there is only one true genealogy.

The evolutionary history of each group, or taxon, of plants or animals is a mystery that is not easy to solve. Nonetheless, cacti are excellent candidates for phylogenetic studies because large amounts of information have been amassed about them. In addition, numerous researchers are applying new methods in an effort to understand how the species are related.

Reconstructing Phylogenies

Philosophies of Classifying Organisms

Developing classification systems has been a human preoccupation for over 2000 years. Initially people used simple methods to name and identify animals and plants as well as to store general facts about them. For example, it was important to know whether an animal had a backbone (a vertebrate) or no backbone (an invertebrate) or whether a plant had flowers (an angiosperm) or seed-bearing cones (a gymnosperm). For flowering plants, people distinguished between species with leaves having parallel veins (monocotyledons) and those with leaves having netlike veins (dicotyledons). It was easy to be overwhelmed by the differences observed between organisms, which were many, but observers were looking for similarities in order to group species in ways that made them easier to remember. When a group of species became too big and heterogeneous, however, it often became too general to be useful, so subcategories were invented to recognize major differences within the group. Consequently, classification of organisms has always been a process involving a delicate balance between emphasis of similarities on the one hand and recognition of differences on the other.

The art of classification had to become a science when evolution became accepted as a biological fact. It was not enough to know that two organisms resembled each other in certain ways; rather, biologists desired to know whether two taxonomic groups were closest kin, or what we now call sister taxa, that is, groups that arose from the same common ancestor. Superficially this may appear to be a purely academic exercise, but actually it is quite practical. Sister taxa usually share a great many structural, chemical, and genetic similarities. Hence, a phylogenetic model of relationships is, in the vast majority of cases, an excellent way to organize information; and it also enables a researcher to study how structure and chemistry have evolved from one species to the next. Not every person who classifies organisms agrees with the phylogenetic approach, but opposition is now quite weak because the general, long-range goal of biology is to gain an understanding of the evolution of organisms.

Even though scientists generally agree that biological classification must have an evolutionary framework, there is still vigorous disagreement on how that goal can be accomplished. Not everyone shares in the optimism that the branching sequence of the species, called cladogenesis (branch formation), can be elucidated and then properly expressed in a system of classification. Nevertheless, if cactologists are going to participate in phylogenetic studies, then they should be informed about the issues and methods of the field.

Characters and Character States

In biology, a feature such as eye color or stamen number is termed a character, and any form of that character, such as brown eye color and five stamens, is termed a character state. When two species look alike, it is because they have many similar character states, a condition that usually means that they are genetically similar and probably share a common ancestry. To produce a phylogeny, the researcher analyzes the distribution of character states to quantify the relative genetic similarity between the species.

The choice of characters to be used in an analysis is of the utmost importance. Characters have to be consistent and dependable; therefore, we should not use features that are unpredictable or that are strongly affected by environmental factors. For example, in cacti many spine features are often useless for phylogenetic analysis because spine length, number, and even color can vary greatly within a population and throughout the range of the species (Fig. 5.5). The worker must also be absolutely certain that a character state present in one species is homologous with one present in other species; that is, in both species the character is coded for by the same genes, which are inherited from a common ancestor. If the genes are not the same, major problems result. For example, the inferior ovary in many plant families is appendicular in nature, whereas that of Cactaceae is receptacular. Thus, the two types of inferior ovary are certainly not homologous structures; instead, they are convergent structures (Chapter 2).

For each character there are two or more character states. For the taxon under scrutiny the investigator usually wants to know which is most primitive, or plesiomorphic, and which are derived, or apomorphic. These are strictly relative terms, however. For example, because betalains occur only in cacti and its closest relatives (Chapters 9 and 11), the presence of betalains is an apomorphic character state for the dicotyledons; but the presence of betalains is plesiomorphic in the cactus family, in which all species have this class of pigments. Decisions on plesiomorphic versus apomorphic are often difficult to make when qualitative differences, such as the length or color of structures, are being judged, and consequently disagreements over the primitive and derived status of a character state are very common in systematic literature.

Analyzing Character States

Once a list of the species has been made and the character states have been recorded for all species on the list, the next step is to determine which characters should be used to elucidate taxonomic and evolutionary relationships. There are three common approaches used in analyzing characters: (1) phenetics, (2) evolutionary systematics, and (3) cladistics, or phylogenetic systematics. Each of the three approaches emphasizes different information in the data set, and phenetics and cladistics typically rely upon computer analyses.

Phenetics is a method of cluster analysis and is not a form of phylogenetic analysis; it uses all characters in the data set to determine the *overall* similarity between species, that is, which two species appear to be the most similar. Consider an example of four species, A through D, for which we have data on 100 characters. Species A and B share 95 of their character states (95% similarity), species C and D share 80 character states, and the two pairs share only 50 of them. Figure 10.1 shows how these results can be diagramed as a phenogram. This diagram does not tell the reader which species originated first, however, and it does not guarantee that the branching pattern mimics the speciation events.

In evolutionary systematics the investigator uses only a portion of the characters available in drawing a phylogeny. A phylogeny is assembled by following the evolution of a particular structure, such as the evolution of flower shape (Fig. 10.2) or of the skull and tusks of elephants (proboscidians). In a phylogeny the taxa are arranged so that the one with the

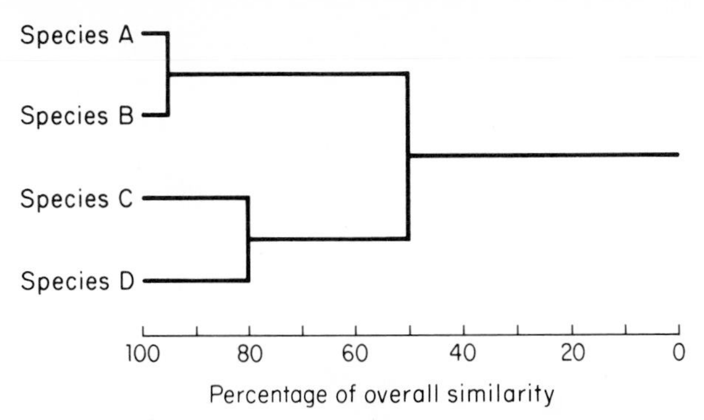

Figure 10.1. Phenogram for four hypothetical species, of which species A and B share 95% of all features analyzed, species C and D share 80% of all features analyzed, and the four taxa share 50%.

most primitive features is placed at the base of the tree and those mostly highly derived are located at the tips of the branches. This type of phylogeny depicts the stages, or grades, of evolution but often does not precisely represent the genealogy of the species.

The third approach is popularly called cladistics; and cladistics has recently dominated the literature because it has several highly favorable attributes. In Latin, *clade* means branch, so cladistics studies the branching sequences of evolution. In cladistic analyses all characters are used, but species are grouped solely on the basis of synapomorphies, that is, shared derived character states (*syn* means together or shared). The diagram produced from this analysis is called a cladogram (Fig. 10.3); and in a cladogram, each branch must be identified by one or more synapomorphies that occur in all species on that branch. A synapomorphy that identifies an evolutionary branch (clade) should be unique, that is, it should not occur anywhere else in the entire group of species.

The methodology of cladistics is to follow the origin of new traits. Consider the case in which a new sulfur-containing compound appears in a plant species; that compound will then be passed on to all subsequent species that arise from the initial population with the sulfur-containing chemical. Although the derived character state may not be important for the success of that clade, it certainly can be used as a tag for identifying that new branch. In contrast, the methodology of phenetics combines the information of the derived character states with those retained from the primitive ancestor of the entire evolutionary tree. Consequently, the information on the origin of branches is potentially lost or is masked by the shared primitive character states (symplesiomorphies).

Evolutionary systematics differs significantly from both phenetics and cladistics because it uses a fairly small portion of the data set. This is called weighting, which means that some characters are selected to carry more weight than others, and this weighting is done before there is a rigorous search for alternative phylogenies. Therefore, by choosing the "important" characters ahead of time, an author can impose a particular phylogeny on the reader. In contrast, the other approaches do not weight characters at the start, and they search for the solution that explains all the observations in the least complicated, that is, the most parsimonious, way.

The difference between evolutionary systematics and cladistics or phylogenetic systematics is a deep philosophical question: How does a researcher choose the correct solution? For instance, there can be 13 possible configurations for the phylogenetic relationships of three taxa (Fig. 10.4); and if the number of taxa is greater, for example, ten or more, the number of possible phylogenetic configurations staggers the mind. For classifying organisms, the goal of the scientist should be to ascertain which of these models do not fit very well, so they can be discarded, and to determine which are the most parsimonious and therefore presumably describe the data with the greatest fidelity. Using the assumption of parsimony requires the investigator to consider all of the most logical solutions, whether there be one or one hundred; and the researcher is forced to demonstrate that the one or several phylogenetic models that are chosen are more acceptable than are those that have been discarded. Adopting this philosophy is supposed to remove some of the subjective and intuitive aspects of phylogenetic reconstructions that characterized many of the published phylogenies of the past.

The tone of our remarks reflects our general confidence in cladistic methods for attempting phylogenetic reconstructions, because cladistics presently seems to offer the best tools for expressing the genealogy of organisms. Nonetheless, there are still certain problems with this approach. An investigator needs a long list of characters to determine the branching sequence of large taxa, and for many groups the data base is presently insufficient. Many cladograms have places where more than two lines radiate from a single point because the branching events cannot be resolved (Fig. 10.5). We can be frustrated, therefore, to discover after many hours of computer analysis that completely resolved phylogenetic models may still be difficult to obtain. Regardless of these and other technical problems in doing cladistic analysis, the analytical methods of cladistics have enabled biologists to break away from the earlier, mostly intuitive evolutionary reconstruc-

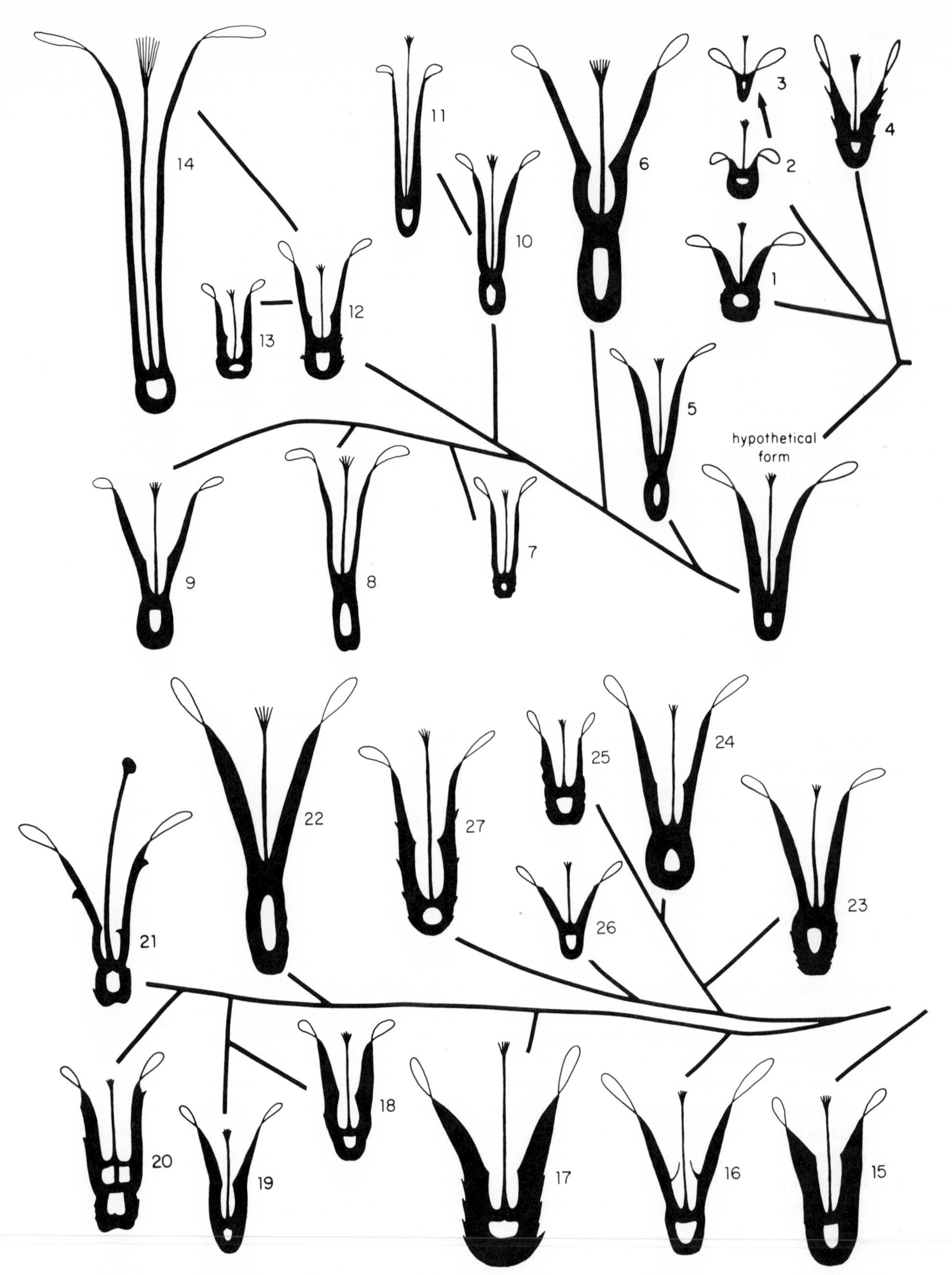

Figure 10.2. A hypothetical representation of flower evolution in the tribe Pachycereeae, showing the way an evolutionary systematist might determine phylogeny by describing changes in one feature. Here changes in flower size and shape and in locule shape are hypothesized for the subtribe Stenocereinae (upper) and the subtribe Pachycereinae (lower); but only about one third of the species have been illustrated. 1, *Polaskia chende;* 2, *P. chichipe;* 3, *Myrtillocactus* spp.; 4, *Escontria chiotilla;* 5, *Stenocereus hystrix;* 6, *S. pruinosus;* 7, *S. beneckei;* 8, *S. quevedonis;* 9, *S. thurberi;* 10, *S. standleyi;* 11, *S. alamosensis;* 12, *S. stellatus;* 13, *S. treleasei;* 14, *S. gummosus;* 15, *Lemaireocereus hollianus* (may not belong to this subtribe); 16, *Cephalocereus* spp.; 17, *Mitrocereus fulviceps;* 18, *Neobuxbaumia tetetzo;* 19, *N. mezcalensis;* 20, *N. polylophus;* 21, *N. euphorbioides;* 22, *Carnegiea gigantea;* 23, *Pachycereus pringlei;* 24, *P. weberi;* 25, *P. marginatus;* 26, *Lophocereus* spp., 27, *Backebergia militaris.*

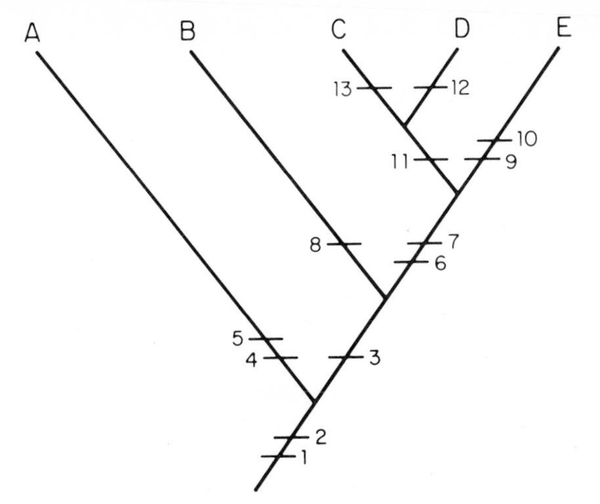

Figure 10.3. Hypothetical cladogram that models the most likely branching sequence of five species, A through E. To understand this cladogram, we have shown the 13 synapomorphies that were used to define the branching events. Two synapomorphies (1 and 2) distinguish this group of species from all others in the family. Once the group originated, species A diverged from the others after developing two synapomorphies (4 and 5), whereas the other populations had synapomorphy 3. Next, species B diverged after developing synapomorphy 8, and the remainder developed synapomorphies 6 and 7. Subsequently, five more synapomorphies help to resolve the branching sequence of the most recent species C–E.

tions and to register some successes in nearly every major group of organisms.

Phylogenetic Studies and the Classification of Cacti

Even though there are several ways to conduct phylogenetic research, such studies are rare in the cactus literature and very detailed phylogenetic studies have only recently been attempted on cacti. Most of the literature on the Cactaceae has been heavily biased toward describing and naming the species, a field of research called alpha taxonomy. It is always the first phase of phylogenetic research, because the species must be thoroughly collected and their features cataloged before quantitative and qualitative analyses can proceed.

As we discussed earlier in this chapter, a goal of modern biological classification is to arrange the species in a way that reflects their evolutionary history and pattern of descent. But the species that are living today are only a small percentage of the species that have lived on the earth; and when the biologist arranges the species into larger units, such as genera, tribes, and families, a category is invented to include all species that presumably arose from the same ancestor, for example, the first population of cactus plants. Anyone can see the problems in doing this. Some workers will conclude that the genus should include only the very closest relatives, perhaps 5 species, whereas another worker may place many more species in the same genus, perhaps 20 species. Both classifications cluster the species that descended from one ancestral population, but one author has a broader definition of a genus than another. In other words, the researcher makes subjective decisions about where the breaks occur between groups of species.

From an evolutionary point of view, only the species and the smaller populations of the species are capable of evolving and are, therefore, "real" evolutionary units. For example, in a species one individual is usually capable of mating and producing fertile offspring with any individual of the opposite sex in that species. In contrast, any taxonomic unit

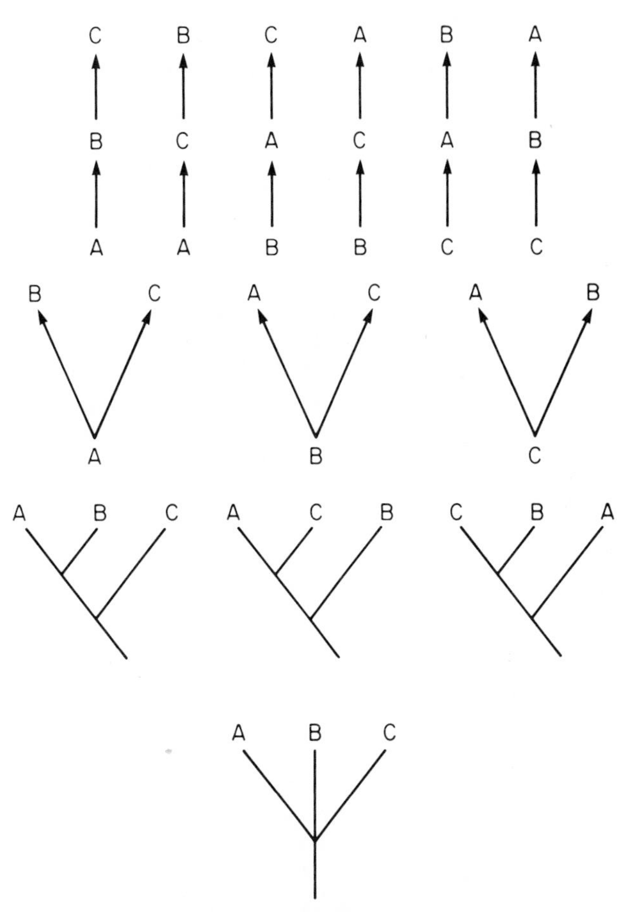

Figure 10.4. Diagrams of the 13 possible ways that three species, A, B, and C, can be related to each other: one species could have given rise to another and that to another *(top row)*; one extant species could have given rise independently to each of the other two *(second row)*; or each of the species could be a product of evolutionary divergence from ancestors that are no longer extant *(third and fourth rows)*. (Redrawn from Cracraft, 1974.)

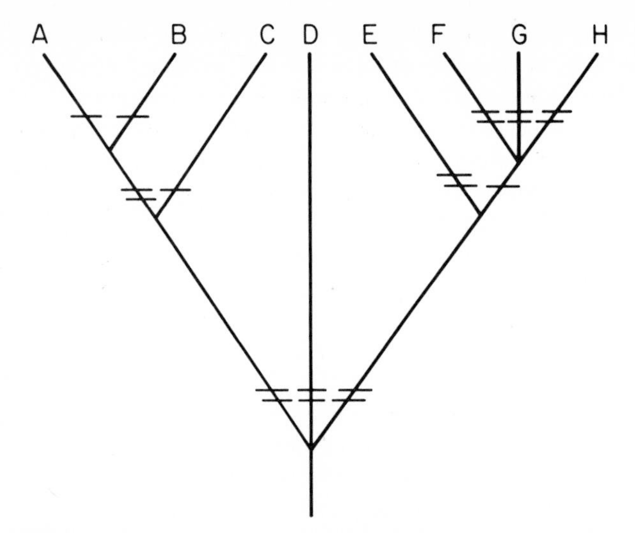

Figure 10.5. Cladogram of eight species (A to H), in which there are two unresolved trichotomies. The sequence of branching at the base of the cladogram and the sequence on the left for the events that leads to the formation of species F, G, and H cannot be determined because of insufficient information. The horizontal lines indicate that there are one or two unique synapomorphies that characterize each evolutionary branch in the diagram.

above the level of the species is usually not a reproductive unit; instead, it is an "artificial" category that has been devised on the basis of convenience and is used to place the species into pigeonholes. Consequently, there are no concrete guidelines to tell us how large or how small a genus, tribe, or family should be; and taxonomists have struggled for over two centuries to compose the best systems for showing the relationships of the species.

Early Systems for Classifying the Cacti

Because the cacti are native organisms of the New World, exclusive of some populations of *Rhipsalis,* European knowledge of the cacti was scanty in 1753 when Linnaeus officially described 22 species (Chapter 1). Nonetheless, the morphological diversity of these species was so great that a classification with several genera was immediately required. The real surge in knowledge of cacti came in the next century, when botanical exploration penetrated to the semiarid and arid inland habitats of the Western Hemisphere. By the end of the nineteenth century, over 500 species (currently recognized) had been discovered. It was at the turn of the century, after a significant percentage of the family had been described, that authors felt a strong need to produce a broad revision of the genera.

The first detailed classification of the cacti was published at the turn of the century by Karl Schumann, a German cactologist. Even though many generic names were available then, Schumann recognized only 21 genera, which he classified into the three subfamilies currently in use. The genera of subfamily Cactoideae were grouped into three tribes, Echinocacteae, which included the columnar cacti, Mamillarieae, and Rhipsalideae (Table 10.1). His tribes were not phylogenetically sound, but he was rather progressive in recognizing many of the generic names that had been proposed in that century.

Soon after Schumann completed his generic treatment of the cacti, Britton and Rose began their famous cactus investigations, discussed briefly in Chapter 1, which led to the familial monograph that is still the cornerstone of cactus classification. The greatest revisions were made in the largest subfamily, Cactoideae, which they called tribe Cacteae, and in which they recognized 114 genera in 8 subtribes and subdivided *Cereus,* the columnar cacti, into many smaller, more narrowly defined genera. Some of these segregate genera were based on names proposed in 1905 by Alwyn Berger, curator of the botanical gardens in La Mortola, Ventimiglia, Italy, who had suggested that *Cereus* be subdivided into 18 subgenera.

Table 10.1. The classification of cactus genera proposed by Schumann in 1898.

SUBFAMILY CEREOIDEAE (=CACTOIDEAE)
TRIBE ECHINOCACTEAE
Cereus Mill. (=most columnar cacti)
Pilocereus Lem. (now part of *Cephalocereus)*
Cephalocereus K. Schum. (narrowly defined)
Phyllocactus Link (=*Epiphyllum)*
Epiphyllum Pfeiff.
Echinopsis Zucc.
Echinocereus Engelm.
Echinocactus Link & Otto
Melocactus Link & Otto
Leuchtenbergia Hook. & Fisch.
TRIBE MAMILLARIEAE
Mamillaria Haw. *(=Mammillaria)*
Pelecyphora Ehrenb.
Ariocarpus Scheidw.
TRIBE RHIPSALIDEAE
Pfeiffera Salm-Dyck *(=Rhipsalis)*
Hariota DC. *(=Rhipsalis)*
Rhipsalis Gaertn.
SUBFAMILY OPUNTIOIDEAE
TRIBE OPUNTIEAE
Opuntia Haw.
Nopalea Salm-Dyck
Pterocactus K. Schum.
SUBFAMILY PEIRESKIOIDEAE (=PERESKIOIDEAE)
TRIBE PEIRESKIEAE
Maihuenia Phil.
Peireskia (=Pereskia)

It is obvious that Britton and Rose had a concept of the genus that was different from that of Berger; Britton and Rose would be called liberal or "splitters," whereas Berger would be called conservative or a "lumper." Certainly these terms are relative, inasmuch as some of the more recent cactologists recognized so many genera that they make Britton and Rose appear to be conservative. Nonetheless, conservative cactologists have suggested that Britton and Rose ushered in a period of liberalism in cactus taxonomy, in which groups of columnar cacti were simply raised to generic rank so that they would be easier to remember. Obviously, there was a need to have some additional names because *Cereus* had accumulated a long list of species and had become a very heterogeneous taxon. But a major driving force for subdividing this genus was an intellectual urge to identify each cluster of species that descended from a single common ancestor, that is, a monophyletic taxon. Neither Britton and Rose nor Berger were successful in discovering all of the natural lineages of cacti, but it is a great credit to Britton and Rose that over 50 of the 79 generic names authored by them are still widely regarded as phylogenetically sound taxa, although, as we shall discuss, some of these have been greatly redefined. Consequently, in cactus systematics the generic concept, although never strictly defined, often corresponds fairly closely to that presented by Britton and Rose.

After Britton and Rose, cactus systematics seemed to proceed simultaneously in three different directions. One group of workers, primarily those who studied the Mexican and West Indian cacti, followed the lead set by Britton and Rose. At the same time cactologists began to collect the diverse but relatively poorly known cactus floras of South America. These new collections greatly changed the data base and caused workers to rethink the genera described by Britton and Rose. Gradually at first and then in a torrent, cactus enthusiasts, most notably Curt Backeberg of Germany, proposed many new genera and hundreds of species and varieties from South America as well as revised names of North American genera. For example, Backeberg published about 80 new generic names, but today only 15 to 20 of these have much chance of surviving as monophyletic genera or subgenera. Backeberg and cactus horticulturalists of that period were very impressed by the variability of the cacti, so they chose to emphasize fairly minor and often plastic features of these plants. The generic classification of Backeberg, such as his last system, published in 1966, is the best-known example of a liberal system. In Backeberg's system very minor differences between species were weighted very heavily. The third taxonomic path was taken in North America by Lyman Benson, who began studies of the native cacti north of Mexico. He chose to reclassify all species of columnar cacti in *Cereus* and his classification is presently the most conservative.

There are some strong differences of opinion among present cactologists over liberal versus conservative systems of classification. Some of the disagreements between authors return us to the conflict between emphasizing similarities and recognizing differences. What cactologists must keep in view is that the generic concept in Cactaceae must be roughly equivalent to that used in other families of dicotyledons. In this sense, in the most liberal systems of classification, many genera are too narrowly defined and are therefore unintelligible to other biologists. On the other hand, a conservative generic classification of the cacti is also unacceptable if this system does not have perfectly monophyletic taxa and does not permit one to reconstruct the evolution of the genera. A solution to the problem of liberal versus conservative classification systems is the development of a phylogeny of the species.

Buxbaum's Phylogenetic Classifications

The classification of genera used for this book (Table 1.2) is quite similar to the classification devised by the Austrian cactologist Franz Buxbaum. He deserves full credit for initiating a movement to produce a phylogenetic system of classification for the cacti. Beginning around 1950, Buxbaum published detailed descriptions and high-resolution illustrations of cactus reproductive features, including excellent views of whole flowers, longitudinal sections through a flower, and enlargements of small portions within the flowers, for example, nectaries, funiculi, and stigma lobes; of immature and mature fruits; and of whole seeds, seed coats, and embryos dissected out of the seeds. While making these drawings, he spent much time examining the cellular nature of these structures and therefore uncovered many apomorphic character states that had not been observed by earlier workers. Using these derived character states, Buxbaum drastically rearranged the genera and sometimes the species within genera to match more closely the distribution of the apomorphic features, particularly using the flowers and seeds. By today's standards his phylogenetic studies were largely intuitive; but many of the characters he used to determine relationships are now becoming popular characters for highly structured phylogenetic studies.

Buxbaum joined with other students of South American cacti in recognizing that there are major taxonomic differences between the dominant groups in North and South America. His novel classification of the subfamily Cactoideae helped to clarify where the tribes occur. For example, Buxbaum determined that there are three tribes, Cacteae, Echinocereeae, and Pachycereeae, centered around Mexico and four tribes, Browningeae, Cereeae, Notocacteae, and Trichocereeae, that are South American. Tribe Hylocereeae, which has many epiphytic species in addition to its terrestrial species, occurs mainly in North America and the West Indies, with some species in northern South America; and Leptocereeae primarily occur in northwestern South America, Central America, and the West Indies. Certainly Britton and Rose had greatly confused the phylogenetic picture, because in many places they had classified North and South American genera in the same subtribe; and they had not recognized that the features showing similarities on the two continents were convergent, that is, the same basic characteristics had evolved independently in two unrelated taxa.

The first tribal classification of subfamily Cactoideae by Buxbaum appeared in 1958, and revisions of that classification were published through the mid-1970s. Over this period he shifted genera between tribes as he accumulated more information on cacti and eventually made phylogenetic diagrams of the genera within each tribe.

Phylogenetic Studies of Pachycereeae — An Example

It is easy to find fault with Buxbaum's generic phylogenies because they were done with very incomplete sets of data. Looking at his work from a positive viewpoint, however, we can examine the genera included in a tribe by Buxbaum and attempt to construct a totally revised phylogeny of the species. This is what Gibson and co-workers have done for the columnar cacti in tribe Pachycereeae.

Pachycereeae currently includes about 70 species, most of which are the large columnar cacti of Mexico but some of which occur as far south as northern Venezuela. No one has yet discovered a single synapomorphy that would permit a phylogeneticist to define this tribe in a strict cladistic sense, and therefore no one can categorically say that other genera or even another subtribe should be included. One derived feature that occurs in all species concerns the nature of the wood skeleton, which consists of a ring of parallel, discrete rods (Figs. 3.29 and 3.37–3.39).

When Buxbaum first proposed this tribe, he accepted only six genera as definite members of the Pachycereeae; but after conducting a thorough morphological study, including a small book on the phylogeny of the tribe published in 1961 and a series of informal generic descriptions, Buxbaum eventually decided to include fourteen genera. Two apomorphic features were used to add the last two genera to Pachycereeae. During his phylogenetic studies, Buxbaum had discovered some idioblastic, rectangular, pigment cells in the epidermis of the funiculus in some of the species (Figs. 10.6 and 10.7). In the fruit these pigmented cells develop into large, spherical, reddish cells, which are called pearl cells (German, *Perlzellen*) because they look like beads on a colorless string, the funiculus. These structures, which help color the fruit pulp, apparently do not occur in any other groups of cacti or in any other angiosperms. Both *Machaerocereus* and *Rathbunia* were removed from other tribes and placed into tribe Pachycereeae because they have these pearl cells. Moreover, Buxbaum noted that species of *Machaerocereus,* like numerous species of Pachycereeae, have abundant stem triterpenes (Chapter 9).

Buxbaum published his final generic phylogeny of Pachycereeae in 1975; in this system he recognized four subtribes (Fig. 10.8). But many illogical aspects in his phylogeny remained and still had to be resolved. Gibson and Horak made lists of the species that have pearl cells, that have triterpenes, that have tetrahydroisoquinoline alkaloids, and that have certain seed features, so that a test could be conducted on the ability of Buxbaum's phylogeny to explain the distributions of these unusual apomorphic features. The model failed all preliminary tests of parsimony, a result that signified that a major overhaul of the model was required.

Apomorphic character states were used as tags to identify the evolutionary branches and to relate one evolutionary branch to the other species groups. This technique produced a completely different phylogenetic model (Fig. 10.9). Two subtribes, instead of four, were recognized. Subtribe Stenocereinae included all species known to have abundant glycosidic triterpenes, pearl cells, and fairly small, rough seeds. Within this subtribe, all species containing silica bodies in the skin (Figs. 9.21 and 9.22), including the species of *Machaerocereus* and *Rathbunia,* became part of a single genus called *Stenocereus.* The species that had tiny seeds and lacked silica bodies but had other apomorphic vegetative and reproductive features became a second group of species in subtribe Stenocereinae and were classified in the genera *Polaskia, Myrtillocactus,* and *Escontria* (Fig. 5.33). Those species having or suspected of having tetrahydroisoquinoline alkaloids or calcium oxalate crystals in the skin and lacking certain apomorphic character states of subtribe Stenocereinae were placed in subtribe Pachycereinae.

Rearrangement of other species within Pachycereeae by using synapomorphies revealed that two

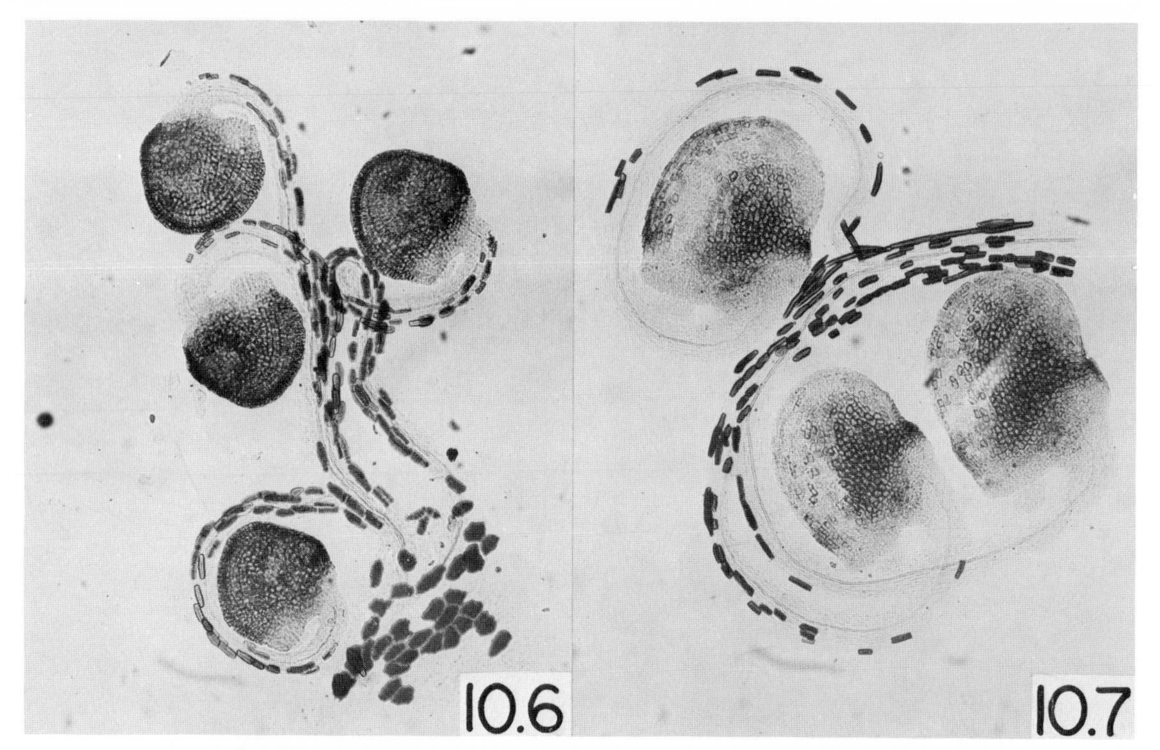

Figures 10.6 and 10.7. Whole mounts of the funiculus and ovules from flower buds before they open; dark, rectangular pigment cells can be observed in the epidermis of the funiculus. (From Gibson and Horak, 1978.)

10.6. *Myrtillocactus cochal.*
10.7. *Stenocereus alamosensis.*

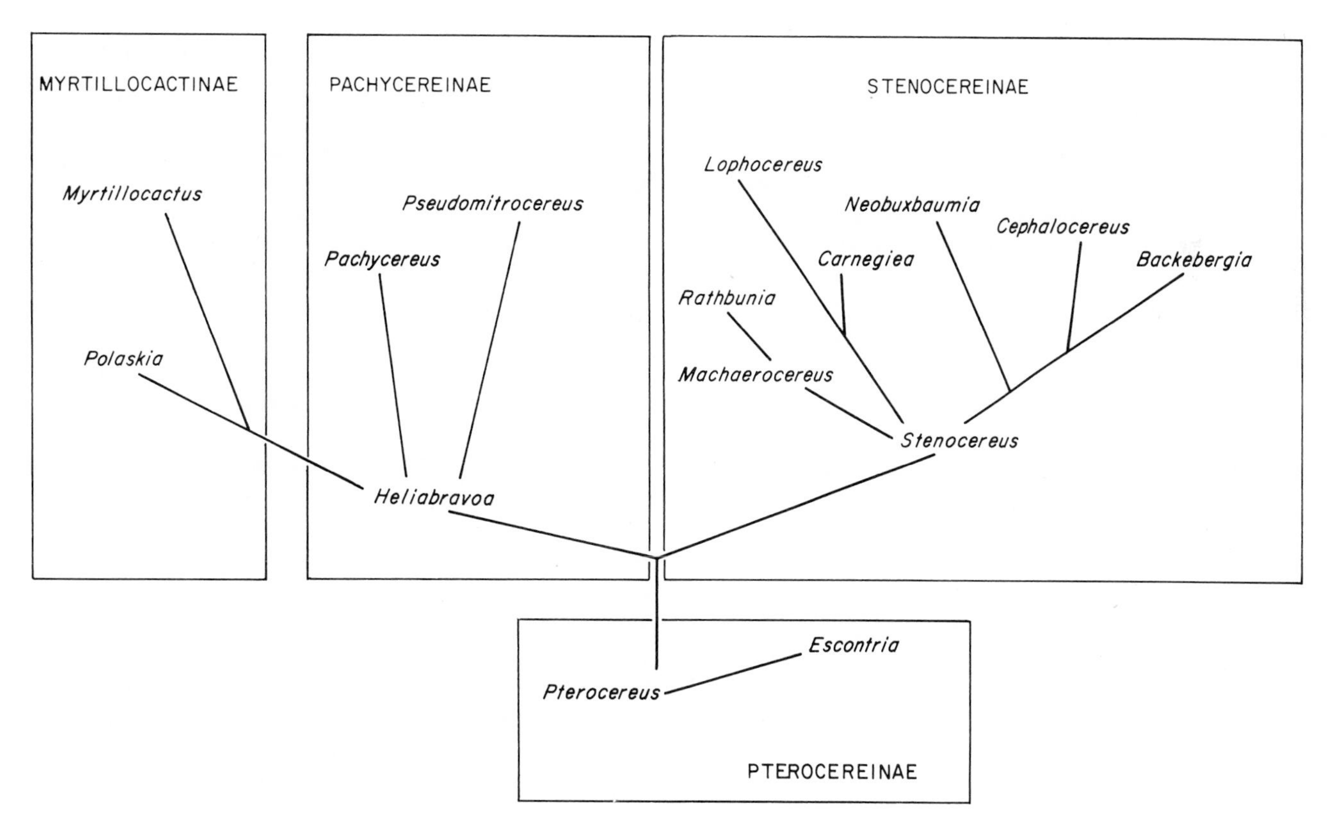

Figure 10.8. Buxbaum's 1975 generic phylogeny of his tribe Pachycereeae, in which he recognized four subtribes. (From Gibson and Horak, 1978.)

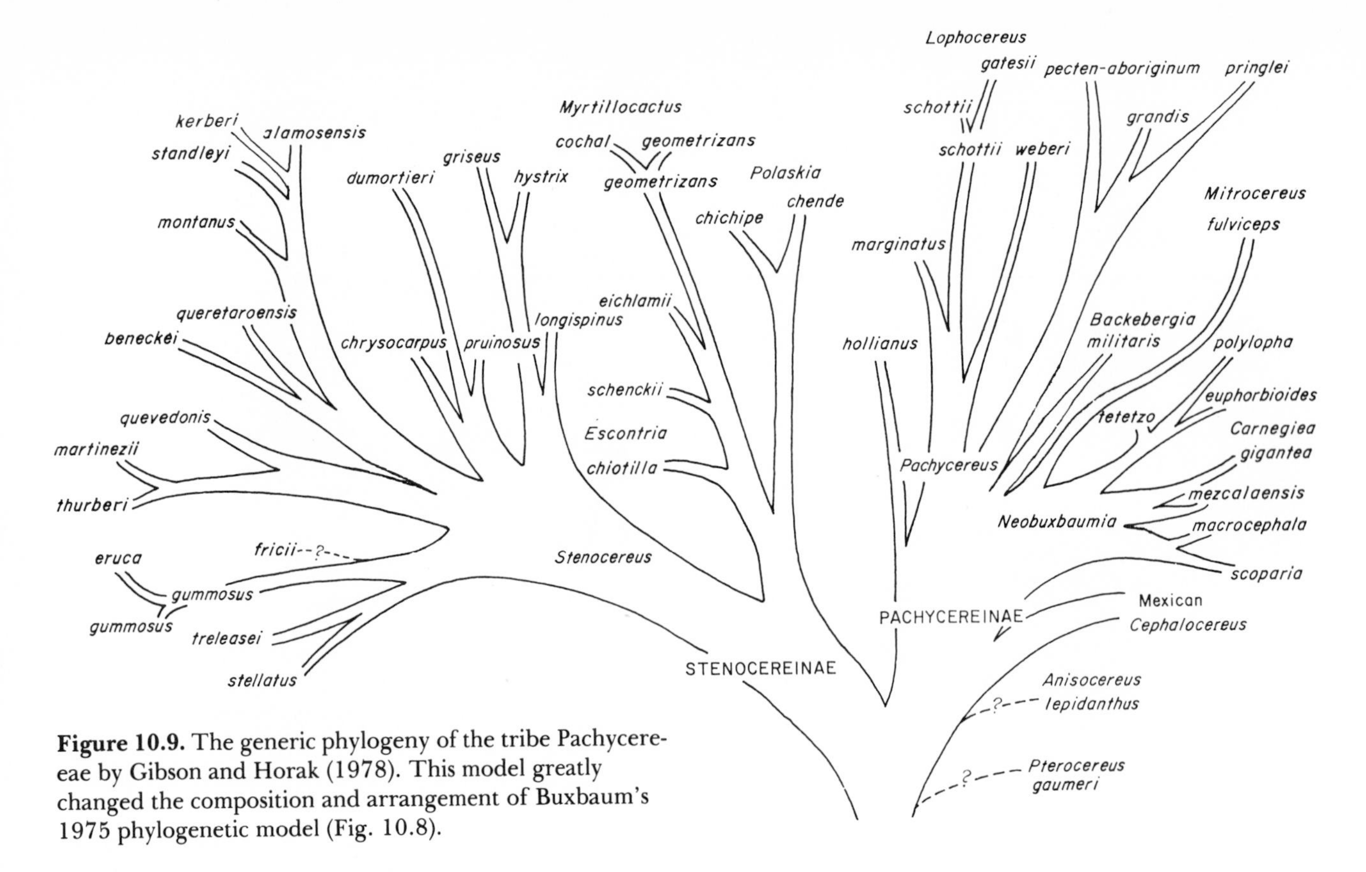

Figure 10.9. The generic phylogeny of the tribe Pachycereeae by Gibson and Horak (1978). This model greatly changed the composition and arrangement of Buxbaum's 1975 phylogenetic model (Fig. 10.8).

Britton and Rose taxa, *Lemaireocereus* and *Pachycereus,* were artificial or form genera. A form genus is, as its name implies, a category into which species with the same external form are placed but in which not all of the species are kin. *Lemaireocereus* was used for columnar cacti with spine clusters on the fruit. *Pachycereus* had been used by Britton and Rose for species with very thick stems, dense, brownish apical pubescence, and often long golden bristles at the apex and on the fruit (Chapter 5). Since Britton and Rose had made their revisions, a number of species had been transferred back and forth between *Lemaireocereus, Pachycereus,* and other genera segregated from them. Once the genus *Lemaireocereus* had been dismembered and reconstructed, the core of its species remained as the genus *Stenocereus,* augmented by *Machaerocereus* and *Rathbunia,* but the type species (the namesake of the genus), *L. hollianus,* which lacks silica bodies, triterpenes, and pearl cells, had to be removed from the subtribe. Also removed were those former species of *Lemaireocereus* that had abundant alkaloids. The genus *Pachycereus* became redefined as a genus rich in alkaloids and having large, glossy, black seeds, stems that blacken rapidly when cut, and many synapomorphies of stem anatomy, areoles, and flower structure.

The initial study by Gibson and Horak was not really a cladistic study because the strict methods of cladistics were not used, but the philosophy of using shared derived character states was successfully employed to analyze relationships. Since then, cladistic studies have been completed for Pachycereeae and lend strong support for most of the initial decisions; these studies also provide more precise models of speciation. A simple cladogram of the lineage for the rathbunias can be used to illustrate this latter point (Fig. 10.10). Three species, *Stenocereus standleyi* and the two former species of *Rathbunia, S. alamosensis* and *S. kerberi,* are very close relatives, and of these, *S. alamosensis* is a very specialized evolutionary product and ranges the farthest north (Sonoran Desert). The cladogram in Figure 10.10 indicates that *S. standleyi* and the ancestor of the rathbunias diverged first, and then the two rathbunias were produced from the common ancestor. *Stenocereus standleyi* and *S. kerberi* are morphologically very similar in most respects. The major difference is the following: the rathbunias have narrow, tubular, diurnal (day-opening) red flowers that are pollinated by hummingbirds (Fig. 1.34), whereas *S. standleyi* has funnelform, nocturnal (night-opening) pinkish white flowers that appear to be pollinated by nocturnal moths. The speciation event separating these two probably reflects selection for the two different forms by pollinators. The moth-pollinated species seems to have arisen along the coastline of western Mexico, whereas the hummingbird-pollinated species seem to have arisen

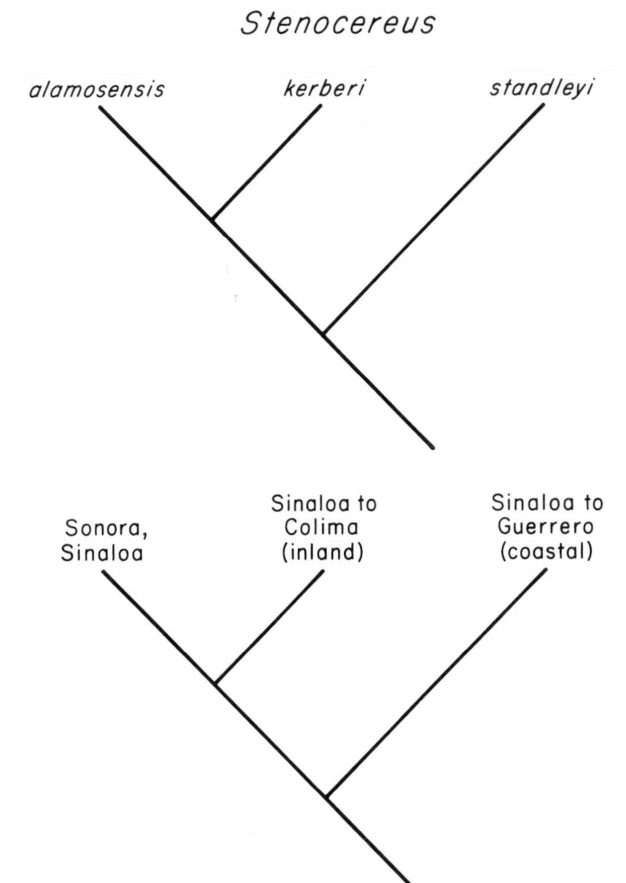

Figure 10.10. Cladogram of three Mexican species of *Stenocereus: S. standleyi, S. kerberi,* and *S. alamosensis.*

somewhat inland in the volcanic hills and mountains, where hummingbirds frequently are found. The final speciation event, producing two hummingbird-pollinated species, could have resulted from a subdivision or geographic isolation separating a desert-adapted species from a species initially inhabiting thorn forest.

Pollination Biology and the Evolution of Flower Structure

In the preceding section, we described a situation in which closest relatives have quite different flowers because they use different types of pollinators; this situation recurs again and again throughout the cactus family. Figure 10.2 shows the diversity of flower types in Pachycereeae, which is a splendid example of adaptive radiation of flower designs in a single country, Mexico. Each flower accommodates the specifications of its particular pollinator: if the pollinator is active during the day, a flower must open after dawn to use that pollinator; and if the pollinator to be used is active at night (for example, bats and many moths), then the flower must open after dusk.

The features in the following list are normally used to describe floral morphology and are strongly influenced by the type of pollinator.

Number of flowers per areole and per plant
Position and orientation of flowers
Time of anthesis (opening)
Length of floral tube to the nectary
Width and shape of the perianth – floral tube
Color of perianth parts
Spination and vesture (pubescence) at anthesis
Thickness of floral tube and thickness of perianth parts
Number of stamens
Length of filaments
Exsertion or insertion of anthers
Exsertion or insertion of stigma lobes
Number of stigma lobes
Size of nectary chamber
Degree of exposure of the nectary
Odor of flower

Each feature in this list can be modified to match the requirements of the pollinator, if the biological link between them is very strong. For example, a bat-pollinated flower typically has a wide entrance, sturdy walls and a strong style, copious pollen, a large, well-concealed nectary, nocturnal anthesis, and a fairly unpleasant odor. At the same time, a bat flower does not require a broad and colorful perianth to attract other animals, and the exposed perianth may be reduced so that other animals that need a landing platform or visual attraction are discouraged from using it.

The narrowly tubular flower, like that of the rathbunias which we discussed earlier and which is pollinated principally by hummingbirds, is another common design that appears throughout the cactus family. These flowers occur where hummingbirds can hover without being injured but in a way that prevents other potential pollinators from landing. Consequently, in the many unrelated species of hummingbird-pollinated cacti, such as in the platyopuntias (the nopaleas), the tropical epiphytes of *Nopalxochia,* the montane species of *Borzicactus* from South America, and the lowland desert rathbunias of North America, the length of the floral tube, the position of the flowers on the plant, and even the chemical nature of the nectar are remarkably similar. Likewise, those species pollinated by night-active hawkmoths, such as many of the epiphytes in *Epiphyllum,* the South American terrestrial cacti of *Echinopsis,* and the North American lowland species *Stenocereus gummosus* of the Sonoran Desert, have the very long and broad, sweetly scented, white flowers that appeal to these insects. Neither the flowers pollinated by hummingbirds nor the ones pollinated by hawkmoths have much development of spines at anthesis, lest their pollinator be injured by them.

Convergence of floral morphology to utilize bats, hummingbirds, nocturnal moths, and other animals has been a major contributor to the confusion of cactus classification, and in the past unrelated plants with convergent flower forms have been classified in the same taxon rather than in their appropriate tribes and subtribes. For example, Berger classified a number of unrelated plants as heliocerei because their flowers open during the daytime or as nyctocerei because their flowers open at night. When one temporarily ignores the features of the flower and concentrates on features that are really unchangeable and have no real role in pollination biology, then many systematic issues can be resolved. For example, in Middle America the terrestrial cacti with night-blooming flowers in the genus *Nyctocereus* are closely related to both the day-blooming species in *Heliocereus* that are pollinated by insects and the day-blooming species with red flowers in *Nopalxochia* and *Aporocactus* that are visited by hummingbirds.

The reason for using animals as pollinators should be clear but still deserves comment. First, to produce seeds, pollen grains from the anthers of one flower need to be transferred to the tip of a stigma on another flower of a different plant. This transfer can be accomplished by the wind, but there is a better chance of getting many pollen grains on the stigma of a single flower if the pollen grains are carried by an animal, especially one that visits many flowers of the same species. Having many pollen grains on a stigma permits these species to produce many seeds in a single fruit (one pollen grain is needed for each seed). The majority of animal-pollinated flowers produce many seeds, whereas most wind-pollinated species have only one or a few seeds per flower. Second, cacti generally are outcrossers. Evolutionists have repeatedly observed that such cross pollination is used by the majority of plant species, presumably because it insures that the offspring have maximum variability of genetic information. Cacti use animals to perform that transfer because they are relatively reliable and efficient, and the animal is rewarded with pollen or nectar or both. Inasmuch as the anthers and stigma lobes of most cactus flowers are fully exposed, probably very few species are pollinated by a single animal species; and now the wide-ranging honey bee, which is a native of the Old World, has intervened in many natural situations. Even where native bees are the principal pollinators, one often observes individuals of several different families of bees working on the same plant.

One way to ensure that cross pollination occurs is to have only male flowers on some individuals and only female flowers on others, a condition known as dioecy. Dioecy appears to be rare in cacti, but there are some species with partial separation of sexes. For example, in *Mammillaria dioica* (Fig. 5.22) from California and Baja California, there are plants that are functionally female because they are male sterile, that is, the male part is nonfunctional. In this case, the pollen grains appear to form within the anther, but the pollen cannot be released because a sheath of tissue surrounds the pollen and the thickened anther wall fails to split open. Species that have evolved from *M. dioica,* such as *M. estebanensis* and *M. multidigitata,* also share that characteristic. *Selenicereus innesii,* which was recently discovered on St. Vincent in the West Indies, is gynodioecious; that is, it has some flowers with functional male and female parts as well as some flowers with only female parts.

Self-pollination or autogamy (self-marriage), in which pollen from the anthers is directly transferred to the stigma of the same flower or the same plant, has been found a number of times in cacti, but there are very few studies that document the extent of self-pollination in cacti. The most extreme case is called cleistogamy (closed marriage), where the flower never opens. Examples of cleistogamy are found in several groups, especially those with very small flowers—most notably, in *Frailea* and *Melocactus.*

Also present in a great many cactus flowers are some wide-ranging species of *Carpophilus* beetles. For a long time cactologists assumed that these beetles were able to transfer pollen to the stigma lobes and thereby cause self-pollination. And because these beetles occur in many cactus flowers, they were assumed to be important pollinating agents. Verne Grant at the University of Texas and a co-worker, Walter A. Connell from the University of Delaware, showed, however, that these beetles are not effective pollinators because they do not visit the stigma lobes.

Important Characters for Phylogenetic Studies

Scattered throughout this book are examples of characters found in various parts of cacti that are potentially valuable characters for use in phylogenetic analyses. Some of these deserve special mention because they have already provided cactologists with powerful information for evaluating the evolutionary histories of cactus lineages.

Seeds. Some of the most elegant discoveries of cactus structure have been made on the seed, especially the ultrastructure of the seed coat. Following the studies by Buxbaum, who studied seeds with only a low-power dissecting microscope, observations using high-magnification photographs generated

Figures 10.11 – 10.14. Seed types of the subfamily Cactoideae.
10.11. *Notocactus succinus* (Notocacteae); a helmet-shaped seed with a very wide hilum. The rough seed coat consists of cells with bulging (convex) outer walls. ×40.
10.12. *Stenocereus stellatus* (Pachycereeae); a seed with a strongly indented hilum and a very rough seed coat. ×25.
10.13. *Blossfeldia liliputana* (Notocacteae); one of the smallest cactus seeds (less than 0.5 mm). The globose seed has a large strophiole *(left)*, and each cell of the seed coat produces a trichome, some of which have been shed. ×55.
10.14. *Stenocereus alamosensis;* a seed that is turned to show the hilum face-on *(upper right)*. The cells of this seed coat are flat or even somewhat concave. ×25.

from the scanning electron microscope (SEM) have revealed that many groups of cacti have distinguishing apomorphic seed features.

Even though SEM studies of cactus seeds have only been published since the early 1970s, enough is now known about seed ultrastructure to compile an atlas as large as this book. From this large database we have selected illustrations of seeds that demonstrate many of the features that have been found and are useful for phylogenetic analysis (Figs. 1.32*F*, 2.15, 10.11 – 10.32). Figures 10.11 through 10.14 show variations in seed shape, overall cellular texture of the seed coat (testa), position of the hilum (point of attachment to the funiculus), and nature of swellings in the region of the hilum (Figs. 10.11 – 10.14; the technical terms for describing these features can be found in the references listed at the end of the chapter). The shape of the seed is closely related to the shape of the embryo within the seed. Seeds with a prominent raphe (a backbonelike ridge; Figs. 10.13 and 10.15 – 10.20) tend to have relatively nonsucculent embryos, whereas those that are very swollen have a succulent hypocotyl (Figs. 3.48 and 3.49). For phylogenetic purposes some of the most important macroscopic features of the seeds are the swellings in the region of the hilum. For example, in the tiny, nearly spherical seeds of *Blossfeldia* (Fig. 10.13) we observe a large swelling of the upper funiculus; this corky structure (strophiole) also occurs in other species of tribe Notocacteae that are believed to be closely related to *Blossfeldia.* This species also has prominent hairs on the surface of the seed, one per cell (Figs. 10.13 and 10.32); and simple but shorter hairs, or papillae, are found on seeds of the closely related species of *Frailea* and in some species of *Notocactus.* Consequently, the occurrence of two apomorphic seed features in genera with many other striking similarities in plant structures is conclusive evidence of their close evolutionary relationship. Presence of a strophiole cannot be used to group species because there are species that clearly belong to other tribes, such as the Cacteae, that have a

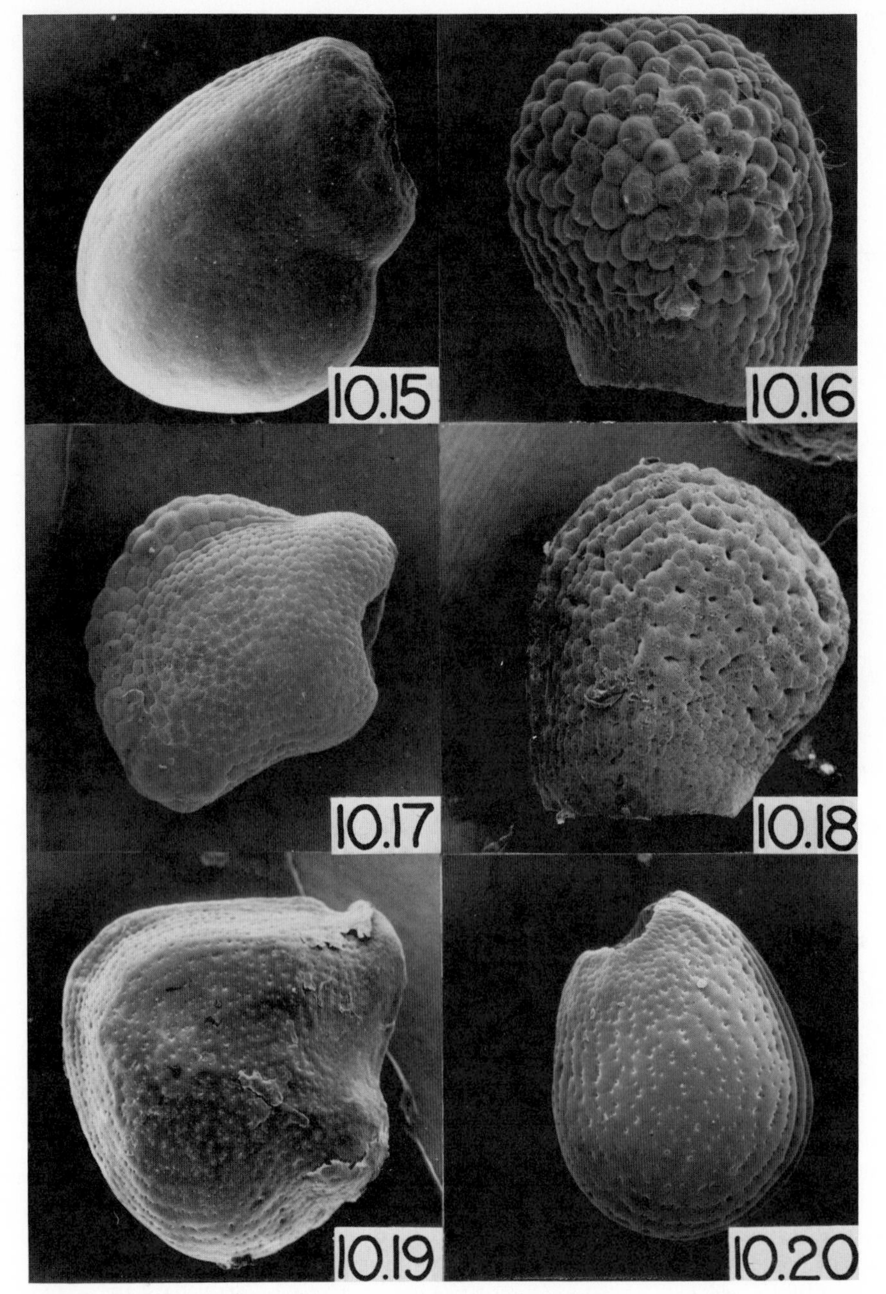

Figures 10.15–10.20. Seeds of the subfamily Cactoideae, showing a range of shapes and textures.
10.15. *Carnegiea gigantea* (Pachycereeae); the seed coat is fairly smooth and glossy. ×25.
10.16. *Echinocereus pentalophus* (Echinocereeae); this seed appears rough because all cells have strongly convex outer walls. ×30.
10.17. *Harrisia rogelii* (Hylocereeae); the seed has a conspicuous raphe, which has larger cells than on the rest of the seed coat. ×30.
10.18. *Polaskia chichipe* (Pachycereeae); the cells have convex outer walls, and between groups of cells deep pits are formed. ×30.
10.19. *Lemaireocereus hollianus* (Leptocereeae?); the seed is squarish, with a broad hilum, a prominent raphe, and minute pits on the surface. ×20. (From Gibson et al., 1985.)
10.20. *Cleistocactus areolatus* (Trichocereeae); an oval seed with a narrow hilum, longitudinal channels on the raphe, and a minutely pitted surface. ×25.

strophiole on the seed but are quite dissimilar in seed, stem, flower, and fruit structure to the Notocacteae.

Viewing the testa at high magnification, we find many other features that can be used for evaluating phylogenetic relationships (Figs. 10.21–10.31). Most noteworthy are the shapes of the cells and the nature of the cuticle covering the testa. We can readily see that many cactus seeds have minute cuticular (waxy) striations (striae) covering the surface of each cell and even crossing the boundaries between the cells. Some species have very fine and distinct striae (Figs. 10.21 and 10.22), some fine and low or indistinct (Figs. 10.27 and 10.28), some coarse and widely spaced (Figs. 10.23 and 10.25), and some distinct around the cell margins but indistinct in the center. Conversely, other species have seed surfaces with minute "bumps" that seem to arise below the cuticle (Fig. 10.29) or seed surfaces that are essentially smooth (Figs. 10.30 and 10.32).

Seed characters are now being widely used to make judgments about the placement of genera in the tribes of subfamily Cactoideae. In Stenocereinae, discussed earlier, the striated seed coat ultrastructure (Figs. 10.12, 10.18, 10.21, 10.23, 10.25, 10.27, and 10.28) is quite similar within the four genera of the subtribe and structurally quite different from the smooth seed coats found in subtribe Pachycereinae. Edward Anderson and co-workers at Whitman College have also used seed coat morphology as one of many apomorphic features in evaluating the generic composition and relationships within tribe Cacteae. The seeds of *Discocactus* closely resemble those of *Gymnocalycium,* an observation that helps us to justify classification of them in Notocacteae. Both of these genera have seeds that are quite unlike the seeds of *Melocactus,* which was formerly placed next to *Discocactus* because the cephalium of *Melocactus* superficially resembles the floriferous shoot tip of *Discocactus.* (Chapter 5). Instead, *Melocactus* has seeds that are quite similar to those of *Buiningia,* which is a genus of tribe Cereeae.

Pollen grain morphology. A number of workers have examined cactus pollen and described its structure (Figs. 1.33*B* and 10.33). Beat Leuenberger and Werner Rauh at the University of Heidelberg examined the ultrastructure of pollen in Cactaceae. Leuenberger found that a single basic type of pollen grains appears in all cacti. Specifically, a cactus pollen grain is roughly spherical with an outer wall (exine) covered with small projections called spinules and depressions called punctae. The spinules may be high to low, equilateral to isolateral in shape, and very abundant to very sparse or absent; the punctae may be wide to narrow and may also be surrounded by a rim. Many species have three elongate apertures, called colpae (Figs. 1.33*B* and 10.33*B*), which are thin areas through which the pollen grain may germinate; but to confuse matters, species have 9, 12, 15, 18, or more colpae.

The study of pollen grains showed no correlation between the size of the pollen grain or the nature of the outer wall sculpturing and the type of pollinator. Also, the study did not find quantitative ways to use pollen grain morphology for phylogenetic reconstructions. Leuenberger was able to show that the variations in pollen grain structure are fairly uniform in each tribe (recognized by Buxbaum), and on this basis some genera that have been misassigned to a tribe could be identified if the pollen grain structure were inconsistent with that of the rest of the tribe.

Vegetative features. In addition to the studies by Bailey on the leaf-bearing cacti and by Gibson and co-workers on Pachycereeae, there have been a number of recent studies that have used vegetative synapomorphies, such as crystals, sclereids, mucilage structure, skin structure, and so forth, in evaluating the relationships of species. The distribution patterns of crystals and of mucilage cells and laticifers have shown the greatest promise for future study.

Around the same time that Leuenberger was studying pollen, he, Wilhelm Barthlott, and other workers at the University of Heidelberg were also doing ultrastructural studies of other cactus features, such as spines (Chapter 5) and epidermis. For spines, several features have been used as synapomorphies for systematic purposes. For example, the spines of Cacteae have prominent papillae or trichomes, whereas those of many Notocacteae have small apical projections on the cells. The downward-pointing cell bases on the spines and glochids of Opuntioideae, when combined with the synapomorphic features of pollen grains with many apertures and large holes in the exine and an aril on the seed, provide a clear way to define the subfamily and show that it has been a separate evolutionary lineage from the other two subfamilies.

Cactus systematics and phylogeny have come a long way since Linnaeus, and they still have a long way to go. Within the last 10 years we have seen a major shift in the systematics of these organisms because new characters are being used for analysis. The features that cactologists use to identify and name species are those that are easy to see with a low-magnification lens; but experience has now taught us that many of the external features, such as growth habit, ribbing, spination, and flower shape, are insufficient by themselves for determining phylogenetic relationships. In fact, many of the traditional features are very misleading because each of these structures has been evolving in parallel ways in many parts of the cactus family. Features that

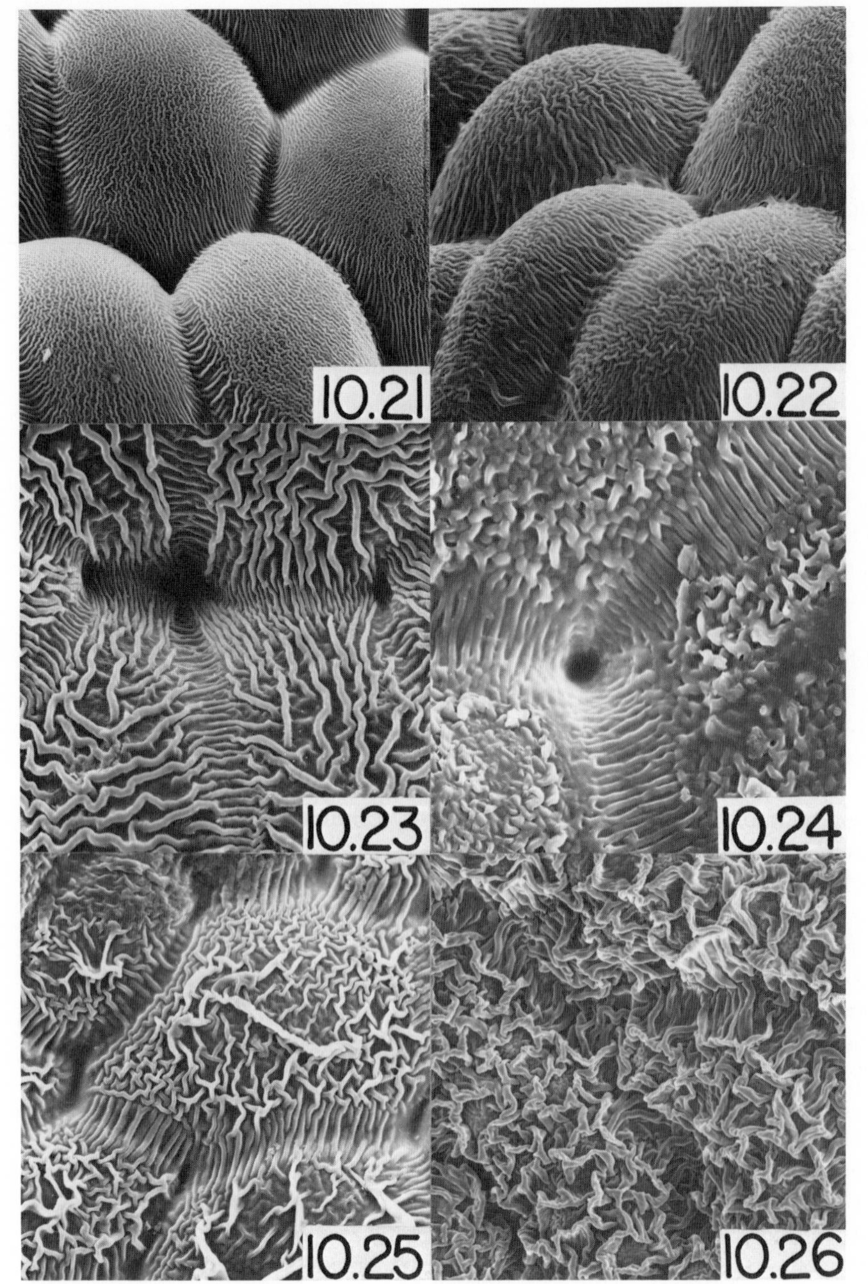

Figures 10.21–10.26. Cuticular striations on the seed coat.
10.21. *Stenocereus quevedonis* (Pachycereeae); numerous, fine striae cover each bulging cell of the seed coat. ×530. (From Gibson et al., 1985.)
10.22. *Notocactus succinus* (Notocacteae); another example of a seed coat with numerous, fine striae on each cell of the seed (see Fig. 10.11). ×400.
10.23. *Polaskia chichipe* (Pachycereeae); close-up of the coarse cuticular striations and deep intercellular pits (see Fig. 10.18). ×1000. (From Gibson et al., 1985.)
10.24. *Harrisia rogelii* (Hylocereeae); close-up of the interconnected cuticular deposits of the seed coat (see Fig. 10.17). Notice the deep pit at the junction of three cells and the long straight striae that cross the margins of the cells. ×1000.
10.25. *Stenocereus treleasei* (Pachycereeae); the coarse, branched cuticular deposits cover each cell, and long, unbranched striae cross the margins of the cells. ×530.
10.26. *Neoporteria* sp. (*Horridocactus,* Notocacteae); a seed coat with a very rough surface formed by irregularly branched cuticular deposits. ×200.

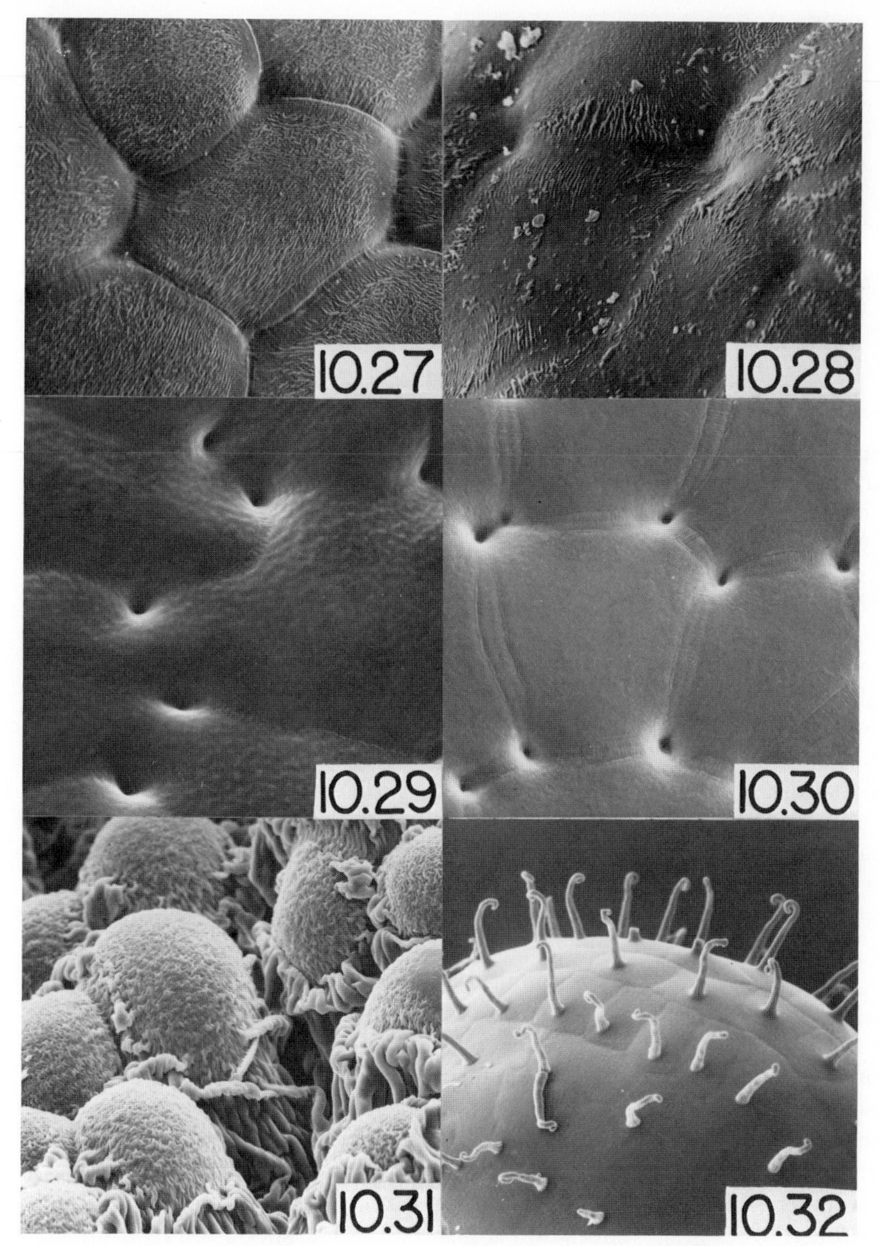

Figures 10.27–10.32. Surfaces of cactus seeds.
10.27 and 10.28. *Stenocereus thurberi* (Pachycereeae); very fine, low striae cover each cell in a seed from near Ciudad Obregon, Sonora, Mexico (Fig. 10.27), but the striae are less obvious on some of the seeds (Fig. 10.28). ×520.
10.29. *Pachycereus marginatus* var. *marginatus* (Pachycereeae); a columnar cactus from Hidalgo, Mexico that has fairly smooth seeds with minute bumps under the surface and deep pits at the junctions of the cells. ×520.
10.30. *Cleistocactus areolatus* (Trichocereeae); another fairly smooth seed coat (Fig. 10.20) with interesting boundaries between the cells and deep pits at the junctions of the cells. ×400.
10.31. *Neoporteria taltalensis* (Notocacteae); an interesting seed coat with low, close, and highly interconnected cuticular deposits on the dome of each cell and very thick, coarse striae that descend from the dome like a skirt. ×585.
10.32. *Blossfeldia liliputana* (Notocacteae); close-up of one of the most photographed cactus seeds (see Fig. 10.13), which has a single-celled trichome that arises from the cell; each trichome has a crozier at the tip. One of the cells has a pair of trichomes, and a few of the cells lack trichomes. ×200.

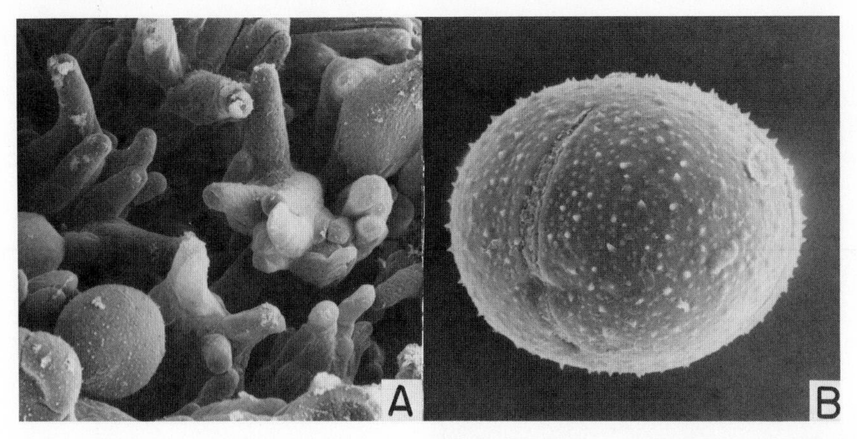

Figure 10.33. Scanning electron photomicrographs of pollen grains of *Stenocereus gummosus,* or agria. ***(A)*** Pollen grain *(lower left),* 0.08 mm in diameter attached to the fingerlike cells (papillae) of the stigma. ***(B)*** Pollen grain in its natural condition (fixed in glutaraldehyde), showing one of the three slitlike furrows (apertures) through which germination can occur and numerous pointed spicules. Compare this figure with Figure 1.33*B*, which was drawn from a pollen grain treated with strong chemicals to remove all soft material and thereby expose the fine structure of the wall.

seem to reveal more about phylogeny, such as the fine features of the stem, the surface sculpturing of a seed, and the chemistry of the stem, must be studied with special techniques or special microscopes. Very careful analytical studies of apomorphic features, combined with studies on the pollination biology, will give us much of the information we need to reconstruct phylogenies of these organisms.

Speciation, Hybridization, and Polyploidy

Speciation is the process by which one population of organisms evolves into two distinct species. A subset of the original species becomes genetically and ecologically or geographically separated from the ancestral populations and becomes an independent evolutionary branch. Through this process of multiplication and diversification, many millions of species have been formed; but over the same period many millions of species have become extinct, leaving incredible gaps in our knowledge of the species. More often than not, information on the origin of the species is fragmentary or nonexistent because scientists cannot identify the ancestors of present-day species.

Modes of Plant Speciation

Many researchers have attempted to explain how one species splits to form two species. In the process they have discovered that speciation can occur via several different pathways, or modes, by which the newly evolving species becomes genetically different from the parental population and then proceeds on its own evolutionary path.

One common mode of speciation is called allopatric speciation by geographic subdivision, or simply geographic speciation. In this case, an ancestral population becomes subdivided into two or more geographic regions, for example, by the formation of a mountain range that separates one population of a species from the rest of the species. The term *allopatric* refers to the condition in which the organisms occur in geographically different places. When a portion of the original population is isolated from the related populations, it may develop genetic differences. If the genetic differences are great enough, then the isolated population may form a new species. Various models have been proposed to show how the isolated species genetically changes and how it can become distinct from the original species. A special version of this model of geographic speciation is called allopatric speciation by disjunct founders. In this case, one or several individual plants, usually seeds, are transported long distances and become established in a geographically isolated location, where the new population may evolve and become a new species. The third common mode of plant speciation, called sympatric speciation by polyploidy, occurs when individuals in a population acquire additional sets of chromosomes that prevent them from mating with plants in the original population; thus, these polyploids can become a new evolutionary branch even though they are growing right next to plants of the original population.

In our discussion of cactus phylogeny and in particular the phylogeny of the rathbunias, we invoked a model of geographic speciation. In this example, plants growing in one geographic region used one type of pollinator and adapted to its specifications whereas those growing in another geographic region used a different type of pollinator and adapted to a different set of specifications. Likewise, plants isolated in desert habitats developed ways of coping with drier and hotter environmental conditions.

When the distributions of cacti are carefully recorded, it becomes clear that closest relatives typically are geographically separated, often by hundreds or even thousands of kilometers; therefore, it is probable that many, if not most, of the speciation events in the Cactaceae occurred in allopatry, when part of the ancestral population was separated from the rest of its relatives.

Sources of Variability in Cactus Species

Individual species of cacti often exhibit a wide range of morphological variability, some of which is related to environmental gradients. In Chapter 5 we showed that the northern populations of saguaro have denser spine coverage at the apex than do southern populations and that the denser coverage helps prevent lethal cold temperatures at the shoot tip. In Chapter 6 we observed that number and shape of ribs can influence PAR interception, and in Chapter 7 we discussed cold-tolerant populations of *Coryphantha vivipara* that inhabit cool temperate areas of the northwestern United States. Finally, in Chapter 8 we reviewed how tree opuntias in the Galápagos Archipelago have different growth habits on different islands, differences that relate to differences in the height of the surrounding vegetation. The obvious interpretation of these observations is that natural selection has operated on local populations to "fine tune" the heritable features, helping individuals in the population to adapt to local environmental conditions. At the same time, the environment has played an important role in modifying plant architecture in local populations. For example, the environment can control the growth habits of platyopuntias and columnar cacti growing on different slopes or the juvenile and adult shoot designs of epiphytes.

Minor variations of spination and flower color can occur within species. For example, in Chapter 5 we introduced *Opuntia violacea* to show that there can be varieties, forms, and races within a species that have an average set of spine features different from spine features found in other populations. These variations in spination are "collectibles." The minor variations, which may not influence the plant's tolerance to high or low temperatures, often wind up in cactus nurseries with or without special latinized names. Flower color can also vary greatly from one isolated population to another, as it does, for example, in the small, globular cacti of the Chihuahuan Desert, where a single variety of cactus may have three or four different floral colors, even though the same pollinators may be used. Local distinctiveness of populations can be caused by inbreeding, which is the local exchange of genetic materials between very closely related plants and which permits a certain group of structural features to become fixed, or uniform, throughout the local population. Hence, some of the variability in a species is caused by the geographic isolation of individual populations, because in that case there is no genetic exchange with other plants in the species.

Although variability is very great in some species of cacti, it cannot be easily related to environmental gradients or spatial isolation. Some species, especially platyopuntias, are so highly variable throughout their range that they appear to be experiencing major changes in their genetic makeup. Recent research has shown that this is indeed the case. Some of the major changes occurring in cacti are those occurring in the number of chromosomes. Most species of cacti are diploid and have 11 pairs, or sets, of chromosomes (=22 chromosomes)—humans have 23 pairs (=46 chromosomes). Other species are polyploid; that is, they have more than two sets of chromosomes. For example, some cacti have 44 chromosomes (tetraploid) or 66 chromosomes (hexaploid). Variability also occurs because individuals from different species are able to exchange genetic material and produce hybrids, which have some characteristics of each parent. In many cases, when hybrids are formed between individuals of two different species, the offspring are polyploids; and this arrangement can lead to new species.

Hybrid Platyopuntias of Southern California

Although hybridization of cacti has been recognized throughout this century, studies of hybridization in natural populations only really got started in the 1960s. The catalyst for modern studies was a study by David L. Walkington at Rancho Santa Ana Botanic Garden and Lyman Benson at Pomona College in which they sorted out the complex and very recent history of hybridization in the platyopuntias of southern California. They determined by structural and chemical studies that plants of "*Opuntia occidentalis*" (Figs. 10.34 and 10.36), which inhabit open and disturbed hillsides in southwestern California, are really hybrids. *Opuntia occidentalis* arose from natural crosses between the native platyopuntias, mostly *O. littoralis* var. *vaseyi* (Fig. 8.40), and a spiny form of *O. ficus-indica* called "*Opuntia megacantha*" (Figs. 10.34 and 10.35). The hybrid looks somewhat like each parent.

When two plant species are crossed, the offspring, which are called the first filial or F_1 generation, typically have characteristics that are intermediate between those of the two parents. For example, crossing a plant with long spines and one with

Figures 10.34–10.36. Hybridization in platyopuntias of southern California.
10.34. Population of platyopuntias near Moorpark, Ventura County, consisting of thick clumps of the hybrid *"Opuntia occidentalis"* and two plants of the tall parent, *O. megacantha (in foreground).*
10.35. *Opuntia megacantha,* a spiny form of the mission fig, *O. ficus-indica,* that occurs in native vegetation near Moorpark, Ventura County.
10.36. Site of a recent fire through chaparral vegetation, in which one of the hybrid clumps of *"O. occidentalis"* was badly burned on the edge but survived in the center. A large clump of the hybrid platyopuntia also survived *(upper right-hand corner).*

extremely short spines may produce offspring with spines of intermediate length. Likewise, an intermediate characteristic would result from crosses between species with large and small cladodes and fruits. When the individuals of the F_1 generation become old enough to produce flowers, new crosses will occur between the hybrids and the parents. Some crosses will be between the hybrid F_1 plants ($F_1 \times F_1$), some between a hybrid and one parent ($F_1 \times P_A$), and some between a hybrid and the other parent ($F_1 \times P_B$). In successive years each plant can breed with either parent or a variety of hybrids, and the end result is a highly variable population called a hybrid swarm.

Opuntia occidentalis is a hybrid swarm resulting from crosses between *O. littoralis* and spiny *O. ficus-indica* at sites where the two parents came into contact. These hybrid swarms arose independently many times in different locations throughout southern California, and especially around the old missions and ranches. The spineless and spiny forms of Indian fig, *O. ficus-indica,* were brought to southern California only after the Franciscan fathers founded the missions in the late 1700s. The cladodes *(nopales)* and fruits *(tunas)* were used as food, and the mucilage of these platyopuntias was also used as a binding material for adobe bricks used in building the missions and homes.

Hybrid swarms occur in many plant species; but in the majority of them, the hybrid populations are restricted in distribution and often disappear soon after they are formed because the hybrids do not grow as well as the parents in native habitats. However, for *O. occidentalis* the reverse is true because the hybrid populations have become more successful than the parents in southern California. In this region, which has chaparral vegetation that is frequently burned by fires, the native species cannot invade chaparral because they are intolerant of fires. *Opuntia occidentalis* can survive fires (Fig. 10.36) because it forms dense thickets, via vegetative propagation of the cladodes. A thicket may be over 2000 m² (about ½ acre) in size, and so the edges but not the center will be burned. Consequently, hybrid populations, which now reproduce largely by vegetative means (apomixis), are spreading in areas where fires have eliminated the parental species.

Cholla Hybrids in Arizona

Chollas are common shrubs of many plant communities in warm, dry areas of the western United States and Mexico. In some habitats only a single species occurs, whereas in others three or four species may coexist on the same hillside. The factors controlling the regional and local distribution of each species have not been studied very much; but a consequence of living on the same hillside is that species flowering at the same time may produce hybrid offspring. So far, hybridization of chollas have been found in several areas, but the best-studied examples occur in Arizona.

Two widespread, shrubby species common in Arizona are the jumping cholla or chain-fruited cholla, *O. fulgida* (Fig. 10.37*A*), a hexaploid (66 chromosomes) that is very abundant in low-elevation desert habitats of the Sonoran Desert, and the cane cholla, *O. spinosior* (Fig. 10.38*A*), a diploid (22 chromosomes) that characteristically occurs in hill or mountain habitats such as chaparral, pinyon–juniper woodland, and desert grassland. *Opuntia fulgida* is a large shrub that has a main trunk and numerous heavy branches. The most distinctive features of this species are the smooth, green, pear-shaped fruits, which form in pendent chains of three or more (Fig. 10.37*B*). In contrast, *O. spinosior* is a smaller, more densely branched species with solitary, yellow, cone-shaped fruits that are covered with very prominent tubercles (Fig. 10.38*B*). Occasionally the two species occur in the same habitat; consequently, hybridization can occur. The hybrids are intermediate in growth habit and vegetative features, and they have moderately tuberculate, green, olive-shaped fruits that are borne either singly or in short chains of only two fruits (Fig. 10.39).

In 1971 Verne and Karen Grant at the University of Texas described one of these hybrid populations from Pinal County, southeast of Phoenix, Arizona, as a new clonal microspecies, *Opuntia kelvinensis.* A clonal microspecies is defined as a plant population that has a definite and narrow geographic range and that reproduces mostly, if not exclusively, by vegetative propagation. In the case of *O. kelvinensis,* which commonly occurs with *O. fulgida* and occasionally with *O. spinosior,* the Grants determined that well-established populations of these hybrids around Kelvin and Sacaton, Arizona, usually have no seeds in the fruits (rarely one or two seeds) and very little viable pollen. Because the plant reproduces by rooting of joints and fruits, this microspecies can be very abundant locally.

Intergeneric hybrids between *Opuntia fulgida* and *O. spinosior* were also studied by Donald J. Pinkava and Lyle A. McGill at Arizona State University, using techniques to study chromosomes. Because typical plants of jumping cholla are hexaploid and typical cane cholla are diploid, when hybrids are formed between these two species, some peculiar genetic combinations result. For example, in a single population we find hexaploids in which either one-third or two-thirds of the genetic information comes from the diploid *O. spinosior* and diploids in which one-half of the genetic information comes from the hexaploid *O. fulgida;* we may also find triploids and tetraploids in the same population. Consequently, if scientists are going to give Latin binomials to hybrids that represent microspecies, they must be certain that the microspecies is discrete both morphologically and cytogenetically.

Opuntia spinosior is also involved in the formation of natural hybrids with other shrubby chollas, such as *O. imbricata, O. acanthocarpa,* and *O. versicolor.* For example, the Grants determined that near Tucson *O. spinosior* × *O. versicolor* has produced hybrid swarms in which the plants reproduce by seeds (sexually) instead of by asexual, vegetative propagation. These two species are able to form fertile offspring because they are both diploids; thus, they have no problems

Figures 10.37–10.39. Cholla hybrids in central Arizona.
10.37. *Opuntia fulgida,* jumping cholla.
(A) Growth habit with many chains of fruit.
(B) Close-up of the long chains of green, pear-shaped fruits.
10.38. *Opuntia spinosior,* cane cholla.
(A) View of the stems.
(B) Close-up of a solitary, yellow, tuberculate fruit.
10.39. *Opuntia kelvinensis,* a plant collected near Kelvey, Arizona.
(A) General nature of the plant, which is fairly similar to *O. fulgida.*
(B) Close-up of the short chains of green, olive-shaped fruits.

of mismatched chromosome number, and the same insects visit and pollinate the flowers of both species. Normally, hybrids swarms of *O. spinosior* × *O. versicolor* do not form because the parental species inhabit different plant communities. Of course, as more cholla populations are studied where two or more species are sympatric, more hybrid populations will be discovered.

Opuntia phaeacantha and Its Relatives

Opuntia phaeacantha (Fig. 10.40) occurs from South Dakota in the north to Chihuahua, Mexico, in the south and from California to eastern Texas. This wide-ranging species and its closest relatives are an interesting and complex taxonomic unit that can be used to illustrate how the various aspects of hybridization and polyploidy have created variability in platyopuntias. The seven species in this group all have fleshy fruits and are closely related: *O. littoralis* (Fig. 8.40), mentioned earlier as an important plant of southern California; *O. martiniana* of northwestern Arizona; *O. curvospina* of southern Nevada and northwestern Arizona; *O. chlorotica* (Fig. 8.38), the widespread pancake pear of semiarid regions from the southern United States and northern Mexico; *O. lindheimeri,* another widespread and variable species; *O. edwardsii* from the Edwards Plateau of Texas; and *Opuntia phaeacantha,* which is the most variable species and has three well-marked varieties, one of

Figure 10.40. Low *(A)* and erect *(B)* forms of *Opuntia phaeacantha* from western Texas.

which, *discata*, the Engelmann prickly pear, is sometimes treated as a separate species.

This group includes diploid, tetraploid, and hexaploid species and even species in which different chromosome numbers occur within the species. *Opuntia chlorotica* is a highly fertile, diploid species ($2n = 22$) that reproduces in the wild by seeds. *Opuntia martiniana, O. edwardsii,* and *O. curvospina* are tetraploid species ($4n = 44$). *Opuntia phaeacantha* is generally hexaploid ($6n = 66$) with occasional tetraploid populations, whereas *O. lindheimeri* has numerous tetraploid and hexaploid populations with one report of a diploid population on the coastline of the Gulf of Mexico. Bruce D. Parfitt and Pinkava at Arizona State University, determined that the tetraploid *O. curvospina* is a morphological intermediate and undoubtedly arose as an interspecific hybrid between the diploid *O. chlorotica* and the hexaploid *O. phaeacantha* var. *major,* both of which are sympatric with *O. curvospina.*

Malcomb G. McLeod and Pinkava analyzed the morphology and cytology of hexaploid *Opuntia phaeacantha* in the Southwest and came to the conclusion that this species probably arose independently in a number of places via interspecific hybridization between *O. chlorotica* as one parent and any of several other species of platyopuntias for the other parent. This multiple origin of *O. phaeacantha* varieties and populations helps to explain why there are both tetraploid and hexaploid races in the species.

Within Texas, hybridization between species has produced numerous hybrid populations, including hybrid swarms, a number of microspecies, and situations where some genetic features are being transferred from one species into another. The Grants have identified numerous hybrid populations, especially hybrid swarms of *O. edwardsii* × *O. phaeacantha* var. *major;* and in six of these populations there has also been some additional hybridization of the *edwardsii-phaeacantha* hybrid with *O. lindheimeri,* a cross that has resulted in offspring known as trihybrids. They have also found several clonal microspecies that are very similar to *O. lindheimeri* and extremely limited in distribution. Two of these occur on Enchanted Rock on the Edwards Plateau; one lives around vernal pools and in rock crevices on the broad top of Enchanted Rock in an area of about 330 m × 150 m, whereas the other microspecies occurs only in liveoak savannah in an area about 330 m × 120 m. Verne and Karen Grant speculated that each population probably arose from a single cladode following some hybridization event. Apomixis — in this case reproduction by cladodes — has been a wonderful way to preserve evolutionary experiments in the Cactaceae.

From the preceding discussion on hybrid platyopuntias, we can see a very complex and dynamic picture of speciation in Cactaceae. A lineage starting as a diploid gives rise to tetraploid and hexaploid races and species. Any of these, if they flower at the same time, can exchange pollen and produce diploid or polyploid hybrids. If a hybrid is a sterile triploid, produced by the fusion of a haploid (from a diploid) and a diploid (from a tetraploid or diploid) sex cell, somewhere in the population there may be a spontaneous doubling of the chromosomes to form a hexaploid plant, which can multiply by vegetative means. The hexaploid individual can then form sexual offspring in two ways when it forms triploid (half of hexaploid) pollen grains and eggs. First, the triploid pollen grain can fuse with a triploid egg to form a new hexaploid plant. Second, a triploid pollen grain or egg can fuse with a haploid sex cell (half of diploid) from a diploid plant, thus producing a tetraploid offspring. Whenever a plant can propagate vegetatively, any of these may persist in a population, even if the individual is sterile.

How many platyopuntia species groups have arisen and diversified following the preceding scenario? At this point, a few records exist of polyploidy in other platyopuntias, but much of the relevant genetic and structural information is lacking. Especially in Mexico, which is rich in species, we desperately need extensive cytogenetic information to work out the phylogeny of the species. Some of the species groups probably arose via hybridization and polyploidy, whereas others probably were produced by geographic speciation events.

Polyploidy in Cactoideae

Polyploidy and interspecific hybridization have been observed much less frequently in Cactoideae, where the vast majority of species examined to date are diploids ($2n = 22$). Occasional chromosome counts of 18 and 24 have been found in diploid cacti. Tetraploids have been found in about 15 genera of Cactoideae, although they are apparently infrequent. Tetraploids are relatively common in *Echinocereus* and in *Mammillaria,* which not only has tetraploids but also the highest chromosome count, $24n$ (24 sets!). *Blossfeldia liliputana* appears to be a hexaploid. At this time it appears that polyploidy has not substantially influenced speciation in this subfamily.

An interesting story on interspecific — actually intergeneric — hybridization has been described from Baja California, where the tetraploid *Bergerocactus emoryi,* a low shrub with hundreds of slender stems, has formed natural hybrids with the small arborescent diploid *Myrtillocactus cochal,* yielding a triploid, ×*Myrtgerocactus lindsayi.* The hybrid has been found only a few times and is intermediate in features between the two parents. *Bergerocactus* has also formed natural hybrids with cardón, *Pachycereus pringlei*, another tetraploid and the largest columnar cactus of that region, to yield an intergeneric hybrid called ×*Pachgerocereus orcuttii*. Both of these strange plants appear to be evolutionary dead ends because they are sterile. A few other naturally occurring interspecific hybrids are suspected, but they are considered to be quite rare, such as crosses between *Stenocereus dumortieri* and *Myrtillocactus geometrizans* in central Mexico and *Espostoa* and *Haageocereus* in Peru. Hybridization may also be a factor in the variations that occur in some of the small cacti of South America in Notocacteae.

The irony of all this is that people who cultivate cacti have made many successful crosses between diploid species within a genus and even between species of distantly related genera in Cactaceae. This is truly remarkable, because in most plant families intergeneric hybrids have never been obtained. The ease with which cacti can cross shows that there are few genetic barriers between cactus species that would limit the exchange of genes.

Intergeneric hybrids are probably rare in nature because two species that may live together often do not flower at the same time and do not use the same pollinators. There have been several suggestions in the literature that "strange" combinations of features found in a species of a genus may have been produced via intrageneric or intergeneric hybridization. Some of this could have been said about the species of *Lemaireocereus* described earlier in this chapter, which appeared to be a mixture of characteristics of several genera. However, when phylogenetic studies revealed the real affinities of each species, the need to invoke intergeneric hybridization vanished. Consequently, we must work out the phylogenetic relationships of genera by applying appropriate methods before we can conclude with certainty anything about the role of hybridization in the diversification of Cactoideae.

SELECTED BIBLIOGRAPHY

Anderson, E. F., and N. H. Boke. 1969. The genus *Pelycyphora* (Cactaceae): resolution of a controversy. *American Journal of Botany* 56:314–326.

Anderson, E. F., and S. M. Skillman. 1984. A comparison of *Aztekium* and *Strombocactus*. *Systematic Botany* 9:42–49.

Barthlott, W. 1981. Epidermal and seed surface characters of plants: systematic applicability and some evolutionary aspects. *Nordic Journal of Botany* 1:345–355.

Barthlott, W., and G. Voit. 1979. Mikromorphologie der Samenschalen und Taxonomie der Cactaceae: ein raster-elektronenmikroscopischer Überblick. *Plant Systematics and Evolution* 132:205–229.

Bemis, W. P., J. W. Berry, and A. J. Deutschman, Jr. 1972. Observations on male sterile mammillariae. *Cactus and Succulent Journal* (Los Angeles) 44:256.

Benson, L. 1982. *The cacti of the United States and Canada.* Stanford University Press, Stanford.

Benson, L., and D. L. Walkington. 1965. The southern Californian prickly pears—invasion, adulteration, and trial-by-fire. *Annals of the Missouri Botanical Garden* 52:262–273.

Berger, A. 1905. A systematic revision of the genus *Cereus*. *Report of the Missouri Botanical Garden* 16:57–86.

——— 1929. *Kakteen.* Verlag von Eugen Ulmer, Stuttgart.

Britton, N. L., and J. N. Rose. 1909. The genus *Cereus* and its allies in North America. *Contributions of the U.S. National Herbarium* 12:413–437.

——— 1919–1923. *The Cactaceae.* 4 vols. Publications of the Carnegie Institute of Washington no. 248, Washington, D.C.

Brown, J. H., and A. C. Gibson. 1983. *Biogeography.* C. V. Mosby, St. Louis.

Buxbaum, F. 1958. The phylogenetic division of the subfamily Cereoideae, Cactaceae. *Madroño* 14:177–206.

——— 1968. The phylogenetic position of the genus *Machaerocerus* Britton et Rose. *Cactus and Succulent Journal* (Los Angeles) 40:195–199.

Cracraft, J. 1974. Phylogenetic models and classification. *Systematic Zoology* 23:71–90.

Faegri, K., and L. van der Pijl. 1979. *The principles of pollination ecology,* 3rd ed. Pergamon Press, New York.

Gibson, A. C. 1982. Phylogenetic relationships of Pachycereeae. Pp. 3–16, in *Ecological genetics and evolution: The cactus-yeast-*Drosophila *model system*, ed. J. S. F. Barker and W. T. Starmer. Academic Press, Sydney.

Gibson, A. C., and K. E. Horak. 1978. Systematic anatomy and phylogeny of Mexican columnar cacti. *Annals of the Missouri Botanical Garden* 65:999–1057.

Gibson, A. C., K. C. Spencer, R. Bajaj, and J. L. McLaughlin. 1985. The ever-changing landscape of cactus systematics. *Annals of the Missouri Botanical Garden*. In press.

Grant, V. 1979. Character coherence in natural hybrid populations in plants. *Botanical Gazette* 140:443–448.

Grant, V., and W. A. Connell. 1979. The association between *Carpophilus* beetles and cactus flowers. *Plant Systematics and Evolution* 133:99–102.

Grant, V., and K. A. Grant. 1971a. Dynamics of clonal microspecies in cholla cactus. *Evolution* 25:144–155.

——— 1971b. Natural hybridization between the cholla cactus species *Opuntia spinosior* and *Opuntia versicolor*. *Proceedings of the National Academy of Sciences, U.S.A.* 68:1993–1995.

——— 1979a. The pollination spectrum in the southwestern American cactus flora. *Plant Systematics and Evolution* 132:29–37.

——— 1979b. Pollination of *Echinocereus fasciculatus* and *Ferocactus wislizenii*. *Plant Systematics and Evolution* 132:85–90.

——— 1979c. Systematics of the *Opuntia phaeacantha* group in Texas. *Botanical Gazette* 140:199–207.

——— 1979d. Hybridization and variation in the *Opuntia phaeacantha* group in central Texas. *Botanical Gazette* 140:208–215.

——— 1979e. Pollination of *Opuntia basilaris* and *O. littoralis*. *Plant Systematics and Evolution* 132:321–325.
——— 1980. Clonal microspecies of hybrid origin in the *Opuntia lindheimeri* group. *Botanical Gazette* 141:101–106.
Grant, V., K. A. Grant, and P. D. Hurd, Jr. 1979. Pollination of *Opuntia lindheimeri* and related species. *Plant Systematics and Evolution* 132:313–320.
Grant, V., and P. D. Hurd, Jr. 1979. Pollination of the southwestern opuntias. *Plant Systematics and Evolution* 133:15–28.
Gregory, M. 1981. References to scanning electron microscope photographs of seeds of Cactaceae. *Cactus and Succulent Journal* (London) 43:114–116.
Jarvis, C. E. 1981. Pollen morphology in the subtribe Borzicactinae F. Buxb. (Cactaceae). *Cactus and Succulent Journal* (London) 43:109–113.
Kurtz, E. B., Jr. 1963. Pollen morphology of the Cactaceae. *Grana Palynologica* 4:367–372.
Leuenberger, B. E. 1974. Testa surface characters of Cactaceae. Preliminary results of a scanning electron microscope study. *Cactus and Succulent Journal* (Los Angeles) 46:175–180.
——— 1976a. Die Pollenmorphologie der Cactaceae und ihre Bedeutung für die Systematik. *Dissertationes Botanicae* 31:1–321.
——— 1976b. Pollen morphology of the Cactaceae. *Cactus and Succulent Journal* (London) 38:79–94.
McLeod, M. G. 1975. A new hybrid fleshy-fruited prickly-pear in California. *Madroño* 23:96–98.
Moran, R. 1962a. *Pachycereus orcuttii*—a puzzle solved. *Cactus and Succulent Journal* (Los Angeles) 34:88–94.
——— 1962b. The unique *Cereus. Cactus and Succulent Journal* (Los Angeles) 34:184–188.
Parfitt, B. D. 1980. Origin of *Opuntia curvospina* (Cactaceae). *Systematic Botany* 5:408–418.
Pinkava, D. J., L. A. McGill, and R. C. Brown. 1973. Chromosome numbers in some cacti of western North America. II. *Brittonia* 25:171–176.
Pinkava, D. J., L. A. McGill, T. Reeves, and M. G. McLeod. 1977. Chromosome numbers in some cacti of western North America. III. *Bulletin of the Torrey Botanical Club* 104:105–110.
Pinkava, D. J., and M. G. McLeod. 1971. Chromosome numbers in some cacti of western North America. *Brittonia* 23:77–94.
Pinkava, D. J., and B. D. Parfitt. 1982. Chromosome numbers in some cacti of western North America. IV. *Bulletin of the Torrey Botanical Club* 109:121–128.
Porsch, O. 1938. Das Bestäubungsleben der Kakteenblüte. *Jahrbuch der Deutsche Kakteen-Gesellschichte* (I):1–80.
Ross, R. 1981. Chromosome counts, cytology, and reproduction in the Cactaceae. *American Journal of Botany* 68:463–470.
Rowley, G. 1980. Pollination syndromes and cactus taxonomy. *Cactus and Succulent Journal (London)* 42:95–98.
Schumann, K. 1898–1902. *Gesamt beschreibung der Kakteen*. J. Neumann, Neudamm.
Tsukada, M. 1964. Pollen morphology and identification. II. Cactaceae. *Pollen y Spores* 6:54–84.
Walkington, D. L. 1966. Morphological and chemical evidence for hybridization in some species of *Opuntia* occurring in southern California. Ph.D. diss. Claremont Graduate School and University Center, Claremont.
Weedin, J. F., and A. M. Powell. 1978. Chromosome numbers in Chihuahuan Desert Cactaceae. *American Journal of Botany* 65:531–537.
Wiley, E. O. 1981. *Phylogenetics, the theory and practice of phylogenetic systematics*. Wiley-Interscience, New York.

11

Evolutionary Relationships

Our curiosity about the evolutionary relationships of the cacti does not stop at the boundary of that family. The Cactaceae is, to be sure, a highly specialized family of dicotyledons that must be more closely related to some families than to others. Botanists want to determine which plants are the closest living relatives to cacti so that we can understand where they originated and then speculate on the factors and processes that initiated this particular plant design. Many attempts have been made to unravel this mystery, and as we shall see, the most success has been obtained by studying how the Cactaceae and its closest relatives, collectively called the centrosperms, all share a group of derived, and often peculiar, features.

One way to begin this endeavor is to figure out how the primitive cactus may have looked. For cacti we already accept that the ancestor was a plant similar in appearance to a *Pereskia* (Chapter 2). When this has been determined, then we can search outside that family for plants sharing the same basic characteristics.

The Hypothetical Ancestor

Some evolutionists invent a hypothetical ancestor as an exercise. To this organism is assigned all the features needed to evolve into all of the descendant species. To invent this organism, the worker must figure out what the original state was for each character. Hundreds of features can be evaluated: Was the ancestor a perennial or an annual? Did it live in the desert? Did it have spines and areoles? Were the ancestral flowers solitary or clustered on an inflorescence? Was the first member of this family a CAM plant? These and scores of other questions can be asked, ranging from the level of the whole organism to that of the molecule. When the character state occurs in all species, the choice is simple; but when two or more character states are present in the family, the investigator must logically choose the one that can be most simply modified to yield the other states. Our creation is truly hypothetical because no such plant having the primitive form of every character probably ever existed; but the ancestor of a family is presumed to have possessed most of the proposed features, and that may be good enough to solve the mystery.

The reader may wish to return to Chapter 2 and review the features found in present-day species of *Pereskia* (Fig. 11.1). Some of the data presented in Chapter 2 may now take on new and important meaning.

Vegetative Morphology and Anatomy

On the basis of comparisons with living species, the following picture of the ancestor for the Cactaceae can be drawn. The original plant was most certainly a woody perennial with relatively large, simple, petiolate leaves arranged in a helically alternate pattern. In other words, the ancestor was a leafy shrub or tree. We conclude this because the least specialized pereskias are erect, woody plants, as are the primitive leaf-bearing species of Opuntioideae and the least specialized Cactoideae. The only cacti that are herbaceous, never forming a conspicuous woody cylinder, have highly bizarre growth forms, such as plants that creep along the ground, geophytes, some highly reduced epiphytes, and some very tiny solitary or caespitose plants. The ancestral wood skeleton consisted of a solid cylinder without conspicuous, unlignified primary rays, and a number of architectural designs of the holes (rays) eventually evolved. Their leaves had a short petiole and a single, ovate blade with entire margins and pinnate venation. The leaf blade was relatively thin, probably had stomates on

Figure 11.1. *Pereskia sacharosa,* a leaf-bearing species having many of the features that would be required in the hypothetical ancestor of the cactus family.

both sides, and was the chief carbon-assimilating (photosynthetic) organ of the plant. Our hypothetical ancestor was a C_3 plant. In the axil of the leaf an areole had developed in place of an axillary bud; this axillary short shoot produced spines the first year and new leaves and possibly more spines in subsequent years. This plant grew in a seasonally dry, short-tree forest, was leafless during the dry months (drought deciduous), and flowered after the leaves were fully expanded.

The stem of this hypothetical plant developed from an apical meristem that had a special type of cellular zonation (Chapter 2); the stem was cylindrical and relatively thin and lacked any development of cortical succulence. These stems stayed green for only a short period and were not important photosynthetic organs. The stem epidermis had few, if any, stomates and was quickly replaced by periderm (Fig. 11.2), which had cork cells but lacked special hard, mechanical cells. The cortex was thin with no collenchymatous hypodermis and probably no mucilage cells. Its vascular system was an open system, and the vascular supply to each leaf arose from only one bundle in the stem and split one or more times before entering the petiole (Chapter 3). The first-formed phloem produced some fibers. The sieve elements of the phloem contained colorless plastids within which was a special accumulation of protein. This protein occurred as a fibrous ring next to a central, globular proteinaceous crystal (Chapter 9). The first-formed xylem of this plant was highly specialized, consisting of vessel elements with simple perforation plates. The pith was narrow to moderately wide (up to 3 cm), but the tissue was firm and lacked any mucilage cells or additional vascular bundles.

A single vascular cambium formed a complete cylinder, producing secondary xylem to the inside and phloem to the outside. The elongate initials of the vascular cambium were arranged in lateral tiers that were diagonally arranged (Fig. 2.7). Some phloem cells differentiated as thick-walled, elongate, mechanical cells that looked like fibers but were really sclereids, which had heavily lignified cell walls. The xylem (Fig. 2.6) was fully lignified, and consisted of short, narrow (highly specialized) vessel elements, short libriform wood fibers, axial wood parenchyma, and specialized vascular rays (Chapter 8).

The seed germinated by the formation of an operculum. When the seed became swollen with water, a crack formed along the dorsal side of the seed; thus, the root emerged from the region of the seed nearest to the point of attachment (the hilum-micropylar end). The two cotyledons were broad and very large and thin when fully expanded. The roots produced by the seedling were shallow, fibrous, and eventually became woody.

Reproductive Structures

The ancestral cactus flower was produced from a newly formed areole at the tip of the current year's growth. It was probably a solitary flower borne on a short petiole (Fig. 11.1), but it could also have been borne on a simple inflorescence. Leaves and areoles covered the outside of the flower because the ovary

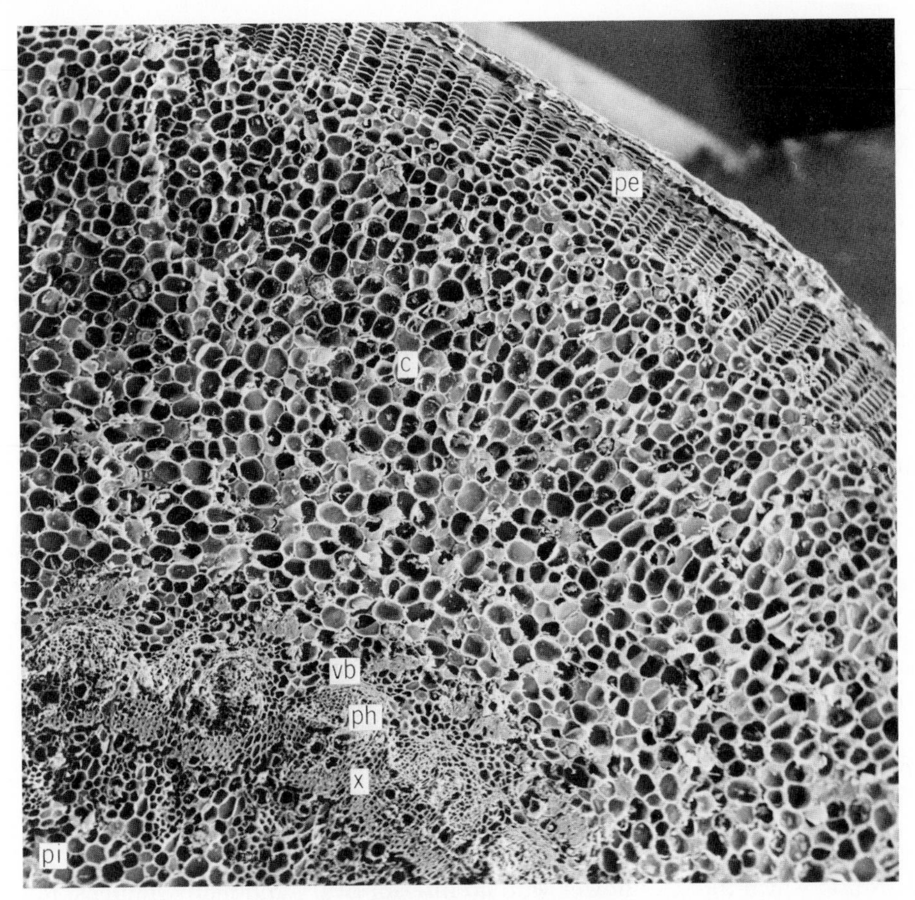

Figure 11.2. *Pereskia sacharosa;* scanning electron photomicrograph of a young stem cross section. This stem, which has a conspicuous cortex (c), cannot be an important photosynthetic organ because the epidermis has been replaced by periderm (pe), consisting of files of thin-walled cork cells. Toward the center of the stem are the vascular bundles (vb) with xylem (x) and phloem (ph), which surround an unspecialized pith (pi). There are no mucilage cells in the tissues.

had become somewhat sunken into the receptacle-stem tissue (Chapter 2). Spines were not produced on the outside of the flower or the fruit. The bracts present on the ovary were fairly similar in size and shape to the perianth parts, which were colored by a lavender to purple betalain pigment (Chapter 9). Anthocyanin pigments were not present in the reproductive or vegetative structures of this plant. There were many separate perianth parts, which were arranged helically.

Many free stamens were produced in a centrifugal order (inside to outside) in the flower. Each stamen had a fairly long and slender filament with a single vascular bundle, and the anther, which had four chambers, dehisced longitudinally to release the abundant pollen grains. Each pollen grain was 0.07 to 0.09 mm in diameter, had three long apertures for germination (tricolpate), and was covered with short, conelike spinules interspersed with small holes (punctae) that each had an encircling rim. The pollen grain contained three nuclei (pollen grains of many angiosperms have only two nuclei when they are released).

The female part of the flower was initiated as a set of 4 to 12 superior and separate carpels (apocarpous) arranged in a ring, which grew together during development to form a single ovary with a large central chamber (locule), but the stigmas remained free. Inside the ovary longitudinal ridges occurred at the places where the carpels fused, along which the ovules were produced. Placentation was fundamentally axile (attached to the inner side of the ovary) although it later appeared to be parietal, when during the growth of the fruit the ovaries became sunken in the stem tissues. The ovary ripened into a thick-walled, fleshy berry that was reinforced with many lignified sclereids.

Each ovule was campylotropous, that is, the long axis of the ovule was positioned at right angles to the stalk (funiculus) (Fig. 11.3). The ovule had two integuments and a fleshy nucellus (crassinucellar). A starchy tissue called endosperm was produced during

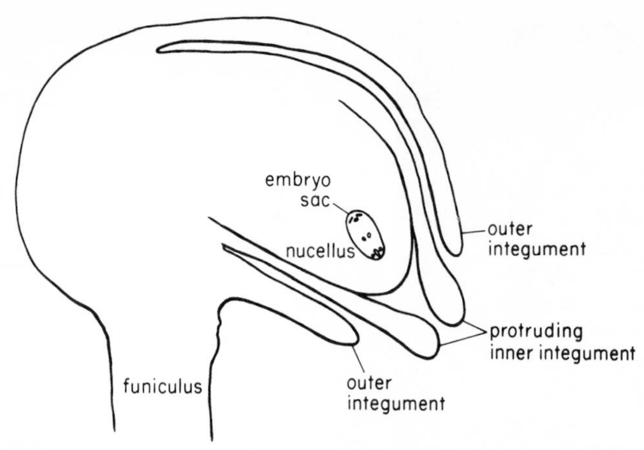

Figure 11.3. Diagram of a campylotropous ovule, such as that found in species of *Opuntia.* The ovule has been cut lengthwise to show the important regions. One major centrospermous feature is the protrusion of the inner integument beyond the outer integument. The embryo sac, in which the embryo will form, has eight nuclei, a condition that occurs in all centrosperms and in nearly 90% of all angiosperms. The nucellus, which surrounds the embryo sac, is relatively thick (crassinucellar).

the early phase of embryo development but was used up in the growth of the embryo. Food storage in the seed was provided by the perisperm, a tissue produced by the mother tissues and around which the curved embryo was wrapped. The seed coat was black, smooth, and glossy, lacking distinguishing features and also lacking an aril. Seeds of the ancestral cactus were probably a few millimeters in length. It is hard to speculate on the pollination biology and fruit or seed dispersal of the ancestral cactus because of lack of data on what animals might have been present with the ancestors of the cacti.

Relatives

All modern systems of biological classification are arranged as a hierarchy. As we have already described (Chapters 1 and 10), all species that are morphologically similar and descended from a single ancestral population are clustered into larger groups called genera, and genera are clustered into subtribes and tribes, and tribes into subfamilies and families. Thus, in Cactaceae there are three subfamilies that contain about 121 genera and 1600 species. Many other taxonomic levels are also available in this hierarchy, not only between the categories of species and family but also above the level of family. So far in this book we have used only the taxonomic categories below the level of the family. But in our study of evolutionary relationships we are searching for groups that are higher and larger than the family, namely, a suborder and an order. Hence, we need to identify all of the families that have evolved from the same common ancestor as the cacti.

Early Systems of Classifying the Cactaceae

As described in Chapter 10, the goal of modern systematic biology is to develop classification systems that show phylogeny, the actual evolutionary sequence of the organisms. The operational word here is *evolutionary;* because Charles Darwin did not publish his theory of evolution until 1859, phylogenetic studies only began after that date and consequently, the search for the relatives of cacti only began in earnest after that publication. But even though there is little reason to evaluate carefully the earliest systems, students can learn from the mistakes that have been made in these primitive classification systems and can determine how to spot them in the future.

The earliest attempts to place the cactus family into a classification system used features of the flower that were very easy to observe but highly artificial. For example, Linnaeus classified cacti as plants that had one ovary and many stamens attached to the calyx. Classified next to the cacti were members of the rose family (Rosaceae). Other taxonomists grouped cacti with families that have a flower with an inferior ovary and many stamens, for example, myrtles (Myrtaceae) or the ice plants (Aizoaceae). But perhaps the most familiar and widely accepted way of classifying the cactus family was to place cacti into a group called the Parietales, which included passion flowers (Passifloraceae), begonias (Begoniaceae), and gourds (Cucurbitaceae), all of which have parietal placentation.

In the 1980s it is obvious to systematic botanists that features such as numerous stamens, inferior ovary, and parietal placentation have independently arisen many times in completely unrelated groups of angiosperms (convergent evolution). But in the formative years of biology, convergent evolution was not commonly accepted or well understood. Moreover, with regard to cacti, we now know that the characters previously used to classify them were very misleading. Rather than studying the leaf-bearing cacti, which have a superior ovary composed of numerous carpels and axile placentation, workers studied the highly specialized species. Nonetheless, even in this century a number of classification systems, such as one published by Charles Bessey in 1915, placed cacti next to the families with parietal placentation. Traditions are sometimes hard to break.

Centrosperms

The order of plants to which cacti belong is called the centrosperms. This name was invented in 1876

Table 11.1. Families assigned to the centrospermous order (Caryophyllales or Chenopodiales) during the last 120 years. [a]

Family	Number of species in family	Distribution
Achatocarpaceae (achatocarpus family)	8	Texas and northwestern Mexico to Argentina
Aizoaceae (ice plant family)	2000	Mostly southern Africa, but some species in Australia and southern South America and naturalized throughout the Northern Hemisphere
Amaranthaceae (amaranth family)	ca. 900	Cosmopolitan
Basellaceae (basella family)	20	Southern United States to tropical and Andean South America, West Indies, tropical Africa, and Madagascar
Cactaceae (cactus family)	1550	New World from southern Canada to Patagonia, and one genus *(Rhipsalis)* in the Old World tropics; naturalized elsewhere
Caryophyllaceae (carnation or pink family)	1750	Cosmopolitan
Chenopodiaceae (goosefoot family)	ca. 1400	Cosmopolitan (includes *Dysphania*)
Didiereaceae (didierea family)	11	Madagascar
Halophytaceae (halophytum family)	1	Argentina
Molluginaceae (carpet-weed family)	ca. 100	Tropical and subtropical regions, especially Africa
Nyctaginaceae (four o'clock family)	290	Nearly cosmopolitan, but mostly in the New World
Phytolaccaceae (pokeweed family)	ca. 120	Mostly New World, but also present in the Old World, such as in Africa, and Madagascar (incl. *Agdestis, Barbeuia, Gisekia, Petiveria,* and *Stegnosperma*)
Portulacaceae (purslane or portulaca family)	ca. 600	Cosmopolitan (incl. *Hectorella* and *Lyallia*)

a. Families not presently assigned to the centrosperms: Bataceae, Begoniaceae, Crassulaceae, Cynocrambaceae, Elatinaceae, Frankeniaceae, Gyrostemonaceae, Hydrostachyaceae, Plumbaginaceae, Podostemonaceae, Polygonaceae, Rhabdodendraceae, Salicaceae, Tamaricaceae, and Theligonaceae.

by August W. Eichler, a German taxonomist, who identified a group of plants featuring a campylotropous ovule and a peripheral embryo that surrounds the central nutritive perisperm. Whereas our present definition of this group is more precise and uses many more features, these two original features are still generally correct.

Table 11.1 lists the families and common names of plants that we will be discussing throughout this chapter, including families that are properly classified as centrosperms as well as those families that had been classified as centrosperms but which have subsequently been removed from this category. This table also summarizes information on the size and distribution of each centrospermous family. Our goal here is to identify the core families—the members of the order that have appeared in the majority of the classification systems.

Before Eichler, the strong evolutionary relationships of the centrosperms were recognized by the German Alexander Braun. In 1864 Braun created an order called the Caryophyllinae. This included eight families: Aizoaceae, Amaranthaceae, Cactaceae (as Opuntiaceae), Caryophyllaceae, Chenopodiaceae, Nyctaginaceae, Phytolaccaceae, and Portulacaceae. When Eichler inaugurated the ordinal name Centrospermae for these same species, he added the knotweeds (Polygonaceae) and considered adding the begonias, one of the families with parietal placentation. Two years later he revised his classification by deleting Cactaceae from the order, adding Theligonaceae, and including Basellaceae as part of the Portulacaceae.

From these initial papers, a seemingly endless parade of systematic schemes began, ultimately encompassing dozens of attempts to define the families in the centrospermous order and to determine the interfamilial relationships within that order. The principal classification schemes are summarized in Table 11.2, which illustrates the following points.

Table 11.2. Selected classifications of the centrospermous order,[a] 1864–1981.

Family	Braun (1864)	Eichler (1878)	Hallier (1912)	Bessey (1915)	Wettstein (1935)	Melchior (1964)	Cronquist (1968)	Thorne (1968)	Takhtajan (1969)	Airy-Shaw (1973)	Cronquist (1981)	Thorne (1981)
Families currently included in the centrospermous order												
Aizoaceae	+	+	+	+	+	+	+	+	+	+	+	+
Amaranthaceae	+	+	+	+	+	+	+	+	+	+	+	+
Basellaceae	•	•	+	+	•	+	+	+	+	+	+	+
Cactaceae	+[b]	—	+	+	+	—	+	+	+	+	+	+
Caryophyllaceae	+	+	+	+	+	+	+	+	+	+	+	+
Chenopodiaceae	+	+	+	+	+	+	+	+	+	+	+	+
Didiereaceae	—	—	+	—	—	+	+	+	+	+	+	+
Halophytaceae	•	•	•	•	•	•	•	•	+	+	•	+
Molluginaceae	•	•	•	•	•	+	+	•	+	•	+	+
Nyctaginaceae	+	+	+	+	+	+	+	+	+	+	+	+
Phytolaccaceae (includes Achatocarpaceae, *Stegnosperma*)	+	+	+	+	+	+	+	+	+	+	+	+
Portulacaceae	+	+	+	+	+	+	+	+	+	+	+	+
Families not currently included in the centrospermous order												
Bataceae	—	—	—	+	—	—	+	—	+	+	—	—
Gyrostemonaceae	—	—	—	—	—	+	•	+	+	+	—	—
Plumbaginaceae	—	—	+	—	—	—	+	—	+[c]	+	—	—
Polygonaceae	—	+	+	+	+[c]	—	+	+	+[c]	+	+[c]	—
Theligonaceae	—	+	—	+	•	—	—	+	—	+	—	—
Others	—	—	1	6	—	—	—	—	1	6	—	—

a. + indicates that the family is classified in the order; • indicates that the species of that taxon are classified in the order but in another family; — indicates that the family was not mentioned.
b. Treated as a separate order, Cactales, closely related to the centrosperms.
c. Treated as a family closely related to the centrospermous order.

First, all systems have included seven of the core families that were originally listed by Braun in 1864 and by Eichler in 1878, but cacti were not always included. Second, some families were very reluctantly accepted as members of this order, most notably Cactaceae and Didiereaceae, because superficially they do not resemble the other families. Third, some of the families that are not closely related to centrosperms, such as the salt-tolerant succulent Bataceae, remained classified in the order until very recent times. Finally, there has been much difference of opinion on how to classify certain genera, such as *Stegnosperma* and *Halophytum,* which are often treated as monogeneric families, and *Dysphania,* which is now commonly treated as a member of the Chenopodiaceae. Table 11.2 does not show how certain genera have been shifted from one centrospermous family to another, such as *Gisekia,* which recently was transferred from Molluginaceae to Phytolaccaceae. Fortunately, no exchange of species has been needed for the Cactaceae, which is a very precisely defined family.

Shared Derived Features of Centrosperms

As discussed in Chapter 10, the most reliable, but certainly not guaranteed, way to assign phylogenetic relationships is to determine which taxa share unusual and highly derived features. For example, if a pair of plant families shares four specialized characteristics that appear in no other plant family and if these features are totally unrelated in function and development, it is unlikely that this unique combination could have arisen independently more than once. Thus, the probability that these families are closely related is very high. The more shared derived traits that exist between two taxa, the higher is the probability of the relatedness of the taxa.

Over the last 100 years, a number of unusual features have been found in centrospermous families, a finding that sheds much light on the order. These features not only show us what families belong in the order but also indicate how the families should be clustered within the order. Specifically, the features that have been most useful can be summarized as five categories.

Ovule and seed. Since Eichler, characteristics of the ovule and the mature seed have yielded excellent data on the families of centrosperms. Data collected by the Indian embryologist Panchanan Maheshwari showed systematists that a great many embryological features are shared by centrosperms—and collectively it is an impressive list.

- Ovules campylotropous (other types are present but are secondarily derived)
- Nucellus massive (crassinucellar) and strongly curved
- Anther tapetum glandular; its cells are di- and tetranucleate; periplasmodium absent
- Ovules with two integuments; inner integument protruding; protrusion forms the micropyle and approaches the funiculus
- Hypodermal archesporial cell cuts off one cell
- Polygonum-type embryo sac (monosporic, eight nuclei) derived from the chalazal megaspore of the tetrad
- Divisions of microspore mother cell simultaneous
- Pollen grains trinucleate (three nuclei) when released from the anther
- Perisperm instead of endosperm as the chief storage region
- Embryo strongly curved and peripheral, surrounding the perisperm

First, the campylotropous ovule contains a very thick nucellus (nurse tissue of the embryo). Second, the micropyle of the ovule is formed by the swollen tip of the inner integument, which protrudes from the ovule. Third, the embryo is strongly curved and located adjacent to the wall of the seed (peripheral). And fourth, this strongly curved embryo surrounds the central starchy tissue, which is the perisperm. Perisperm, which forms from the female nucellar tissue, is uncommon in the angiosperms, which normally form endosperm by the fusion of two female cells and one male cell at the time of fertilization. Even if highly specialized species in a family deviate somewhat from this design, for example, by having succulent embryos with little perisperm, the primitive and typical members of each family have the necessary features.

Gross anatomy. Plant anatomists discovered that many species of the centrosperms have a form of stem growth termed anomalous secondary thickening by successive cambia. This ponderous term simply means that many centrosperms do not have the usual dicotyledonous type of secondary growth. Instead of having one cylindrical vascular cambium throughout its lifetime, a plant forms a succession of cambia, and each one functions for only a short time. In these plants a cambium develops and produces secondary xylem to the inside and secondary phloem to the outside; but the cambium becomes entrapped between the tissues and a new cambium forms close to the outside of the stem (Figs. 11.4 and 11.5). The mature stem, therefore, has regions of phloem located within the wood. This feature has evolved several times in dicotyledons, but most examples are found in the families presently classified in the centrospermous order. Not all families of centrosperms have anomalous secondary thickening,

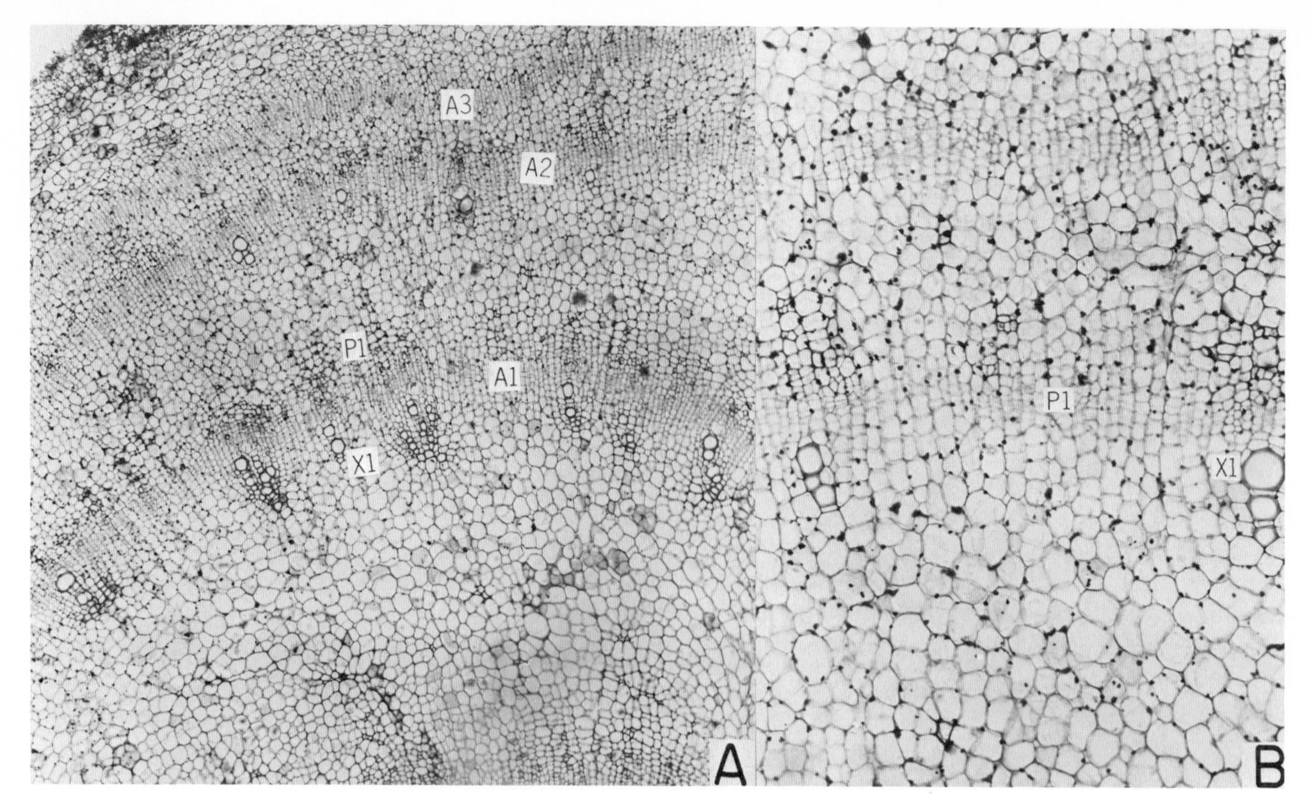

Figure 11.4. Anomalous secondary thickening in the root of a sugar beet (*Beta vulgaris*, Chenopodiaceae), a centrosperm.
(A) Low-magnification view of the root in cross section; several bands of anomalous secondary growth (A1 – A3) are visible. The oldest band (A1) has produced numerous vessels in the xylem (X1) and sieve tubes in the phloem (P1); the younger band (A2), which originated outside A1, has relatively little xylem and phloem; and the youngest band (A3) has not yet produced functional vessels in the xylem ×80.
(B) Closer view of the outer two anomalies (A2 and A3), showing the vessels with thickened walls in the xylem of A2 (X1) and sieve tubes in the phloem (P1). ×320.

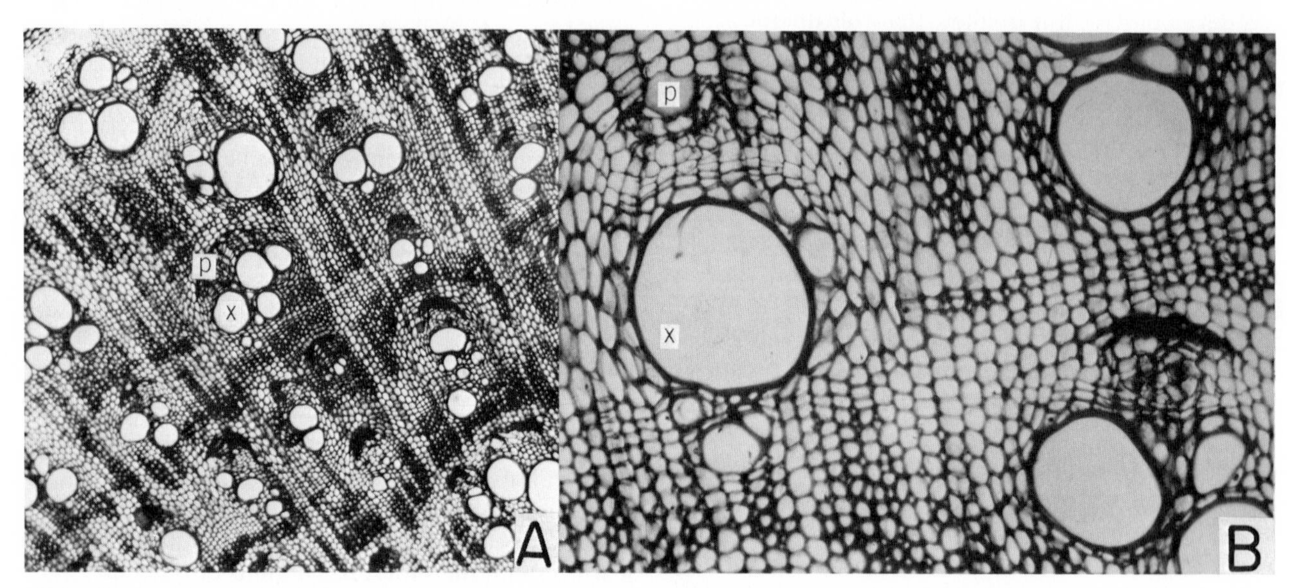

Figure 11.5. Cross sections of an old stem of *Bougainvillea spectabilis*, which has anomalous secondary thickening by successive cambia.
(A) At low magnification, the xylem (x) and phloem (p) appear to be organized in bundles, although they are formed from a layer of cambial initials. The bundlelike nature of this stem is termed foraminate because the stem appears to have small openings where the phloem occurs. ×40.
(B) At four times the magnification of *A*, the cells of the xylem and phloem are visible. ×160.

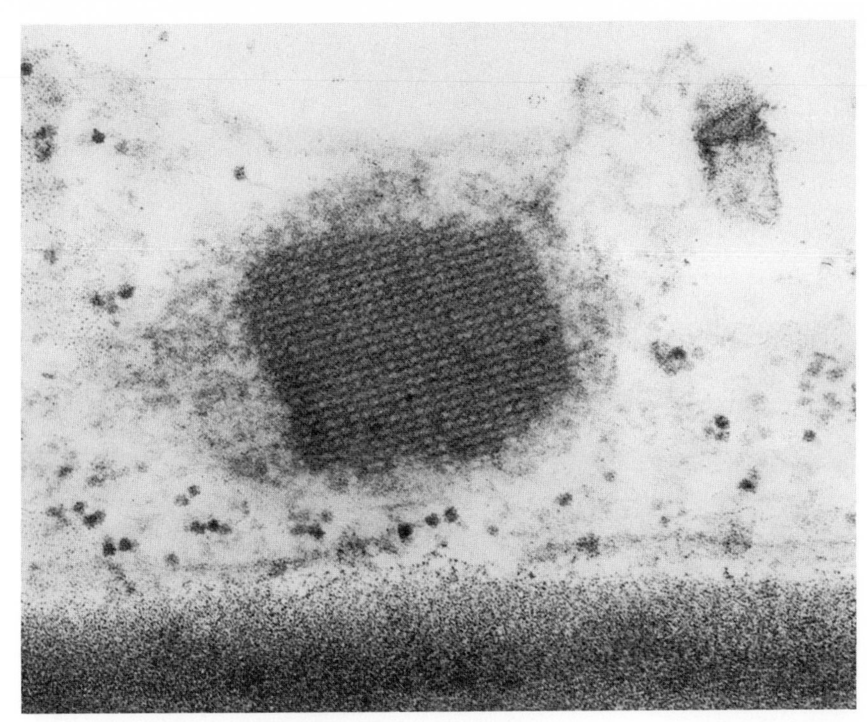

Figure 11.6. Polygonal, proteinaceous inclusion in a sieve-tube element leucoplast of *Stenosperma cubense* (Stegnospermaceae). The gridlike structure is characteristic of crystalline proteins. Compare this figure to Figure 9.13, which shows a globular protein lacking a crystalline structure. ×76,500.

however; the cacti, the portulacas, the basellas, and the didiereas do not have this feature.

Betalains. As discussed in Chapter 9, the water-soluble pigments in the flowers, fruits, stems, and spines of cacti are betalains, a unique class of nitrogen-containing compounds (Figs. 9.1 and 9.2). Betalains are found only in centrospermous families. Two families of centrosperms, Caryophyllaceae (the carnations) and Molluginaceae (as currently defined), have no betalains but instead have the water-soluble anthocyanin pigments found in all other orders of angiosperms.

Pollen grains. The structure of pollen grains in dicotyledons is quite diverse, and it would take a book to describe the important variations that occur from family to family. We can simplify things by stating that the basic structure of pollen grains is fairly specific for each family and each order. This is especially true for centrosperms, which all share the same basic pollen grain architecture: three apertures or colpae (Figs. 1.33*B* and 10.33*B*), outer wall (exine) with small spinules and punctae, and three nuclei. The number of apertures increases to 6, 9, 12, 15, and so forth in several families and is exceedingly high in Chenopodiaceae and Amaranthaceae (ChenAm pollen), which is literally covered with round apertures, much like the dimples of a golf ball. In Cactaceae, some of the pollen grains of *Opuntia* look like ChenAm pollen.

Plastid ultrastructure in the sieve-tube elements. In Chapter 9 we discussed the inclusions that have been discovered in the colorless plastids of phloem sieve-tube elements in cacti (Fig. 9.13). In angiosperms there are six major categories of sieve-tube element inclusions, one with starch only and five that have protein present. Just as all monocotyledons have a particular arrangement of proteinaceous inclusions in the plastid, termed Type PII, all species of centrosperms, and only centrosperms, have another unique pattern, termed Type PIII. In this type, a protein globule and a peripheral ring of fibrous protein is present in each plastid. The majority of centrosperms have plastids identical to those found in cacti (Type PIII c f), which has a large, central, protein globule. In Caryophyllaceae, Stegnospermaceae (Fig. 11.6), and *Limeum* of Molluginaceae, however, the central protein is a polygonal crystal; and in all species of Chenopodiaceae and Amaranthaceae a central protein crystal is absent.

A Search for the Sister Taxon of Cactaceae

The cactus family has finally found a permanent home in the centrospermous order, because phylogeneticists have become convinced that they belong to the same group with the betalain-containing families, which also share many other synapomorphies. Nonetheless, major problems still confront an-

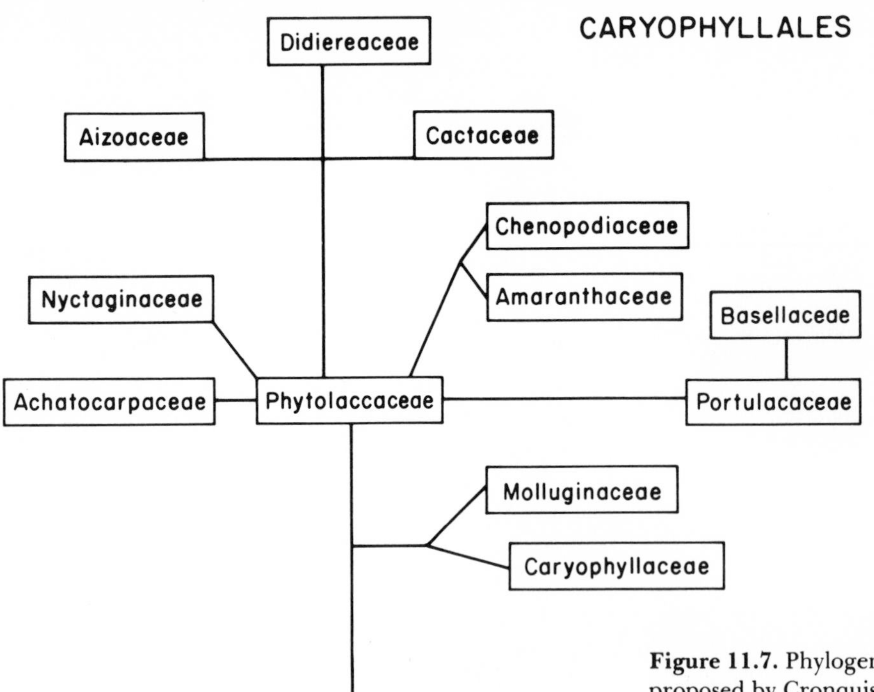

Figure 11.7. Phylogeny of the centrospermous families proposed by Cronquist (1981).

giosperm phylogeneticists: What is the family most closely related to the Cactaceae (its sister taxon)? What is the correct phylogeny for the families of this order? What family of dicotyledons is most closely related to the centrosperms? Certainly this book is not the proper forum for presenting primary phylogenetic research on these topics, but we can discuss each of these questions, which are all very tightly linked, in a narrative way.

To begin with, we can quickly rule out some of the centrospermous families as candidates for being the sister taxon of the cacti. Removed from contention are Chenopodiaceae and Amaranthaceae, which not only have a unique subtype of sieve-tube-element plastid that lacks a central protein and a complex pollen grain (ChenAm pollen), but also have a long list of adaptations of flowers, fruits, and vegetative parts that are too highly specialized to bear any close resemblance to the cacti. Likewise, for many good reasons, authors have abandoned any hope of finding tight relationships between Cactaceae and Nyctaginaceae. Nyctaginaceae, which includes bougainvillea, often have opposite leaves, highly derived forms of anomalous secondary thickening, and flowers and fruits too highly specialized to be compared with the pereskias. Cacti also cannot be placed next to Caryophyllaceae, which have different sieve-tube-element plastids and anthocyanins and do not have betalains.

Left for consideration are eight families and some satellite genera of these families that are sometimes recognized as separate families. For convenience, and based on our current knowledge of their structural relationships, we will treat these as three units: (1) Phytolaccaceae, including *Gisekia,* Molluginaceae, Stegnospermaceae, *Phaulothamnus,* and *Barbeuia;* (2) Aizoaceae and Halophytaceae; and (3) Portulacaceae, including *Hectorella* and *Lyallia,* Didiereaceae, and Basellaceae. Part of each of these groups has been proposed by some past author as the sister taxon of Cactaceae.

Phytolaccaceae have been treated as the family closest to the ancestor of the centrosperms by many authors, most recently by Arthur Cronquist, a world-renowned phylogeneticist, whose 1981 system of classification is used by many botanists. In his latest familial classification of the order (Fig. 11.7), which he called the Caryophyllales, Cronquist placed the Phytolaccaceae, including *Stegnosperma, Gisekia,* and *Barbeuia,* in the middle of the betalain families. This model fits the pattern of evolutionary systematics discussed in Chapter 10. Cronquist suggested that Phytolaccaceae itself as well as its immediate ancestors, gave rise to four lineages, one of which produced the cacti, the ice plants, and the didiereas.

In all places where Phytolaccaceae has been accepted as representing the ancestral stock, authors have emphasized two things. First, the pokeweeds (Phytolaccaceae) have all of the typical features of the betalain-producing families, and they also have general shoot and flower characteristics found in many dicotyledonous families. Second, the ovary of some species is formed from a ring of separate carpels (apocarpy), an arrangement that is considered

Figure 11.8. A leafy shoot of *Phytolacca dioica,* or ombu, a large tree that is cultivated at the Huntington Botanical Garden in San Marino, California. The leaf is composed of an ovate blade and a long petiole (up to 10 cm), and the leaves are arranged alternately on the stem. In the background is the outer bark of this tree.

primitive in angiosperms as a whole compared with the fusion of carpels early in their development (syncarpy). These reasons are, of course, equally true for our hypothetical ancestor of Cactaceae, which is based on *Pereskia.* There are many similarities between our hypothetical ancestor of cacti and the primitive features of Phytolaccaceae, including seed morphology, but a very important synapomorphy stands between them. Phytolaccaceae (Fig. 11.8) and *Stegnosperma* (Fig. 11.9) have anomalous secondary thickenings, as do the Nyctaginaceae, Chenopodiaceae, Amaranthaceae, and Aizoaceae, whereas cacti have the characteristic wood of dicotyledons. This difference is certainly not trivial, because thickening by successive cambia is not only a derived feature but also involves major changes, both in developmental processes and even in the organization of the primary vascular system. Especially for the latter reason, there is much circumstantial evidence supporting the conclusion that once anomalous secondary thickening occurs in a taxon it cannot easily be lost.

The origin of anomalous secondary thickening in this order has received relatively little attention, but the research on *Stegnosperma* by Karl E. Horak at the University of Arizona has shed considerable light on the subject. Horak studied stem anatomy of *Stegnosperma* because it has a single vascular cambium and produces typical dicotyledonous wood and bark for several years before shifting to anomalous secondary thickening in older stems and roots. By studying the inception of anomalies, Horak discovered that the development of the first anomaly in the stem is associated with the formation of a major branch. Likewise, in the root the development of the anomaly is evident with the appearance of a major branch root.

Figure 11.10 shows the anomalous portion of the woody stem of *Stegnosperma,* which consists of lignified tissues, xylem, and parenchyma, interrupted by regions of phloem. Normally phloem does not form within xylem. In highly specialized genera of centrosperms, xylem and phloem formed in anomalies appear as discrete bundles (Fig. 11.5), which with great effort can be traced from the tip of the stem

Figure 11.9. *Stegnosperma cubense,* a weak-stemmed shrub that grows near beaches in tropical America. The alternate leaves are simple and entire, and the fruits are borne at the end of lateral branches.

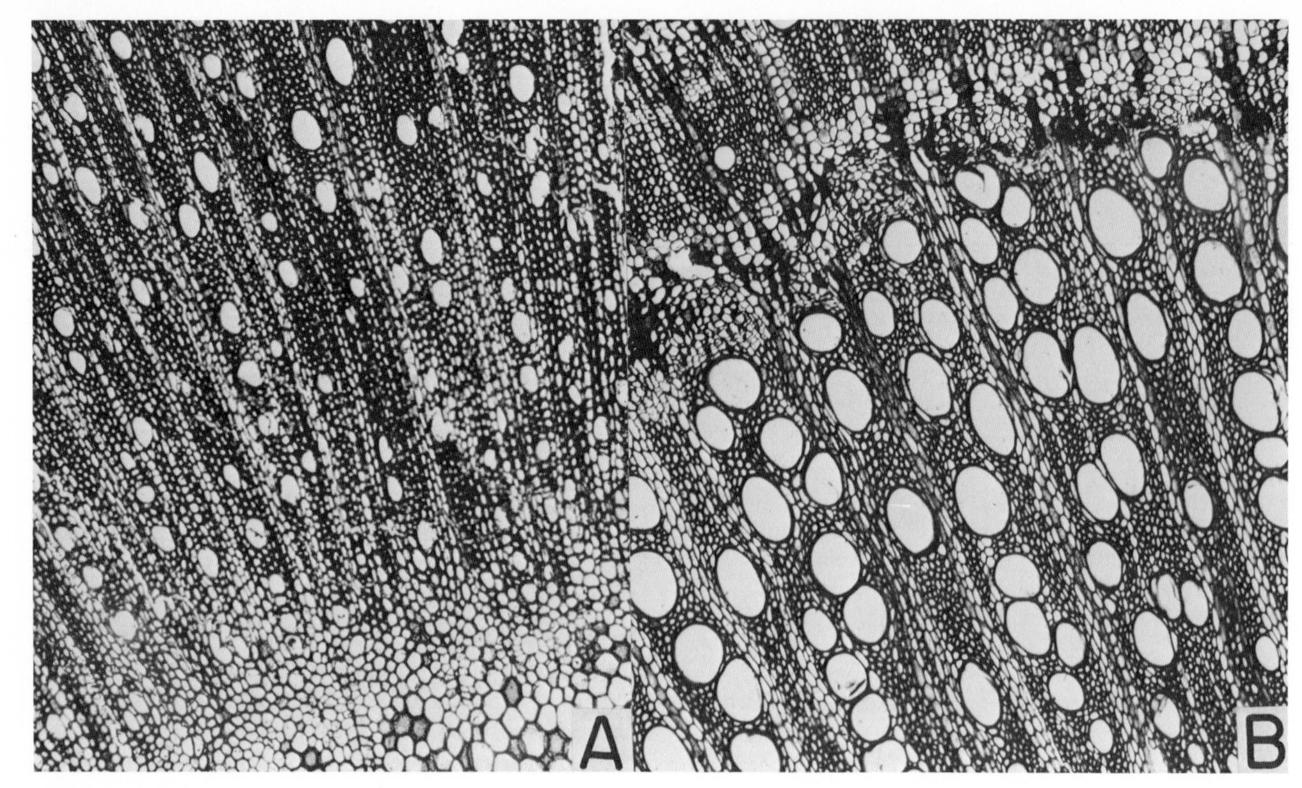

Figure 11.10. Stem cross sections of *Stegnosperma cubense*. ***(A)*** Low-magnification view of the inner region of the woody stem, which has normal secondary thickening. ***(B)*** Outer region of the same woody stem, which has anomalous secondary thickening bands of phloem between bends of wood (secondary xylem). ×45.

right down into the root. No one has experimentally studied this vascular arrangement in sufficient detail to appreciate its physiological significance. Of course, it might just be a developmental oddity, but because there is a direct connection between stems and individual roots through the bundlelike organization of the wood, it may also be a structural design to improve sugar transport in the phloem or, less likely, a strategy to control water flow. Whatever the causes and effects, there are reasons to treat anomalous secondary thickening as a specialization that did not include the Cactaceae, Portulacaceae, and Didiereaceae, all of which have a perfectly normal vascular cambium throughout their lifetimes. This observation forces one to conclude that the phytolacs, which do have anomalous secondary thickening, cannot be the direct ancestors of the Cactaceae but instead must be treated as an evolutionary branch distinct from the cactus lineage.

Aizoaceae is a family that continues to be suggested as the sister family of Cactaceae (Figs. 8.25 and 11.11). In 1984, James E. Rodman at the National Science Foundation and a number of co-workers at Yale University came to this conclusion after a long phenetic and cladistic analysis of the centrospermous families. The synapomorphic features occurring in both the Cactaceae and Aizoaceae but absent in most of the other centrospermous families were (1) Crassulacean acid metabolism, (2) floral hypanthium (Chapter 2), and (3) centrifugal order of stamen initiation (starting from next to the ovary and proceeding outward toward the petals). Stamen number, which is high in both families, was scored as plesiomorphic. Presence of a floral hypanthium was the single most important synapomorphy showing the close relationship between these two families

Figure 11.11. *Lampranthus* sp., one of the commonly cultivated genera of the ice plant family (Aizoaceae), has flowers with many separate petals and many stamens.

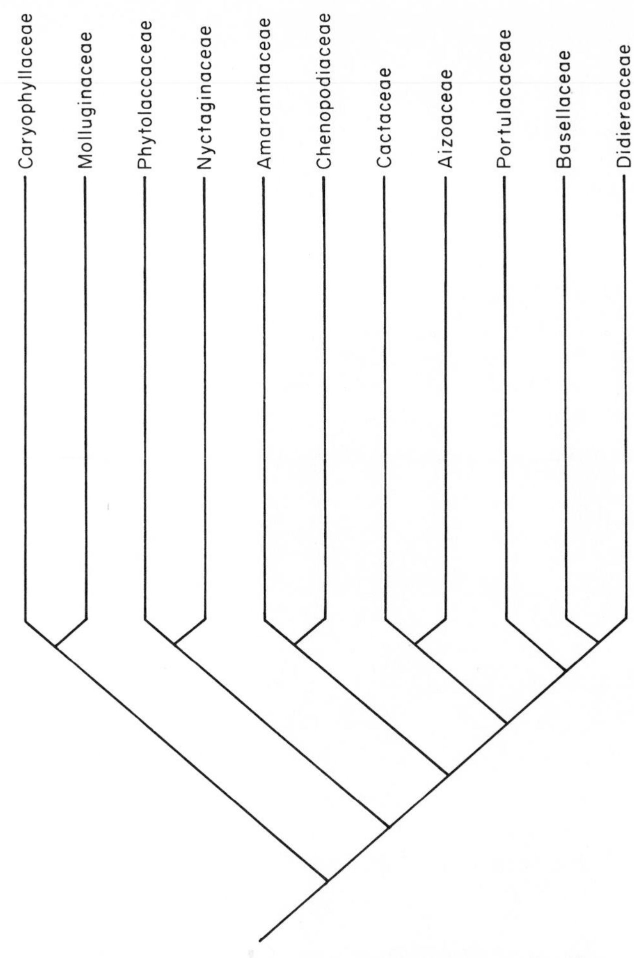

Figure 11.12. Phylogeny of the centrospermous families. In this phylogenetic model, favored by Rodman and co-workers (1984), Cactaceae are considered to be most closely related to Aizoaceae; the character that they share is the presence of a fleshy hypanthium, which developed into an inferior ovary.

(Fig. 11.12); and these authors proposed phylogenetic models to reflect that observation.

Crassulacean acid metabolism should not really be used as an important synapomorphic feature because it is not present in the most primitive-looking (non-succulent) members of the families, although many of the enzymes may be. This situation suggests that CAM has evolved independently in each of these families when succulence evolved. And certainly CAM is not a single feature for phylogenetic analysis, rather, it is a group of features. The similarities in flower structure, including the formation of an inferior ovary with many stamens, are suspected by structural botanists to be a case of convergent evolution because the details of how the ovary is formed and how stamens are initiated appear to be different in the two families. The extremely high numbers of stamens in both families are specializations. Consequently, on developmental grounds the most important characters for showing affinities of Cactaceae and Aizoaceae may be convergent and not homologous.

Halophytum ameghinoi, the only species of the Halophytaceae, was not included in the analysis by Rodman and co-workers because many details about this narrowly endemic plant of Argentina are missing. Traditional wisdom placed this genus in the Chenopodiaceae; but because *Halophytum* does not have either the ChenAm pollen or the sieve-tube-element plastid type and because it also seems to lack the anomalous secondary thickening that all Chenopodiaceae exhibit from the earliest growth of the plant, this assignment is probably incorrect. After describing the wood (secondary xylem) of this annual desert species, Gibson suggested that future workers should compare *Halophytum* with primitive Aizoaceae, which have many similarities and which in some annuals may not form anomalies in the first year.

Another place to look for relationships of *Halophytum* and the cacti is close to the portulaca family. The principal apomorphic features present in *Pereskia* are present in Portulacaceae. Although people in the temperate area are accustomed to thinking of this family as herbs, for example, the widely cultivated species of *Portulaca* (Figs. 11.13 and 11.14), there are some woody species, described as shrubs and half-shrubs with normal wood, such as the commonly cultivated shrub, *Portulacaria afra* (Fig. 11.15). Also in Portulacaceae there are some species with a semi-inferior ovary. In Portulacaceae many of the evolutionary changes documented for cacti have also been observed, for example, variations in seed coat morphology; but clearly the present-day Portulacaceae are more highly derived that our hypothetical ancestor of the Cactaceae.

The Portulacaceae include two aberrant woody species, *Hectorella caespitosa,* a high elevation caespitose plant (cushion plant) of New Zealand and the strikingly similar *Lyallia kerguelensis* from Kerguelen Island in an isolated region of the Indian Ocean. Cited as a very close relative of Portulacaceae are the Didiereaceae (Figs. 11.16 and 11.17), an endemic family of Madagascar comprising desert-dwelling, succulent, small or large shrubs. The Basellaceae, a small family of vines, is presently considered to be a close relative of, if not a derivative from, Portulacaceae.

This group of families related to the portulacas includes many divergent types of growth habits; nonetheless, when the cladistic and phenetic analyses were done by Rodman and co-workers, they formed a distinctive cluster and also were placed next to Cactaceae and Aizoaceae in all of their trials. Interestingly, long ago an interfamilial graft was successful between *Pereskiopsis* (Cactaceae) and *Didierea.* Grafting success has traditionally been used as a reliable index of relationship because distantly related forms

Figures 11.13 – 11.15. Species of the portulaca family (Portulacaceae), which appears to be closely related to the cactus family (Cactaceae).
11.13. Flowers of a species of *Portulaca* from Brazil, which has small, fleshly leaves.
(A) Top view of an open flower.
(B) Side view of another flower and of an ovary that has started to develop into a fruit.
11.14. A different species of *Portulaca,* with fleshy leaves and long hairs arising from the axillary buds.
11.15. *Portulacaria afra,* a shrub from Africa that has succulent leaves arranged in pairs (opposite) along the stem.

Figures 11.16 and 11.17. Shoot design of the Didiereaceae, a family that appears to be closely related to the cacti. Didiereaceae have spines and long shoot–short shoot organization.

11.16. *Alluaudia humbertii.*
11.17. *Didierea trollii.*

have tissue rejection and such grafts fail; however, grafts have been successful between other families, such as between succulent plants of *Sedum* (Crassulaceae) and nonsucculent nightshades of *Solanum* (Solanaceae), which are not closely related.

Because the characters used to classify Aizoaceae next to Cactaceae may be convergent and therefore not derived from a common ancestor, the next group to consider seriously must be the portulaca alliance. Combining Cactaceae with this alliance produces an evolutionary branch that has normal secondary growth and distinguishes it from the species with anomalous secondary thickening.

We have not presented a new phylogeny of the families, but we expect that soon someone will use more synapomorphies to provide better phylogenetic models of the order that will be extremely useful in working out the details of evolution for the families. Meanwhile, it looks as if the betalain families diverged from the anthocyanin families. From this betalain stock, the cactus family originated near the beginning of the betalain-producing plant groups, diverging before the origin of anomalous secondary thickening.

Biogeography

The study of the distributions of animals and plants is called biogeography. A biogeographer not only documents where species occur but also attempts to uncover the historical and ecological reasons why they occur where they do. Consequently, this is a great field for those who like solving mysteries.

One type of mystery is disjunction. When populations of a taxon, for example, a species, a genus, or a family, are geographically isolated from each other, they are said to be disjunct. An example of a broad disjunction is the occurrence of several species of creosotebush *(Larrea)* in deserts of South America and the presence of *L. tridentata* in Mexico and the United States. We would expect to find that closely related species live fairly close to each other, because they arose from a common ancestor, so the observation that close relatives live on opposite sides of a continent or on opposite sides of the world leads us to ask questions about how they achieved such a distribution.

A substantial discontinuity in the range of a taxon may have been caused by several different events. First, a seed of an organism could have been carried from one location over inhospitable territory to another site, where a new population became established; this is called long-distance dispersal. Second, at some time in the past, a population could have covered a large area; but when the climate changed most populations died out, leaving isolated populations, for example, only on mountaintops. In a third case, called vicariance, creation of a geographic barrier, such as the formation of a mountain range or a lake or the splitting apart of continents, subdivided the ancestral species and thus gave each of the smaller populations a chance to experience speciation. Biogeographers have recently invested much effort in determining the effects that the breakup of the world continents, that is, continental drift, has had on the diversification of life on those continents.

Geographic Origin

Because we lack a concrete phylogeny of the centrospermous families or of Cactaceae, there is really no scientific way to determine whether geographic and geologic events were responsible for the origins of these families. Nevertheless, a few tidbits of knowledge that are based on our preliminary discussions on the probable sister taxa of the cactus family can be examined.

The cacti have been used a number of times as evidence of a disjunct distribution of a plant group that was formed by the separation of Africa and South America 125 million years ago.* At first a few people, most notably Alfred Wegener, the German meteorologist who was the father of the continental drift theory, used *Rhipsalis* as an example of a plant present in both South America and Africa, considering this as evidence that the continents split apart. This example was thoroughly discredited by biologists, who noted that the species involved, *R. baccifera* (Fig. 1.24), is widespread in the New World and the Old and because birds feed on the small, sticky, whitish fruits and then deposit the seeds with their feces a long distance from where the fruits were eaten. There are other examples of fleshy-fruited plants occurring in both Africa and South America as a result of bird transport.

Once phylogeneticists had identified the centrosperms as a natural order, investigators wanted to know whether the Cactaceae, whose primitive species are concentrated in northeastern South America and the West Indies, are most closely related to an Old World or a New World family. Given our previous discussion of familial affinities, we can see that there are three possible scenarios. If the Phytolaccaceae is a family more primitive than the cacti, then the mystery is easily solved, because phytolacs are mostly a family of the New World tropics, where the cacti are native. *Stegnosperma* (Fig. 11.9), which sometimes is classified as a separate family and sometimes in the Phytolaccaceae, occurs mostly in Mexico and the West Indies; and the Phaulothamnaceae (two genera), classified next to the phytolacs, are also New World plants.

* For a review of biogeography and the geologic history of the world, see J. H. Brown and A. C. Gibson, *Biogeography* (St. Louis: C. V. Mosby, 1983).

If Aizoaceae is the closest family to Cactaceae, the mystery takes on a different plot. The ice plants are mostly an African family. They have speciated tremendously in southern Africa, but some species have been transported to other parts of the world in very recent times. Some writers enjoyed using Cactaceae and Aizoaceae as a "good" example of vicariance caused by the breakup of Gondwanaland, the ancient Southern Hemisphere supercontinent that consisted of present-day South America, Africa, Madagascar, India, New Zealand, Australia, New Guinea, and Antarctica.

A third scenario, which has not been carefully considered, is the model of Cactaceae-Portulacaceae-Didiereaceae-Basellaceae. Once again we have a New World–Old World disjunction that is just as interesting as the one hypothesized for cacti and ice plants. Remember that the woody members of the Portulacaceae occur in South Africa, Kerguelen Island, and New Zealand, which were once part of the Gondwanaland landmass. Add to this the Didiereaceae on Madagascar, another piece of the Gondwanaland puzzle, and we would be forced to consider the possibility that portions of the Cactaceae-Portulacaceae-Didiereaceae lineage were stranded and then diverged on five different pieces of land from Gondwanaland. Meanwhile, the Basellaceae evolved in the New World from a portulacaceous or another centrospermous ancestor.

Let us next consider another aspect of this biogeographic mystery. We shall assume that all of the betalain-producing families, both those with normal secondary growth and those with the anomalies, have been in the Southern Hemisphere for quite a while. The Chenopodiaceae, which have a fossil pollen record dating back at least 60 million years, are exceedingly well developed in Africa and Australia; and the Amaranthaceae in southwestern Australia have evolved a remarkable group of species. In contrast, the anthocyanin-producing Caryophyllaceae is best developed in the Northern Hemisphere, and this family could have been evolving in a separate region from the betalain taxa. Because the Chenopodiaceae are at least 60 million years old and probably derived from other betalain-producing plants, the age of the order must predate the fossil record and must, therefore, have begun in the Cretaceous, which ended 65 million years ago. The period when the pieces of Gondwanaland were separating from one another started around 125 million years ago and was certainly finished by 45 million years ago.

One possible explanation for the distribution pattern of betalain-producing centrosperms is that the ancestors of the cacti and of the phytolacs were present in South America and Africa around the time or soon after Africa and South America were separated by a substantial ocean in the South Atlantic. This disjunction could have been a vicariance of large, widespread initial populations or could have been achieved via long-distance dispersal. Our hypothesis also suggests that the cacti differentiated from the ancestral group following geographic separation, which would have occurred since 100 million years ago. Likewise, the Didiereaceae of Madagascar, which closely resemble Portulacaceae first and then Cactaceae, would have differentiated on Madagascar independent from the population that gave rise to the Cactaceae. The development of similar vegetative characteristics in Cactaceae and Didiereaceae, including short shoots and the wood structure, would be a case of parallel evolution, whereby they started out from a similar plant design and developed a similar shoot structure but very different structures of their flowers and fruits.

Center of Origin and Patterns of Cactus Radiations

One common pastime of biogeographers has been to determine the center of origin, or birthplace, for a group. To do this, early biogeographers developed a list of rules for determining where the center might be. Researchers hoped that they could get the correct answer by responding to a series of simple questions about the group, such as where the most species or the most primitive species occur, assuming that most species stay near their birthplace or that at least the ones with the primitive features do. Many, if not all, of the criteria used by early biogeographers contained serious flaws in logic, and today workers have essentially abandoned the search for a single, static birthplace and prefer instead to study the geographic spread and fragmentation of taxa through time.

All of this notwithstanding, the cactus phylogeneticist Buxbaum, discussed in Chapter 10, made a case for using the West Indies as the center of origin for the cacti, because, he reasoned, some species with putatively primitive features occur there. A number of cactologists have repeated this idea, but we feel that Buxbaum's choice of the West Indies is not correct. First, the current West Indies were not always where they are today. About 90 million years ago the Greater Antilles (Cuba, Jamaica, Hispaniola, and Puerto Rico) were situated to the west, where they constituted Central America (Fig. 11.18*A*). At that time the Greater Antilles were a chain of large islands between Mexico of North America and Colombia of South America. The movements of the earth's crustal plates, as part of continental drift, caused these islands to be displaced eastward from their position in Central America, eventually to reside in the Gulf of Mexico, while a new Central America was being formed (Fig. 11.18*B–D*). The Lesser Antilles, which presently form the eastern boundary of the Caribbean region, arose fairly recently from the floor of the seas from volcanic activity (Fig. 11.18*D*). Consequently, if the origin of the Cactaceae occurred before 90 million years ago, then the West Indies could not have been the place, at least not the islands presently residing in the middle of the Gulf of Mexico.

We can also ask whether cacti with very primitive features still occur in the West Indies. For example, let us examine the leafy cacti of the genus *Pereskia.* Species of the group to which *P. sacharosa* belongs (Table 2.1) occur exclusively in northern and eastern South America, especially in Brazil. The second group has a broad range: one group of species passes through Central America into southern Mexico; several species occur in northern South America; one group of species occurs in the Andes of Peru and Bolivia; and one species, the very widespread *P. aculeata,* occurs around the Caribbean area, including the West Indies, northern South America, west coastal Central America, and Mexico. The third portion of the genus has two species in northern South America and two other species, considered to be derived taxa, that occur in the Greater Antilles. Taking these distributions at face value, we would have to conclude that the West Indies have received its few pereskias from northern South America, that is, from seasonally dry areas in Venezuela and neighboring Brazil.

The analysis can be extended to the subfamily Opuntioideae, which must be an early offshoot of the leafy cacti. Here all of the primitive clades of this subfamily are absent from the West Indies. A species of the leaf-bearing *Quiabentia* and the hummingbird-pollinated *Tacinga* occur in Brazil; whereas other species of *Quiabentia* and the genera *Austrocylindropuntia* and *Tephrocactus* occur in western and southern areas of South America. Likewise, the clades of North American Opuntioideae with cylindrical (not flattened) stems show no primitiveness in the West Indies. The leafy species of *Pereskiopsis* are found in Mexico and Guatemala; and only one cylindropuntia, *O. caribaea,* occurs in the West Indies, whereas all the rest of them — nearly 40 species — occur in Mexico and United States. Among the groups of platyopuntias, most taxa are restricted to either North or South America and only a few of them occur in the West Indies. Of these, several are taxa that occur widely around the Caribbean region, and only one, the dominant arborescent consoleas, is endemic to the islands. Once again, for the West Indies there is

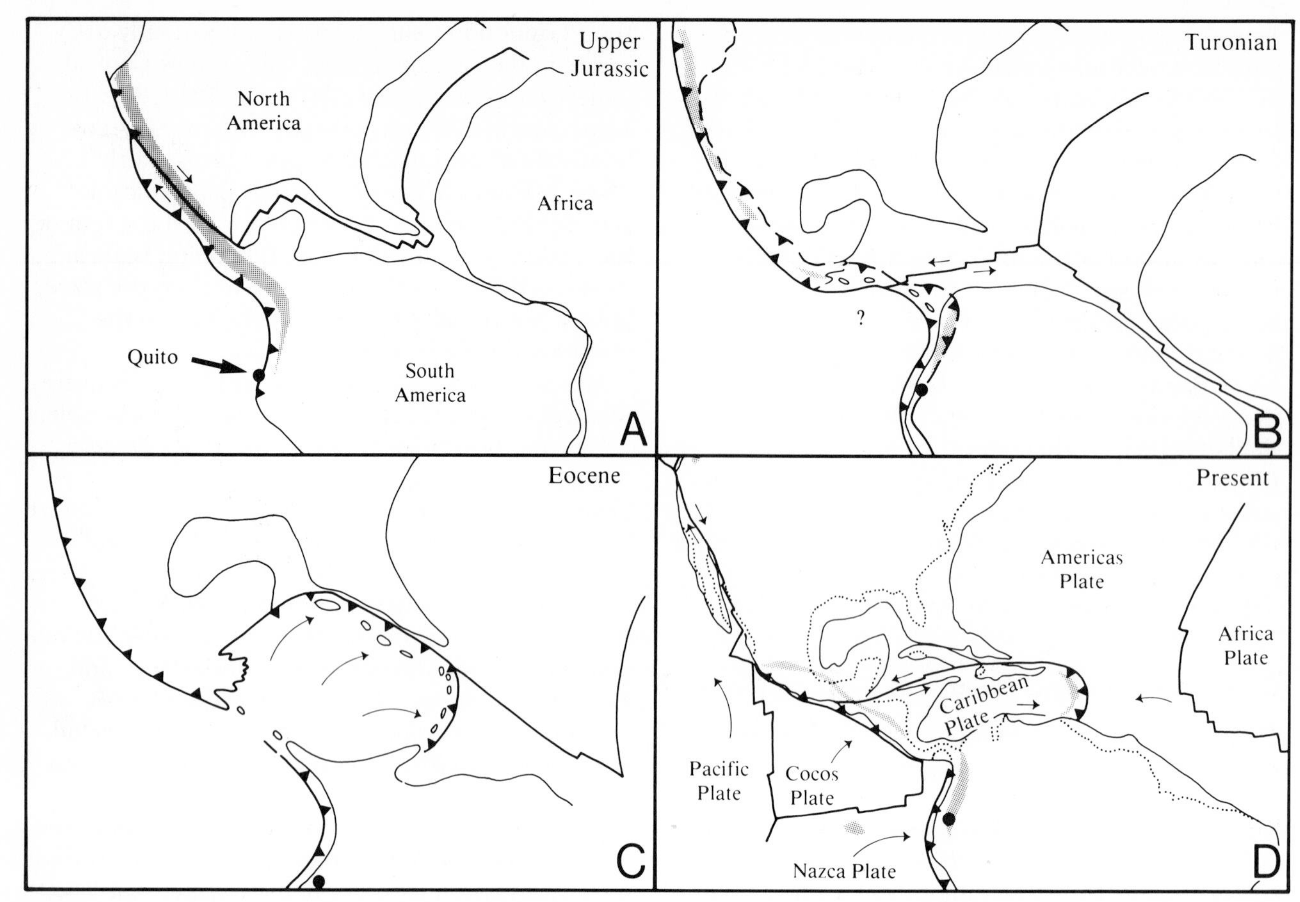

Figure 11.18. Schematic maps *(A–D)* of the geologic history of Central America and the West Indies from the Upper Jurassic (150 million years ago) to the Turonian (90 million years ago) to the Eocene (45 million years ago) to the present. In the Jurassic *(A)* the Gulf of Mexico did not exist, and North and South America were much closer together. By the Turonian *(B)* the Gulf of Mexico had begun to form as North and South America started to separate via spreading of the seafloor between the two continents; Central America was not solid but was a series of islands located between southern Mexico and Colombia. In the Turonian the islands between Mexico and Colombia began their displacement northeastward to become the Greater Antilles in the West Indies. Movement of islands into the Gulf of Mexico left a vacancy in the region of Central America during the Eocene *(C)*. Subsequent movements of oceanic plates in the eastern Pacific *(D)* caused a new Central America to form, mostly within the last six million years. (Reproduced by permission from J. H. Brown and A. C. Gibson, *Biogeography,* St. Louis, 1983, The C. V. Mosby Co.)

a pattern of receiving derived clades from the mainland areas and not being the center of origin for the major opuntioid lineages.

Evolution of the Tribes of Cactoideae

As mentioned in Chapter 10 and as noted in Table 1.2, each tribe of Cactoideae tends to be restricted in distribution to a portion of the Western Hemisphere. There are exceptions, of course, such as that previously described for *Rhipsalis.* Yet, even though cacti generally have juicy fruits, long-distance dispersal of most tribes of Cactoideae has apparently not been effective in causing these cacti to range widely between the continental areas or even within a continent. For example, the cactus floras in western South America on either side of the Andean Cordillera are extremely different, a fact indicating that they have not been able to cross the Andes since the great height of the mountain ranges was established.

Because there are prominent geographic and climatological barriers in the range of the cactus family, it is not surprising to find that many of the genera and even the tribes are limited by these barriers. Without dwelling on the details of each, we shall indicate the distributions of the tribes of the subfamily Cactoideae (Table 1.2).

Leptocereeae. Tribe Leptocereeae, which is presumed to show the primitive features of the subfamily, occurs in northwestern South America, Central America, and the West Indies. About half of the species occur in the West Indies, but there are mainland species with primitive features in Bolivia and Peru. Two genera, *Neoraimondia,* a distinctive genus with very long-lived flowering shoots (Fig. 5.35), and *Armatocereus,* the species of which are distributed as far north as Costa Rica, have evolved on the western slopes of the Andes. *Lemaireocereus hollianus* from southern Mexico and *Jasminocereus* and *Brachycereus* from the Galápagos Archipelago may also belong to this tribe.

Browningeae. Tribe Browningeae consists of several small genera found in Peru and northern Chile. These species, which have well-developed, overlapping bracts covering the ovary and floral tube, are intermediate in some features between Leptocereeae and the next tribe, Trichocereeae.

Trichocereeae. In South America, tribe Trichocereeae has evolved many species, particularly on the western slopes of the Andean Cordillera. As illustrated earlier (Figs. 8.1 – 8.8), the diversity of growth habits in this group is very broad. This tribe is characterized by having long silky hairs on the areoles of the floral tube. Within this tribe the species have evolved floral designs that use a variety of animal pollinators.

Pachycereeae. As discussed in Chapter 10, tribe Pachycereeae has radiated in Mexico to form the familiar species of columnar cacti. The species with plesiomorphic features all reside in southern Mexico, especially in the Valley of Tehuacán in Puebla, and lines of morphological specialization seem to diverge from that region, mostly northward but also into Central America and the Caribbean region. Major structural changes in this tribe have occurred in growth habit, floral design (Fig. 10.2), rib design, and internal stem anatomy.

Hylocereeae. Tribe Hylocereeae, which is poorly defined, includes terrestrial and epiphytic cacti. The primitive growth habits and vegetative forms occur in genera such as *Harrisia,* which is widespread in the West Indies (to Florida and the Bahamas), and other woody groups, such as *Dendrocereus* of the Greater Antilles and *Acanthocereus,* which is very widespread but especially common in the Caribbean region. From terrestrial forms have evolved plants that root in the ground but climb up trees or clamber over rocks, and these in turn evolved into many types of epiphytes (Chapters 6 and 8). Epiphytes have adaptations for coping with low light and damp environments. The epiphytes have evolved as a diverse group of genera, with greatest concentrations in Central America and western Mexico, on islands in the West Indies, and to a lesser degree in northernmost South America. Two species are native to the Amazon Basin, and *Mediocactus* occurs in Peru. The geophytic species of *Peniocereus* (Fig. 11.19) are presumably derived from *Acanthocereus.*

Figure 11.19. *Peniocereus rosei,* a geophytic cactus species from Mexico, with a greatly enlarged root and slender stems.

Cereeae. The Cereeae are a group of highly specialized columnar cacti, which have evolved mainly in eastern South America, especially in Brazil. The tribe generally lacks spine clusters and even can lack areoles on the floral tube and ovary. Specializations in this tribe include the evolution of *Melocactus* (Figs. 5.30 and 5.31), which is a barrel cactus that forms a true cephalium; and this genus has spread throughout the West Indies, northern and northwestern South America, and Central America to southernmost Mexico.

Echinocereeae. Tribe Echinocereeae presently includes the mound-forming hedgehog cacti of North America (*Echinocereus;* Fig. 1.9) and the geophytic wilcoxias *(Wilcoxia).* No one has been able to determine whether *Bergerocactus emoryi* of coastal Baja California and adjacent southern California belongs here or in another tribe, for example, Hylocereeae.

Cacteae. Cacteae, the largest tribe of the subfamily, occurs mainly in North America, especially in Mexico, although a handful of species live in dry habitats of Central America, northern South America, and the West Indies. This tribe lacks tall columnar forms, although some of the tallest columnar forms of *Ferocactus* may qualify as columnars. Much of the evolution of this tribe has involved modifications of tubercles and areoles, for example, the specialization in *Mammillaria* of axillary rather than areolar flowering meristems. Flowers in many genera lack areoles on the pericarpel and the floral tube.

Notocacteae. Many of the species of Cactoideae found in southern South America belong to tribe Notocacteae, which is closely related to tribe Trichocereeae. Notocacteae has a few columnar forms in *Corryocactus,* but mostly it has small growth habits, including several types of barrel cacti, low caespitose plants, and many globular forms with brightly colored flowers as well as a variety of epiphytes

(*Rhipsalis* and its close relatives). Although much of the evolution in this tribe has occurred east of the Andes, an important radiation of this tribe also occurred in Chile.

Within the next decade cactologists can expect to see a phylogenetic classification of the cactus tribes and genera. Already many areas of research are adding substantial data that can be used in phylogenetic analyses. One by one the genera are being scrutinized closely. As this process progresses, some of the genera in Table 1.2 will become redefined to include a different list of species, and probably some of the generic names will also have to change. As the names change, our perceptions of the evolutionary histories of the groups of cacti also change and become more accurate; consequently scientists are then able to say more about how the genera and species have arisen.

The Evolution of Cactus Characteristics—An Overview

In describing the biology of an organisms, from simple multicellular ones, such as an earthworm or a moss, to complex ones, such as humans or cacti, the tendency is to concentrate on the pieces, losing sight of the whole. Now is the time to pull some of the pieces together to reconstruct the evolution of cacti.

Origin of the First Cactus

The hypothetical cactus ancestor that was described at the beginning of this chapter can be used to make comparisons with other nonsucculent dicotyledons, that is, to locate a family of dicotyledons that most closely resembles cacti and that could be the sister taxon of the centrospermous order. Many families have been suggested as candidates, for example, the Dilleniaceae, the Theaceae, and the Schizandraceae, but none has yet emerged as the winner. At the very least we would expect the nonsucculent sister taxon would be a woody plant with *Pereskia*-like secondary xylem, helically arranged simple leaves with C_3 photosynthesis, centrospermous pollen grain and seed morphology, sieve-tube-element plastids with proteinaceous inclusions, and flowers with separate petals and some separate carpels. Betalains are optional for the sister taxon, but it probably could not have anomalous secondary thickening because this trait developed within the centrospermous order.

In addition, once we have defined the hypothetical ancestor, we can easily picture which apomorphic features set the stage for the evolutionary divergence of the cactus family. The list of these features is fairly short: precocious development of spine-bearing axillary buds that formed spine-bearing short shoots (areoles); a shoot apex that has visually discernible zones rather than a clear tunica–corpus organization; and a flower in which the carpels are starting to be covered by stem tissues. These are the three specializations that distinguish *Pereskia* from any other family of dicotyledons. In the first year the areolar meristem of a cactus produces spines from leaflike primordia; but in some of the pereskias, which have well-developed short or spur shoots, leaves are produced on the spiny areole during subsequent years.

Development of long shoot–short shoot organization is not, of course, unique to cacti; it is frequently observed in woody plants of seasonally dry tropical to warm temperate habitats. Plants with short shoots are often drought deciduous, producing leaves rapidly at the beginning of a growing season, usually immediately after the first effective rainfall, and then shedding leaves weeks or months later when the soil dries out. We can therefore assume that the ancestors of cacti also lived in seasonally dry habitats having a drought-deciduous vegetation of medium-sized shrubs or small trees. In many places around the tropics this type of vegetation is called tropical deciduous thorn forest or thornscrub (Fig. 11.20); and as these names imply, the plants are armed with thorns or spines to deter foraging by large herbivores. Once again, the cacti fit this model because spines emerge precociously on the young cactus shoots, which would be eaten without that type of protection. Interestingly, the flowers and fruits of pereskias tend to lack any armature, a fact indicating that spination on reproductive structures followed shoot spination.

The early evolution of the cactus shoot involved three interrelated events. First, leaves were reduced in size and at the same time became more fleshy, and very early in evolution a stage was reached in which the leaves were reduced to ephemeral and minute structures (Fig. 11.21). Second, as leaves became reduced in size, succulent tissues were formed in the stem cortex and pith. Third, the plant shifted from C_3 photosynthesis in the ancestral cacti to CAM in the succulent ones.

A decrease in leaf area can lower leaf temperature for exposed leaves and thus keep the leaf temperature close to that of the surrounding air. One reason to keep the leaf temperature low is that temperatures above about 50°C are usually lethal to cells. If a leaf cannot transpire enough to keep itself relatively cool, as would occur during water stress, the leaf protein may "cook," that is, denature and become nonfunctional, and the leaf may then fall off. Another important reason for having low leaf temperature is to reduce daytime transpiration. Referring back to

Figure 11.20. Typical landscape of tropical deciduous forest in Puebla in southern Mexico, showing a stand of the solitary columnar cactus, *Neobuxbaumia mezcalaensis.* Columnar species of Mexico are commonly found in these habitats, as are many small cacti and even a species of *Pereskia.*

Equation 1.1, we observe that the rate of water leaving unit area of a leaf or stem is equal to the stomatal conductance for water vapor times the drop in water vapor concentration from the plant tissue to the air. Also, as discussed in Chapter 1, the saturation water vapor concentration is very dependent on temperature. Consequently, when leaf temperature is high, the water vapor concentration from the internal leaf tissues to the air is also usually high.

Figure 11.21. A young, leaf-bearing joint of a cholla, *Opuntia acanthocarpa,* in southern Arizona.

Development of water-storage tissues in the leaf or the stem can help avoid water stress. The water storage of a plant is increased not only by producing more cells but by increasing the size of the cell and especially by increasing the size of the vacuole in each of the cells. Coincidentally, a large vacuole is a primary requirement for CAM, in which large quantities of malate are accumulated each night and lost each day from the vacuole (Chapter 4). All land plants apparently have PEP carboxylase, but a CAM plant has to produce greater amounts of it, regulate it differently, and shift stomatal opening to the nighttime.

Decreases in leaf size and the acquisition of CAM are adaptations for living in a hot, dry climate where drought and water stress are common. This is not the only strategy for life under these conditions, but CAM offers certain distinct advantages, such as nocturnal stomatal opening, which avoids periods when temperatures are high, and the storage of a considerable volume of water per unit of transpiring area. The major disadvantage of CAM is that growth rate and primary production during the most favorable period can be low compared with typical leafy plants growing under the same conditions, in part because of the more complicated method of processing CO_2, with its intermediate storage of malate. Moreover, because of water storage and support tissues, the volume fraction of a cactus stem occupied by chloroplasts, where photosynthesis takes place, is less than the chloroplast volume fraction for a typical C_3 leaf.

Evolution of the Succulent Cactus Growth Form

Cactologists commonly recognize their plants by the presence of ribs or tubercles and large succulent tissues. In the early stages of cactus evolution, however, neither ribs nor tubercles were present. For example, in the species of *Pereskiopsis, Quiabentia, Tacinga, Austrocylindropuntia,* and *Maihuenia* that have smaller leaves than the pereskias and succulence in the stem, tubercles are not present. Therefore, because these are the genera that have retained many primitive features, we may assume that the origin of tubercles and certainly ribs appeared later in the evolution of the cactus family and actually only appeared when the leaf blade became vestigial.

For a plant population that is evolving as a stem succulent with CAM, large leaves may be a detriment because they could intercept much of the PAR needed by the stem. On the other hand, loss of leaves might cause an increase of stem succulence because growth substances produced in the leaf, which exert many controls on stem development, cannot be made. For example, growth substances, especially auxin, that are produced during the period of rapid blade expansion, appear to influence xylem development in nonsucculent plants; when leaves are removed, wood fiber development is suppressed and parenchyma cells are formed in the place where fibers would have occurred. No one has examined what happens to shoot development when the young leaf primordia of *Pereskia* are removed.

Stem spination (Fig. 11.22) has been one of the features of cacti that has had a complex evolutionary history, even though its general structure and development are fairly uniform throughout the family. Assuming that spines were first formed to prevent excessive herbivore damage to the shoots, we can also assume that this function has remained except in a few highly specialized small cacti. Spines have also had the detrimental effect of decreasing PAR interception by the stem but have been useful in moderating the stem temperatures, thereby avoiding lethal temperatures. Research has not been done on many types of spines to determine the relative importance of spines for temperature control versus that required for defense of the stems, flowers, and fruits from herbivores.

Figure 11.22. *Espostoa melanostele,* a plant with much apical spination and pubescence; this species grows near the equator on the dry, lowland, western slopes of the Andean Cordillera in Peru.

In Chapter 6 we discussed the effects of the development of tubercles and ribs on PAR interception, and hence on the productivity of the stem, and on volume-to-surface ratios. Tubercles are formed by the peripheral meristem, a special layer of subsurface dividing cells that nonsucculent centrospermous plants do not have; and cactologists suspect that ribs are formed from another special and deeper primary thickening meristem. What causes these regions to undergo cell division has not been studied but in general meristems of this kind have been shown to be strongly influenced by hormones, especially gibberellins or cytokinins. For example, gibberellins, which influence the development of spine primordia on the areole (Chapter 5), also stimulate the formation of collenchyma. But changes in tissue pH can also evoke an increase in collenchyma. In contrast, auxins have been shown to promote periderm formation and the development of strongly lignified cells in the xylem and phloem, all of which cactus shoots generally lack.

When lower than normal levels of auxins are used in plant experiments, all types of interesting anatomical changes occur, for example, a delay in the completion of the vascular cylinder (= large primary rays?), reduction in the amount of secondary xylem produced per year, changes in the lengths and diameters of wood cells, and development of long shoot–short shoot organization. All this is excitingly suspicious for cacti. A strong possibility exists that the reduction in leaf size triggered changes in hormonal balance in the shoot and then set into motion a long series of anatomical consequences, including the development of areoles with spines, ribs, increased succulence, and modifications of the wood. Although few studies have addressed the issue of hormone activity and cell division, in studies on apices of the hedgehog cactus *Echinocereus engelmannii,* Mauseth has shown that cytokinin induces the formation of the pith–rib meristem. Needed now is a series of studies on the actual hormonal levels of cactus tissues and their sensitivities to hormones and other potential growth-inducing substances.

Growth Habit

Growth habit is described by such parameters as total plant height, length and thickness of the trunk, degree of erectness of the plant and individual branches, and the nature of branch formation (Chapter 8). The mechanism for producing each design is related to the mechanical strength of the stem, namely, the architectural design of the woody skeleton, the amount of wood accumulated in each portion of the plant, the weight of the stem being supported, the cellular properties of the wood (Fig. 11.23), and the amount and distribution of succulence, including perhaps the nature of the ribbing.

From an ecological point of view (Chapters 7 and 8), cactus growth habit does not appear to be random but instead seems to be a direct response to conditions in its natural environment. The overall growth habit has evolved over long periods of time, but it is commonly modified from site to site as the cactus species responds in evolutionary and ecological ways to its surroundings. Many of the early data support the hypothesis that shrubby and arborescent cacti have been responding mainly to competition for light with other plants by increasing their overall height, by increasing growth rate to decrease the time that they are shaded, and to modify ribbing to maximize the effectiveness of PAR interception. In cold parts of its range, changes in growth form and habit may also reduce the risk of freezing.

Other environmental factors that have yet to be systematically considered are aspects of water storage, wind damage, and nutrient limitations. At the same time, biotic factors such as pollination requirements and seed dispersal must be considered and entered into the equation. For example, cacti that are bat-pollinated need to have flowers on shoot tips above the neighboring vegetation so that they can be located easily and be accessible to the bats. It is

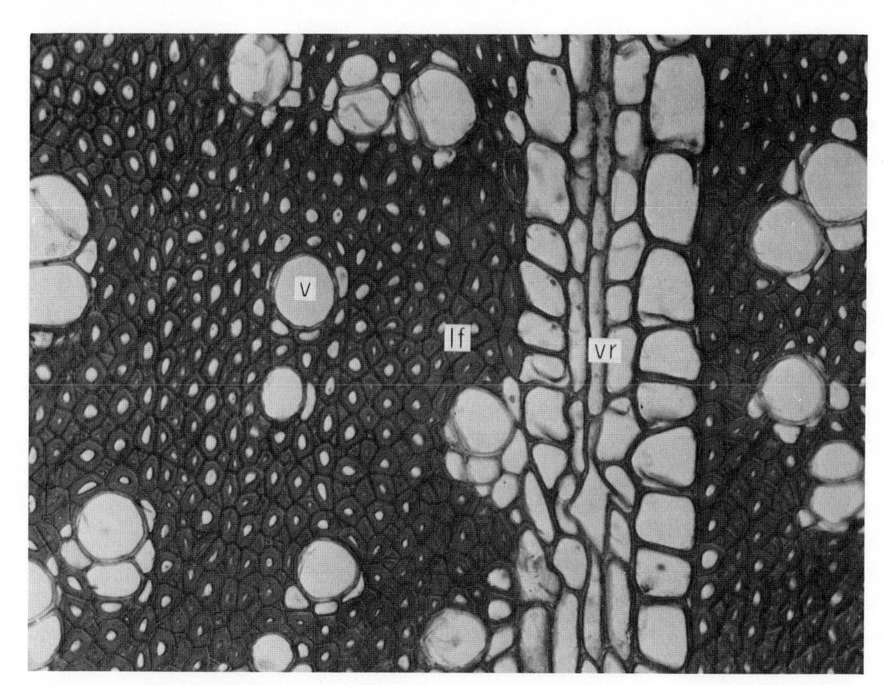

Figure 11.23. Cross section of the wood of the solitary columnar cactus, *Neobuxbaumia mezcalaensis* (see Fig. 11.20), from southern Mexico. This is one of the hardest cactus woods because the cells have relatively thick, lignified cell walls. The three conspicuous components of this wood are the vessels (v), the very thick-walled libriform wood fibers (lf), and the vascular ray (vr). The width of this section is about 1 mm.

Figure 11.24. A community of different cactus growth habits in the Valley of Tehuacán, Puebla, Mexico. Note the large barrel cacti in the foreground and the caespitose plant to the left.

intriguing to imagine that reproductive issues, such as how many seeds are produced on a single plant, may be strategies that are in conflict with vegetative issues, such as PAR interception or water storage (Fig. 11.24). Indeed, this is an area where in-depth comparative studies should be done on growth habits in a single genus to tell us what factors have caused growth habit to evolve in so many different directions.

As studies on growth form and growth habit mature, there will have to be concomitant studies by laboratory scientists on the biochemistry of the stem, studies investigating such important subjects as cellular heat tolerance — which in cacti is among the highest known for vascular plants — and the importance of secondary compounds for reducing herbivory. Recognizing that biology is divided into at least three levels of study relevant to our present considerations — organismal, cellular, and molecular — it is encouraging to see one group in which the discoveries in each area can have significant impacts on the others.

We hope that the reader now has a greater appreciation of the complexity of cacti. Perhaps some of the innocence of cactus watching has been displaced by deeper inquiries into the nature of a cactus plant — its structure, development, physiology, and evolution. Gone also may be some of the old myths about cactus biology, which can no longer by supported by the data collected by scientists over the last 25 years. Replacing the old myths are new ideas — perhaps new myths — that must be tested again and again to determine their ability to explain the facts about cacti. Today the explanations are appealing, but tomorrow, when more data are available and the methods for analyzing biological processes are more sophisticated, today's explanation may have to be discarded.

There is excitement associated with scientific discovery, because as one question is answered a new set of questions appears on the horizon to challenge the investigator. Certainly this book, in reviewing many aspects of cactus biology, has answered some questions that have puzzled cactologists for centuries. But in the process of answering a few questions, we have left unanswered an incredible number of equally interesting questions, questions that need to capture the interest of a new generation of cactophiles. A primer is, as the name implies, the first and not the final chapter of the learning experience.

SELECTED BIBLIOGRAPHY

Airy Shaw, H. K. 1973. *A dictionary of the flowering plants and ferns,* 8th ed. Cambridge University Press, Cambridge.

Bedell, H. G. 1980. A taxonomic and morphological re-evaluation of Stegnospermaceae (Caryophyllales). *Systematic Botany* 5:419–431.

Behnke, H.-D. 1976a. Ultrastructure of sieve-element plastids in Caryophyllales (Centrospermae), evidence for delimitation and classification of the order. *Plant Systematics and Evolution* 126:31–54.

——— 1976b. Sieve-element plastids of *Fouquieria, Frankenia* (Tamaricales), and *Rhabdodendron* (Rutaceae), taxa sometimes allied with Centrospermae (Caryophyllales). *Taxon* 25:265–268.

——— 1981. Sieve-element characters. *Nordic Journal of Botany* 1:381–400.

——— 1983. Sieve-element plastids of Caryophyllales: additional investigations with special reference to the Caryophyllaceae and Molluginaceae. *Plant Systematics and Evolution* 142:109–115.

Behnke, H.-D, and W. Barthlott. 1983. New evidence from ultrastructural and micromorphological fields in angiosperm classification. *Nordic Journal of Botany* 3:43–66.

Behnke, H.-D., T. J. Mabry, P. Neumann, and W. Barthlott. 1983. Ultrastructural, micromorphological and phytochemical evidence for a "central position" of *Macarthuria* (Molluginaceae) within the Caryophyllales. *Plant Systematics and Evolution* 143:151–161.

Bessey, C. E. 1915. The phylogenetic taxonomy of flowering plants. *Annals of the Missouri Botanical Garden* 2:109–164.

Braun, A. 1864. Übersucht des natürlichen Systems. Pp. 22–67 in *Flora der Provinz Brandenburg, der Altmark und des Herzogthums Magdeburg,* ed. P. Ascherson. August Hirschwald, Berlin.

Bregman, R., and F. Bouman. 1983. Seed germination in Cactaceae. *Botanical Journal of the Linnean Society* (London) 84:357–374.

Brown, G. K., and G. S. Varadarajan. 1985. Studies in Caryophyllales I: re-evaluation of classification of Phytolaccaceae. *Systematic Botany* 10:49–64.

Brown, J. H., and A. C. Gibson. 1983. *Biogeography.* C. V. Mosby, St. Louis.

Burret, F., P. Lebreton, and B. Voirin. 1983. Les aglycones flavoniques de cactees: distribution, signification. *Journal of Natural Products (Lloydia)* 45:687–693.

Cronquist, A. 1968. *The evolution and classification of flowering plants.* Houghton Mifflin, Boston.

——— 1981. *An integrated system of classification of flowering plants.* Columbia University Press, New York.

Dahlgren, R. M. T. 1980. A revised system of classification of the angiosperms. *Botanical Journal of the Linnean Society* 80:91–124.

Eckardt, T. 1976. Classical morphological features of centrospermous families. *Plant Systematics and Evolution* 126:5–25.

Ehrendorfer, F. 1976a. Chromosome numbers and differentiation of centrospermous families. *Plant Systematics and Evolution* 126:27–30.

——— 1976b. Closing remarks: systematics and evolution of centrospermous families. *Plant Systematics and Evolution* 126:99–105.

Eichler, A. W. 1878. *Blüthendiagramme,* Part 2. W. Engelmann, Leipzig.

Esau, K., and V. I. Cheadle. 1969. Secondary growth in *Bougainvillea. Annals of Botany* 33:807–819.

Gershenzon, J., and T. J. Mabry. 1983. Secondary metabolites and the higher classification of angiosperms. *Nordic Journal of Botany* 3:5–34.

Gibson, A. C. 1978. Rayless secondary xylem of *Halophytum. Bulletin of the Torrey Botanical Club* 105:39–44.

Hallier, H. 1912. L'origine et le système phylétique des angiospermes exposés à l'aide de leur arbre généalogique. *Archives Néerlandica des Sciences Exactes Naturelles,* sér. 3, B, 1:146–234.

Horak, K. E. 1981a. Anomalous secondary thickening in *Stegnosperma* (Phytolaccaceae). *Bulletin of the Torrey Botanical Club* 108:189–197.

——— 1981b. The three-dimensional structure of vascular tissues in *Stegnosperma* (Phytolaccaceae). *Botanical Gazette* 142:545–549.

Mabry, T. J. 1974. Is the order Centrospermae monophyletic? Pp. 275–285 in *Chemistry in botanical classification,* ed. G. Bendz and J. Santesson. Academic Press, New York.

——— 1976. Pigment dichotomy and DNA–RNA hybridization data for centrospermous families. *Plant Systematics and Evolution* 126:79–94.

——— 1977. The order Centrospermae. *Annals of the Missouri Botanical Garden* 64:210–220.

Mabry, T. J., P. Neuman, and W. R. Philipson. 1978. *Hectorella:* a member of the betalain-suborder Chenopodiineae of the order Centrospermae. *Plant Systematics and Evolution* 130:163–165.

Maheshwari, P. 1950. *An introduction to the embryology of angiosperms.* McGraw-Hill, New York.

Mauseth, J. D. 1984. Introduction to cactus anatomy. Part 10. Flowers and sex. *Cactus and Succulent Journal* (Los Angeles) 56:212–216.

Melchior, H. 1964. *A. Englers Syllabus der Pflanzenfamilien,* Vol. II, 12th ed. (2). Borntraeger, Berlin-Nikolassee.

Nowicke, J. W. 1969. A palynotaxonomic study of the Phytolaccaceae. *Annals of the Missouri Botanical Garden* 55:294–363.

——— 1976. Pollen morphology in the order Centrospermae. *Grana* 15:51–77.

Nowicke, J. W., and J. J. Skvarla. 1977. Pollen morphology and the relationships of the Plumbaginaceae, Polygonaceae, and Primulaceae to the order Centrospermae. *Smithsonian Contributions in Botany* 37:1–64.

Rodman, J. E., M. K. Oliver, R. R. Nakamura, J. U. McClammer Jr., and A. H. Bledsoe. 1984. A taxonomic analysis and revised classification of Centrospermae. *Systematic Botany* 9:297–323.

Sarmiento, G. 1975. The dry plant formations of South

America and their floristic connections. *Journal of Biogeography* 2:233–251.
Skvarla, J. J., and J. W. Nowicke. 1976. Ultrastructure of pollen exine in centrospermous families. *Plant Systematics and Evolution* 126:55–78.
——— 1982. Pollen fine structure and relationships of *Achatocarpus* Triana and *Phaulothamnus* A. Gray. *Taxon* 31:244–249.
Takhtajan, A. 1969. *Flowering plants: origin and dispersal.* Smithsonian Institution Press, Washington.
Thorne, R. F. 1968. Synopsis of a putatively phylogenetic classification of the flowering plants. *Aliso* 6 (4):57–66.
——— 1981. Phytochemistry and angiosperm phylogeny: a summary statement, including a synopsis of the class Angiospermae (Annonopsida). Pp. 233–295 in *Phytochemistry and angiosperm phylogeny,* ed. D. A. Young and D. S. Seigler. Praeger, New York.
Wettstein, R. 1935. *Handbuch der systematischen Botanik,* Vol. II. Franz Deuticke, Leipzig.

Glossary

Index

Glossary

abaxial Directed away from the axis; on a leaf, usually the lower side. (Contrast with adaxial.)
abiotic Pertaining to the nonliving components of the environment, for example, temperature, water, minerals, and solar radiation. (Contrast with biotic.)
absorptance The fraction of radiant energy absorbed by a surface.
accessory vascular bundle In cacti, the minor vascular bundle that arises from an axial bundle or a trace and vascularizes the succulent regions of the cortex, especially in the ribs and tubercles.
acropetal development Development that occurs sequentially from the base to the tip. (Contrast with basipetal development.)
adaptation A feature that substantially improves the ability of an organism to survive and to leave more offspring relative to that of other forms.
adaxial Directed toward the axis; on a leaf, usually the upper side. (Contrast with abaxial.)
adnate Fused to something that is different.
adventitious root A root that originates on a stem or leaf instead of on other roots.
agglutination The massing together of particles or cells.
aglycone A molecular form not containing a sugar.
alkaloid A bitter-tasting, basic (alkaline) compound with at least one nitrogen atom that is usually incorporated in a carbon ring.
allopatric Referring to populations or species that occur in geographically different (nonoverlapping) places. (Contrast with sympatric.)
allopatric speciation by disjunct founders The formation of a new species occurring when one or several individuals start a geographically separate population.
allopatric speciation by geographic subdivision The formation of a new species occurring when a large population is subdivided into two or more smaller, geographically separate populations; geographic speciation.
alpha taxonomy The branch of biology that describes and gives legal, latinized, scientific names to species.
alternate pitting The arrangement of pits in diagonal rows on a vessel or a tracheid.
anatomy The study of tissue structure, usually at the microscopic level.
angiosperm A group of seed plants in which the seed forms within a mature ovary (fruit).
anisotropic See birefringent.
annular Forming a ring.
anomalous secondary thickening by successive cambia An increase in stem or root diameter in which the initial vascular cambium functions for a short time before it is superceded by new cambia that arise in the parenchyma of the phloem or cortex.
anther The pollen-bearing part of a stamen.
anthesis The time when the flower is fully expanded.
anthocyanin A water-soluble, red, purple, or blue flavonoid pigment occurring in vacuoles.
anticlinal Referring to the orientation of a cell wall or a plane of cell division perpendicular to the nearest surface. (Contrast with periclinal.)
aperture (of a pollen grain) A thin, depressed region in the wall of a pollen grain through which germination occurs; a colpa.
apex (adj. **apical**) The tip of an organ.
apical dominance The condition whereby the shoot tip strongly inhibits the growth of axillary buds.
apical meristem The topmost, domelike tip of a stem, root, flower, or leaf primordium in which the initial cells of the organ are formed.
apical zonation In cacti, the special organization of initials in the apical meristem as groups of cells (zones) rather than as discrete cell layers.
apocarpy (adj. **apocarpous**) The condition in which the female part of the flower is composed of two or more carpels that are not fused. (Contrast with syncarpy.)
apomixis The reproduction of a plant by any means other than through the fusion of male and female sex cells.
apomorphy (adj. **apomorphic**) In cladistics, a derived character state. (Contrast with plesiomorphy.)
appendicular model (of the inferior ovary) A model of the origin of the inferior ovary in which the outer structures of the flower (especially the hypanthium) are appended to (fused with) the ovary. (Contrast with receptacular model.)
appressed Pressed closely, or flattened against, the axis or substrate.
arborescent Being treelike.

archesporial cell A cell in a flower that gives rise to the haploid sex cells and surrounding nutritive tissue.
areolar meristem The apical meristem of an areole; comparable in form and function to a shoot apical meristem but destined to form spine primordia and a flower bud.
areole An axillary bud that immediately produces a cluster of spine primordia.
aril An outgrowth that develops in the region of the hilum and partially or entirely envelops the seed.
ascending receptacular system In the cactus flower, the portion of the vascular bundles that passes upward through the pericarpel tissue before the vascular tissues bend down toward the ovary.
asymmetrical areole An areole on which the production of spines is largely or entirely on one side. (Contrast with radial areole.)
autogamy See self-pollination.
auxin An acidic plant hormone that characteristically has either an indole or a naphthalene structure.
axial Referring to a structure oriented with its longest side parallel to the main axis of a plant organ, usually vertical.
axial bundle The vertical bundle of the vascular cylinder that gives rise to the major vascular bundles of the leaf.
axial wood parenchyma Parenchyma cells in the axial (vertical) system of the secondary xylem.
axil The upper (adaxial) angle between the leaf and a stem.
axile placentation The condition in which the position of the placenta and the ovules is along a central axis of the ovary. (Contrast with parietal placentation.)
axillary bud A bud that forms in the axil of a leaf; a lateral bud.

bark A general term used for the thick outer covering of a plant; this structure is composed principally of dead cork cells. See also periderm.
barrel cactus A general term for a cactus with a single thick, ribbed columnar stem 0.5 to 2 m in maximum height.
barrier In biogeography, any biotic or abiotic feature that totally or partially restricts the movement of organisms or their genes from one population or locality to another.
basal cell (of a trichome) The lowermost cell of a trichome, which is located adjacent to the original epidermis and divides repeatedly, thereby forming cells to the outside that increase trichome length.
basal meristematic zone (of a spine primordium) The region of actively dividing, meristematic cells at the base of a spine primordium; the cells in this region produce cells to the upper side during spine formation.
basipetal development Development that occurs sequentially from the tip to the base. (Contrast with acropetal development.)
berry A fleshy fruit that has a soft fruit wall (pericarp) and, usually, many small seeds.
betalain A vacuolar pigment having a skeleton with 15 carbon atoms and 1 nitrogen atom; in cacti, the chemical typically responsible for yellow (betaxanthin) to red or purple (betacyanin) colors.
binomial The two-part, latinized scientific name of an organism; the parts are the genus and the species names.
biogeography The study of the geographic distributions of organisms, both past and present.
biomass The amount of living matter of an organism, population, or community.
biosynthesis (adj. **biosynthetic**) The manufacture of organic molecules by living organisms.
biotic Pertaining to the components of the world that are living or came from living organisms. (Contrast with abiotic.)
birefringent Having the property of appearing optically different (a dark form and a light form) under polarized light when the beam of light is rotated; anisotropic. (Contrast with isotropic.)
boundary layer The unstirred layer of air or water adjacent to any solid surface.
brachysclereid A short, spheroidal cell with a thick, heavily lignified cell wall; cells of this type are most commonly found in parenchymatous tissues, such as stem cortex and pith, and in fruit walls or seed coats.
bract A modified leaf that occurs on a flower or inflorescence.
bridge In plant anatomy, the relatively short, diagonal vascular bundle that connects two vertical (axial) vascular bundles.

C_3 plant The typical green plant in which carbon dioxide is taken up during the day and incorporated into a three-carbon molecule (3-phosphoglycerate), which is then used in the synthesis of sugars; stomates are closed at night.
cactologist A specialist who studies cacti.
cactophile A person who loves cacti.
caespitose Having numerous stems that arise from a common base and form a dense, low, often hemispherical mound.
calciphile A plant that grows most favorably in a soil rich in calcium carbonate (limestone).
callus A tissue composed of thin-walled parenchyma cells that form either as a result of an injury or in tissue culture.
calyx The outermost (lowermost) series of sterile, modified leaves on a flower; a collective term for all sepals in a flower.
CAM See Crassulacean acid metabolism.
cambium A meristem that produces cells to the inside and the outside by periclinal divisions. See also cork cambium and vascular cambium.
campylotropous Having a ovule that is oriented perpendicular to the funiculus.
canopy The crown or upper portion of a plant or plant community.
carbohydrate Any of a large group of compounds produced in plants by photosynthesis and containing carbon, hydrogen, and oxygen, for example, sugars, starches, and cellulose.
carotenoid A yellow, orange, or red pigment found in plant plastids and associated with photosynthesis.
carpel A female, ovule-bearing structure of an angiospermous flower; usually (and in cacti) one of several units that have been fused laterally to form the pistil.

carpel primordium A mound of meristematic tissue that gives rise to a carpel, that is, part of the pistil.
cauline bundle A vertical (axial) vascular bundle that is found in the stem and never produces leaf traces.
cell membrane See plasmalemma.
Cell Theory The widely accepted biological theory, proposed by Schleiden and Schwann, that the cell is the fundamental unit of living organisms and that cells come from preexisting cells.
cellulose The long-chained carbohydrate that constitutes most of a plant cell wall.
cell wall The nonliving structure that bounds a plant cell and is composed largely of cellulose.
center of origin The place in which a particular taxon most likely originated.
central mother cell zone (of the apical meristem) The core of initials, the cells of which seldom divide and do not directly produce a mature plant tissue.
centrifugal In anatomy, developing in a sequence from inside to out.
centrosperm The common name for any species in the order Chenopodiales (Centrospermae); these species generally have an embryo that curves around perisperm.
cephalium In cacti, a highly specialized type of shoot terminus in which internodes are lacking and all areoles on the structure produce flowers, which are protected by a dense cover of spines or trichomes.
cereoid Pertaining to any plant with ribbed, succulent, columnar stems, as in the form genus *Cereus.*
chalaza The region of an ovule where the funiculus is attached. (Contrast with micropyle.)
character Any feature or attribute of an organism.
character state One of several alternative forms of a character, for example, the ancestral form or one of several derived forms.
ChenAm pollen The type of pollen grains that are found in the families Chenopodiaceae and Amaranthaceae and that have many apertures and a highly sculptured, reticulate exine.
chlorenchyma The type of parenchyma cells that have many chloroplasts and are thereby specialized for photosynthesis.
chlorophyll The green pigment produced by plants to capture light energy for photosynthesis.
chloroplast The green, chlorophyll-containing organelle (plastid) in which photosynthesis occurs in a plant cell.
cholla The common name for a group of closely related species of *Opuntia* that occur in North America and the West Indies and have jointed, cylindrical, tuberculate stems.
chromoplast A plastid that contains carotenoids.
clade An evolutionary branch in a phylogeny, especially one based on genealogical (cladistic) relationships.
cladistics The method of reconstructing the phylogeny of a taxon by analyzing shared derived characteristics to determine the precise sequence of evolutionary branching; phylogenetic systematics. (Contrast with evolutionary systematics.)
cladode A flattened photosynthetic stem, characteristic of platyopuntias; a phylloclad or pad.
cladogenesis The evolutionary process that produces a series of branching events.
cladogram A line diagram that is derived from cladistic analysis and shows the branching sequence (genealogy) of the taxa.
cleistogamy The process in which self-pollination occurs within a flower that never opens.
clonal microspecies A very narrowly restricted (endemic) species that reproduces exclusively by cloning, usually vegetatively by the rooting of shoots, for example, joints, or the spread of underground shoots.
closed vascular system The vascular organization of a stem in which the vertical (axial) vascular bundles are interconnected by diagonal bridges. (Contrast with open vascular system.)
cohort In ecology, the set of individuals that started life together, usually in the same season.
cold acclimatization See cold hardening.
cold hardening The process by which a plant tolerates colder temperatures as a result of previous exposure to cold temperatures; cold acclimatization.
collenchyma A type of cells that are living and have unevenly thickened cell walls, rich in pectin and hemicellulose but lacking lignin.
colpa See aperture.
columnar Having the appearance of a column.
companion cell A living parenchyma cell that is found in the phloem and is adjacent to a sieve-tube element; cells of this type provide the energy for sugar transport in the sieve-tube element.
compensation See light compensation or PAR compensation.
competition An interaction between organisms that is mutually detrimental to both participants.
compound ovary An ovary that formed evolutionarily and usually developmentally from the fusion of two or more carpels; a multicarpellate ovary.
conservative In systematics, having the tendency to lump populations or species into fewer, larger taxa. (Contrast with liberal.)
contact parastichy A spiral that can be drawn to connect a series of leaves that are adjacent to and somewhat vertical of each other.
continental drift A model of continental formation and movement, first proposed by Alfred Wegener, in which the continents were once united and have become independent structures that have been displaced over the surface of the earth.
convection A movement of molecules by fluid motion, for example, the movement of air molecules during the heat exchange process at a plant surface.
convergent evolution The development of two or more species with strong superficial resemblances from totally unlike and unrelated ancestors; convergence.
cork cambium The layer of cells that occurs in bark and forms cork cells toward the outside; phellogen.
cork cell A cell that is dead at maturity and has a thick wall impregnated with cork (suberin); the principal cell type of the outer bark.
corolla The collective term for the petals of a flower.
corpus The core of meristematic cells in the shoot tip, internal to the tunica; these cells divide in many planes to give rise to the cortex, vascular tissues, and pith of a stem.
cortex The tissue located beneath the epidermis in any

plant organ, usually composed mostly of soft-walled parenchyma cells; in leaves, also called mesophyll.

cotyledon An embryonic leaf of a seed plant.

crassinucellar Having a very thick nucellus around the embryo sac.

Crassulacean acid metabolism (CAM) The basic physiological process of succulent plants in which carbon dioxide is taken up in the dark and incorporated into organic acids; the stomates are generally closed during the daytime, when sugars are made by breaking down the organic acids.

cross pollination The transfer of pollen grains from one individual to the receptive stigma of another, genetically dissimilar individual.

cross section The view or piece obtained when a long axis is cut at right angles.

crypsis A form of mimicry in which an organism is unnoticed by prey or a predator because its form and coloration match that of the background.

crystal sand A term describing the numerous, minute, rhomboidal crystals of calcium oxalate found in certain plant cells.

cuticle The nonliving layer of wax that covers the epidermis of aerial plant parts and makes the plant surface relatively impervious to water.

cutin A complex fatty substance that forms the wax of the cuticle.

cyathium The specialized inflorescence of the genus *Euphorbia;* this inflorescence appears to be a single flower but actually consists of male and female flowers and colorful bracts.

cylindropuntia The common name of species of the subfamily Opuntioideae that have jointed, cylindrical stems. Usually these species are classified in the genus *Opuntia;* rarely, as a segregate genus.

cytogenetics The study of chromosomes and their behavior in cells and populations.

cytokinin A plant hormone that characteristically has a purine or purinelike structure.

cytosol The liquid portion of the cell protoplast, occurring outside the vacuole and excluding all cell organelles.

daughter cell One of the two identical cells that form when a cell divides; a sister cell.

decarboxylation The removal of a carboxyl radical (—COOH) from an organic molecule, a reaction leading to the release of carbon dioxide.

deciduous Referring to a plant that sheds all of its leaves at approximately the same time and thus has bare branches part of the year; not evergreen.

decumbent Referring to stems that lie on the ground but have ascending shoot tips.

dehiscence The act of splitting along a line to discharge the contents, as in the dehiscence of a fruit to release seeds or of an anther to release pollen grains.

dehydration The loss of water. (Contrast with rehydration.)

derivative A cell that is produced by division of a meristematic cell and begins to specialize as part of the plant body. (Contrast with initial.)

derived Pertaining to any character state of an organism that is more specialized than and has originated from another character state; specialized from a primitive condition; apomorphic.

dermatogen (of the apical meristem) The layer of cells that covers the apical meristem and gives rise to the epidermis.

determinate growth The pattern of shoot formation in which discrete (of a predetermined size) shoot units are produced regardless of the environmental conditions, as in joints of chollas and cladodes of most platyopuntias. (Contrast with indeterminate growth.)

dichotomy A condition or situation in which branching into two equal parts or two similar choices occurs.

dicotyledon An angiosperm that has two cotyledons on the embryo and the potential to form xylem and phloem from a vascular cambium.

dictyosome An organelle that is composed of stalked, flattened membranes and is active in the manufacture and secretion of cellular products such as pectin and mucilage.

diffusion The random, thermal motion of particles leading to net movement toward regions of lower concentration.

dihydroisoquinoline An alkaloid similar to and possibly the biosynthetic precursor of the tetrahydroisoquinolines.

dimorphic Having two distinct forms.

dioecy (adj. **dioecious)** The condition in which a species has individuals with either only female or only male reproductive systems. (Contrast with monoecy.)

diol An alcohol with two hydroxyl groups (—OH).

diploid Having two sets of chromosomes ($2n$).

disjunct Referring to a taxon whose range is geographically isolated from that of its closest relative.

disjunction A discontinuous range of a taxon (for example, a species or a genus) in which at least two closely related populations are separated by a wide distance; a gap in the distribution of a taxon.

diterpene A 20-carbon molecule that typically has 2 to 4 carbon rings, for example, a gibberellin.

diurnal Occurring during the daytime. (Contrast with nocturnal.)

dorsal bundle A vascular bundle (trace) that passes through the adaxial side of a carpel.

drought-deciduous Referring to a plant having leaves that are shed for at least one season in response to drought.

druse An aggregate crystal of calcium oxalate in which the crystals tend to radiate from a central point.

ecology The study of the abundance and distribution of organisms.

ecophysiology The study of the growth and distribution of organisms by determining physiological responses to environmental conditions.

embryo The young vegetative plant that occurs in a seed.

embryo sac The haploid female plant (female gametophyte), which is located in the ovule.

endemic Referring to a taxon that is naturally restricted to a particular (specified) geographic area, for example, a state, island, or country.

endosperm The nutritive tissue in the seed of an angiosperm; tissue derived from the fusion of one haploid male nucleus (sperm nucleus) and two or more haploid female nuclei.

entire Having borders without incisions or indentations.

enzyme An organic molecule that speeds up (catalyzes) a chemical reaction.

epicotyl The shoot part of an embryo or a young plant that is positioned above the attachment of the cotyledon.

epicuticular wax A powderlike wax that occurs on the surface of the cuticle.

epigyny (adj. **epigynous)** The condition in which the flower has an inferior ovary and the perianth and stamens are attached above the pistil. (Contrast with hypogyny.)

epiphyte A plant not rooted in soil but usually living on another plant. An epiphyte derives its moisture and nutrients from atmospheric precipitation and the surrounding organisms.

epithelium The layer of parenchyma cells that lines a plant cavity in which a secondary plant product is stored.

evolution Any irreversible change in the genetic composition of a population.

evolutionary systematics The method of reconstructing the phylogeny of a taxon by analyzing the evolution of major features along with the distribution of both shared primitive and shared derived characteristics. (Contrast with cladistics.)

exine The hard, resistant, waxy, outer wall of a pollen grain, which is usually highly sculptured.

extant Living at this time. (Contrast with extinct.)

extinct No longer living. (Contrast with extant.)

extracellular Pertaining to anything occurring within a plant but in the regions between living protoplasts, for example, in cell walls or intercellular air spaces.

extrastelar Pertaining to tissues outside the stele, especially the cortex.

family A taxonomic category above the rank of genus, tribe, and subfamily.

fasciated Having a crested or monstrosely shaped shoot.

fascicular cambium The region of the vascular cambium that forms within a vascular bundle. (Contrast with interfascicular cambium.)

fastigiate Having erect structures that are more or less parallel with each other.

fenestrate Having the appearance of small windows.

fertile In plants, having reproductive structures.

fertilization The fusion of one male sex cell (sperm nucleus) and one female sex cell (egg).

Fibonacci angle The angle between two successive leaf primordia, as viewed from above. In a plant with helically alternate phyllotaxy, the angle is approximately 137.5°.

Fibonacci sequence See Fibonacci summation series.

Fibonacci summation series The mathematical series in which each succeeding number is the sum of the previous two: 1, 1, 2, 3, 5, 8, 13, 21, 34, . . . ; the Fibonacci sequence.

filament The stalk of a stamen that bears the anther.

flavonoid A compound that is found in most plants and has 3 carbon rings (15 carbon atoms) with an oxygen atom in the middle ring.

floral apex The growing tip of a flower; this structure forms the perianth, stamens, and carpels.

floral tube A cylindrical to funnel-shaped structure that bears at its top the sepals and petals and positions the perianth away from the base of the pistil.

floriferous Bearing flowers.

flower The reproductive shoot of an angiosperm; this structure consists of sterile parts (perianth) and parts containing sex organs (stamen or pistil or both).

foraminate Having perforations (holes), as in wood (secondary xylem) with patches of phloem, which collapse to form holes as the stem dries.

form genus An artificial (unnatural) genus, which includes unrelated species because they share a particular morphological feature, for example, ribbed, columnar stems.

frost plasmolysis The process in which a cell protoplast pulls away from the cell wall because water has been lost from the cell and has frozen in the extracellular regions.

fruit The ripened ovary of an angiosperm; sometimes a fruit also includes other reproductive structures.

funiculus (adj. **funicular)** The stalk of an ovule.

fusiform Having a long spindlelike outline.

fusiform initial A vertically elongate cell that is found in the vascular cambium and gives rise to the vertically elongate cells of the secondary xylem and phloem. (Contrast with ray initial.)

fusiform parenchyma cell (of wood) The elongate parenchyma cell that is found in highly specialized secondary xylem, is nucleated, and lacks a thick, lignified cell wall.

gas exchange The release and uptake of gases such as carbon dioxide and water vapor.

gas-phase conductance A parameter indicating the ease of movement of molecules in a gas phase, such as air.

genetic spiral See parastichy.

genus (pl. **genera)** A taxonomic category for classifying the species derived from a common ancestor; a rank beneath tribe and subtribe.

geographic speciation See allopatric speciation by geographic subdivision.

geophyte An herbaceous perennial that has a large underground storage stem or root, which enables the plant to survive long periods of dormacy.

germination The emergence of the embryo from a seed.

gibberellin A class of plant hormones with an acidic, diterpene structure (4 carbon rings), for example, gibberellic acid; often associated with stimulating cell division and elongation.

glabrous Being smooth and devoid of trichomes. (Contrast with pubescent.)

glandular Having or secreting a sticky substance.

glaucous Having a gray to white or blue surface, usually because of the presence of epicuticular wax.

globular Roughly spherical in shape; globose.

glochid A short, very thin, deciduous spine that has barbs and is easily dislodged from the areole; a characteristic feature of the subfamily Opuntioideae.

glucoside An organic molecule with an attached glucose; a type of glycoside.

glycoside An organic molecule with an attached sugar.

grade A level of organization, for example, the succulence in cortex, which may or may not have evolved only once in a taxon.

graft The union of parts of two plants, accomplished by matching the respective tissues.

groundmass The typical cells that constitute a tissue.

ground tissue The general term for cortex, pith, and medullary rays.

growth habit The general architecture of a plant, including size, degree of erectness, and manner of producing branches and roots.

guard cells (two) A pair of cells that flank the stomate; the cells that control stomatal opening and closing by changes in their shape in response to changes in their turgor pressure (hydrostatic pressure).

gymnosperm A cone-bearing seed plant, which lacks true flowers.

gynodioecious Having two types of flowers in a population, those with both male and female organs and those with only male organs (staminate).

gypsophile A plant that grows most favorably in a soil rich in calcium sulfate (gypsum).

hallucinogen (adj. **hallucinogenic)** A psychoactive compound that produces unreal sense perceptions.

haploid Having one set of chromosomes ($1n$).

hardening See cold hardening; heat hardening.

hardiness The ability of a plant to withstand low or high temperatures.

heat convection coefficient A parameter relating heat movement to differences in air temperature.

heat hardening The process by which a plant tolerates hotter temperatures as a result of previous exposure to hot temperatures.

hectare The measure equal to 10,000 square meters; 2.47 acres.

helical Having the form of a helix.

herbaceous Having shoots that lack conspicuous amounts of wood (secondary xylem).

herbarium A museum in which dried, flattened, scientific specimens of plants are stored in cabinets.

herbivory The condition in which an animal feeds mostly or entirely on plants.

hermaphroditic Having both female and male reproductive structures in the same individual.

hexaploid Having six sets of chromosomes ($6n$).

hilum The depression on a seed where the funiculus was attached.

homologous In systematics, pertaining to two or more organisms or features that have arisen from the same genetic sequence on a chromosome.

hormone An organic molecule synthesized by an organism that controls its rate of growth and metabolism.

hybrid Any individual formed by the fusion of two genetically dissimilar cells.

hybridization The production of offspring by parents of two different species or dissimilar populations.

hybrid swarm A population in which there is a confusing mixture of two parental species and various types of hybrids involving the two parents.

hydrolyze To break down a compound by adding water, as occurs during removal of a sugar from an organic compound by treatment with an acid.

hydrophilic Having a strong affinity for water.

hydrostatic pressure A physical condition in a liquid where differences lead to flow; positive hydrostatic pressures (turgor pressures) are caused by water uptake into plant cells.

hypanthium A disk-shaped structure that surrounds the ovary and to which sepals, petals, and stamens are attached.

hypocotyl The part of the embryonic stem below the cotyledon and above the radicle.

hypocotyl–root axis The general term for the embryonic plant axis below the cotyledon; so called because a sharp boundary between the hypocotyl and radical cannot be observed.

hypodermis (adj. **hypodermal)** The outermost region of the cortex of certain stems in which cells in the outer one or more layers have thick cell walls.

hypogyny (adj. **hypogynous)** The condition in which the flower has a superior ovary. (Contrast with epigyny.)

idioblast A cell that differs markedly in form and function from the surrounding cells of a tissue, for example, a mucilage cell or a crystal-bearing cell in the cortex.

inactivation temperature The temperature at which biological activity is lost.

inbreeding The condition in which very closely related individuals mate for generations instead of mating with dissimilar individuals.

indehiscent Staying closed, not splitting open.

indeterminate growth The pattern of shoot formation in which shoot growth is tightly controlled by environmental conditions; thus, growth continues indefinitely when the conditions are favorable. (Contrast with determinate growth.)

inferior ovary In a flower, the condition in which the ovary is positioned beneath (inferior to) the other floral parts; epigyny. (Contrast with superior ovary.)

inflorescence A reproductive stalk with two or more flowers.

inhibition The cessation or retardation of a process.

initial A meristematic cell that divides to give rise to two cells, one of which (the initial) remains in the meristem and the other of which (the derivative) becomes part of the plant body. (Contrast with derivative.)

initial zone In cactus shoot apices, a region of the meristem that contains initials, which do not become part of the plant body.

integument The cell layers that envelop the nucellus in the ovule and later become the seed coat; in angiosperms, this structure consists of two parts (outer and inner integuments).

intercalary meristem A platelike region of dividing cells, typically located at the base of an organ (for example, a leaf, spine, stamen, or ovary) between tissues that are no longer meristematic.

intercellular air space The space between cells that is filled with gases.

interfamilial Pertaining to a feature that occurs in several, closely related families.
interfascicular cambium The region of the vascular cambium that forms in the medullary rays. (Contrast with fascicular cambium.)
intergeneric Pertaining to a feature that occurs in several, closely related genera.
internode The region of a stem between nodes. (Contrast with node.)
interspecific Pertaining to a feature of or event between more than one species. (Contrast with intraspecific.)
intraspecific Pertaining to a feature of or event within a single species. (Contrast with interspecific.)
ionic milieu The particular concentration of ions in a certain region, such as that within a cell.
isotropic Having the property of appearing optically the same under polarized light when the beam of light is rotated. (Contrast with birefringent.)

joint In cacti, a shoot segment that arises abruptly from an old stem areole and is clearly demarcated by a relatively narrow base; a determinate shoot; in platyopuntias, a cladode.
juvenilism In evolutionary biology, the condition in which the features found only in a young individual are retained by the adult.

lamina See leaf blade.
lateral bud See axillary bud.
latex A fluid, often white but also colored or colorless, that exudes from a cut plant surface; a complex fluid that often contains rubber (not in cacti) and other secondary plant products, for example, alkaloids.
laticifer The latex-bearing tube or cavity of a plant.
leaf-bearing cactus A species of cactus that has relatively broad, flat, large leaves; a species of *Pereskia, Pereskiopsis,* or *Quiabentia.*
leaf blade The broad, expanded portion of a leaf where photosynthesis occurs; the lamina.
leaf primordium A mound of meristematic tissue that gives rise to a leaf.
leaf succulent A plant in which the principal succulent tissue occurs in the leaves rather than in the stem.
leaf trace A major vascular bundle that vascularizes the leaf and arises from the vascular cylinder of the stem.
liberal In systematics, having the tendency to subdivide populations or species into numerous, smaller taxa. (Contrast with conservative.)
libriform (wood) fiber The typical elongate, narrow, and thick-walled cell of most dicotyledonous woods; the pits on the lateral walls of such a cell are very narrow.
light compensation (point) The light level at which there is no net exchange of carbon dioxide, that is, photosynthesis balances respiration; PAR compensation point.
lignin A complex, long-chained molecule found in cell walls that makes them extremely hard.
lipid An organic substance having the general property of a fat.
liquid-phase conductance A parameter indicating the ease of movement of molecules in a liquid phase, such as water.
lobed Having deep, rounded indentations along the edge of a flattened structure, as in a leaf or rib.
locule A cavity or compartment of an ovary.
long-distance dispersal The process by which an organism colonizes a favorable, distant habitat by crossing over inhospitable environments.
long shoot On a plant with long shoot–short shoot organization, a portion of the shoot with relatively long, conspicuous internodes.
long shoot–short shoot organization The arrangement of the shoot into a stem that has relatively long internodes and upon which very short branches form because their internodes are very short or absent.
longwave radiation The energy that is radiated from an object as a result of its surface temperature; wavelengths are in the infrared spectrum, that is, wavelengths longer than the red part of the spectrum.
lumen (of a cell) The space within the cell wall.
lupane Any of a series (subclass) of triterpenes.

macronutrient A nutrient that is present in plant tissues at a fairly high level, for example, nitrogen, phosphorus, and potassium. (Contrast with micronutrient.)
male sterile Having a male structure that does not produce viable sex cells.
marginal meristem (of a leaf) The meristem that is found along the border of a leaf and is responsible for forming the leaf blade.
medullary bundle A vascular bundle that occurs in the pith of a stem.
medullary ray The region of parenchyma cells that separates vascular bundles in the stem vascular cylinder and thereby connects the cortex and the pith.
megaspore The initial haploid cell occurring in a female reproductive organ and giving rise to haploid sex cells.
meristem (adj. **meristematic**) Any plant tissue that is primarily concerned with the formation of new cells.
meristemoid A cell that is located in an older tissue and is meristematic for the short interval during which it produces a special group of derivatives, for example, the subsidiary and guard cells.
mesophyll The cortical region of a leaf; this region is rich in chloroplasts and designed to facilitate light reception and gas movements for photosynthesis.
metabolism (adj. **metabolic**) The chemical activities occurring in a living organism.
microhabitat The local environment, which often determines the success and productivity of an organism.
micronutrient A nutrient that is present in plant tissues at fairly low levels and that has its major functions in proteins. (Contrast with macronutrient.)
micropyle The opening in the integuments of an ovule, usually the entrance for the pollen tube. (Contrast with chalaza.)
microspecies See clonal microspecies.
microspore mother cell A diploid cell that is found in a male organ and divides to give rise to four haploid sex cells.
midvein The central, main vein of a leaf or leaflike structure.
mimicry The striking resemblance of one organism to another organism or to a background (for example, soil

or tree bark); this resemblance often deceives predators or prey.

monocotyledon A type of angiosperm that has one cotyledon on the embryo and does not have a vascular cambium.

monoecious Having separate male and female reproductive structures on a single individual.

monohydroxy Having one hydroxyl group (—OH).

monophyletic Having arisen from a single ancestral population.

monosporic In embryology, the condition in which only one haploid cell of several survives to produce the next structure; a description of the embryo sac development in most angiosperms produced from one haploid cell.

monstrose The general term for a mutant plant form that is lacking the typical geometry of the species.

morphology (adj. **morphological**) The study of plant form, particularly at the level of overall organization.

mortality The production of deaths within a population.

mucilage A sticky, jellylike, water-absorbing substance found in certain plants.

mucilage cell A cell that produces and stores mucilage.

multicarpellate ovary See compound ovary.

multicellular Having more than one cell.

multiseriate Having three or more layers or rows of cells.

necrotic Dying or decaying.

nectary The structure that secretes copious amounts of a sugar solution that acts as a reward for animal visitors; usually present in flowers but sometimes present on fruits, stems, and leaves.

nectary chamber See nectary.

negative hydrostatic pressure See tension.

nocturnal Occurring at night. (Contrast with diurnal.)

node The place on a stem where one or more leaves are attached. (Contrast with internode.)

nonfibrous wood In cacti, the condition in which the secondary xylem lacks wood fibers.

nucellus (adj. **nucellar**) The internal tissue of an ovule within which the embryo sac develops.

nucleus (adj. **nucleate**) The organelle that contains the chromosomes.

nurse plant A plant that provides shelter for a seedling or young plant by ameliorating the local environment.

Oberblatt The upper portion of a developing leaf; this structure normally becomes the petiole, midvein, and leaf blade. (Contrast with Unterblatt.)

obovate Having the outline of an egg, with the base of the structure at the narrow end. (Contrast with ovate.)

oleanane Any of a series (subclass) of triterpenes.

open vascular system The vascular organization of a stem in which the vertical (axial) vascular bundles are not interconnected by diagonal bridges. (Contrast with closed vascular system.)

operculate Opening by means of a lid or cover.

order A taxonomic category for classifying closely related families.

organelle A membrane-bounded body that occurs within a cell and has a special function.

organic acid An organic compound whose protons (H^+) readily dissociate from a carboxyl group (—COOH).

organogenesis The formation of an organ, for example, a leaf, a root, or floral structures.

orthostichy A vertical or nearly vertical line that can be drawn to connect a series of leaves or leaflike structures. (Contrast with parastichy.)

osmotic pressure A property of a solution that is caused by its dissolved solutes and which affects water energy and hence water flow.

outcrossing The exchange of genetic information between two individuals that are not identical, for example, the exchange of pollen between genetically dissimilar plants.

ovary The lower part of a pistil; this structure contains the ovules and develops into a fruit.

ovate Having the outline of an egg, with the base of the structure at the wide end. (Contrast with obovate.)

ovule The structure that contains the embryo sac and eventually matures into a seed.

pad See cladode.

palmate venation A type of venation, principally of leaves, in which several equal, major veins diverge from the petiole at the very base of the leaf blade.

papilla (adj. **papillose**) A minute, nipplelike projection of an epidermal cell.

PAR See photosynthetically active radiation.

paracytic Having a pair of subsidiary cells that have their long axis parallel with the long axis of the guard cells.

parallelocytic Having three of more subsidiary cells that have their long axes parallel with the long axis of the guard cells.

parastichy A spiral that can be drawn to connect a sequential series of leaves or leaflike structures, youngest to oldest; a genetic spiral. (Contrast with orthostichy.)

paratracheal (wood parenchyma) Being directly attached to a vessel.

PAR compensation (point) The PAR level at which there is no net exchange of carbon dioxide, that is, the level at which photosynthesis balances respiration.

parenchyma The type of living (nucleate) cells that occur in all plant organs and tissues and perform various physiological functions; any living cell with a thin cell wall.

parietal placentation The condition in which the position of the placenta and the ovules is on the outer wall of the locule. (Contrast with axile placentation.)

parsimony The logical principle that the simplest solution, which involves the fewest logical steps or conditions, should be chosen when trying to choose from two or more conflicting explanations.

pascal A unit of pressure equal to 1 newton per meter squared; 10^5 pascal (Pa) = 0.1 megapascal (MPa) ≅ 1 atmosphere.

pearl cell A special epidermal cell of the funiculus that is pigmented at anthesis and becomes a colored cell of the fruit pulp.

pectin A compound that occurs between plant cells and holds them together as tissues.

pectinate Having projections (spines) that radiate from a line like the teeth of a comb.

pedicel The stalk of a single flower. (Contrast with peduncle.)
peduncle The main stalk of an inflorescence. (Contrast with pedicel.)
pendent Hanging from a fixed point.
perennial Living more than 2 years.
perforation A hole in a cell wall, for example, in the end wall of a vessel element or of a sieve-tube element.
perianth parts The sterile, leaflike parts of a flower; this term describes both the petals and the sepals (perianth) and is usually used when petals are not distinct from sepals.
pericarpel In a cactus flower, the outer covering that encloses the ovary and is actually stem tissue.
periclinal Referring to the orientation of a cell wall or a plane of cell division parallel to the nearest surface. (Contrast with anticlinal.)
periderm The covering of an old stem or root that is produced by the cork cambium; the outer bark.
peripheral Pertaining to a structure that occurs on the edge of another structure.
peripheral meristem In cacti, the layer of meristematic cells that is located inside the protoderm and which produces the cells of the outer cortex (ribs and tubercles); a subprotodermal meristem.
peripheral zone (of the apical meristem) A region of actively dividing cells that lies beneath the dermatogen and produces cells for the stem cortex and vascular cylinder.
periplasmodium A mucilaginous mass of cells in the nucellus of some angiosperms.
perisperm The nutritive tissue that occurs in the seed of certain angiosperms and forms from diploid cells of the ovule, not from a fusion of male and female nuclei.
permeability The overall effect of factors that determine the passage of something across a membrane or other barrier.
petal A sterile, leaflike organ on a flower positioned between the sepals (if present) and the stamens and pistil; usually the showy floral part that attracts pollinators; part of the corolla.
petiole (adj. **petiolate)** The stalk of a leaf.
phellogen See cork cambium.
phenethylamine Any member of a class of alkaloids that have the basic carbon-ring structure of tyramine.
phenetics The study of the overall similarities of organisms. (Contrast with phylogeny.)
phenogram A diagram that expresses the probable phenetic relationships of three or more taxa.
phloem The plant tissue that transports sugars and other organic compounds.
photon A discrete unit of radiation energy that can be absorbed by a photosynthetic pigment molecule.
photosynthesis The process in which carbon dioxide is incorporated into organic compounds in the chloroplast of a plant.
photosynthetically active radiation (PAR) The portion of the light spectrum that can be absorbed by photosynthetic pigments, for example, chlorophyll and carotenoids; about 400 to 700 nm in wavelength.
phylloclad See cladode.
phyllotaxy The pattern of leaf arrangement on the stem.
phylogenetic systematics See cladistics.
phylogeny The evolutionary relationships between an ancestor and its descendants. (Contrast with phenetics.)
physiology The study of cell metabolism and other vital processes of organisms.
pigment cell A cell that contains a large amount of a pigment, such as a carotenoid or betalain.
pinnate Having parts arranged like those of a feather, that is, on both sides of a common axis.
pinnate venation A type of venation, principally of leaves, in which the minor veins diverge in a parallel fashion from a central midvein.
pistil (adj. **pistillate)** The female organ of a flower; this structure consists of a stigma, style, and ovary and is the product of one or more fused carpels.
pit The thin area in a lignified cell wall, as in a vessel element or a primary phloem fiber.
pith The central tissue of a stem, typically composed of soft-walled parenchyma cells.
pith–rib meristem zone (of an apical meristem) The region of meristematic cells that are found in the cactus shoot apex and produce long files of cells for the pith by repeated anticlinal divisions.
plant hair See trichome.
plasmalemma The membrane that delimits the living cell from the cell wall; the plasma membrane or cell membrane.
plasma membrane See plasmalemma.
plastid A structure that lies within a cell and is bounded by a double membrane; it peforms a specific physiological function, for example, a chloroplast is the site of photosynthesis.
platyopuntia The common name for any of a large group of species in the subfamily Opuntioideae that has cladodes; also called a prickly pear, a term that technically refers to the fruit.
plesiomorphy (adj. **plesiomorphic)** In cladistics, a primitive or ancestral character state. (Contrast with apomorphy.)
ploidy The number of sets of chromosomes in a cell.
pollen grain The small, haploid male plant (male gametophyte), which is enclosed in a hard pollen wall.
pollen tube The structure formed to carry the male sex cells from the pollen grain to the embryo sac.
pollination The transfer of pollen grains to a receptive stigma, usually by wind or a flower-visiting animal.
Polygonum-type embryo sac The common type of embryo sac that occurs in angiosperms and contains eight haploid nuclei, three at each end and two in the center.
polyhedral Having many sides.
polymorphic Having three or more distinct forms.
polyploid Having three or more sets of chromosomes.
polysaccharide A long-chained carbohydrate that can be broken down into simple sugars (monosaccharides).
polyterpene A long-chained molecule that can be broken down into smaller organic, terpene compounds.
population The individuals of a species occurring in a particular region or at a specific locality.
precocious Developing at a very early time.
predation The act of feeding on other organisms, which negatively affects the victim.

primary phloem The first-formed phloem of a vascular bundle.
primary phloem fiber A very long, thick-walled cell of the primary phloem in a vascular bundle.
primary ray The wide regions of soft parenchyma in a woody cylinder; this tissue connects the cortex and the pith and is not formed by the vascular cambium.
prismatic crystal An individual crystal that is birefringent under polarized light.
procambium The term applied to cells of vascular bundles when the cells are still meristematic.
procumbent Trailing or lying flat; prostrate.
proteinaceous inclusion A mass of protein within a cell or an organelle.
protein globule A dense, spheroidal mass of protein.
protoderm The name applied to the surface layer of a plant after the cell layer has formed from an apical meristem but before the cells have enlarged to become the epidermis.
protoplast The living unit of the cell, which includes everything within the cell wall.
pseudocephalium The general term for a cluster of very pubescent, flower-bearing areoles that are formed along the sides or top of a cactus stem and are not terminal structures because they do not include the shoot apex. (Contrast with cephalium.)
pseudopalmate venation A leaf venation pattern in which several principal veins arise from near the leaf base; in this venation pattern the midvein is not prominent.
pubescence (adj. **pubescent**) The cover of trichomes. (Contrast with glabrous.)
pulp The soft, parenchymatous tissue that fills the locules of a fruit.
punctae The minute holes or pits in the outer wall (exine) of a pollen grain.

radial Pertaining to a radius, for example, along the plane from the central pith to the bark.
radial areole An areole on which the spines radiate from a central point. (Contrast with asymmetrical areole.)
radiation In physics, the movement of energy by electromagnetic waves; in biology, the evolutionary divergence of a taxon into a number of different forms.
radicle The first root of a plant; this structure typically forms in the embryo and emerges first from the seed during germination.
rain root A fine root that appears shortly after a dormant plant has received abundant moisture.
raphe The prominent ridge of a seed; the long axis of the embryo is beneath and parallel to this structure.
raphides Needle-shaped calcium oxalate crystals, which usually occur in bundles.
ray A general term for a region of parenchyma cells that radiates from the pith.
ray initial A relatively short cell that is found in the vascular cambium and that gives rise to the short cells of the vascular ray. (Contrast with fusiform initial.)
receptacle The flower part that supports the floral organs.
receptacular model (or the inferior ovary) A model of the origin of the inferior ovary in which the ovary has become sunken into the tissues of the receptacle. (Contrast with appendicular model.)
recurrent system In the cactus flower, the portion of the vascular system in which the vascular bundles bend down toward the ovary from the floral tube.
reflectivity (reflectance) The fraction of radiant energy that is reflected by a surface.
rehydration The re-uptake of water. (Contrast with dehydration.)
reproductive Pertaining to any plant part that is formed from the flower or a cluster of flowers; the seed is the final reproductive product. (Contrast with vegetative.)
respiration The process in which energy and carbon dioxide are released when organic compounds are broken down in a cell organelle called a mitochondrion.
reticulate Having a netlike appearance.
rhomboidal Having the general three-dimensional shape of a rhombus.
rib A vertical or nearly vertical ridge that is formed on the side of a stem by the outward displacement of the tubercles.
ribulose-1,5-bisphosphate carboxylase/oxygenase The enzyme that catalyzes the incorporation of carbon dioxide into organic compounds in photosynthesis; abbreviated as Rubisco.
rootcap The domelike mass of cells that covers and protects the root apical meristem.
root hair The fine outgrowth (trichome) of an epidermal cell on a young root; a structure that greatly increases the water-absorbing area of the root system.
root primordium A mound of meristematic tissue that becomes a root.
Rubisco See ribulose-1,5-bisphosphate carboxylase/oxygenase.

saline Containing salt, especially sodium chloride.
sap A nontechnical term for any copious, watery substance obtained from a plant when it is cut.
sapogenin See saponin.
saponin Any of a group of organic compounds that freely produces suds (foam) when mixed with water; a sapogenin.
scandent Climbing over other structures.
scientific name The latinized name of a taxon. See also binomial.
sclereid A type of sclerenchyma cell that is relatively short, dead at maturity, and has an extremely thick, hard cell wall.
sclerenchyma The type of cells that have very thick cell walls generally impregnated with lignin and are often dead at maturity; a fiber or sclereid.
sclerified Having a thick cell wall that is impregnated with lignin.
secondary growth The cells added to a plant by both the vascular cambium and the cork cambium.
secondary phloem The phloem tissue produced by a vascular cambium.
secondary plant product An organic chemical that is made by a plant but is not a product or an intermediate of the biochemical pathways common to most living

organisms; an unusual molecule in a plant, such as a triterpene or mucilage.

secondary xylem The xylem tissue produced by a vascular cambium; wood.

seed A mature ovule, which has a mature embryo and a hard testa.

seed coat See testa.

segregate A taxon that is removed from one group by treating it as a separate but close relative of the original taxon.

self-pollination The transfer of pollen grains from the anther to the stigma of the same plant; autogamy.

sepal A sterile, leaflike, usually nonshowy organ on a flower, located to the outside and below the petals; part of the calyx.

sessile Having no petiole or pedicel; stalkless.

shoot The aboveground vegetative part of a plant, including leaves, stems, and shoot apical meristems.

shoot apical meristem The ultimate growing tip of each shoot, consisting of initials and recent derivatives.

short shoot The highly condensed lateral branch of a shoot with long shoot–short shoot organization.

sieve tube The long, cylindrical, essentially hollow conduit of the phloem; sugars are transported through this structure.

sieve-tube element A unit of a sieve tube.

silica body A deposit of silica (SiO_2) in the protoplast of a plant cell.

simple In plant morphology, pertaining to a leaf that has only one blade.

sister taxa The pair of species, extant or extinct, that are genealogically most closely related.

skin The general term for the tough covering of a cactus stem; the skin includes cuticle, epidermis, and hypodermis.

solitary In morphology, having only a single main axis.

speciation The process in which two or more contemporaneous species evolve from a single ancestral population.

species The fundamental taxonomic category consisting of all organisms that are morphologically and reproductively more similar to each other than to other populations and that come from a single ancestral population.

sperm nucleus A haploid male nucleus (two in a pollen tube) that is discharged to fuse with a haploid female nucleus.

spination The condition of having spines.

spine In cacti, a structure derived from a modified leaf primordium that lacks a blade and usually consists of a hard, tapered axis with a sharp tip.

spine primordium A mound of meristematic tissue that gives rise to a spine; in cacti, modified leaf primordium.

spinule A small spinelike projection on the exine of a pollen grain.

spur shoot A type of short shoot that becomes fairly long because it grows for many years.

stamen (adj. **staminate)** The male organ of a flower, including the filament and anther.

starch A long-chained polysaccharide composed of glucose molecules (monomers).

starch grain A deposit of starch in a plastid.

steady state A situation in which things are not changing with time.

stele In roots, the core of vascular tissues that is bounded on the outside by the cortex.

stem succulent A plant that has most of its succulence in a photosynthetic stem.

sterol A complex alcohol occurring in living organisms, for example, cholesterol.

stigma The upper tip of a pistil, above the style, that receives the pollen and on which the pollen grain germinates; can be subdivided into stigma lobes.

stipule One of a pair of appendages at the base of a petiole in many plants; absent in cacti.

stomatal conductance A measure of the ease with which gas molecules can move through the stomates.

stomate The pore in the epidermis of an aerial organ; gas exchange occurs through the stomate, and its opening is controlled by the surrounding guard cells.

storied wood The specialized condition of a wood in which, when seen in tangential view, the cells are aligned in horizontal tiers.

striae The minute cuticular striations (ridges) on an epidermal surface, for example, on a seed coat.

strophiole The corky swelling of the upper funiculus; this structure may cover part of the seed.

style The cylindrical portion of the pistil that lies between the ovary and the stigma and through which the pollen tube must grow.

subapical Occurring just behind the growing tip.

subfamily The taxonomic category below family.

subprotodermal meristem See peripheral meristem.

subsidiary cell An epidermal cell that is associated with the guard cell and that generally forms from the same meristemoid.

substomatal canal The cylindrical passage that traverses the hypodermis between the stomate and the chlorenchyma cells to permit gas exchange.

substrate The surface in or to which an organism is attached or from which it draws nutrients; a compound used in a chemical reaction.

subtribe A taxonomic category between tribe and genus.

subtropical Being nearly tropical; an area with moderate temperatures but occasional freezing in the poleward areas.

succulence (adj. **succulent)** The condition in which a plant has relatively thick, water-storing tissues, which enable the plant to tolerate drought for extremely long periods.

suffrutescent Having a low habit with a short woody base that forms herbaceous shoots; a diminutive shrub.

sugar residue The general term for a sugar that has been chemically removed (hydrolyzed) from an organic compound.

supercooling The cooling of a substance below the temperature at which freezing usually occurs.

superior ovary In a flower, the condition in which the ovary is located above the attachment of the stamens and perianth; hypogyny. (Contrast with inferior ovary.)

sympatric Referring to populations or species that occur in the same locality and are therefore close enough to mate. (Contrast with allopatric.)

sympatric speciation by polyploidy The formation of a new species occurring when individuals in a population experience a change in ploidy and therefore can no longer breed with the ancestral form.

symplesiomorphy In cladistics, the term used for a primitive character state shared between two or more taxa. (Contrast with synapomorphy.)

synapomorphy In cladistics, the term used for a derived character state shared between two or more taxa. (Contrast with symplesiomorphy.)

syncarpy (adj. **syncarpous**) The condition in which the female part of the flower is composed of a single, compound ovary because two or more carpels have been fused. (Contrast with apocarpy.)

synonym In taxonomy, a scientific name that should not be used because it refers to the same species by another, older scientific name.

systematics The study of the evolutionary relationships between organisms.

tangential section A section, generally of a cylindrical surface, that is longitudinal and at right angles to the radius.

tapetum In the anther, a layer of cells that nourishes the development of the pollen grains.

taxon (pl. **taxa**) A convenient and general term for any taxonomic category, for example, a species, genus, or family.

taxonomy The study of the names of organisms, which often includes the entire process of classification. See also systematics.

tension A negative pressure, such as the negative hydrostatic pressure occurring in the xylem of a transpiring plant.

terete Cylindrical.

terpene An organic compound containing units (monomers) of 10 carbon atoms arranged in two rings; generally found in the aromatic oils of plants.

terrestrial In botany, having its roots in soil.

testa The hard covering of a seed; a seed coat.

tetrad A group of four cells that developed from a single mother cell.

tetrahydroisoquinoline Any of a class of alkaloids that have a nitrogen atom in one of the two carbon rings.

tetraploid Having four sets of chromosomes ($4n$).

thornscrub A type of vegetation found in dry, subtropical regions dominated by shrubs (less than 4 m tall) with spines and thorns.

tonoplast The membrane that surrounds the vacuole.

trace A major vascular bundle that supplies a primordium from the vascular cylinder.

transpiration The loss of water vapor from a plant through the epidermis, especially through the stomates.

transpirational cooling The loss of heat that occurs when water evaporates and leaves a plant.

trichome A surface structure that consists of one or more cells and arises from the epidermis of a plant organ; a plant hair.

trichotomy A condition or situation in which branching into three equal parts or three similar situations occurs.

tricolpate Having three apertures (colpae) in the exine of a pollen grain.

trihybrid The hybrid plant formed from the combination of chromosome sets from three different parents.

trinucleate Having three nuclei in a structure, usually in a cell or celllike structure.

triol An alcohol that has three hydroxyl groups (—OH).

triploid Having three sets of chromosomes ($3n$).

triterpene Any of a class of sterols; these compounds are produced mainly by plants and usually occur as glycosides.

tropical An area near the equator in which the mean annual temperature is moderate and no freezing occurs.

truncate Having an abrupt tip or base that is essentially at right angles to the long axis.

tuber A type of enlarged, underground shoot used as a storage organ for starch.

tubercle (adj. **tuberculate**) An enlarged, succulent projection on a stem; in cacti, a modified leaf base.

tunica The surface layer or layers of a shoot apical meristem in which the cells always divide by forming anticlinal walls.

tunica–corpus organization (of a shoot apical meristem) The organization of meristematic regions in an angiosperm into one or more surface layers (tunica) and an interior core (corpus).

turbinate Having the shape of a top.

type species The species to which the description of a genus of family is legally assigned.

tyramine alkaloid An alkaloid that closely resembles and is derived from the amino acid tyramine; a type of phenethylamine.

uniseriate Having a layer one–cell thick.

Unterblatt The lower portion of a developing leaf; this structure normally becomes the leaf base or, in cacti, a tubercle. (Contrast with Oberblatt.)

vacuole A compartment that is found within a living cell, is bounded by the tonoplast, and contains water, organic acids, ions, pigments, and a variety of water-soluble organic compounds.

vascular bundle A strand composed of xylem and phloem and often a vascular cambium.

vascular cambium A uniseriate cylinder of meristematic cells (initials) that forms between the xylem and phloem to produce, by periclinal divisions, secondary xylem (wood) to the inside and secondary phloem (inner bark) to the outside; this meristem is present in gymnosperms and most dicotyledons, but is absent in monocotyledons.

vascular cylinder The cylinder of vascular bundles formed in a stem of a root.

vascular ray A radial array of parenchyma cells that is produced by ray initials of a vascular cambium and which occurs in the secondary tissues of xylem and phloem.

vascular tissue The cells that are formed from procambium and transport water (xylem) and sugars (phloem) within the plant.

vascular tracheid A short, dead, fusiform cell that occurs in the secondary xylem and has scanty thickenings on the walls but no perforations.

vegetative Pertaining to any plant part that originates from an embryo (or zygote) exclusive of the flower and fruit. (Contrast with reproductive.)

vegetative reproduction The formation of a new individual from a vegetative part of another plant and not from a seed.

venation The arrangement of visible veins (vascular bundles) in a leaf or leaflike structure.

ventral bundle The vascular bundle that passes through the adaxial side of a carpel.

vesicle A small, spherical organelle that is pinched off a dictyosome and which transports the secreted substance to where it will be used.

vessel The long, cylindrical, hollow conduit of the xylem, which transports water in a plant; this structure is composed of many individual cells and is a characteristic of most angiosperms.

vessel element A unit of a vessel; this structure consists of a hollow, dead cell with one or more perforations at each end.

vesselless Having no vessels in the xylem.

vesture (vestiture) See pubescence.

vicariance The subdivision of a species range that results in the formation of two or more allopatric populations.

vital stain A biological stain (dye) that is taken up only by living cells.

volume-to-surface ratio The comparison of organ volume to surface area; this ratio is an estimate of the amount of water available to the plant during drought.

water potential A measure of the energy of water; water spontaneously tends to move toward regions of lower water potential.

water-use efficiency An expression of the amount of carbon dioxide fixed by a plant per unit of water lost by transpiration.

water vapor conductance A measure of the ease with which water vapor molecules can move out across plant surfaces.

weighting In systematics, the placing of special emphasis on the importance of selected characters in the analysis of evolutionary relationships.

wood fiber A long, narrow, thick-walled, nonconducting cell of a wood, for example, a libriform wood fiber.

wood parenchyma The parenchyma cells, generally with lignified walls, that occur in secondary xylem.

wood skeleton The common name for a dried cylinder of wood that retains geometric holes or lines where soft parenchyma cells were present.

xylem (adj. **xylary**) The water-conducting tissue of a plant.

zygomorphic Having irregular arrangement of floral parts, usually with two symmetrical halves but no radial symmetry.

Index